AF696566

# Television Receivers

# On-Line Updates

Additional updates relating to television in general, and this book in particular, can be found at the *Standard Handbook of Video and Television Engineering* web site:

**www.tvhandbook.com**

The tvhandbook.com web site supports the professional video community with news, updates, and product information relating to the broadcast, post production, consumer, and business/industrial applications of digital video.

Check the site regularly for news, updated chapters, and special events related to video engineering. The technologies encompassed by *Television Receivers* arc changing rapidly, with new developments each month. Changing market conditions and regulatory issues are adding to the rapid flow of news and information in this area.

Specific services found at **www.tvhandbook.com** include:

- **Video Technology News**. News reports and technical articles on the latest developments in digital television, both in the U.S. and around the world. Check in at least once a month to see what's happening in the fast-moving area of digital television.
- **Television Handbook Resource Center**. Check for the latest information on professional and broadcast video systems. The Resource Center provides updates on implementation and standardization efforts, plus links to related web sites.
- **tvhandbook.com Update Port**. Updated material for *Television Receivers* is posted on the site regularly. Material available includes updated sections and chapters in areas of rapidly advancing technologies.
- **tvhandbook.com Book Store**. Check to find the latest books on digital video and audio technologies. Direct links to authors and publishers are provided. You can also place secure orders from our on-line bookstore.

In addition to the resources outlined above, detailed information is available on other books in the McGraw-Hill Video/Audio Series.

# Television Receivers

Jerry C. Whitaker, Editor

**McGraw-Hill**
New York Chicago San Francisco
Lisbon London Madrid Mexico City Milan
New Delhi San Juan Seoul Singapore
Sydney Toronto

Cataloging-in-Publication Data is on file with the Library of Congress.

McGraw-Hill

A Division of The McGraw-Hill Companies

Copyright © 2001, by The McGraw-Hill Companies, Inc. All rights reserved. Printed in the United States of America. Except as permitted under the United States Copyright Act of 1976, no part of this publication may be reproduced or distributed in any form or by any means, or stored in a data base or retrieval system, without the prior written permission of the publisher.

1 2 3 4 5 6 7 8 9 0 DOC/DOC 0 6 5 4 3 2 1 0

**ISBN 0-07-138042-6**

*The sponsoring editor for this book was Steve Chapman and the production supervisor was Pamela Pelton. The book was set in Times New Roman and Helvetica by Technical Press, Morgan Hill, CA.*

Printed and bound by R. R. Donnelley & Sons Company.

McGraw-Hill books are available at special quantity discounts to use as premiums and sales promotions, or for use in corporate training programs. For more information, please write to the Director of Special Sales, McGraw-Hill, Two Penn Plaza, New York, NY 10121-2298. Or contact your local bookstore.

Information contained in this work has been obtained by The McGraw-Hill Companies, Inc. ("McGraw-Hill") from sources believed to be reliable. However, neither McGraw-Hill nor its authors guarantees the accuracy or completeness of any information published herein, and neither McGraw-Hill nor its authors shall be responsible for any errors, omissions, or damages arising out of use of this information. This work is published with the understanding that McGraw-Hill and its authors are supplying information, but are not attempting to render engineering or other professional services. If such services are required, the assistance of an appropriate professional should be sought.

This book is printed on recycled, acid-free paper containing a minimum of 50% recycled, de-inked fiber.

# Contents

## Chapter 13: ATSC DTV Receiver Systems 227



## Chapter 14: Consumer Video and Networking Issues 265

# Contributors

Oktay Alkin
Edward W. Allen
K. Blair Benson
Carl Bentz
B. W. Bomar
J. A. Chambers
William Daniel
L. E. Donovan
Donald G. Fink
Susan A. R. Garrod
R. A. Hedler
L. H. Hoke, Jr.
Scott Johnson
Charles A. Kase
J. D. Knox
Kishore Manghnani
L. L. Maninger
D. E. Manners
D. Stevens McVoy
W. G. Miller
R. A. Momberger
Chandy Nilakantan
John Norgard
R. J. Peffer
Skip Pizzi
Ken Polhmann
Wilbur L. Pritchard
J. D. Robbins
Ulrich L. Rohde
Donald L. Say
Joseph L. Stern
S. Tantaratana
Daniel R. von Recklinghausen
J. G. Zahnen
Rodger E. Ziemer

Some material in this book has been adapted from the McGraw-Hill *Standard Handbook of Video and Television Engineering*, 3rd edition. Used with permission. All rights reserved.

# Preface

The familiar—and ubiquitous—television receiver has gone through significant and far reaching changes within the past few years. Once a relatively simple, single purpose device intended for viewing over-the-air broadcasts, the television set has become the focal point of entertainment and information services in the home. What began as an all-in-one-box receiver has evolved into a suite of devices intended to serve a viewing public that demands more selections, more flexibility, simpler control, and better pictures and sound. The TV set is—in fact—going through the same metamorphosis that audio systems did in the 1970s. The "console stereo" of the 1960s evolved from a multipurpose, albeit inflexible, aural entertainment system into the *component* scenario that is universal today for high-end audio devices.

The move to such a component approach to television is important because it permits the consumer to select the elements and features that suit his or her individual needs and tastes. Upgrade options also are simplified. By separating the display device and its related circuitry from the receiver makes it possible to upgrade from, for example, an NTSC receiver to an NTSC/DTV-compliant receiver without sacrificing the display—where most of the cost is concentrated and where most of the really significant advancements are likely to be seen in the coming years. Cable and satellite systems also play into this component scenario, as new services are rolled out to consumers.

While the arguments in favor of a component approach to video devices are compelling, the tradeoff is complexity and—of course—cost. Consumers have clearly indicated that they want entertainment devices to be simple to operate and simple to install, and price is always an issue. Here again, the integration of audio devices points the way for video manufacturers. There is ample evidence that consumers will pay a premium if a given device or system gives them what they want.

Apart from consumer television, videoconferencing is a related area of receiver system development. Although not strictly a receiver (in many cases, it is really a computer), emerging desktop videoconferencing systems and other interactive services promise to bring visual communication to businesses with the ease of a phone call. Long hamstrung by bandwidth bottlenecks, the era of high-speed real-time networking has—at last—made desktop videoconferencing possible, and more importantly, practical.

### Digital Television

The driving force for new developments in television receiving systems is DTV—and more importantly, HDTV. Stations coast-to-coast are broadcasting regular HDTV programming and beginning to fashion new business strategies made possible by this new technology. While broadcasters are the most visible early adopters of DTV, many other industries are gearing up and—indeed—are already making commercials, television programs, and motion pictures using HDTV equipment made possible by the ATSC DTV standard. Put side-by-side, the breadth of the HDTV applications now finding commercial success is staggering—from NASA to postproduction to medical imaging. Television receivers play a critical role in the transformation to digital.

This is an exciting time for video professionals and for consumers. Our new digital television system, the product of more than a decade of work by scientists and engineers around the world, is on the air. It works. It makes beautiful pictures. It provides capabilities never before possible. Indeed, the ATSC DTV system offers features thought to be little more than fantasy just 10 years ago.

The launch of every new technology has been fraught with growing pains. Consider what television engineers of the 1950s faced when their management decided that it was good business to convert to color. In 1954, getting a black-and-white signal on the air was an accomplishment, let alone holding the frequency and waveform tolerances sufficiently tight that the NTSC color system would look halfway decent. And then there were the receivers: big, expensive, and unstable. Round picture tubes with a long necks. The sets commandeered living rooms across America. And service these things? There were two dozen convergence adjustments alone in the RCA CTC-15 chassis, the first design that really worked—and it was released nearly a decade after the first color broadcasts began.

These problems were, in the final analysis, only details. Obstacles to be overcome. Challenges to be met. Such is the case with DTV. If the history of color television tells us anything, it is that a well-conceived standard will enable enormous improvements in end-to-end performance as receiver manufacturers in partnership with transmission system engineers fine-tune the overall system.

The ATSC digital television standard is, of course, just technology—a technical scheme that permits certain tasks to be accomplished. As such, it is neutral insofar as intrinsic benefits. DTV will, instead, become whatever we choose make it.

At this point in the transition to DTV, many challenges exist. Many more will be uncovered as the early-adopters learn by doing. The bottom line, however, is that DTV is the future of the professional video industry.

### About the Book

*Television Receivers* examines the technologies involved in the reception, decoding, and display of video signals. The underlying technologies that go into a modern receiver are examined, and examples are provided of typical uses. New developmental efforts also are explained and the benefits of the work are outlined.

This publication is directed toward technical and engineering personnel involved in the design, specification, installation, and maintenance of video receiver systems. The basic principles of analog and digital television reception and display are discussed, with emphasis on how the underlying technologies influence the ultimate applications.

The author has made every effort to cover the subject of television receivers comprehensively. Extensive references are included at the end of each chapter to direct readers to sources of additional information. The included CD-ROM contains additional reference information, particularly with regard to digital television standards.

### Technology for Tomorrow

Advances in imaging resolution and clarity are the driving forces behind many new product offerings, particularly in the consumer electronics business. Sophisticated set-top boxes capable of receiving and decoding HDTV broadcasts from terrestrial DTV stations, satellite services, and cable feeds—in addition to receiving conventional television signals—are available and affordable. On the display side, tremendous advances in performance and cost have been made in flat panel displays and in wide-screen HDTV rear-projection displays.

It is the goal of *Television Receivers* to bring these diverse concepts and technologies together in an understandable form.

*Jerry C. Whitaker*

June, 2001

For

*Larry Whitaker*

my favorite older brother

Chapter

1

# The Electromagnetic Spectrum

**John Norgard**

## 1.1 Introduction

The electromagnetic (EM) spectrum consists of all forms of EM radiation—EM waves (radiant energy) propagating through space, from DC to light to gamma rays. The EM spectrum can be arranged in order of frequency and/or wavelength into a number of regions, usually wide in extent, within which the EM waves have some specified common characteristics, such as characteristics relating to the production or detection of the radiation. A common example is the spectrum of the radiant energy in white light, as dispersed by a prism, to produce a "rainbow" of its constituent colors. Specific frequency ranges are often called *bands*; several contiguous frequency bands are usually called *spectrums*; and sub-frequency ranges within a band are sometimes called *segments*.

The EM spectrum can be displayed as a function of frequency (or wavelength). In air, frequency and wavelength are inversely proportional, $f = c/\lambda$ (where $c \approx 3 \times 10^8$ m/s, the speed of light in a vacuum). The MKS unit of frequency is the Hertz and the MKS unit of wavelength is the meter. Frequency is also measured in the following sub-units:

- Kilohertz, 1 kHz = $10^3$ Hz
- Megahertz, 1 MHz = $10^6$ Hz
- Gigahertz, 1 GHz = $10^9$ Hz
- Terahertz, 1 THz = $10^{12}$ Hz
- Petahertz, 1 PHz = $10^{15}$ Hz
- Exahertz, 1 EHz = $10^{18}$ Hz

Or for very high frequencies, *electron volts*, 1 ev ~ $2.41 \times 10^{14}$ Hz

Wavelength is also measured in the following sub-units:

- Centimeters, 1 cm = $10^{-2}$ m
- Millimeters, 1 mm = $10^{-3}$ m
- Micrometers, 1 μm = $10^{-6}$ m (microns)
- Nanometers, 1 nm = $10^{-9}$ m

- Ångstroms, 1 Å = $10^{-10}$ m
- Picometers, 1 pm = $10^{-12}$ m
- Femtometers, 1 fm = $10^{-15}$ m
- Attometers, 1 am = $10^{-18}$ m

## 1.2 Spectral Sub-Regions

For convenience, the overall EM spectrum can be divided into three main sub-regions:

- *Optical spectrum*
- *DC to light spectrum*
- *Light to gamma ray spectrum*

These main sub-regions of the EM spectrum are next discussed. Note that the boundaries between some of the spectral regions are somewhat arbitrary. Certain spectral bands have no sharp edges and merge into each other, while other spectral segments overlap each other slightly.

### 1.2a Optical Spectrum

The optical spectrum is the "middle" frequency/wavelength region of the EM spectrum. It is defined here as the visible and near-visible regions of the EM spectrum and includes:

- The *infrared (IR)* band, circa 300 µm–0.7 µm (circa 1 THz–429 THz)
- The *visible light* band, 0.7 µm–0.4 µm (429 THz–750 THz)
- The *ultraviolet (UV)* band, 0.4 µm–circa 10 nm (750 THz–circa 30 PHz), approximately 100 ev

These regions of the EM spectrum are usually described in terms of their wavelengths.

Atomic and molecular radiation produce radiant light energy. Molecular radiation and radiation from hot bodies produce EM waves in the IR band. Atomic radiation (outer shell electrons) and radiation from arcs and sparks produce EM waves in the UV band.

#### Visible Light Band

In the "middle" of the optical spectrum is the visible light band, extending approximately from 0.4 µm (violet) up to 0.7 µm (red), i.e. from 750 THz (violet) down to 429 THz (red). EM radiation in this region of the EM spectrum, when entering the eye, gives rise to visual sensations (colors), according to the spectral response of the eye, which responds only to radiant energy in the visible light band extending from the extreme long wavelength edge of red to the extreme short wavelength edge of violet. (The spectral response of the eye is sometimes quoted as extending from 0.38 µm (violet) up to 0.75 or 0.78 µm (red); i.e., from 789 THz down to 400 or 385 THz.)This visible light band is further subdivided into the various colors of the rainbow, in decreasing wavelength/increasing frequency:

- Red, a primary color, peak intensity at 700.0 nm (429 THz)

- Orange
- Yellow
- Green, a primary color, peak intensity at 546.1 nm (549 THz)
- Cyan
- Blue, a primary color, peak intensity at 435.8 nm (688 THz)
- Indigo
- Violet

## IR Band

The IR band is the region of the EM spectrum lying immediately below the visible light band. The IR band consists of EM radiation with wavelengths extending between the longest visible red (circa 0.7 µm) and the shortest microwaves (300 µm–1 mm), i.e., from circa 429 THz down to 1 THz–300 GHz.

The IR band is further subdivided into the "near" (shortwave), "intermediate" (midwave), and "far" (longwave) IR segments as follows [1]:

- *Near* IR segment, 0.7 µm up to 3 µm (429 THz down to 100 THz)
- *Intermediate* IR segment, 3 µm up to 7 µm (100 THz down to 42.9 THz)
- *Far* IR segment, 7 µm up to 300 µm (42.9 THz down to 1 THz)
- Sub-millimeter band, 100 µm up to 1 mm (3 THz down to 300 GHz). Note that the sub-millimeter region of wavelengths is sometimes included in the very far region of the IR band.

EM radiation is produced by oscillating and rotating molecules and atoms. Therefore, all objects at temperatures above absolute zero emit EM radiation by virtue of their thermal motion (warmth) alone. Objects near room temperature emit most of their radiation in the IR band. However, even relatively cool objects emit some IR radiation; hot objects, such as incandescent filaments, emit strong IR radiation.

IR radiation is sometimes incorrectly called "radiant heat" because warm bodies emit IR radiation and bodies that absorb IR radiation are warmed. However, IR radiation is not itself "heat". This radiant energy is called "black body" radiation. Such waves are emitted by all material objects. For example, the background cosmic radiation (2.7K) emits microwaves; room temperature objects (293K) emit IR rays; the Sun (6000K) emits yellow light; the Solar Corona (1 million K) emits X rays.

IR astronomy uses the 1 µm to 1 mm part of the IR band to study celestial objects by their IR emissions. IR detectors are used in night vision systems, intruder alarm systems, weather forecasting, and missile guidance systems. IR photography uses multilayered color film, with an IR sensitive emulsion in the wavelengths between 700–900 nm, for medical and forensic applications, and for aerial surveying.

---

1. Some reference texts use 2.5 mm (120 THz) as the breakpoint between the near and the intermediate IR bands and 10 mm (30 THz) as the breakpoint between the intermediate and the far IR bands. Also, 15 mm (20 Thz) is sometimes considered as the long wavelength end of the far IR band.

### UV Band

The UV band is the region of the EM spectrum lying immediately above the visible light band. The UV band consists of EM radiation with wavelengths extending between the shortest visible violet (circa 0.4 µm) and the longest X rays (circa 10 nm), i.e., from 750 THz—approximately 3 ev—up to circa 30 PHz—approximately 100 ev.[2]

The UV band is further subdivided into the "near" and the "far" UV segments as follows:

- *Near*" UV segment, circa 0.4 µm down to 100 nm (circa 750 THz up to 3 PHz, approximately 3 ev up to 10 ev)
- *Far* UV segment, 100 nm down to circa 10 nm, (3 PHz up to circa 30 PHz, approximately 10 ev up to 100 ev)

The far UV band is also referred to as the *vacuum UV band*, since air is opaque to all UV radiation in this region.

UV radiation is produced by electron transitions in atoms and molecules, as in a mercury discharge lamp. Radiation in the UV range is easily detected and can cause florescence in some substances, and can produce photographic and ionizing effects.

In UV astronomy, the emissions of celestial bodies in the wavelength band between 50–320 nm are detected and analyzed to study the heavens. The hottest stars emit most of their radiation in the UV band.

## 1.2b DC to Light

Below the IR band are the lower frequency (longer wavelength) regions of the EM spectrum, subdivided generally into the following spectral bands (by frequency/wavelength):

- *Microwave* band, 300 GHz down to 300 MHz (1 mm up to 1 m). Some reference works define the lower edge of the microwave spectrum at 1 GHz.
- *Radio frequency (RF)* band, 300 MHz down to 10 kHz (1 m up to 30 Km)
- *Power (PF)/telephony* band, 10 kHz down to dc (30 Km up to ∞)

These regions of the EM spectrum are usually described in terms of their frequencies.

Radiations whose wavelengths are of the order of millimeters & centimeters are called *microwaves*, and those still longer are called radio frequency (RF) waves (or *Hertzian waves*).

Radiation from electronic devices produces EM waves in both the microwave and RF bands. Power frequency energy is generated by rotating machinery. Direct current (dc) is produced by batteries or rectified alternating current (ac).

### Microwave Band

The microwave band is the region of wavelengths lying between the far IR/sub-millimeter region and the conventional RF region. The boundaries of the microwave band have not been definitely fixed, but it is commonly regarded as the region of the EM spectrum extending from about 1 mm up to 1 m in wavelengths, i.e. from 300 GHz down to 300 MHz. The microwave band is further sub-divided into the following segments:

---

2. Some references use 4, 5, or 6 nm as the upper edge of the UV band.

- *Millimeter* waves, 300 GHz down to 30 GHz (1 mm up to 1 cm); the EHF band. (Some references consider the top edge of the millimeter region to stop at 100 GHz.)
- *Centimeter* waves, 30 GHz down to 3 GHz (1 cm up to 10 cm); the SHF band.

The microwave band usually includes the UHF band from 3 GHz down to 300 MHz (from 10 cm up to 1 m).

Microwaves are used in radar, in communication links spanning moderate distances, as radio carrier waves in television broadcasting, for mechanical heating, and cooking in microwave ovens.

## Radio Frequency (RF) Band

The RF range of the EM spectrum is the wavelength band suitable for utilization in radio communications extending from 10 kHz up to 300 MHz (from 30 Km down to 1 m). (Some references consider the RF band as extending from 10 kHz to 300 GHz, with the microwave band as a subset of the RF band from 300 MHz to 300 GHz.)

Some of the radio waves in this band serve as the carriers of low-frequency audio signals; other radio waves are modulated by video and digital information. The *amplitude modulated* (AM) broadcasting band uses waves with frequencies between 550–1640 kHz; the *frequency modulated* (FM) broadcasting band uses waves with frequencies between 88–108 MHz.

In the U.S., the Federal Communications Commission (FCC) is responsible for assigning a range of frequencies to specific services. The International Telecommunications Union (ITU) coordinates frequency band allocation and cooperation on a worldwide basis.

Radio astronomy uses radio telescopes to receive and study radio waves naturally emitted by objects in space. Radio waves are emitted from hot gases (*thermal radiation*), from charged particles spiraling in magnetic fields (*synchrotron radiation*), and from excited atoms and molecules in space (*spectral lines*), such as the 21 cm line emitted by hydrogen gas.

## Power Frequency (PF)/Telephone Band

The PF range of the EM spectrum is the wavelength band suitable for generating, transmitting, and consuming low frequency power, extending from 10 kHz down to dc (zero frequency), i.e., from 30Km up in wavelength. In the US, most power is generated at 60 Hz (some military and computer applications use 400 Hz); in other countries, including Europe, power is generated at 50 Hz.

## Frequency Band Designations

The combined microwave, RF (Hertzian Waves), and power/telephone spectra are subdivided into the specific bands given in Table 1.1, which lists the international radio frequency band designations and the numerical designations. Note that the band designated (12) has no commonly used name or abbreviation.

The radar band often is considered to extend from the middle of the HF (7) band to the end of the EHF (11) band. The current US Tri-Service radar band designations are listed in Table 1.2.

An alternate and more detailed sub-division of the UHF (9), SHF (10), and EHF (11) bands is given in Table 1.3.

Several other frequency bands of interest (not exclusive) are listed in Tables 1.4–1.6.

**Table 1.1 Frequency Band Designations**

| Description | Band Designation | Frequency | Wavelength |
|---|---|---|---|
| Extremely Low Frequency | ELF (1) Band | 3 Hz up to 30 Hz | 100 Mm down to 10 Mm |
| Super Low Frequency | SLF (2) Band | 30 Hz up to 300 Hz | 10 Mm down to 1 Mm |
| Ultra Low Frequency | ULF (3) Band | 300 Hz up to 3 kHz | 1 Mm down to 100 Km |
| Very Low Frequency | VLF (4) Band | 3 kHz up to 30 kHz | 100 Km down to 10 Km |
| Low Frequency | LF (5) Band | 30 kHz up to 300 kHz | 10 Km down to 1 Km |
| Medium Frequency | MF (6) Band | 300 kHz up to 3 MHz | 1 Km down to 100 m |
| High Frequency | HF (7) Band | 3 MHz up to 30 MHz | 100 m down to 10 m |
| Very High Frequency | VHF (8) Band | 30 MHz up to 300 MHz | 10 m down to 1 m |
| Ultra High Frequency | UHF (9) Band | 300 MHz up to 3 GHz | 1 m down to 10 cm |
| Super High Frequency | SHF (10) Band | 3 GHz up to 30 GHz | 10 cm down to 1 cm |
| Extremely High Frequency | EHF (11) Band | 30 GHz up to 300 GHz | 1 cm down to 1 mm |
| — | (12) Band | 300 GHz up to 3 THz | 1 mm down to 100 μ |

**Table 1.2 Radar Band Designations**

| Band | Frequency | Wavelength |
|---|---|---|
| A Band | 0 Hz up to 250 MHz | ∞ down to 1.2 m |
| B Band | 250 MHz up to 500 MHz | 1.2 m down to 60 cm |
| C Band | 500 MHz up to 1 GHz | 60 cm down to 30 cm |
| D Band | 1 GHz up to 2 GHz | 30 cm down to 15 cm |
| E Band | 2 GHz up to 3 GHz | 15 cm down to 10 cm |
| F Band | 3 GHz up to 4 GHz | 10 cm down to 7.5 cm |
| G Band | 4 GHz up to 6 GHz | 7.5 cm down to 5 cm |
| H Band | 6 GHz up to 8 GHz | 5 cm down to 3.75 cm |
| I Band | 8 GHz up to 10 GHz | 3.75 cm down to 3 cm |
| J Band | 10 GHz up to 20 GHz | 3 cm down to 1.5 cm |
| K Band | 20 GHz up to 40 GHz | 1.5 cm down to 7.5 mm |
| L Band | 40 GHz up to 60 GHz | 7.5 mm down to 5 mm) |
| M Band | 60 GHz up to 100 GHz | 5 mm down to 3 mm |
| N Band | 100 GHz up to 200 GHz | 3 mm down to 1.5 mm |
| O Band | 200 GHz up to 300 GHz | 1.5 mm down to 1 mm |

**Table 1.3 Detail of UHF, SHF, and EHF Band Designations**

| Band | Frequency | Wavelength |
|---|---|---|
| L Band | 1.12 GHz up to 1.7 GHz | 26.8 cm down to 17.6 cm |
| LS Band | 1.7 GHz up to 2.6 GHz | 17.6 cm down to 11.5 cm |
| S Band | 2.6 GHz up to 3.95 GHz | 11.5 cm down to 7.59 cm |
| C(G) Band | 3.95 GHz up to 5.85 GHz | 7.59 cm down to 5.13 cm |
| XN(J, XC) Band | 5.85 GHz up to 8.2 GHz | 5.13 cm down to 3.66 cm |
| XB(H, BL) Band | 7.05 GHz up to 10 GHz | 4.26 cm down to 3 cm |
| X Band | 8.2 GHz up to 12.4 GHz | 3.66 cm down to 2.42 cm |
| Ku(P) Band | 12.4 GHz up to 18 GHz | 2.42 cm down to 1.67 cm |
| K Band | 18 GHz up to 26.5 GHz | 1.67 cm down to 1.13 cm |
| V(R, Ka) Band | 26.5 GHz up to 40 GHz | 1.13 cm down to 7.5 mm |
| Q(V) Band | 33 GHz up to 50 GHz | 9.09 mm down to 6 mm |
| M(W) Band | 50 GHz up to 75 GHz | 6 mm down to 4 mm |
| E(Y) Band | 60 GHz up to 90 GHz | 5 mm down to 3.33 mm |
| F(N) Band | 90 GHz up to 140 GHz | 3.33 mm down to 2.14 mm |
| G(A) | 140 GHz p to 220 GHz | 2.14 mm down to 1.36 mm |
| R Band | 220 GHz up to 325 GHz | 1.36 mm down to 0.923 mm |

**Table 1.4 Low Frequency Bands of Interest**

| Band | Frequency |
|---|---|
| Sub-sonic band | 0 Hz–10 Hz |
| Audio band | 10 Hz–10 kHz |
| Ultra-sonic band | 10 kHz and up |

**Table 1.5 Applications of Interest in the RF Band**

| Band | Frequency |
|---|---|
| Longwave broadcasting band | 150–290 kHz |
| AM broadcasting band | 550–1640 kHz (1.640 MHz), 107 channels, 10 kHz separation |
| International broadcasting band | 3–30 MHz |
| Shortwave broadcasting band | 5.95–26.1 MHz (8 bands) |
| VHF TV (Channels 2 - 4) | 54–72 MHz |
| VHF TV (Channels 5 - 6) | 76–88 MHz |
| FM broadcasting band | 88–108 MHz |
| VHF TV (Channels 7 - 13) | 174–216 MHz |
| UHF TV (Channels 14 - 69) | 512–806 MHz |

## 1.2c Light to Gamma Rays

Above the UV spectrum are the higher frequency (shorter wavelength) regions of the EM spectrum, subdivided generally into the following spectral bands (by frequency/wavelength):

- *X ray* band, approximately 10 ev up to 1 Mev (circa 10 nm down to circa 1 pm), circa 3 PHz up to circa 300 EHz
- *Gamma ray* band, approximately 1 Kev up to ∞ (circa 300 pm down to 0 m), circa 1 EHz up to ∞

These regions of the EM spectrum are usually described in terms of their photon energies in electron volts. Note that the bottom of the gamma ray band overlaps the top of the X ray band.

It should be pointed out that *cosmic "rays"* (from astronomical sources) are not EM waves (rays) and, therefore, are not part of the EM spectrum. Cosmic "rays" are high energy charged particles (electrons, protons, and ions) of extraterrestrial origin moving through space, which may have energies as high as $10^{20}$ ev. Cosmic "rays" have been traced to cataclysmic astrophysical/cosmological events, such as exploding stars and black holes. Cosmic "rays" are emitted by supernova remnants, pulsars, quasars, and radio galaxies. Comic "rays" that collide with molecules in the Earth's upper atmosphere produce secondary cosmic "rays" and gamma rays of high energy that also contribute to natural background radiation. These gamma rays are sometimes called "cosmic" or *secondary* gamma rays. Cosmic rays are a useful source of high-energy particles for certain scientific experiments.

Radiation from atomic inner shell excitations produces EM waves in the X ray band. Radiation from naturally radioactive nuclei produces EM waves in the gamma ray band.

### X Ray Band

The X ray band is further sub-divided into the following segments:

- *Soft* X rays, approximately 10 ev up to 10 Kev (circa 10 nm down to 100 pm), circa 3 PHz up to 3 EHz
- *Hard* X rays, approximately 10 Kev up to 1Mev (100 pm down to circa 1 pm), 3 EHz up to circa 300 EHz

Because the physical nature of these rays was first unknown, this radiation was called X rays. The more powerful X rays are called hard X rays and are of high frequencies and, therefore, are more energetic; less powerful X rays are called soft X rays and have lower energies.

X rays are produced by transitions of electrons in the inner levels of excited atoms or by rapid deceleration of charged particles (*Brehmsstrahlung* or breaking radiation). An important source of X rays is *synchrotron radiation*. X rays can also be produced when high energy electrons from a heated filament cathode strike the surface of a target anode (usually tungsten) between which a high alternating voltage (approximately 100 kV) is applied.

X rays are a highly penetrating form of EM radiation and applications of X rays are based on their short wavelengths and their ability to easily pass through matter. X rays are very useful in crystallography for determining crystalline structure and in medicine for photographing the body. Because different parts of the body absorb X rays to a different extent, X rays passing through the body provide a visual image of its interior structure when striking a photographic plate. X rays are dangerous and can destroy living tissue. They can also cause severe skin burns. X rays are useful in the diagnosis and non-destructive testing of products for defects.

**Table 1.6 Applications of Interest in the Microwave Band (up to 40 GHz):**

| Application | Frequency |
|---|---|
| Aero Navigation | 0.96–1.215 GHz |
| GPS Down Link | 1.2276 GHz |
| Military COM/Radar | 1.35–1.40 GHz |
| Miscellaneous COM/Radar | 1.40–1.71 GHz |
| L-Band Telemetry | 1.435–1.535 GHz |
| GPS Down Link | 1.57 GHz |
| Military COM (Troposcatter/Telemetry) | 1.71–1.85 GHz |
| Commercial COM & Private LOS | 1.85–2.20 GHz |
| Microwave Ovens | 2.45 GHz |
| Commercial COM/Radar | 2.45–2.69 GHz |
| Instructional TV | 2.50–2.69 GHz |
| Military Radar (Airport Surveillance) | 2.70–2.90 GHz |
| Maritime Navigation Radar | 2.90–3.10 GHz |
| Miscellaneous Radars | 2.90–3.70 GHz |
| Commercial C-Band SAT COM Down Link | 3.70–4.20 GHz |
| Radar Altimeter | 4.20–4.40 GHz |
| Military COM (Troposcatter) | 4.40–4.99 GHz |
| Commercial Microwave Landing System | 5.00–5.25 GHz |
| Miscellaneous Radars | 5.25–5.925 GHz |
| C-Band Weather Radar | 5.35–5.47 GHz |
| Commercial C-Band SAT COM Up Link | 5.925–6.425 GHz |
| Commercial COM | 6.425–7.125 GHz |
| Mobile TV Links | 6.875–7.125 GHz |
| Military LOS COM | 7.125–7.25 GHz |
| Military SAT COM Down Link | 7.25–7.75 GHz |
| Military LOS COM | 7.75–7.9 GHz |
| Military SAT COM Up Link | 7.90–8.40 GHz |
| Miscellaneous Radars | 8.50–10.55 GHz |
| Precision Approach Radar | 9.00–9.20 GHz |
| X-Band Weather Radar (& Maritime Navigation Radar) | 9.30–9.50 GHz |
| Police Radar | 10.525 GHz |
| Commercial Mobile COM (LOS & ENG) | 10.55–10.68 GHz |
| Common Carrier LOS COM | 10.70–11.70 GHz |
| Commercial COM | 10.70–13.25 GHz |
| Commercial Ku-Band SAT COM Down Link | 11.70–12.20 GHz |
| DBS Down Link & Private LOS COM | 12.20–12.70 GHz |
| ENG & LOS COM | 12.75–13.25 GHz |
| Miscellaneous Radars & SAT COM | 13.25–14.00 GHz |
| Commercial Ku-Band SAT COM Up Link | 14.00–14.50 GHz |
| Military COM (LOS, Mobile, &Tactical) | 14.50–15.35 GHz |
| Aero Navigation | 15.40–15.70 GHz |
| Miscellaneous Radars | 15.70–17.70 GHz |
| DBS Up Link | 17.30–17.80 GHz |
| Common Carrier LOS COM | 17.70–19.70 GHz |
| Commercial COM (SAT COM & LOS) | 17.70–20.20 GHz |
| Private LOS COM | 18.36–19.04 GHz |
| Military SAT COM | 20.20–21.20 GHz |
| Miscellaneous COM | 21.20–24.00 GHz |
| Police Radar | 24.15 GHz |
| Navigation Radar | 24.25–25.25 GHz |
| Military COM | 25.25–27.50 GHz |
| Commercial COM | 27.50–30.00 GHz |
| Military SAT COM | 30.00–31.00 GHz |
| Commercial COM | 31.00–31.20 GHz |

### Gamma Ray Band

The gamma ray band is sub-divided into the following segments:

- *Primary* gamma rays, approximately 1 Kev up to 1 Mev (circa 300 pm down to 300 fm), circa 1 EHz up to 1000 EHz
- *Secondary* gamma rays, approximately 1 Mev up to ∞ (300 fm down to 0 m), 1000 EHz up to ∞

Secondary gamma rays are created from collisions of high energy cosmic rays with particles in the Earth's upper atmosphere.

The primary gamma rays are further sub-divided into the following segments:

- *Soft* gamma rays, approximately 1 Kev up to circa 300 Kev (circa 300 pm down to circa 3 pm), circa 1 EHz up to circa 100 EHz
- *Hard* gamma rays, approximately 300 Kev up to 1 Mev (circa 3 pm down to 300 fm), circa 100 EHz up to 1000 EHz

Gamma rays are essentially very energetic X rays. The distinction between the two is based on their origin. X rays are emitted during atomic processes involving energetic electrons; gamma rays are emitted by excited nuclei or other processes involving sub-atomic particles.

Gamma rays are emitted by the nucleus of radioactive material during the process of natural radioactive decay as a result of transitions from high energy excited states to low energy states in atomic nuclei. Cobalt 90 is a common gamma ray source (with a half-life of 5.26 years). Gamma rays are also produced by the interaction of high energy electrons with matter. "Cosmic" gamma rays cannot penetrate the Earth's atmosphere.

Applications of gamma rays are found both in medicine and in industry. In medicine, gamma rays are used for cancer treatment, diagnoses, and prevention. Gamma ray emitting radioisotopes are used as tracers. In industry, gamma rays are used in the inspection of castings, seams, and welds.

## 1.3 Bibliography

Collocott, T. C., A. B. Dobson, and W. R. Chambers (eds.): *Dictionary of Science & Technology.*

*Handbook of Physics*, McGraw-Hill, New York, N.Y., 1958.

Judd, D. B., and G. Wyszecki: *Color in Business, Science and Industry*, 3rd ed., John Wiley and Sons, New York, N.Y.

Kaufman, Ed: *IES Illumination Handbook*, Illumination Engineering Society.

Lapedes, D. N. (ed.): *The McGraw-Hill Encyclopedia of Science & Technology*, 2nd ed., McGraw-Hill, New York, N.Y.

Norgard, John: "Electromagnetic Spectrum," *NAB Engineering Handbook*, $9^{th}$ ed., Jerry C. Whitaker (ed.), National Association of Broadcasters, Washington, D.C., 1999.

Norgard, John: "Electromagnetic Spectrum," *The Electronics Handbook*, Jerry C. Whitaker (ed.), CRC Press, Boca Raton, Fla., 1996.

Stemson, A: *Photometry and Radiometry for Engineers*, John Wiley and Sons, New York, N.Y.

*The Cambridge Encyclopedia*, Cambridge University Press, 1990.

*The Columbia Encyclopedia*, Columbia University Press, 1993.

*Webster's New World Encyclopedia*, Prentice Hall, 1992.

Wyszecki, G., and W. S. Stiles: *Color Science, Concepts and Methods, Quantitative Data and Formulae*, 2nd ed., John Wiley and Sons, New York, N.Y.

Chapter

# 2

# Propagation

**William Daniel, Edward W. Allen, Donald G. Fink**

## 2.1 Introduction

The portion of the electromagnetic spectrum currently used for radio transmissions lies between approximately 10 kHz and 40 GHz. The influence on radio waves of the medium through which they propagate is frequency-dependent. The lower frequencies are greatly influenced by the characteristics of the earth's surface and the ionosphere, while the highest frequencies are greatly affected by the atmosphere, especially rain. There are no clear-cut boundaries between frequency ranges but instead considerable overlap in propagation modes and effects of the path medium.

In the U.S., those frequencies allocated for television-related use include the following:

- 54–72 MHz: TV channels 2–4
- 76–88 MHz: TV channels 5–6
- 174–216 MHz: TV channels 7–13
- 470–806 MHz: TV channels 14–69
- 0.9–12.2 GHz: nonexclusive TV terrestrial and satellite ancillary services
- 12.2–12.7 GHz: direct satellite broadcasting
- 12.7–40 GHz: direct satellite broadcasting

## 2.2 Propagation in Free Space

For simplicity and ease of explanation, propagation in space and under certain conditions involving simple geometry, in which the wave fronts remain coherent, may be treated as *ray propagation*. It. should be kept in mind that this assumption may not hold in the presence of obstructions, surface roughness, and other conditions which are often encountered in practice.

For the simplest case of propagation in space, namely that of uniform radiation in all directions from a point source, or *isotropic radiator*, it is useful to consider the analogy to a point source of light, The radiant energy passes with uniform intensity through all portions of an imaginary spherical surface located at a radius $r$ from the source. The area of such a surface is $4\pi r^2$

and the power flow per unit area $W = P_t / 4\pi r^2$, where $P_t$ is the total power radiated by the source and $W$ is represented in $W/m^2$. In the engineering of broadcasting and of some other radio services, it is conventional to measure the intensity of radiation in terms of the strength of the electric field $E_o$ rather than in terms of power density $W$. The power density is equal to the square of the field strength divided by the impedance of the medium, so for free space:

$$W = \frac{E_o^2}{120\pi} \qquad (2.1)$$

and:

$$P_t = \frac{4\pi r^2 E_o^2}{120\pi} \qquad (2.2)$$

or:

$$P_t = \frac{r^2 E_o^2}{30} \qquad (2.3)$$

Where:
$P_t$ = watts radiated
$E_o$ = the free space field in volts per meter
$r$ = the radius in meters

A more conventional and useful form of this equation, which applies also to antennas other than isotropic radiators, is:

$$E_o = \frac{\sqrt{30\, g_t\, P_t}}{r} \qquad (2.4)$$

where $g_t$ is the power gain of the antenna in the pertinent direction compared to an isotropic radiator.

An isotropic antenna is useful as a reference for specifying the radiation patterns for more complex antennas but does not in fact exist. The simplest forms of practical antennas are the *electric doublet* and the *magnetic doublet*, the former a straight conductor that is short compared with the wavelength and the latter a conducting loop of short radius compared with the wavelength. For the doublet radiator, the gain is 1.5 and the field strength in the equatorial plane is:

$$E_o = \frac{\sqrt{45\, P_t}}{r} \qquad (2.5)$$

For a half-wave dipole, namely, a straight conductor one-half wave in length, the power gain is 1.64 and:

$$E_o = \frac{7\sqrt{P_t}}{r} \tag{2.6}$$

From the foregoing equations it can be seen that for free space:

- The radiation intensity in watts per square meter is proportional to the radiated power and inversely proportional to the square of the radius or distance from the radiator.
- The electric field strength is proportional to the square root of the radiated power and inversely proportional to the distance from the radiator.

## 2.2a Transmission Loss Between Antennas in Free Space

The maximum useful power $P_r$ that can be delivered to a matched receiver is given by [1]:

$$P_r = \left(\frac{E\lambda}{2\pi}\right)^2 \frac{g_r}{120} \tag{2.7}$$

Where:
$E$ = received field strength in volts per meter
$\lambda$ = wavelength in meters, $300/F$
$F$ = frequency in MHz
$g_r$ = receiving antenna power gain over an isotropic radiator

This relationship between received power and the received field strength is shown by scales 2, 3, and 4 in Figure 2.1 for a half-wave dipole. For example, the maximum useful power at 100 MHz that can be delivered by a half-wave dipole in a field of 50 dB above 1 µV/m is 95 dB below 1 W.

A general relation for the ratio of the received power to the radiated power obtained from Equations (2.4) and (2.7) is:

$$\frac{P_r}{P_t} = \left(\frac{\lambda}{4\pi r}\right)^2 g_t g_r \left(\frac{E}{E_o}\right)^2 \tag{2.8}$$

When both antennas are half-wave dipoles, the power-transfer ratio is:

$$\frac{P_r}{P_t} = \left(\frac{1.64\lambda}{4\pi r}\right)^2 \left(\frac{E}{E_o}\right)^2 = \left(\frac{0.13\lambda}{r}\right)^2 \left(\frac{E}{E_o}\right)^2 \tag{2.9}$$

and is shown on scales 1 to 4 of Figure 2.2. For free-space transmission, $E/E_o = 1$.

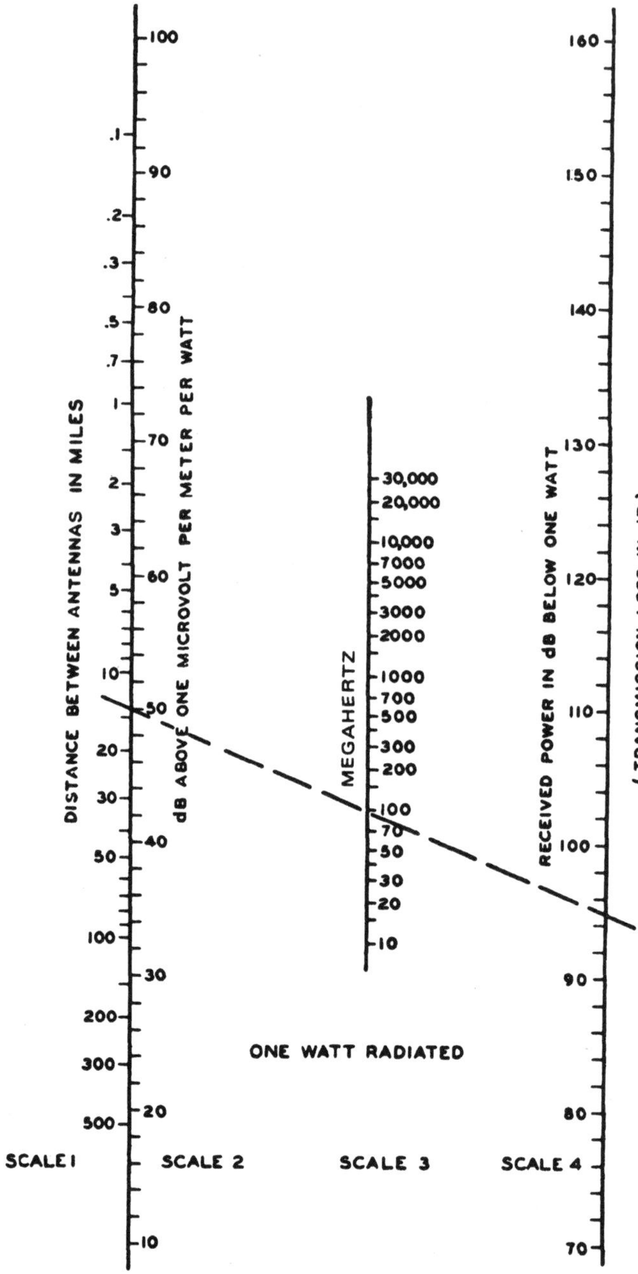


**Figure 2.1** Free-space field intensity and received power between half-wave dipoles. (*From* [2]. *Used with permission.*)

When the antennas are horns, paraboloids, or rnultielement arrays, a more convenient expression for the ratio of the received power to the radiated power is given by the following:

$$\frac{P_r}{P_t} = \frac{B_t B_r}{(\lambda r)^2}\left(\frac{E}{E_o}\right)^2 \tag{2.10}$$

where $B_t$ and $B_r$ are the effective areas of the transmitting and receiving antennas, respectively. This relation is obtained from Equation (2.8) by substituting as follows:

$$g = \frac{4\pi B}{\lambda^2} \tag{2.11}$$

This is shown in Figure 2.2 for free-space transmission when $B_t = B_r$. For example, the free-space loss at 4000 MHz between two antennas of 10 ft$^2$ (0.93 m$^2$) effective area is about 72 dB for a distance of 30 mi (48 km).

## 2.3 Propagation Over Plane Earth

The presence of the ground modifies the generation and propagation of radio waves so that the received field strength is ordinarily different than would be expected in free space [3, 4]. The ground acts as a partial reflector and as a partial absorber, and both of these properties affect the distribution of energy in the region above the earth.

### 2.3a Field Strengths Over Plane Earth

The geometry of the simple case of propagation between two antennas each placed several wavelengths above a plane earth is shown in Figure 2.3. For isotropic antennas, for simple magnetic-doublet antennas with vertical polarization, or for simple electric-doublet antennas with horizontal polarization the resultant received field is [4, 5]:

$$E = \frac{E_o d}{r_1} + \frac{E_o d R e^{j\Delta}}{r_2} = E_o\left(\cos\theta_1 + R\cos\theta_2\, e^{j\Delta}\right) \tag{2.12}$$

For simple magnetic-doublet antennas with horizontal polarization or electric-doublet antennas with vertical polarization at both the transmitter and receiver, it is necessary to correct for the cosine radiation and absorption patterns in the plane of propagation. The received field is:

$$E = E_o\left(\cos^3\theta_1 + R\cos^3\theta_2\, e^{j\Delta}\right) \tag{2.13}$$

Where:

$E_o$ = the free-space field at distance $d$ in the equatorial plane of the doublet

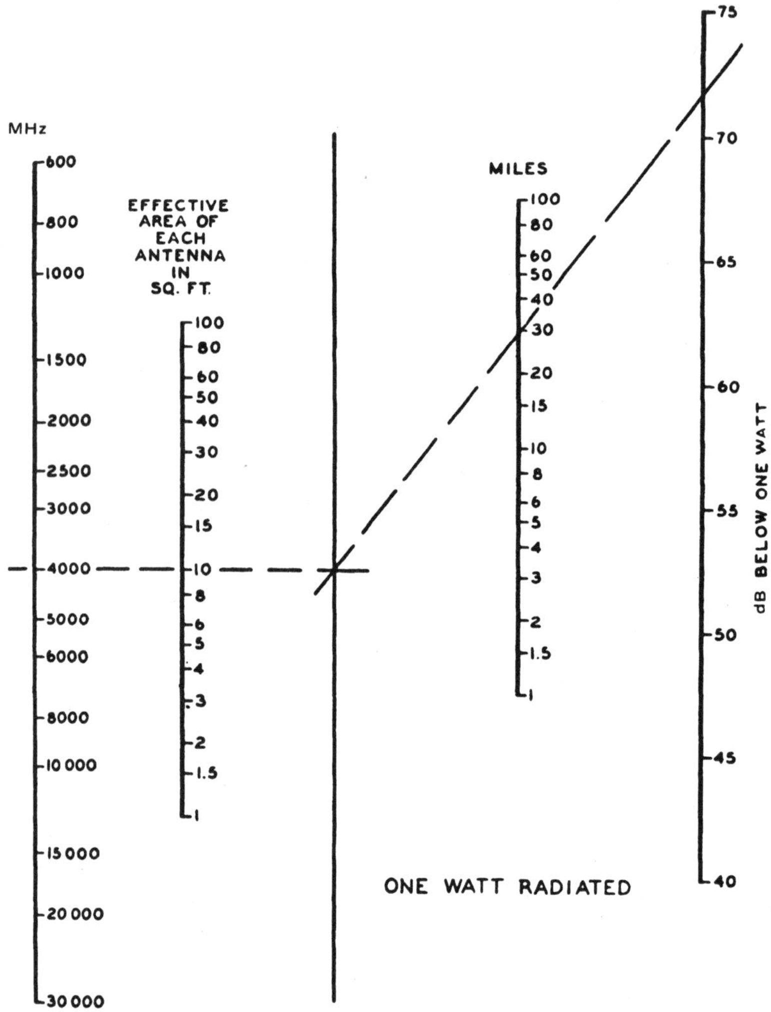


**Figure 2.2** Received power in free space between two antennas of equal effective areas. (*From* [2]. *Used with permission.*)

$R$ = the complex reflection coefficient of the earth
$j$ = the square root of $-1$
$e^{j\Delta} = \cos \Delta + j \sin \Delta$
$\Delta$ = the phase difference between the direct wave received over path $r_1$ and the ground-reflected wave received over path $r_2$, which is due to the difference in path lengths

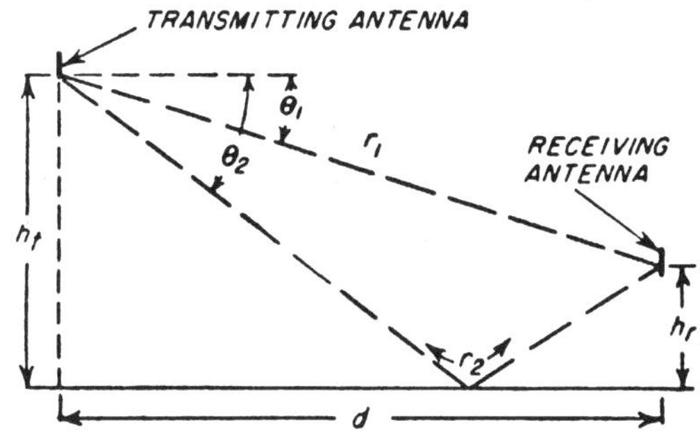


**Figure 2.3** Ray paths for antennas above plane earth. (*From* [2]. *Used with permission.*)

For distances such that θ is small and the differences between $d$ and $r_1$ and $r_2$ can be neglected, Equations (2.12) and (2.13) become:

$$E = E_o\left(1 + R e^{j\Delta}\right) \tag{2.14}$$

When the angle θ is very small, $R$ is approximately equal to –1. For the case of two antennas, one or both of which may be relatively close to the earth, a surface-wave term must be added and Equation (2.14) becomes [3, 6]:

$$E = E_o[1 + Re^{j\Delta} + (1 - R)Ae^{j\Delta}] \tag{2.15}$$

The quantity $A$ is the *surface-wave attenuation factor*, which depends upon the frequency, ground constants, and type of polarization. It is never greater than unity and decreases with increasing distance and frequency, as indicated by the following approximate equation [1]:

$$A \cong \frac{-1}{1 + j\left(\frac{2\pi d}{\lambda}\right)(\sin\theta + z)^2} \tag{2.16}$$

This approximate expression is sufficiently accurate as long as $A < 0.1$, and it gives the magnitude of $A$ within about 2 dB for all values of $A$. However, as $A$ approaches unity, the error in phase approaches 180°. More accurate values are given by Norton [3] where, in his nomenclature, $A = f(P,B)\, e^{i\phi}$.

The equation (14.4.15) for the absolute value of field strength has been developed from the successive consideration of the various components that make up the ground wave, but the following equivalent expressions may be found more convenient for rapid calculation:

$$E = E_o\left\{2\sin\frac{\Delta}{2} + j\left[(1 + R) + (1 - R)A\right]e^{j\Delta/2}\right\} \tag{2.17}$$

When the distance $d$ between antennas is greater than about five times the sum of the two antenna heights $h_t$ and $h_r$, the phase difference angle $\Delta$ (rad) is:

$$\Delta = \frac{4\pi h_t h_r}{\lambda d} \tag{2.18}$$

Also, when the angle $\Delta$ is greater than about 0.5 rad, the terms inside the brackets of Equation (2.17)—which include the surface wave—are usually negligible, and a sufficiently accurate expression is given by the following:

$$E = E_o \left( 2 \sin \frac{2\pi h_t h_r}{\lambda d} \right) \tag{2.19}$$

In this case, the principal effect of the ground is to produce interference fringes or *lobes*, so that the field strength oscillates about the free-space field as the distance between antennas or the height of either antenna is varied.

When the angle $\Delta$ is less than about 0.5 rad, there is a region in which the surface wave may be important but not controlling. In this region, sin $\Delta/2$ is approximately equal to $\Delta/2$ and:

$$E = E_o \frac{4\pi h_t' h_r'}{\lambda d} \tag{2.20}$$

In this equation $h' = h + jh_o$, where $h$ is the actual antenna height and $h_o = \lambda/2\pi z$ has been designated as the minimum effective antenna height. The magnitude of the minimum effective height $h_o$ is shown in Figure 2.4 for seawater and for "good" and "poor" soil. "Good" soil corresponds roughly to clay, loam, marsh, or swamp, while "poor" soil means rocky or sandy ground [1].

The surface wave is controlling for antenna heights less than the minimum effective height, and in this region the received field or power is not affected appreciably by changes in the antenna height. For antenna heights that are greater than the minimum effective height, the received field or power is increased approximately 6 dB every time the antenna height is doubled, until free-space transmission is reached. It is ordinarily sufficiently accurate to assume that $h'$ is equal to the actual antenna height or the minimum effective antenna height, whichever is the larger.

When translated into terms of antenna heights in feet, distance in miles, effective power in kilowatts radiated from a half-wave dipole, and frequency $F$ in megahertz, Equation (2.20) becomes the following very useful formula for the rapid calculation of approximate values of field strength for purposes of prediction or for comparison with measured values:

$$E \cong F \frac{h_t' h_r' \sqrt{P_t}}{3 d^2} \tag{2.21}$$

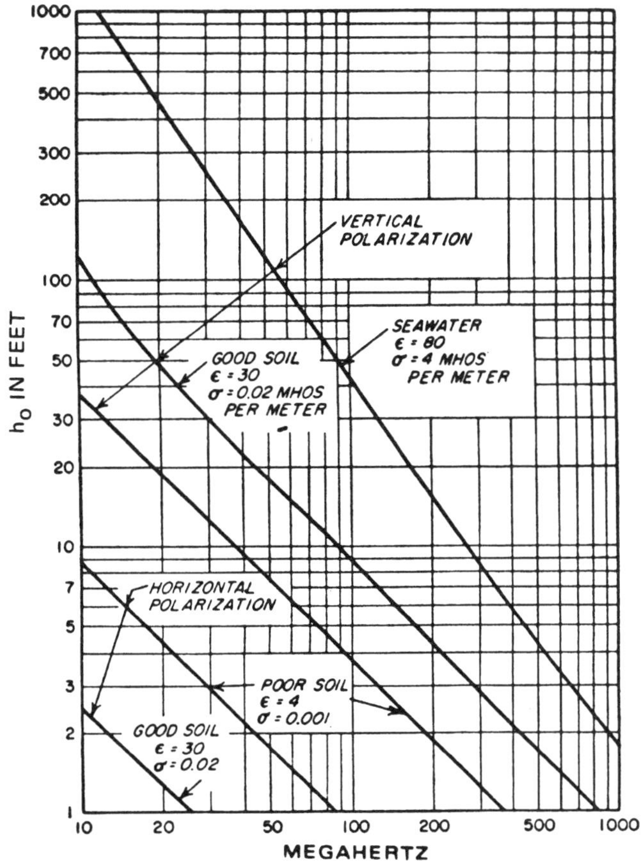


**Figure 2.4** Minimum effective antenna height. (*From* [2]. *Used with permission.*)

## 2.3b Transmission Loss Between Antennas Over Plane Earth

The ratio of the received power to the radiated power for transmission over plane earth is obtained by substituting Equation (2.20) into (2.8), resulting in the following:

$$\frac{P_r}{P_t} = \left(\frac{\lambda}{4\pi d}\right)^2 g_t\, g_r \left(\frac{4\pi h_t'\, h_r'}{\lambda d}\right) = \left(\frac{h_t'\, h_r'}{d^2}\right)^2 g_t\, g_r \tag{2.22}$$

This relationship is independent of frequency, and is shown on Figure 2.5 for half-wave dipoles ($g_t = g_r = 1.64$). A line through the two scales of antenna height determines a point on the unlabeled scale between them, and a second line through this point and the distance scale determines the received power for 1 W radiated. When the received field strength is desired, the power indicated on Figure 2.5 can be transferred to scale 4 of Figure 2.1, and a line through the frequency on scale 3 indicates the received field strength on scale 2. The results shown on Figure 2.5 are valid as long as the value of received power indicated is lower than that shown on Figure 2.3 for free-space transmission. When this condition is not met, it means that the angle Δ is too large for Equation (2.20) to be accurate and that the received field strength or power oscillates around the free-space value as indicated by Equation (2.19) [1].

## 2.3c Propagation Over Smooth Spherical Earth

The curvature of the earth has three effects on the propagation of radio waves at points within the line of sight:

- The *reflection coefficient* of the ground-reflected wave differs for the curved surface of the earth from that for a plane surface. This effect is of little importance, however, under the circumstances normally encountered in practice.
- Because the ground-reflected wave is reflected against the curved surface of the earth, its energy diverges more than would be indicated by the inverse distance-squared law, and the ground-reflected wave must be multiplied by a divergence factor *D*.
- The heights of the transmitting and receiving antennas $h_t'$ and $h_r'$ , above the plane that is tangent to the surface of the earth at the point of reflection of the ground-reflected wave, are less than the antenna heights $h_t$ and $h_r$ above the surface of the earth, as shown in Figure 2.6.

Under these conditions, Equation (2.14), which applies to larger distances within the line of sight and to antennas of sufficient height that the surface component may be neglected, becomes:

$$E = E_o\left(1 + D\,R'\,e^{j\Delta}\right) \tag{2.23}$$

Similar substitutions of the values that correspond in Figures 2.3 and 2.6 can be made in Equations (2.15 through (2.22). However, under practical conditions, it is generally satisfactory to use the plane-earth formulas for the purpose of calculating smooth-earth values. An exception to this is usually made in the preparation of standard reference curves, which are generally calculated by the use of the more exact formulas [1, 4–9]

## 2.3d Propagation Beyond the Line of Sight

Radio waves are bent around the earth by the phenomenon of *diffraction*, with the ease of bending decreasing as the frequency increases. Diffraction is a fundamental property of wave motion, and in optics it is the correction to apply to geometrical optics (*ray theory*) to obtain the more accurate *wave optics*. In wave optics, each point on the wave front is considered to act as a radiating source. When the wave front is coherent or undisturbed, the resultant is a progression of the front in a direction perpendicular thereto, along a path that constitutes the ray. When the front is disturbed, the resultant front can be changed in both magnitude and direction with resulting

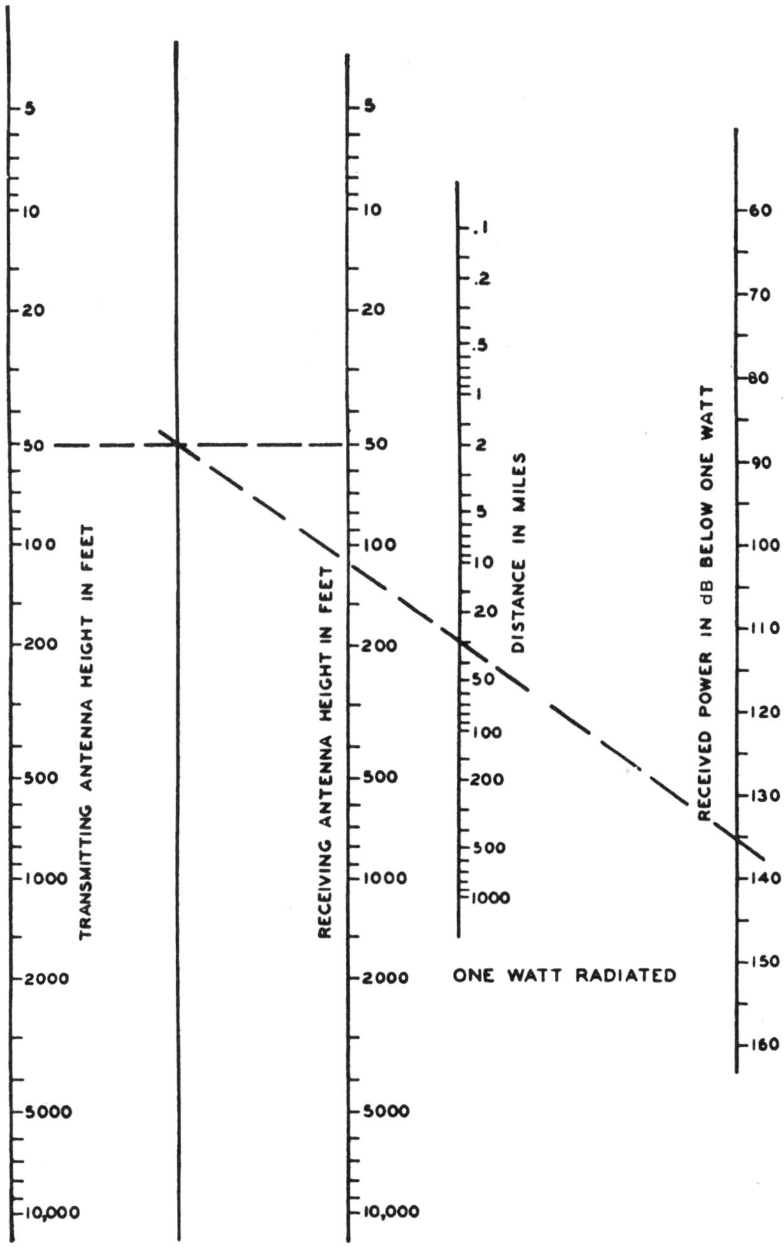


**Figure 2.5** Received power over plane earth between half-wave dipoles. *Notes:* (1) This chart is not valid when the indicated received power is greater than the free space power shown in Figure 2.1. (2) Use the actual antenna height or the minimum effective height shown in Figure 2.4, whichever is the larger. (*From* [2]. *Used with permission.*)

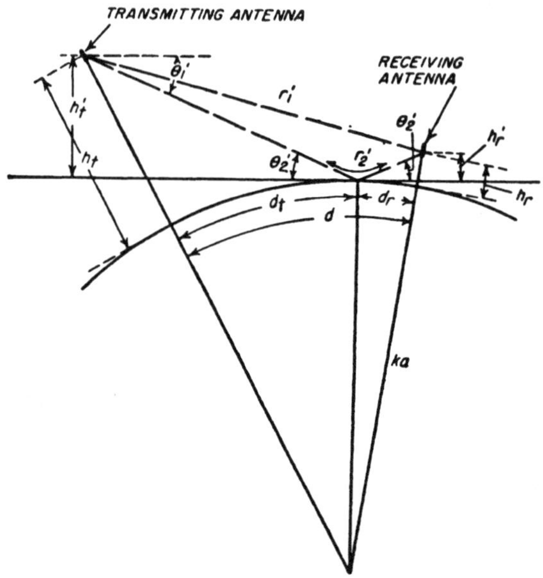


**Figure 2.6** Ray paths for antennas above spherical earth. (*From* [2]. *Used with permission.*)

attenuation and bending of the ray. Thus, all shadows are somewhat "fuzzy" on the edges and the transition from "light" to "dark" areas is gradual, rather than infinitely sharp.

The effect of diffraction around the earth's curvature is to make possible transmission beyond the line of sight, with somewhat greater loss than is incurred in free space or over plane earth. The magnitude of this loss increases as either the distance or the frequency is increased and it depends to some extent on the antenna height.

The calculation of the field strength to be expected at any particular point in space beyond the line of sight around a spherical earth is rather complex, so that individual calculations are seldom made except with specially designed software. Rather, nomograms or families of curves are usually prepared for general application to large numbers of cases. The original wave equations of Van der Pol and Bremmer [6] have been modified by Burrows [7] and by Norton [3, 5] so as to make them more readily usable and particularly adaptable to the production of families of curves. Such curves have been prepared by a variety of organizations. These curves have not been included herein, in view of the large number of curves that are required to satisfy the possible variations in frequency, electrical characteristics of the earth, polarization, and antenna height. Also, the values of field strength indicated by smooth-earth curves are subject to considerable modification under actual conditions found in practice. For VHF and UHF broadcast purposes, the smooth-earth curves have been to a great extent superseded by curves modified to reflect average conditions of terrain.

Figure 2.7 is a nomogram to determine the additional loss caused by the curvature of the earth [1]. This loss must be added to the free-space loss found from Figure 2.1. A scale is included to provide for the effect of changes in the effective radius of the earth, caused by atmospheric refraction. Figure 2.7 gives the loss relative to free space as a function of three distances; $d_1$ is the distance to the horizon from the lower antenna, $d_2$ is the distance to the horizon from the

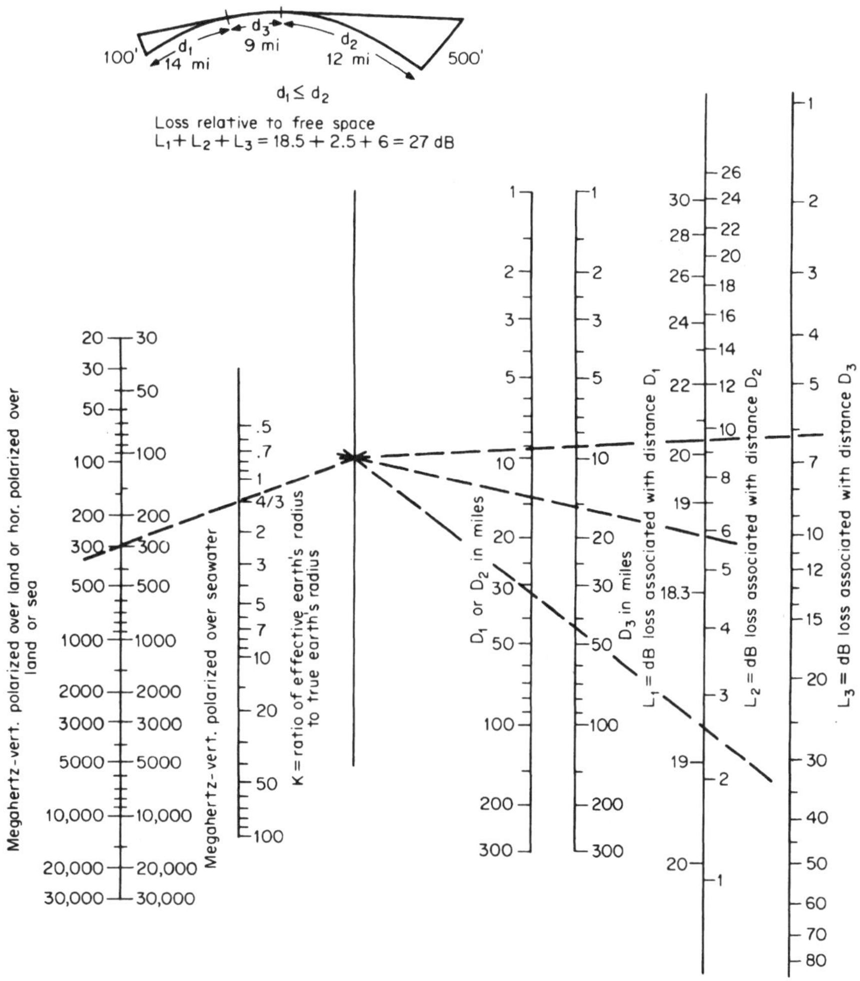


**Figure 2.7** Loss beyond line of sight in decibels. (*From* [2]. *Used with permission.*)

higher antenna, and $d_3$ is the distance between the horizons. The total distance between antennas is $d = d_1 + d_2 + d_3$.

The horizon distances $d_1$ and $d_2$ for the respective antenna heights $h_1$ and $h_2$ and for any assumed value of the earth's radius factor $k$ can be determined from Figure 2.8 [1].

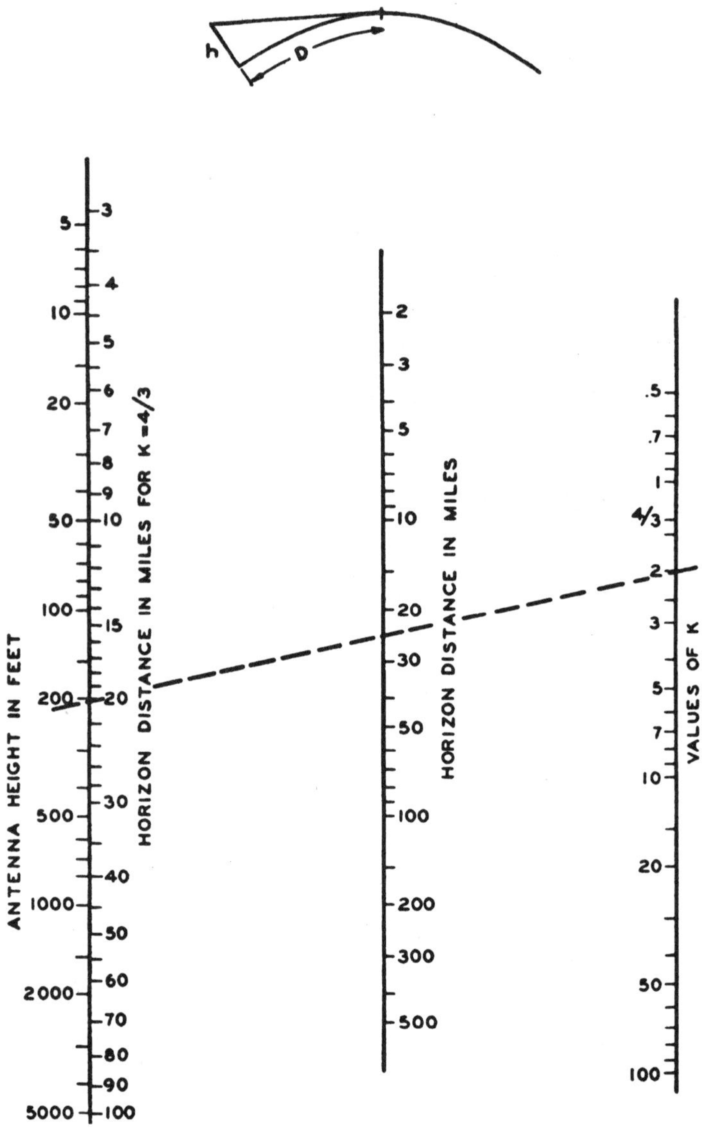


**Figure 2.8** Distance to the horizon. (*From* [2]. *Used with permission.*)

## 2.3e Effects of Hills, Buildings, Vegetation, and the Atmosphere

The preceding discussion assumes that the earth is a perfectly smooth sphere with a uniform or a simple atmosphere, for which condition calculations of expected field strengths or transmission

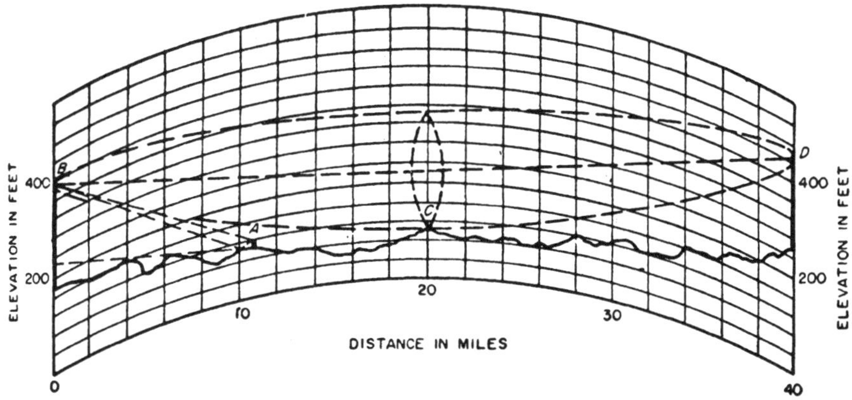


**Figure 2.9** Ray paths for antennas over rough terrain. (*From* [2]. *Used with permission.*)

losses can be computed for the regions within the line of sight and regions well beyond the line of sight, and interpolations can be made for intermediate distances. The presence of hills, buildings, and trees has such complex effects on propagation that it is impossible to compute in detail the field strengths to be expected at discrete points in the immediate vicinity of such obstructions or even the median values over very small areas. However, by the examination of the earth profile over the path of propagation and by the use of certain simplifying assumptions, predictions that are more accurate than smooth-earth calculations can be made of the median values to be expected over areas representative of the gross features of terrain.

## Effects of Hills

The profile of the earth between the transmitting and receiving points is taken from available topographic maps and is plotted on a chart that provides for average air refraction by the use of a 4/3 earth radius, as shown in Figure 2.9. The vertical scale is greatly exaggerated for convenience in displaying significant angles and path differences. Under these conditions, vertical dimensions are measured along vertical parallel lines rather than along radii normal to the curved surface, and the propagation paths appear as straight lines. The field to be expected at a low receiving antenna at $A$ from a high transmitting antenna at $B$ can be predicted by plane-earth methods, by drawing a tangent to the profile at the point at which reflection appears to occur with equal incident and reflection angles. The heights of the transmitting and receiving antennas above the tangent are used in conjunction with Figure 14.4.5 to compute the transmission loss, or with Equation (2.21) to compute the field strength. A similar procedure can be used for more distantly spaced high antennas when the line of sight does not clear the profile by at least the first *Fresnel zone* [10].

Propagation over a sharp ridge, or over a hill when both the transmitting and receiving antenna locations are distant from the hill, may be treated as diffraction over a knife edge, shown schematically in Figure 2.10*a* [1, 9–14]. The height of the obstruction $H$ is measured from the line joining the centers of the two antennas to the top of the ridge. As shown in Figure 2.11, the

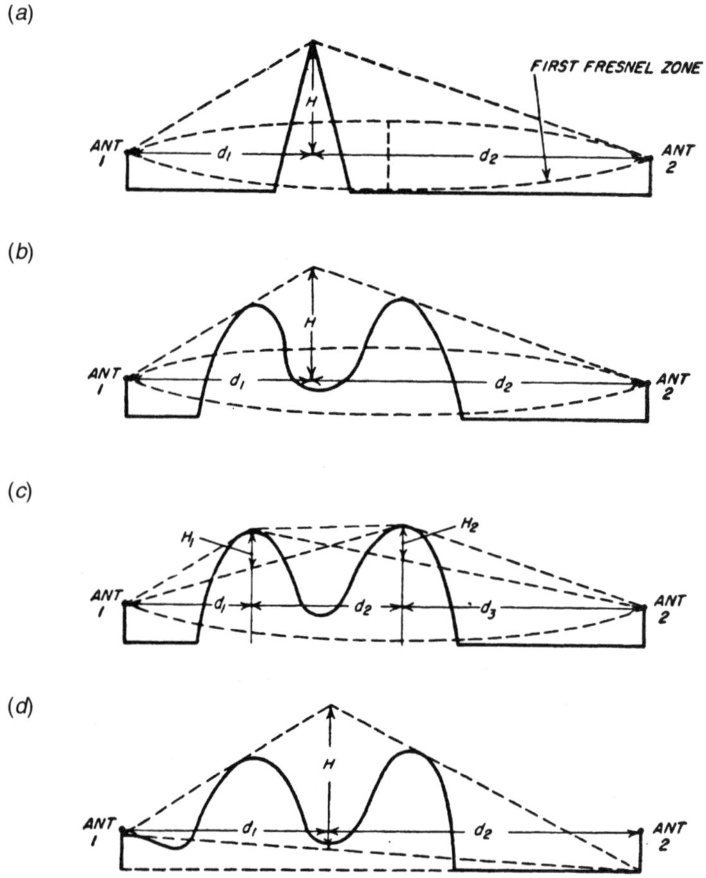


**Figure 2.10** Ray paths for antennas behind hills: (*a–d*), see text. (*From* [2]. *Used with permission.*)

shadow loss approaches 6 dB as $H$ approaches 0—*grazing incidence*—and it increases with increasing positive values of $H$. When the direct ray clears the obstruction, $H$ is negative, and the shadow loss approaches 0 dB in an oscillatory manner as the clearance is increased. Thus, a substantial clearance is required over line-of-sight paths in order to obtain free-space transmission. There is an optimum clearance, called the first Fresnel-zone clearance, for which the transmission is theoretically 1.2 dB better than in free space. Physically, this clearance is of such magnitude that the phase shift along a line from the antenna to the top of the obstruction and from there to the second antenna is about one-half wavelength greater than the phase shift of the direct path between antennas.

The locations of the first three Fresnel zones are indicated on the right-hand scale on Figure 2.11, and by means of this chart the required clearances can be obtained. At 3000 MHz, for example, the direct ray should clear all obstructions in the center of a 40 mi (64 km) path by

*Note*: When accuracy greater than ±1.5 dB is required, values on the $d_1$ scale should be: $d_1 \dfrac{\sqrt{2}}{1 + d_1/d_2}$

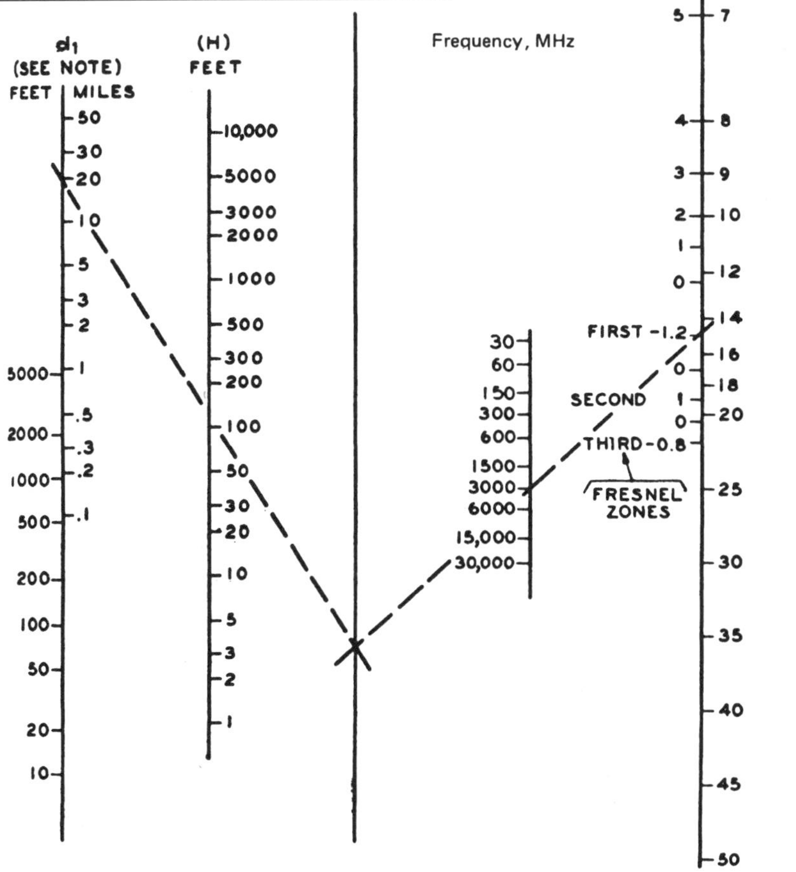


**Figure 2.11** Shadow loss relative to free space. (*From* [2]. *Used with permission.*)

about 120 ft (36 m) to obtain full first-zone clearance, as shown at "C" in Figure 2.9. The corresponding clearance for a ridge 100 ft (30 m) in front of either antenna is 4 ft (1.2 m). The locus of all points that satisfy this condition for all distances is an ellipsoid of revolution with foci at the two antennas.

When there are two or more knife-edge obstructions or hills between the transmitting and receiving antennas, an equivalent knife edge can be represented by drawing a line from each antenna through the top of the peak that blocks the line of sight, as in Figure 2.10*b*.

Alternatively, the transmission loss can be computed by adding the losses incurred when passing over each of the successive hills, as in Figure 2.10*c*. The height $H_1$ is measured from the top of hill 1 to the line connecting antenna 1 and the top of hill 2. Similarly, $H_2$ is measured from the top of hill 2 to the line connecting antenna 2 and the top of hill 1. The nomogram in Figure 2.11 is used for calculating the losses for terrain conditions represented by Figure 2.10*a–c*.

This procedure applies to conditions for which the earth-reflected wave can be neglected, such as the presence of rough earth, trees, or structures at locations along the profile at points where earth reflection would otherwise take place at the frequency under consideration; or where first Fresnel-zone clearance is obtained in the foreground of each antenna and the geometry is such that reflected components do not contribute to the field within the first Fresnel zone above the obstruction. If conditions are favorable to earth reflection, the base line of the diffraction triangle should not be drawn through the antennas, but through the points of earth reflection, as in Figure 2.10*d*. *H* is measured vertically from this base line to the top of the hill, while $d_1$ and $d_2$ are measured to the antennas as before. In this case, Figure 2.12 is used to estimate the shadow loss to be added to the plane-earth attenuation [1].

Under conditions where the earth-reflected components reinforce the direct components at the transmitting and receiving antenna locations, paths may be found for which the transmission loss over an obstacle is less than the loss over spherical earth. This effect may be useful in establishing VHF relay circuits where line-of-sight operation is not practical. Little utility, however, can be expected for mobile or broadcast services [14].

An alternative method for predicting the median value for all measurements in a completely shadowed area is as follows [15]:

1. The roughness of the terrain is assumed to be represented by height *H*, shown on the profile at the top of Figure 2.13.
2. This height is the difference in elevation between the bottom of the valley and the elevation necessary to obtain line of sight with the transmitting antenna.
3. The difference between the measured value of field intensity and the value to be expected over plane earth is computed for each point of measurement within the shadowed area.
4. The median value for each of several such locations is plotted as a function of sq. rt. ($H/\lambda$).

These empirical relationships are summarized in the nomogram shown in Figure 2.13. The scales on the right-hand line indicate the median value of shadow loss, compared with plane-earth values, and the difference in shadow loss to be expected between the median and the 90 percent values. For example, with variations in terrain of 500 ft (150 m), the estimated median shadow loss at 4500 MHz is about 20 dB and the shadow loss exceeded in 90 percent of the possible locations is about 20 + 15 = 35 dB. This analysis is based on large-scale variations in field intensity, and does not include the standing-wave effects that sometimes cause the field intensity to vary considerably within a matter of a few feet.

## Effects of Buildings

Built-up areas have little effect on radio transmission at frequencies below a few megahertz, since the size of any obstruction is usually small compared with the wavelength, and the shadows caused by steel buildings and bridges are not noticeable except immediately behind these obstructions. However, at 30 MHz and above, the absorption of a radio wave in going through an obstruction and the shadow loss in going over it are not negligible, and both types of losses tend

*Note*: When accuracy greater than ±1.5 dB is required, values on the $d_1$ scale should be:

$$d_1 \frac{\sqrt{2}}{1+d_1/d_2}$$

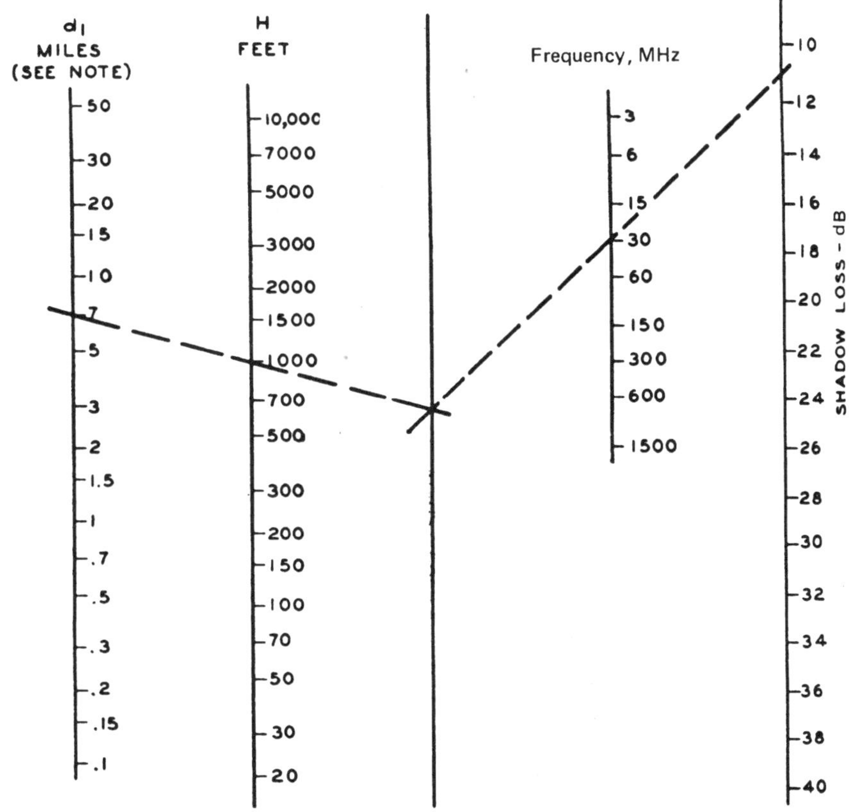


**Figure 2.12** Shadow loss relative to plane earth. (*From* [2]. *Used with permission.*)

to increase as the frequency increases. The attenuation through a brick wall, for example, can vary from 2 to 5 dB at 30 MHz and from 10 to 40 dB at 3000 MHz, depending on whether the wall is dry or wet. Consequently, most buildings are rather opaque at frequencies of the order of thousands of megahertz.

For radio-relay purposes, it is the usual practice to select clear sites; but where this is not feasible the expected fields behind large buildings can be predicted by the preceding diffraction methods. In the engineering of mobile- and broadcast-radio systems it has not been found practi-

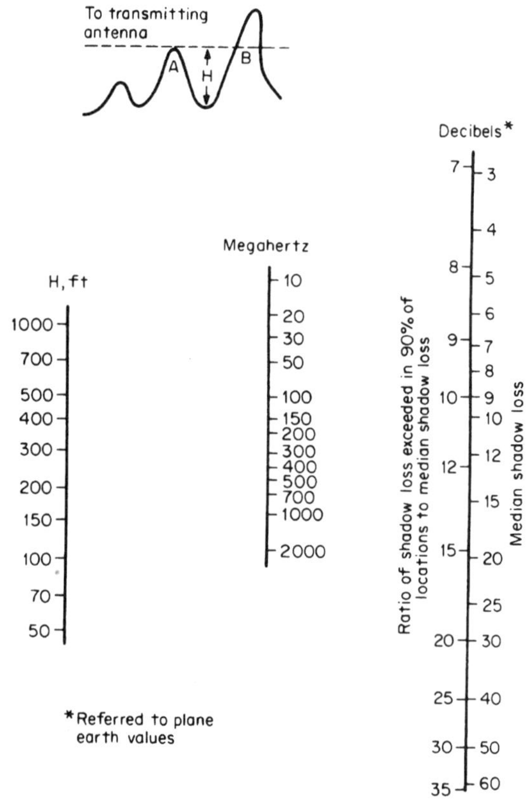


**Figure 2.13** Estimated distribution of shadow loss for random locations (referred to plane-earth values). (*From* [2]. *Used with permission.*)

cal in general to relate measurements made in built-up areas to the particular geometry of buildings, so that it is conventional to treat them statistically. However, measurements have been divided according to general categories into which buildings can readily be classified, namely, the tall buildings typical of the centers of cities on the one hand, and typical two-story residential areas on the other.

Buildings are more transparent to radio waves than the solid earth, and there is ordinarily much more backscatter in the city than in the open country. Both of these factors tend to reduce the shadow losses caused by the buildings. On the other hand, the angles of diffraction over or around the buildings are usually greater than for natural terrain, and this factor tends to increase the loss resulting from the presence of buildings. Quantitative data on the effects of buildings indicate that in the range of 40 to 450 MHz there is no significant change with frequency, or at least the variation with frequency is somewhat less than the square-root relationship noted in the case of hills. The median field strength at street level for random locations in New York City is about 25 dB below the corresponding plane-earth value. The corresponding values for the 10 percent and 90 percent points are about –15 and –35 dB, respectively [1, 15]. Measurements in congested residential areas indicate somewhat less attenuation than among large buildings.

### Effects of Trees and Other Vegetation

When an antenna is surrounded by moderately thick trees and below treetop level, the average loss at 30 MHz resulting from the trees is usually 2 or 3 dB for vertical polarization and negligible with horizontal polarization. However, large and rapid variations in the received field strength can exist within a small area, resulting from the standing-wave pattern set up by reflections from trees located at a distance of as much as 100 ft (30 m) or more from the antenna. Consequently, several nearby locations should be investigated for best results. At 100 MHz, the average loss from surrounding trees may be 5 to 10 dB for vertical polarization and 2 or 3 dB for horizontal polarization. The tree losses continue to increase as the frequency increases, and above 300 to 500 MHz they tend to be independent of the type of polarization. Above 1000 MHz, trees that are thick enough to block vision present an almost solid obstruction, and the diffraction loss over or around these obstructions can be obtained from Figures 2.9 or 2.11.

There is a pronounced seasonal effect in the case of deciduous trees, with less shadowing and absorption in the winter months when the leaves have fallen. However, when the path of travel through the trees is sufficiently long that it is obscured, losses of the above magnitudes can be incurred, and the principal mode of propagation may be by diffraction over the trees.

When the antenna is raised above trees and other forms of vegetation, the prediction of field strengths again depends upon the proper estimation of the height of the antenna above the areas of reflection and of the applicable reflection coefficients. For growth of fairly uniform height and for angles near grazing incidence, reflection coefficients will approach –1 at frequencies near 30 MHz. As indicated by Rayleigh's criterion of roughness, the apparent roughness for given conditions of geometry increases with frequency so that near 1000 MHz even such low and relatively uniform growth as farm crops or tall grass may have reflection coefficients of about –0.3 for small angles of reflection [17].

The distribution of losses in the immediate vicinity of trees does not follow normal probability law but is more accurately represented by Rayleigh's law, which is the distribution of the sum of a large number of equal vectors having random phases.

## 2.3f Effects of the Lower Atmosphere (Troposphere)

Radio waves propagating through the lower atmosphere, or troposphere, are subject to absorption, scattering, and bending. Absorption is negligible in the VHF–UHF frequency range but becomes significant at frequencies above 10 GHz. The index of refraction of the atmosphere, $n$, is slightly greater than 1 and varies with temperature, pressure, and water vapor pressure, and therefore with height, climate, and local meteorological conditions. An exponential model showing a decrease with height to 37 to 43 mi (60 to 70 kin) is generally accepted [18, 19]. For this model, variation of $n$ is approximately linear for the first kilometer above the surface in which most of the effect on radio waves traveling horizontally occurs. For average conditions, the effect of the atmosphere can be included in the expression of earth diffraction around the smooth earth without discarding the useful concept of straight-line propagation by multiplying the actual earth's radius by $k$ to obtain an effective earth's radius, where

$$k = \frac{1}{1 + a(dn/dh)} \tag{2.24}$$

Where:
$a$ = the actual radius of the earth
$dn/dh$ = the rate of change of the refractive index with height

Through the use of average annual values of the refractive index gradient, $k$ is found to be 4/3 for temperate climates.

### Stratification and Ducts

As a result of climatological and weather processes such as *subsidence*, *advection*, and surface heating and radiative cooling, the lower atmosphere tends to be stratified in layers with contrasting refractivity gradients [20]. For convenience in evaluating the effect of this stratification, *radio refractivity* $N$ is defined as $N = (n - 1) \times 10^6$ and can be derived from:

$$N = 77.6\frac{P}{T} + 3.73\times10^5\frac{e}{T^2} \tag{2.25}$$

Where:
$P$ = atmospheric pressure, mbar
$T$ = absolute temperature, K
$e$ = water vapor pressure, mbar

When the gradient of $N$ is equal to –39 $N$-units per kilometer, normal propagation takes place, corresponding to the effective earth's radius $ka$, where $k = 4/3$.

When $dN/dh$ is less than –39 $N$-units per kilometer, *subrefraction* occurs and the radio wave is bent strongly downward.

When $dN/dh$ is less than –157 $N$-units per kilometer, the radio energy may be bent downward sufficiently to be reflected from the earth, after which the ray is again bent toward the earth, and so on. The radio energy thus is trapped in a duct or waveguide. The wave also may be trapped between two elevated layers, in which case energy is not lost at the ground reflection points and even greater enhancement occurs. Radio waves thus trapped or *ducted* can produce fields exceeding those for free-space propagation because the spread of energy in the vertical direction is eliminated as opposed to the free-space case, where the energy spreads out in two directions orthogonal to the direction of propagation. Ducting is responsible for abnormally high fields beyond the radio horizon. These enhanced fields occur for significant periods of time on overwater paths in areas where meteorological conditions are favorable. Such conditions exist for significant periods of time and over significant horizontal extent in the coastal areas of southern California and around the Gulf of Mexico. Over land, the effect is less pronounced because surface features of the earth tend to limit the horizontal dimension of ducting layers [20].

### Tropospheric Scatter

The most consistent long-term mode of propagation beyond the radio horizon is that of scattering by small-scale fluctuations in the refractive index resulting from turbulence. Energy is scattered from multitudinous irregularities in the common volume which consists of that portion of troposphere visible to both transmitting and receiving site. There are some empirical data that show a correlation between the variations in the field beyond the horizon and $\Delta N$, the difference between the reflectivity on the ground and at a height of 1 km [21]. Procedures have been developed for calculating scatter fields for beyond-the-horizon radio relay systems as a function of frequency and distance [22, 23]. These procedures, however, require detailed knowledge of path configuration and climate.

The effect of scatter propagation is incorporated in the statistical evaluation of propagation (considered previously in this chapter), where the attenuation of fields beyond the diffraction zone is based on empirical data and shows a linear decrease with distance of approximately 0.2 dB/mi (0.1 dB/km) for the VHF–UHF frequency band.

## 2.3g Atmospheric Fading

Variations in the received field strengths around the median values are caused by changes in atmospheric conditions. Field strengths tend to be higher in summer than in winter, and higher at night than during the day, for paths over land beyond the line of sight. As a first approximation, the distribution of long-term variations in field strength in decibels follows a normal probability law.

Measurements indicate that the fading range reaches a maximum somewhat beyond the horizon and then decreases slowly with distance out to several hundred miles. Also, the fading range at the distance of maximum fading increases with frequency, while at the greater distances where the fading range decreases, the range is also less dependent on frequency. Thus, the slope of the graph $N$ must be adjusted for both distance and frequency. This behavior does not lend itself to treatment as a function of the earth's radius factor $k$, since calculations based on the same range of $k$ produce families of curves in which the fading range increases systematically with increasing distance and with increasing frequency.

### Effects of the Upper Atmosphere (Ionosphere)

Four principal recognized layers or regions in the ionosphere are the $E$ layer, the $F1$ layer, the $F2$ layer (centered at heights of about 100, 200, and 300 km, respectively), and the $D$ region, which is less clearly defined but lies below the $E$ layer. These *regular* layers are produced by radiation from the sun, so that the ion density—and hence the frequency of the radio waves that can be reflected thereby—is higher during the day than at night, The characteristics of the layers are different for different geographic locations and the geographic effects are not the same for all layers. The characteristics also differ with the seasons and with the intensity of the sun's radiation, as evidenced by the sunspot numbers, and the differences are generally more pronounced upon the $F2$ than upon the $F1$ and $E$ layers. There are also certain random effects that are associated with solar and magnetic disturbances. Other effects that occur at or just below the $E$ layer have been established as being caused by meteors [24].

The greatest potential for television interference by way of the ionosphere is from *sporadic E ionization*, which consists of occasional patches of intense ionization occurring 62 to 75 mi (100 to 120 km) above the earth's surface and apparently formed by the interaction of winds in the neutral atmosphere with the earth's magnetic field. Sporadic $E$ ionization can reflect VHF signals back to earth at levels capable of causing interference to analog television reception for periods lasting from 1 h or more, and in some cases totaling more than 100 h per year. In the U.S., VHF sporadic $E$ propagation occurs a greater percentage of the time in the southern half of the country and during the May to August period [25].

## 2.4 References


1. Bullington, K.: "Radio Propagation at Frequencies above 30 Mc," *Proc. IRE*, pg. 1122, October 1947.
2. Fink, D. G., (ed.): *Television Engineering Handbook*, McGraw-Hill, New York, N.Y., 1957.
3. Eckersley, T. L.: "Ultra-Short-Wave Refraction and Diffraction," *J. Inst. Elec. Engrs.*, pg. 286, March 1937.
4. Norton, K. A.: "Ground Wave Intensity over a Finitely Conducting Spherical Earth," *Proc. IRE*, pg. 622, December 1941.
5. Norton, K. A.: "The Propagation of Radio Waves over a Finitely Conducting Spherical Earth," *Phil. Mag.*, June 1938.
6. van der Pol, Balth, and H. Bremmer: "The Diffraction of Electromagnetic Waves from an Electrical Point Source Round a Finitely Conducting Sphere, with Applications to Radiotelegraphy and to Theory of the Rainbow," pt. 1, *Phil. Mag.*, July, 1937; pt. 2, *Phil. Mag.*, November 1937.
7. Burrows, C. R., and M. C. Gray: "The Effect of the Earth's Curvature on Groundwave Propagation," *Proc. IRE*, pg. 16, January 1941.
8. "The Propagation of Radio Waves through the Standard Atmosphere," Summary Technical Report of the Committee on Propagation, vol. 3, National Defense Research Council, Washington, D.C., 1946, published by Academic Press, New York, N.Y.
9. "Radio Wave Propagation," Summary Technical Report of the Committee on Propagation of the National Defense Research Committee, Academic Press, New York, N.Y., 1949.
10. de Lisle, E. W.: "Computations of VHF and UHF Propagation for Radio Relay Applications," RCA, Report by International Division, New York, N.Y.
11. Selvidge, H.: "Diffraction Measurements at Ultra High Frequencies," *Proc. IRE*, pg. 10, January 1941.
12. McPetrie, J. S., and L. H. Ford: "An Experimental Investigation on the Propagation of Radio Waves over Bare Ridges in the Wavelength Range 10 cm to 10 m," *J. Inst. Elec. Engrs.*, pt. 3, vol. 93, pg. 527, 1946.
13. Megaw, E. C. S.: "Some Effects of Obstacles on the Propagation of Very Short Radio Waves," *J. Inst. Elec. Engrs.*, pt. 3, vol. 95, no. 34, pg. 97, March 1948.
14. Dickson, F. H., J. J. Egli, J. W. Herbstreit, and G. S. Wickizer: "Large Reductions of VHF Transmission Loss and Fading by the Presence of a Mountain Obstacle in Beyond-Line-of-Sight Paths," *Proc. IRE*, vol. 41, no. 8, pg. 96, August 1953.
15. Bullington, K.: "Radio Propagation Variations at VHF and UHF," *Proc. IRE*, pg. 27, January 1950.
16. "Report of the Ad Hoc Committee, Federal Communications Commission," vol. 1, May 1949; vol. 2, July 1950.
17. Epstein, J., and D. Peterson: "An Experimental Study of Wave Propagation at 850 Mc," *Proc. IRE*, pg. 595, May 1953.

18. “Documents of the XVth Plenary Assembly,” CCIR Report 563, vol. 5, Geneva, 1982.

19. Bean, B. R., and E. J. Dutton: “Radio Meteorology,” National Bureau of Standards Monograph 92, March 1, 1966.

20. Dougherty, H. T., and E. J. Dutton: “The Role of Elevated Ducting for Radio Service and Interference Fields,” NTIA Report 81–69, March 1981.

21. “Documents of the XVth Plenary Assembly,” CCIR Report 881, vol. 5, Geneva, 1982.

22. “Documents of the XVth Plenary Assembly,” CCIR Report 238, vol. 5, Geneva, 1982.

23. Longley, A. G., and P. L. Rice: “Prediction of Tropospheric Radio Transmission over Irregular Terrain—A Computer Method,” ESSA (Environmental Science Services Administration), U.S. Dept. of Commerce, Report ERL (Environment Research Laboratories) 79-ITS 67, July 1968.

24. National Bureau of Standards Circular 462, “Ionospheric Radio Propagation,” June 1948.

25. Smith, E. E., and E. W. Davis: “Wind-induced Ions Thwart TV Reception,” *IEEE Spectrum*, pp. 52—55, February 1981.

Chapter

3

# Frequency Sources and References

**Ulrich L. Rohde**

**Jerry C. Whitaker, Editor**

## 3.1 Introduction[1]

A stable frequency reference is an important component of any transmission/reception system. The quartz crystal is the classic device used for frequency generation and control. Quartz acts as a stable high *Q* mechanical resonator. Crystal resonators are available for operation at frequencies ranging from 1 kHz to 300 MHz and beyond.

## 3.2 Characteristics of Crystal Devices

The operating characteristics of a crystal are determined by the *cut* of the device from a bulk "mother" crystal. The behavior of the device depends heavily on the size and shape of the crystal, and the angle of the cut. To provide for operation at a wide range of frequencies, different cuts, vibrating in one or more selected modes, are used.

The properties of a piezoelectric quartz crystal can be expressed in terms of three sets of axes. (See Figure 3.1.) The axis joining the points at the ends of the crystal is known as the *optical axis*, and electrical stresses applied in this direction exhibit no piezoelectric properties. The three axes $X'$, $X''$, and $X'''$, passing through the corners of the hexagon that forms the section perpendicular to the optical axes, are known as the *electrical axes*. The three axes $Y'$, $Y''$, and $Y'''$, which are perpendicular to the faces of the crystal, are the *mechanical axes*.

If a flat section is cut from a quartz crystal in such a way that the flat sides are perpendicular to an electrical axis, as shown in Figure 3.1*b* (*X*-cut), the mechanical stresses along the *Y*-axis of such a section produce electric charges on the flat sides of the crystal. If the direction of these stresses is changed from tension to compression, or vice versa, the polarity of the charges on the

1. Portions of this chapter are based on: Rohde, Ulrich L., and Jerry C. Whitaker: *Communications Receivers: Principles and Design*, 3rd ed., McGraw-Hill, New York, N.Y., 2000. Used with permission.

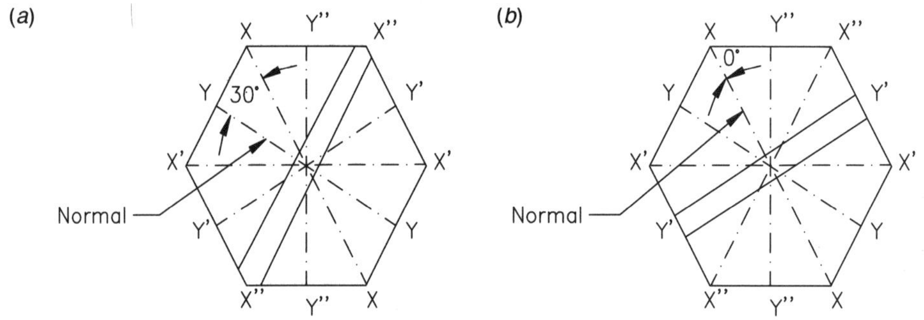


**Figure 3.1** Cross section of a quartz crystal taken in the plane perpendicular to the optical axis: (*a*) *Y*-cut plate, (*b*) *X*-cut plate.

crystal surfaces is reversed. Conversely, if electric charges are placed on the flat sides of the crystal by applying a voltage across the faces, a mechanical stress is produced in the direction of the *Y*-axis. This property by which mechanical and electrical properties are interconnected in a crystal is known as the *piezoelectric effect* and is exhibited by all sections cut from a piezoelectric crystal. Thus, if mechanical forces are applied across the faces of a crystal section having its flat sides perpendicular to a *Y*-axis, as shown in Figure 3.1*a*, piezoelectric charges will be produced because forces and potentials developed in such a crystal have components across at least one of the *Y*- or *X*-axes, respectively.

An alternating voltage applied across a quartz crystal will cause the crystal to vibrate and, if the frequency of the applied alternating voltage approximates a frequency at which mechanical resonance can exist in the crystal, the amplitude vibrations will be large. Any crystal has a number of such resonant frequencies that depend upon the crystal dimensions, the type of mechanical oscillation involved, and the orientation of the plate cut from the natural crystal.

Crystals are temperature-sensitive, as shown in Figure 3.2. The extent to which a device is affected by changes in temperature is determined by its cut and packaging. Crystals also exhibit changes in frequency with time. Such *aging* is caused by one or both of the following:

- Mass transfer to or from the resonator surface
- Stress relief within the device itself

Crystal aging is most pronounced when the device is new. As stress within the internal structure is relieved, the aging process slows.

## 3.2a Frequency Stabilization

The stability of a quartz crystal is inadequate for most commercial and industrial applications. Two common methods are used to provide the required long-term frequency stability:

- Oven-controlled crystal oscillator—The crystal is installed in a temperature-controlled box. Because the temperature in the box is constant, controlled by a thermostat, the crystal remains on-frequency. The temperature of the enclosure usually is set to the *turnover temperature* of the crystal. (The turnover point is illustrated in Figure 3.2.)

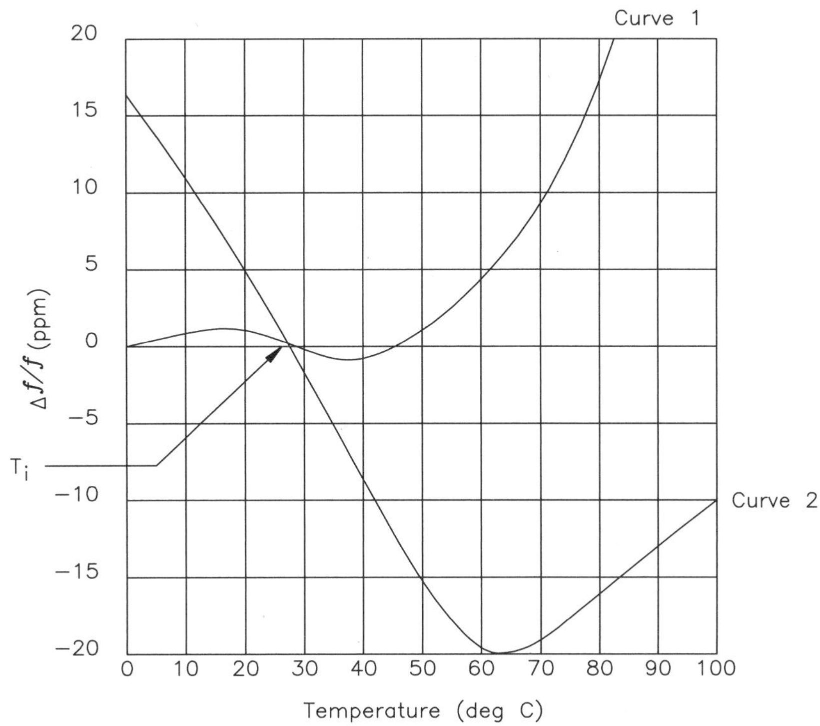


**Figure 3.2** The effects of temperature on two types of AT-cut crystals.

- Temperature-compensated crystal oscillator (TCXO)—The frequency vs. temperature changes of the crystal are compensated by varying a load capacitor. A thermistor network typically is used to generate a correction voltage that feeds a voltage-controlled capacitor (varactor) to retune the crystal to the desired on-frequency value.

## 3.2b Equivalent Circuit of a Quartz Resonator

Because of the piezoelectric effect, the quartz resonator behaves physically as a vibrating object when driven by a periodic electric signal near a resonant frequency of the cavity [1, 2]. This resonant behavior may be characterized by equivalent electrical characteristics, which may then be used to accurately determine the response and performance of the electromechanical system. Dye and Butterworth (1926) [3] developed the equivalent lump electrical network that describes an electrostatic signal driving a mechanically vibrating system (an electromechanical transducer). The equivalent electric network that models the piezo resonator was later developed by Van Dyke (1928), Butterworth, and others, and is shown in Figure 3.3. In the equivalent network, the circuit parameters $R_1$, $L_1$, and $C_1$ represent parameters associated with the quartz crystal, while $C_0$ represents the shunt capacitance of the resonator electrodes in parallel with the container that packages the crystal.

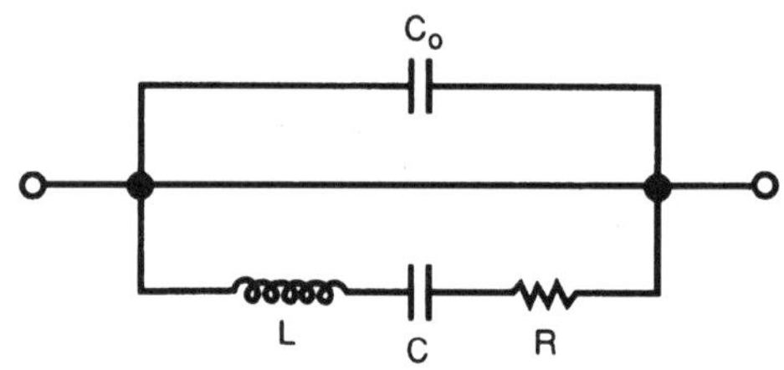


**Figure 3.3** Equivalent network of a piezoelectric resonator. (*From* [1]. *Used with permission.*)

**Table 3.1 Typical Equivalent Circuit Parameter Values for Quartz Crystals** (*After* [1].)

| Parameter | 200 kHz Fundamental | 2 Mhz | 30 Mhz, 3rd Overtone | 50 Mhz, 5th Overtone |
|---|---|---|---|---|
| $R_1$ | 2 kΩ | 100 Ω | 40 Ω | 20 Ω |
| $L_1$ | 27 H | 520 mH | 11 mH | 6 mH |
| $C_1$ | 0.024 pF | 0.012 pF | 0.0026 pF | 0.0005 pF |
| $C_o$ | 9 pF | 4 pF | 6 pF | 4 pF |
| $Q$ | $18 \times 10^3$ | $18 \times 10^3$ | $18 \times 10^3$ | $18 \times 10^3$ |

The inductance $L_1$ is associated with the mass of the quartz slab, and the capacitance $C_1$ is associated with the stiffness of the slab. Finally, the resistance, $R_1$ is determined by the loss processes associated with the operation of the crystal. Each of the equivalent parameters can be measured using standard network measurement techniques. Typical parameters are shown in Table 3.1.

Application of an electric field across the electrodes of the crystal results in both face and thickness shear waves developing within the crystal resonator. For a *Y*-cut crystal, face shear modes occur in the *X-Z* plane, while thickness shear modes develop in the *X-Y* plane.

The vibration is enhanced for electric signals near a resonant frequency of the resonator. Low frequency resonances tend to occur in the face shear modes, while high frequency resonances occur in thickness shear modes. In either case, when the applied electric signal is near a resonant frequency, the energy required to maintain the vibrational motion is small.

## 3.2c Temperature Compensation

Compensation for temperature-related frequency shifts can be accomplished by connecting a reactance, such as a variable voltage capacitor (a varactor) in series with the resonator [1]. The effect of this reactance is to *pull* the frequency to cancel the effect of frequency drift caused by temperature changes. An example of a frequency versus temperature characteristic both before and after compensation is shown in Figure 3.4.

Several techniques for temperature compensation are possible. They include:

- Analog temperature compensation
- Hybrid analog-digital compensation

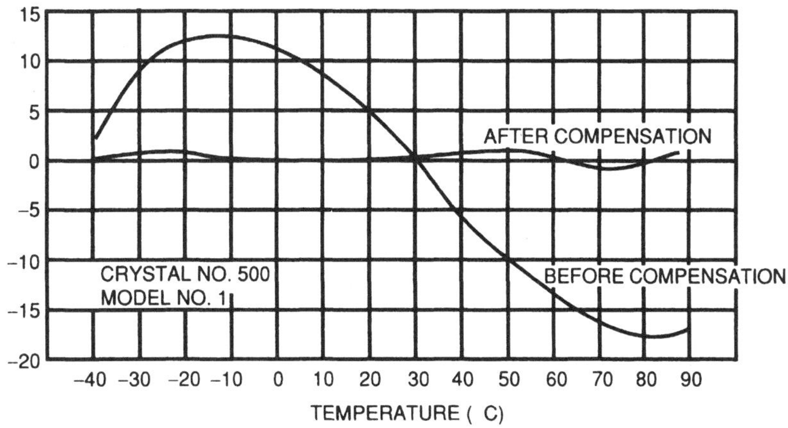


**Figure 3.4** Frequency vs. temperature characteristics for a typical temperature-compensated crystal oscillator. (*After* [2].)

- Digital temperature compensation

For analog compensation, a varactor is controlled by a voltage divider network that contains both thermistors and resistors to manage the compensation. Stability values are in the range 0.5 to 10 ppm. The lowest value requires that individual components in the voltage divider network be adjusted under controlled-temperature conditions. This is difficult over large temperature ranges for analog networks because of the tight component value tolerances required and because of interactions between different analog segments. Analog networks can be segmented in order to increase the independence of the component adjustments required to improve stability.

These problems can be alleviated through the use of some form of hybrid or digital compensation. In this scheme, the crystal oscillator is connected to a varactor in series, as in the analog case. This capacitor provides the coarse compensation (about ±4 parts in $10^7$). The fine compensation is provided by the digital network to further improve the stability. The hybrid analog-digital method is then capable of producing stability values of ±5 parts in $10^8$ over the range –40°C to +80°C.

## 3.2d Stress Compensation

Stress compensation is employed in order to stabilize the crystal resonator frequency against shifts caused by mechanical stresses [1]. These result from temperature changes or from electrode changes, due to differences in the expansion coefficients, which can occur over time. The frequency variations are caused by alteration of the crystal elastic coefficients with stress bias. The method developed for stress compensation is a special crystal cut called the *stress-compensated* or SC-cut. This cut was developed to obtain a resonator whose frequency would be independent of specified stresses.

The SC-cut has some advantages over the more common AT- and BT-cuts. These include a flatter frequency-temperature response, especially at the *turning points*, and less susceptibility to aging because of changes in the mounting stresses. The lower sensitivity to frequency changes means that SC-cut resonators can be frequency-temperature tested more rapidly than their AT- and BT-cut counterparts. A disadvantage of the SC-cut is that it is sensitive to air-damping and must be operated in a vacuum environment to obtain optimum $Q$ factors. The frequency-temperature curve is also roughly 70° higher for the SC-cut over the more common AT-cut, which could also be a disadvantage in some applications.

### 3.2e Aging Effects

Aging in crystal oscillators generally refers to any change over time that will affect the frequency characteristics of the oscillator, or the physical parameters that describe the device; such as motional time constant and equivalent circuit parameters [1]. Some factors that influence aging include the following:

- Surface deterioration
- Surface contamination
- Electrode composition
- Environmental conditions

Deterioration on the surface contributes to strain in the crystal, and this type of strain can impact the operating frequency of the oscillator. More importantly, the surface strain is sensitive to humidity and temperature early in the life of the crystal. One method of combating this surface degradation mechanism is to etch away the strained surface layers. The initial slope of the frequency change versus etch time curve is exponential. However, after the strained surface layers have been removed, the slope becomes linear.

Another surface issue is contamination from foreign materials. The presence of foreign substances will load the oscillator with an additional mass, and this can be reflected in the equivalent circuit parameters. The contaminants can be introduced during fabrication processes, from outgassing of other materials after the device has been packaged, or from contaminants in the environment during storage. The contamination process is often increased in higher temperature environments, and temperature cycling will lead to stability problems for the device.

Aging problems associated with the device electrodes includes delamination, leading to increased resistance or corrosion effects. Typical electrode materials include gold, silver, and aluminum. Of these, silver is the most popular material.

## 3.3 Oscillators

Receivers are seldom single-channel devices, but more often cover wide frequency ranges. In the superheterodyne receiver, this is accomplished by mixing the input signal with an LO signal. The LO source must meet a variety of requirements:

- It must have high spectral purity.

- It must be agile so that it can move rapidly (jump) between frequencies in a time frame that can be as short as a few microseconds.
- The increments in which frequencies can be selected must be small.

Frequency resolution between 1 and 100 Hz is generally adequate below 30 MHz; however, there are a number of systems that provide 0.001-Hz steps. At higher frequencies, resolution is generally greater than 1 kHz.

In most modern receivers, such a frequency source is typically a synthesizer that generates all individual frequencies as needed over the required frequency band. The modern synthesizer provides stable phase-coherent outputs. The frequencies are derived from a master standard, which can be a high-precision crystal oscillator, a secondary atomic standard (such as a rubidium gas cell), or a primary standard using a cesium atomic beam. The following characteristics must be specified for the synthesizer:

- Frequency range
- Frequency resolution
- Frequency indication
- Maximum frequency error
- Settling time
- Reference frequency
- Output power
- Harmonic distortion
- SSB phase noise
- Discrete spurs (spurious frequencies)
- Wide-band noise
- Control interface
- Power consumption
- Mechanical size
- Environmental conditions

Free-running tunable oscillators, once used in radio receivers, have generally been replaced in modern receivers because of their lack of precision and stability. Fixed-tuned crystal-controlled oscillators are still used in second- and third-oscillator applications in multiconversion superheterodyne receivers that do not require single-reference precision. Oscillators used in synthesizers have variable tuning capability, which may be voltage-controlled, generally by varactor diodes.

Synthesizer designs have used mixing from multiple crystal sources and mixing of signals derived from a single source through frequency multiplication and division. Synthesizers may be "direct" and use the product of multiple mixing and filtering or "indirect" and use a *phase-locked-loop* (PLL) slaved to the direct output to provide reduced spurious signals. There are a number of publications describing these and other techniques, the classic texts being [4] through [6]. Most modern receivers operating below 3 GHz use single- or multiple-loop digital PLL synthesizers, although for some applications, direct digital waveform synthesis may be used.

## 3.3a Key Terms

The following characteristics are commonly used to describe oscillator performance:

- **Frequency pushing**. Frequency pushing characterizes the degree to which an oscillator's frequency is affected by its supply voltage. For example, a sudden current surge caused by activating a transceiver's RF power amplifier may produce a spike on the VCO's dc power supply and a consequent frequency jump. Frequency pushing is specified in frequency/voltage form, and is tested by varying the VCO's dc supply voltage (typically ±1 V) with its tuning voltage held constant.
- **Harmonic output power**. Harmonic output power is measured relative to the output power of the oscillator. Typical values are 20 dB or more suppression relative to the fundamental. This suppression can be improved by additional filtering.
- **Output power**. The output power of the oscillator, typically expressed in dBm, is measured into a 50-Ω load. The output power is always combined with a specification for flatness or variation. A typical spec would be 0 dBm ±1 dB.
- **Output power as a function of temperature**. All active circuits vary in performance as a function of temperature. The output power of an oscillator over a temperature range should vary less than a specified value, such as 1 dB.
- **Post-tuning drift**. After a voltage step is applied to the tuning diode input, the oscillator frequency may continue to change until it settles to a final value. This post-tuning drift is one of the parameters that limits the bandwidth of the VCO input.
- **Power consumption**. This characteristic conveys the dc power, usually specified in milliwatts and sometimes qualified by operating voltage, required by the oscillator to function properly.
- **Sensitivity to load changes**. To keep manufacturing costs down, many wireless applications use a VCO alone, without the buffering action of a high reverse-isolation amplifier stage. In such applications, *frequency pulling*, the change of frequency resulting from partially reactive loads, is an important oscillator characteristic. Pulling is commonly specified in terms of the frequency shift that occurs when the oscillator is connected to a load that exhibits a non-unity VSWR (such as 1.75, usually referenced to 50 Ω), compared to the frequency that results with a unity-VSWR load (usually 50 Ω). Frequency pulling must be minimized, especially in cases where power stages are close to the VCO unit and short pulses may affect the output frequency. Such feedback can make-phase locking impossible.
- **Spurious outputs**. The spurious output specification of a VCO, expressed in decibels, characterizes the strength of unwanted and nonharmonically related components relative to the oscillator fundamental. Because a stable, properly designed oscillator is inherently clean, such *spurs* are typically introduced only by external sources in the form of radiated or conducted interference.
- **Temperature drift**. Although the synthesizer is responsible for locking and maintaining the oscillator frequency, the VCO frequency change as a function of temperature is a critical parameter and must be specified. This value varies between 10 kHz/°C to several hundred kHz/°C depending on the center frequency and tuning range.

- **Tuning characteristic**. This specification shows the relationship, depicted as a graph, between the VCO operating frequency and the tuning voltage applied. Ideally, the correspondence between operating frequency and tuning voltage is linear.
- **Tuning linearity.** For stable synthesizers, a constant deviation of frequency versus tuning voltage is desirable. It is also important to make sure that there are no breaks in the tuning range—for example, that the oscillator does not stop operating with a tuning voltage of 0 V.
- **Tuning sensitivity, tuning performance**. This datum, typically expressed in megahertz per volt (MHz/V), characterizes how much the frequency of a VCO changes per unit of tuning-voltage change.
- **Tuning speed**. This characteristic is defined as the time necessary for the VCO to reach 90 percent of its final frequency upon the application of a tuning-voltage step. Tuning speed depends on the internal components between the input pin and tuning diode—including, among other things, the capacitance present at the input port. The input port's parasitic elements determine the VCO's maximum possible modulation bandwidth.

## 3.4 Phase-Locked Loop Synthesizers

The traditional synthesizer consists of a single loop and the step size of the output frequency is equal to the reference frequency at the phase detector. Figure 3.5 shows this classic approach.

Wireless applications with a step size of 200 kHz have made the life of the designer somewhat easier, since such a wide step size reduces the division ratio. As can be seen in Figure 3.5, the simplest form of a digital synthesizer consists of a *voltage-controlled oscillator* (VCO). For PLL applications, the oscillator sensitivity, typically expressed in megahertz per volt (MHz/V), needs to be stated. For high-performance test equipment applications, the VCO is frequently provided in the form of a YIG oscillator. These oscillators operate at extremely high $Q$ and are fairly expensive. The VCO needs to be separated from any load by a post amplifier, which drives a sine-wave-to-logic-waveform translator.

Typically, an ECL line receiver or its equivalent would serve for this function. This stage, in turn, drives a programmable divider and divides the oscillator frequency down to a reference frequency, such as 200 kHz. Assuming an oscillator frequency of 1 GHz, the division ratio would be 1 GHz/200 kHz = 5000. In Figure 3.5, we are looking at a 3 GHz output frequency and a step size determined by the reference source resolution, assuming a fixed division ratio.

The phase detector, actually the *phase/frequency detector* (PFD), is driven by the reference frequency on one side and the oscillator frequency, divided down, on the other side. The PFD is typically designed to operate from a few kilohertz to several tens of megahertz, 50 MHz for example. In our case, this would mean a step size of 50 MHz. Most of these types of phase detectors use MOS technology to keep the levels for on/off voltage high, and their noise contribution needs to be carefully evaluated. While the synthesizer is not locked, the output of the PFD is a dc control voltage with a superimposed ac voltage equal to the difference between the two frequencies prior to lock. Most PFDs have a flip-flop-based architecture, and their output consists of a train of pulses that must to be integrated to produce the control voltage necessary for driving the VCO into the locked condition. This integrator also serves as a loop filter. Its purpose is to suppress the reference frequency and provide the necessary phase/frequency response for stable locking. The basic loop really is a nonlinear control loop that, for the purpose of analysis, is

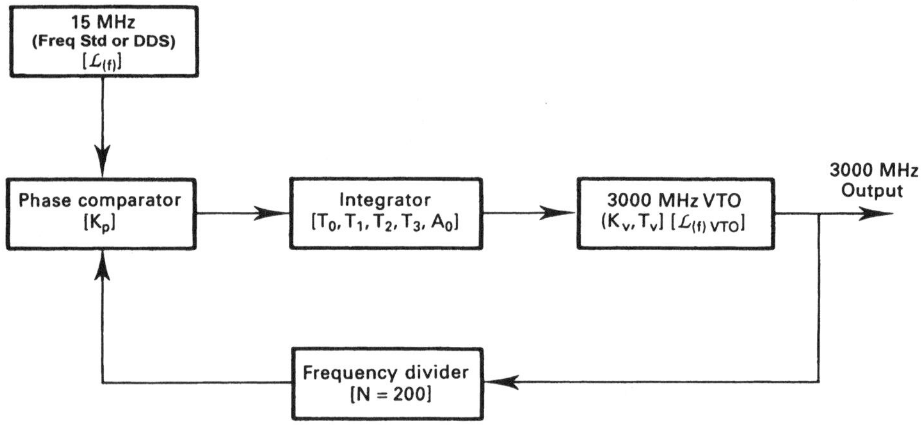


**Figure 3.5** Block diagram of a PLL synthesizer driven by a frequency standard, DDS, or fractional-*N* synthesizer for high resolution at the output. The last two standards allow a relatively low division ratio and provide quasi-arbitrary resolution.

always assumed to be linear or piecewise-linear. The most linear phase detector is a diode ring, but it has a low-level dc output that requires an operational amplifier to shift the level of typically ±0.3 V to the high voltage required for the tuning diode. These values are usually somewhere between 5 and 30 V. The tuning diode itself needs to have the appropriate breakdown voltage and voltage-dependent linearity to provide constant loop gain. In most cases, this is not possible, especially if the division ratio changes by a factor of 2 or more; in this case, it is a wise decision to use coarse steering so that fine tuning shows a reasonably linear performance. These loops are also called Type 2 *second-order loops*. This is because the loop has two integrators, one being the loop filter and other one being the tuning diode. The order of the loop is determined by the filter. Table 3.2 shows circuit and transfer characteristics of several PLL filters.

The filters shown in the table are single-ended, which means that they are driven from the output of a CMOS switch in the phase/frequency detector. This type of configuration exhibits a problem under lock conditions: If we assume that the CMOS switches are identical and have no leakage, initially the output current charges the integrator and the system will go into lock. If it is locked and stays locked, there is no need for a correction voltage, and therefore the CMOS switches will not supply any output. The very moment a tiny drift occurs, a correction voltage is required, and therefore there is a drastic change from no loop gain (closed condition) to a loop gain (necessary for acquiring lock). This transition causes a number of nonlinear phenomena, and therefore it is a better choice to either use a passive filter in a symmetrical configuration or a symmetrical loop filter with an operational amplifier instead of the CMOS switches. Many of the modern PFDs have different outputs to accommodate this. An ill-conditioned filter or selection of values for the filter frequently leads to a very low-frequency type of oscillation, often referred to as *motorboating*. This can only be corrected by adjusting the phase/frequency behavior of the filter.

The Bode diagram is a useful tool in designing the appropriate loop filter. The Bode diagram shows the open-loop performance, both magnitude and phase, of the phase-locked loop.

**Table 3.2 PLL Filter Characteristics**

## Circuit and Transfer Characteristics of Several PLL Filters

| Type | Passive | | Active | |
|---|---|---|---|---|
| | 1 | 2 | 3 | 4 |
| Circuit | 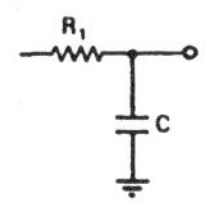 | 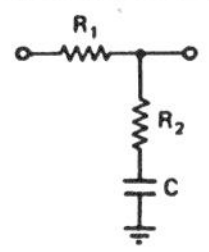 | 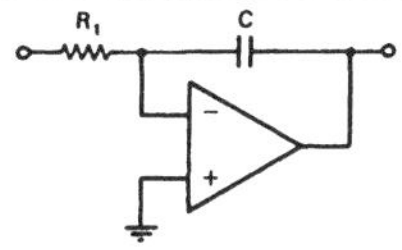 | 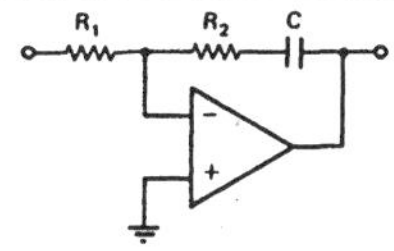 |
| Transfer characteristic | 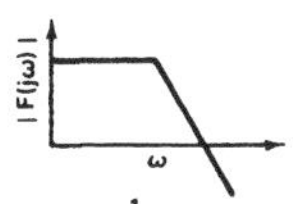 | 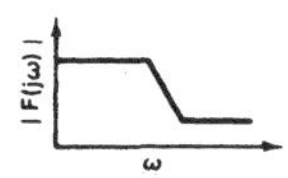 | 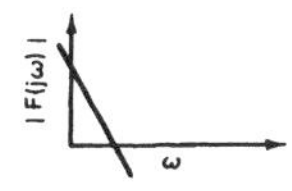 | 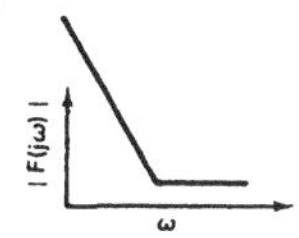 |
| $F(j\omega) =$ | $\frac{1}{1 + j\omega\tau_1}$ | $\frac{1 + j\omega\tau_2}{1 + j\omega(\tau_1 + \tau_2)}$ | $\frac{1}{j\omega\tau_1}$ | $\frac{1 + j\omega\tau_2}{j\omega\tau_1}$ |

$\tau_1 = R_1C,\ \tau_2 = R_2C$

## Implementation of Different Loop Filters

| Passive Lead-Lag | Passive Lead Lag with Pole | Active Integrator | Active Integrator with Pole |
|---|---|---|---|
| 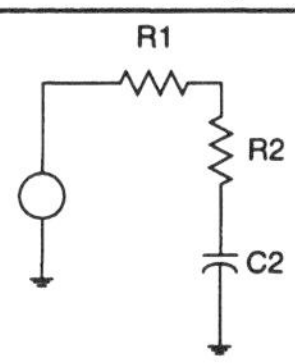 | 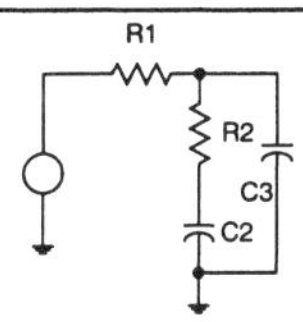 | 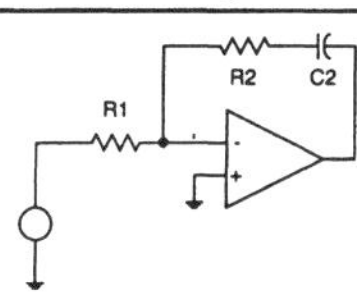 | 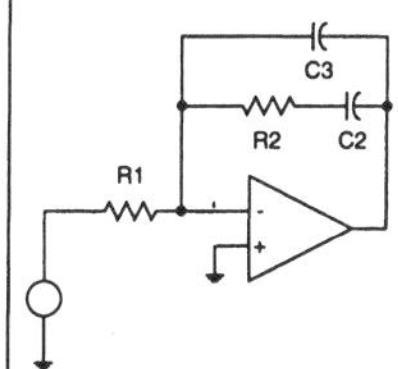 |
| $F(s) = \frac{s\tau_2 + 1}{[s(\tau_1 + \tau_2) + 1]}$ <br> $\tau_1 = R_1C_2; \tau_2 = R_2C_2$ | $F(s) = \frac{s\tau_2 + 1}{[s(\tau_1 + \tau_2) + 1](s\tau_3 + 1)}$ <br> $\tau_1 = R_1C_2; \tau_2 = R_2C_2;$ <br> $\tau_3 = (R_2 \Vert R_1)C_3$ | $F(s) = \frac{s\tau_2 + 1}{s\tau_1}$ <br> $\tau_1 = R_1C_2; \tau_2 = R_2C_2;$ | $F(s) = \frac{s\tau_2 + 1}{s\tau_1(s\tau_3 + 1)}$ <br> $\tau_1 = R_1(C_2 + C_3); \tau_2 = R_2C_2;$ <br> $\tau_3 = R_2(C_3 \Vert C_2)$ |
| Type 1.5, 2nd Order (Low Gain) | Type 1.5, 3rd Order (Low Gain) | Type 2, 2nd Order (High Gain) | Type 2, 3rd Order (High Gain) |

## Recommended Passive Filters for Charge Pumps

| Integrator | Integrator With Poles | Integrator With 2 Poles |
|---|---|---|
| 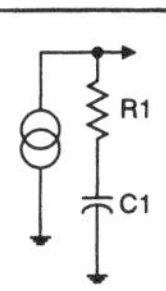 | 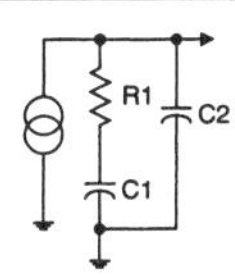 | 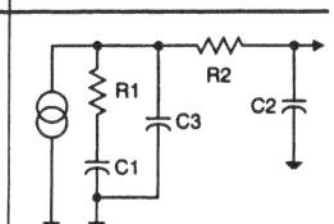 |
| $F(s) = R_1\frac{s\tau_1 + 1}{s\tau_1}$ <br> $\tau_1 = R_1C_1$ | $F(s) = R_1\frac{s\tau_1 + 1}{s\tau_1(s\tau_2 + 1)}$ <br> $\tau_1 = R_1C_1; \tau_2 = R_2(\frac{C_1C_2}{C_1 + C_2})$ | $F(s) = R_1\frac{s\tau_1 + 1}{s\tau_1(s\tau_2 + 1)(s\tau_3 + 1)}$ <br> $\tau_1 = R_1C_1; \tau_2 = R_1\frac{C_1C_3}{C_1 + C_3};$ <br> $\tau_3 = R_2C_2$ |
| Type 2, 2nd Order | Type 2, 3rd Order | Type 2, 4th Order |

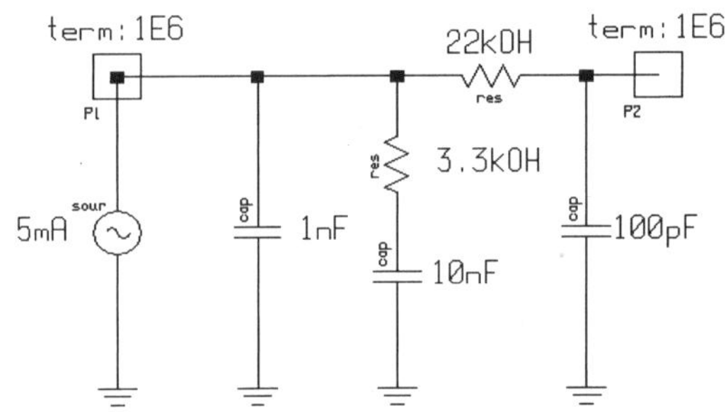


**Figure 3.6** Type 1 high-order loop filter used for passive filter evaluation. The 1-nF capacitor is used for spike suppression. The filter consists of a lag portion and an additional low pass section.

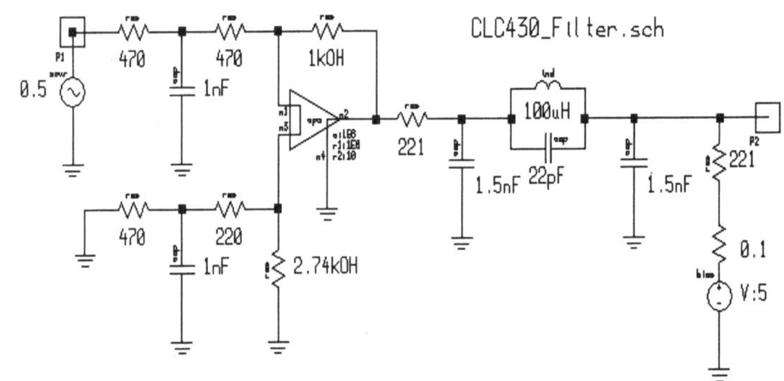


**Figure 3.7** A Type 2 high-order filter with a notch to suppress the discrete reference spurs.

## 3.4a Practical PLL Circuits

Figure 3.6 shows a passive filter that is used for a National LMX synthesizer chip. This chip has a charge-pump output, which explains the need for the first capacitor.

Figure 3.7 shows an active integrator operating at a reference frequency of several megahertz. The notch filter at the output reduces the reference frequency considerably. The notch is at about 4.5 MHz.

Figure 3.8 shows the combination of a phase/frequency discriminator and a higher-order loop filter as used in more complicated systems, such as fractional-division synthesizers.

Figure 3.9 shows a custom-built phase detector with a noise floor of better than –168 dBc/Hz.

## 3.4b Fractional-Division Synthesizers

In conventional synthesizers, the minimum step size is equal to the reference frequency. In order to achieve a finer resolution, we can either manipulate reference as outlined in Figure 3.5, or we

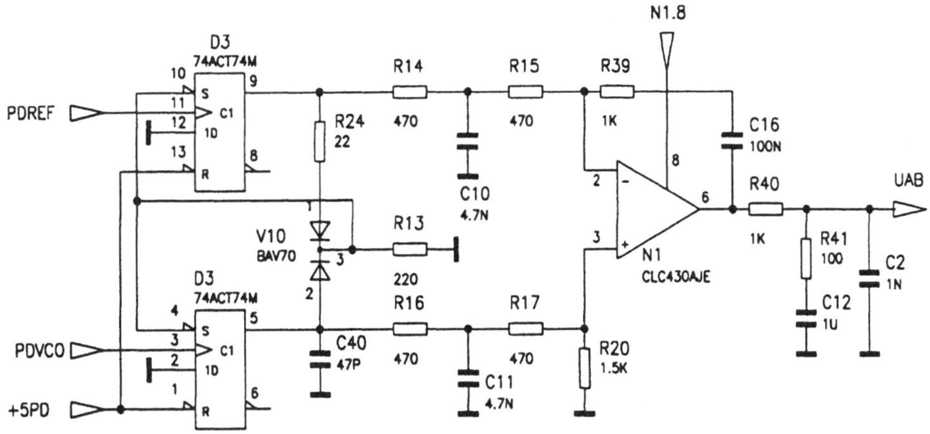


**Figure 3.8** Phase/frequency discriminator including an active loop filter capable of operating up to 100 MHz.

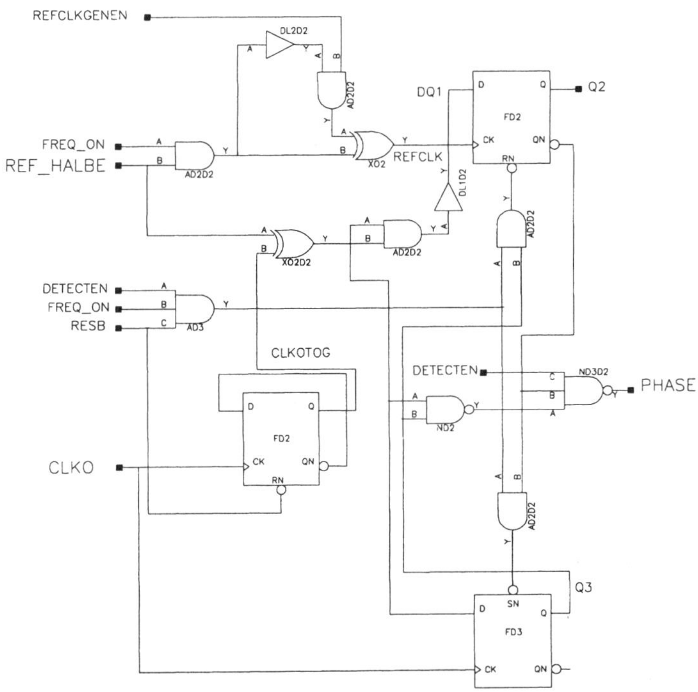


**Figure 3.9** Custom-built phase detector with a noise floor of better than –168 dBc/Hz. This phase detector shows extremely low phase jitter.

can use fractional division. The principle of the fractional-$N$-division synthesizer has been around for some time. In the past, implementation of this technique has been done in an analog system. The Figure 3.5 single loop uses a frequency divider where the division ratio is an integer value between 1 and some very large number, hopefully not as high as 50,000. It would be ideal to be able to build a synthesizer with the 1.25 MHz reference or 50 MHz reference and yet obtain the desired step size resolution, such as 25 kHz. This would lead to a much smaller division ratio and better phase noise performance.

An alternative would be for $N$ to take on fractional values. The output frequency could then be changed in fractional increments of the reference frequency. Although a digital divider cannot provide a fractional division ratio, ways can be found to accomplish the same task effectively. The most frequently used method is to divide the output frequency by $N + 1$ every $M$ cycles and to divide by $N$ the rest of the time. The effective division ratio is then $N + 1/M$, and the average output frequency is given by

$$f_0 = \left( N + \frac{1}{M} \right) f_r \tag{3.1}$$

This expression shows that $f_0$ can be varied in fractional increments of the reference frequency by varying $M$. The technique is equivalent to constructing a fractional divider, but the fractional part of the division is actually implemented using a phase accumulator. This method can be expanded to frequencies much higher than 6 GHz using the appropriate synchronous dividers. The phase accumulator approach is illustrated by the following example.

Consider the problem of generating 899.8 MHz using a fractional-$N$ loop with a 50-MHz reference frequency; 899.8 MHz = 50 MHz ($N - K/F$). The integral part of the division $N$ is set to 17 and the fractional part $K/F$ is 996/1000 (the fractional part $K/F$ is not a integer). The VCO output is divided by 996 × every 1,000 cycles. This can easily be implemented by adding the number 0.996 to the contents of an accumulator every cycle. Every time the accumulator overflows, the divider divides by 18 rather than by 17. Only the fractional value of the addition is retained in the phase accumulator. If we move to the lower band or try to generate 850.2 MHz, $N$ remains 17 and $K/F$ becomes 4/1000. This method of fractional division was first introduced by using analog implementation and noise cancellation, but today it is implemented as a totally digital approach. The necessary resolution is obtained from the *dual-modulus prescaling*, which allows for a well-established method for achieving a high-performance frequency synthesizer operating at UHF and higher frequencies. Dual-modulus prescaling avoids the loss of resolution in a system compared to a simple prescaler. It allows a VCO step equal to the value of the reference frequency to be obtained. This method needs an additional counter and the dual-modulus prescaler then divides one or two values depending upon the state of its control. The only drawback of prescalers is the minimum division ratio of the prescaler for approximately $N^2$.

The dual modulus divider is the key to implementing the fractional-$N$ synthesizer principle. Although the fractional-$N$ technique appears to have a good potential of solving the resolution limitation, it is not free of complications. Typically, an overflow from the phase accumulator, which is the adder with the output feedback to the input after being latched, is used to change the instantaneous division ratio. Each overflow produces a jitter at the output frequency, caused by the fractional division, and is limited to the fractional portion of the desired division ratio.

In our example, we had chosen a step size of 200 kHz, and yet the discrete side bands vary from 200 kHz for $K/F = 4/1000$ to 49.8 MHz for $K/F = 996/1000$. It will become the task of the loop filter to remove those discrete spurious elements. While in the past the removal of discrete spurs was accomplished by analog techniques, various digital methods are now available. The microprocessor has to solve the following equation:

$$N^* = \left( N + \frac{K}{F} \right) = \left[ N(F - K) + (N + 1)K \right] \tag{3.2}$$

For example, for $F_0 = 850.2$ MHz, we obtain

$$N^* = \frac{850.2 \text{ MHz}}{50 \text{ MHz}} = 17.004 \tag{3.3}$$

Following the formula above

$$N^* = \left( N + \frac{K}{F} \right) = \frac{\left[ 17(1000 - 4) + (17 + 1) \times 4 \right]}{1000} = \frac{\left[ 16932\_72 \right]}{1000} = 17.004$$

$$F_{out} = 50 \text{ MHz} \times \frac{\left[ 16932 - 72 \right]}{1000} = 846.6 \text{ MHz} + 3.6 \text{ MHz} = 850.2 \text{ MHz} \tag{3.4}$$

By increasing the number of accumulators, frequency resolution considerably below a step size of 1 Hz is possible with the same switching speed.

## 3.5 Multiloop Synthesizers

To avoid the limitations of the single-loop synthesizer, synthesizers are often designed to employ more than one loop. Figure 3.10 shows a block diagram of a multiloop synthesizer. The first LO, operating from 81.4 to 111.4 MHz, is a two-loop synthesizer using a frequency translation stage. It comprises a 70- to 80-MHz loop, a divider, two frequency translators, and an output loop at the final LO frequency. Two single-loop synthesizers are also used later in the receiver, but our discussion will be confined to the multiloop unit.

A 10-MHz crystal oscillator is used as the standard to which all of the internal oscillator frequencies are locked. A divide-by-100 circuit reduces this to a 100-kHz reference used in both loops of the synthesizer. The 100-kHz reference is further divided by 100 to provide the 1-kHz reference for the 70- to 80-MHz loop. The output of this loop is then further divided by 100 to

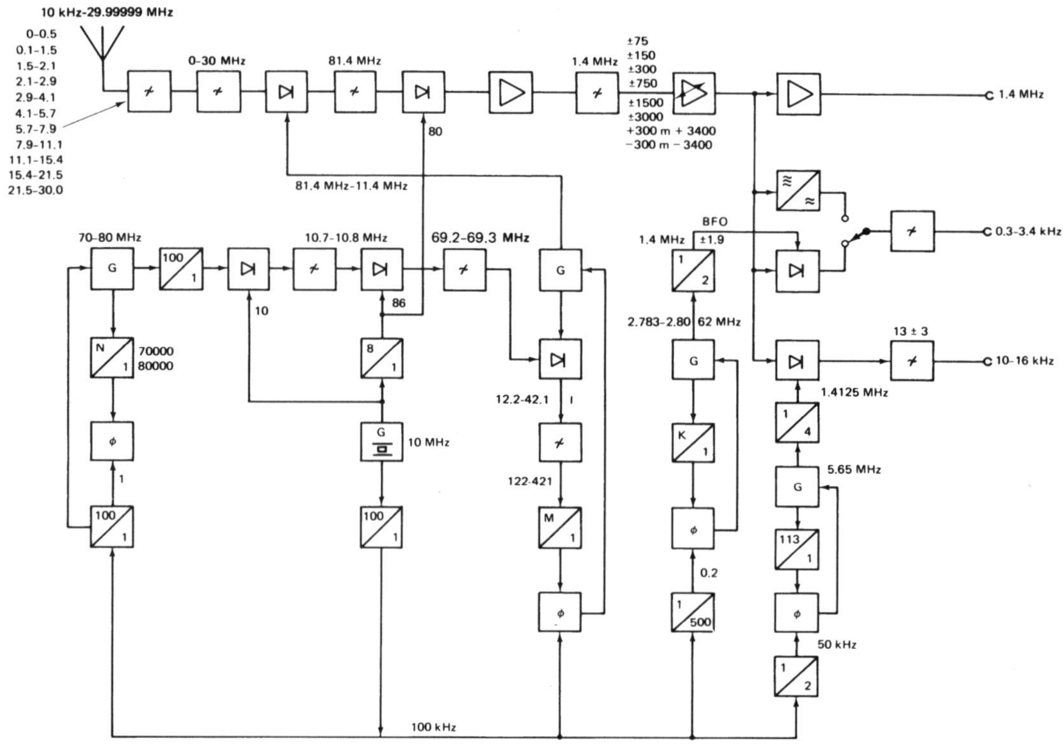


**Figure 3.10** Block diagram of a multiloop synthesizer. (*Courtesy of Rohde & Schwarz.*)

provide 10-Hz steps between 0.7 and 0.8 MHz. This division improves the noise sidebands and spurious signal suppression of the loop by 40 dB.

The 0.7- to 0.8-MHz band is converted to 10.7 to 10.8 MHz by mixing it with the 10-MHz reference. A crystal filter is used to provide the necessary suppression of the two inputs to the mixer. The resultant signal is translated to 69.2 to 69.3 MHz by further mixing with a signal of 80 MHz, the eighth harmonic of the 10-MHz frequency standard. The 80-MHz signal can be generated either by a frequency multiplier and crystal filter or by using an 80-MHz crystal oscillator under phase-lock control of the standard. In the former case, the noise sideband performance of the standard is degraded by 18 dB over the standard. Another possibility is the use of an 80-MHz crystal oscillator standard, followed by a divide-by-8 circuit to produce the 10-MHz internal reference. However, it is not possible to build crystal oscillators at 80 MHz with as good long- and short-term frequency stability as at 10 MHz. Hence, the phase-locked crystal oscillator approach was used to achieve high stability.

The 69.2- to 69.3-MHz output frequency from the mixer, after filtering, is mixed with the final VCO output frequency to produce a signal of 12.2 to 42.1 MHz, which after division by $M$ is used for comparison in the final PLL with the 100-kHz reference. The $M$ value is used to select the 0.1-MHz steps, while the value of $N$ shifts the 70- to 80-MHz oscillator to provide 10-

Hz resolution over the 0.1-MHz band resulting from its division by 100. This synthesizer provides the first oscillator frequency for a receiver with 81.4-MHz first IF, for a band of input frequencies up to 30 MHz, with 10-Hz step resolution.

This multiloop synthesizer illustrates the most important principles found in general-purpose synthesizers. A different auxiliary loop could be used to provide further resolution. For example, by replacing the 10.7- to 10.8-MHz loop by a digital direct frequency synthesizer, quasi-infinite resolution could be obtained by a microprocessor-controlled low-frequency synthesizer. Such a synthesizer has a very fast lock-up time, comparable to the speed of the 100-kHz loop. In the design indicated, the switching speed is determined by the 1-kHz fine resolution loop. For a loop bandwidth of 50 Hz for this loop, we will obtain a settling time in the vicinity of 40 ms. The longest time is needed when the resolution synthesizer is required to jump from 80 to 70 MHz. During this frequency jump, the loop will go out of both phase and frequency lock and will need complete reacquisition. Thus, each time a 100-kHz segment is passed through, there is such a disturbance. The same occurs when the output frequency loop jumps over a large frequency segment. The VCO, operating from 81.4 to 111.4 MHz, is coarse tuned by switching diodes, and some of the jumps are objectionable. The noise sideband performance of this synthesizer is determined inside the loop bandwidth by the reference noise multiplied to the output frequency, and outside the loop bandwidth by the VCO noise. The latter noise can be kept low by building a high-quality oscillator.

Another interesting multiloop synthesizer is shown in Figure 3.11. A 13- to 14-MHz singleloop synthesizer with 1-kHz steps is divided by 10 to provide 100-Hz steps and a 20-dB noise improvement. A synthesizer in this frequency range can be built with one oscillator in a single chip. The output loop, operating from 69.5 to 98 MHz, is generated in a synthesizer loop with 50-kHz reference and 66.6- to 66.7-MHz frequency offset generated by translation of the fine loop frequency. The reference frequencies for the two loops are generated from a 10.7 MHz temperature-compensated crystal oscillator standard. An additional 57.3-MHz TCXO is used to drive a second oscillator in the receiver as well as to offset the fine frequency loop of the synthesizer. An increase in frequency of this oscillator changes the output frequency of the first LO (synthesizer) in the opposite direction to the second LO, thereby canceling the drift error. This drift-canceling technique is sometimes referred to as the *Barlow-Wadley principle.*

## 3.6 Direct Digital Synthesis

*Direct digital frequency synthesis* (DDFS) is an oscillator scheme wherein a digital representation of the desired signal is generated and then applied to a D/A converter to reform the digital representation to an analog waveform. Advances in high-speed microelectronics, particularly the microprocessor, make DDFS practical at frequencies in the very-high-frequency band and below. Systems can be compact, use low power, and provide fine frequency resolution with virtually instantaneous switching of frequencies. DDFS is finding increasing application, particularly in conjunction with PLL synthesizers.

DDFS uses a single-frequency source (clock) as a time reference. One method of digitally generating the values of a sine wave is to solve the digital recursion relation as follows

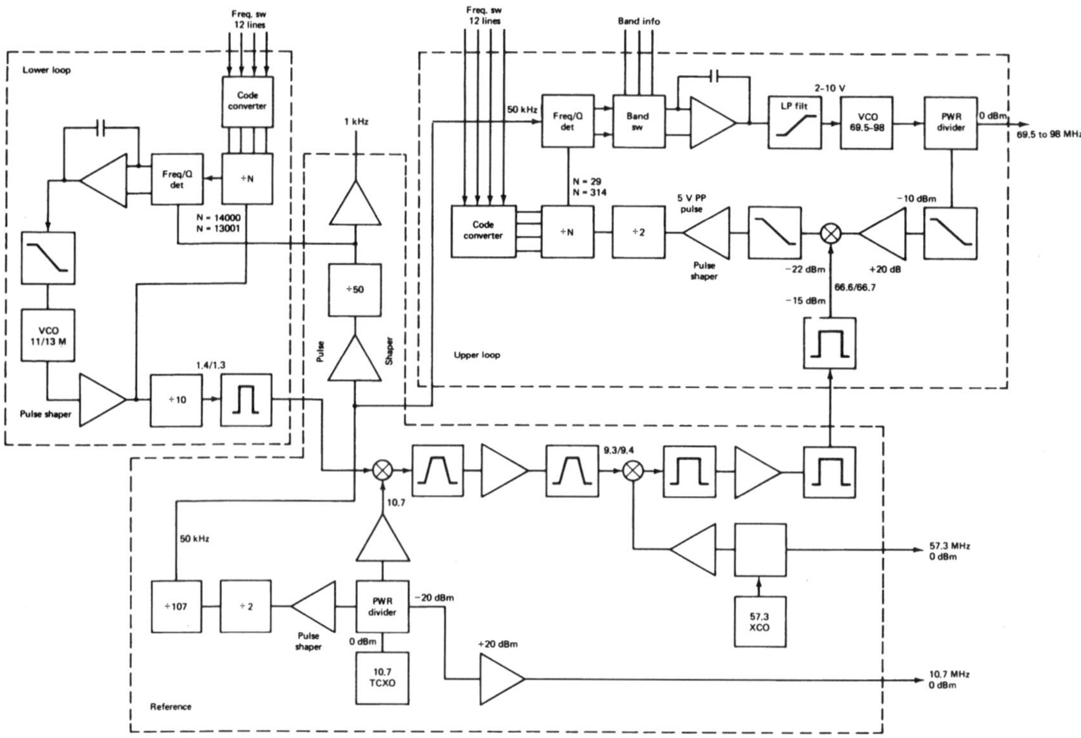


**Figure 3.11** Block diagram of a multiloop frequency synthesizer incorporating a drift-canceling technique. (*Courtesy of Rohde & Schwarz.*)

$$Y_n = [2 \cos (2 \pi f t)] Y_{(n-1)} - Y_{(n-2)} \tag{3.5}$$

This is solved by $Y_n = \cos(2fnt)$. There are at least two problems with this method, however. The noise can increase until a limit cycle (nonlinear oscillation) occurs. Also, the finite word length used to represent 2 cos (2*ft*) places a limitation on the frequency resolution. Another method of DDFS, *direct table lookup*, consists of storing the sinusoidal amplitude coefficients for successive phase increments in memory. The continuing miniaturization in size and cost of ROM make this the most frequently used technique.

One method of direct table lookup outputs the same *N* points for each cycle of the sine wave, and changes the output frequency by adjusting the rate at which the points are computed. It is relatively difficult to obtain fine frequency resolution with this approach, so a modified table lookup method is generally used. It is this method that we describe here. The function cos (2*ft*) is approximated by outputting the function cos (2*fnT*) for $n = 1, 2, 3, \ldots$ , where *T* is the interval between conversions of digital words in the D/A converter and *n* represents the successive sample numbers. The sampling frequency, or rate, of the system is $1/T$. The lowest output frequency

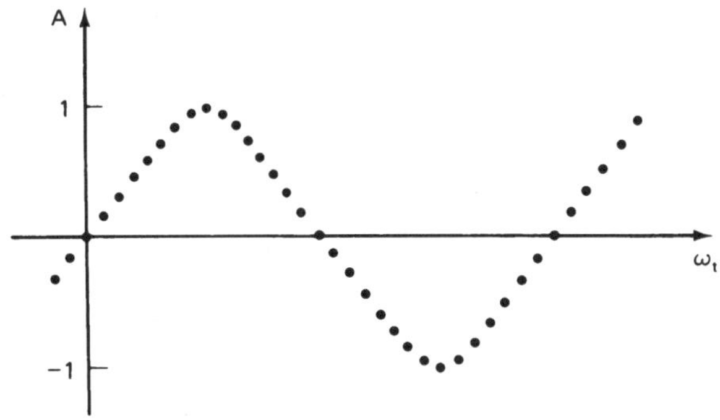


**Figure 3.12** Synthesized waveform generated by direct digital synthesis.

waveform contains $N$ distinct points in its waveform, as illustrated in Figure 3.12. A waveform of twice the frequency can be generated, using the same sampling rate, but outputting every other data point. A waveform $k$ times as fast is obtained by outputting every $k$th point at the same rate $1/T$. The frequency resolution, then, is the same as the lowest frequency $f_L$.

The maximum output frequency is selected so that it is an integral multiple of $f_L$, that is, $f_U = kF_L$. If $P$ points are used in the waveform of the highest frequency, then $N$ (= $kP$) points are used in the lowest frequency waveform. The number $N$ is limited by the available memory size. The minimum value that $P$ can assume is usually taken to be four. With this small value of $P$, the output contains many harmonics of the desired frequency. These can be removed by the use of low-pass filtering at the D/A output. For $P = 4$, the period of the highest frequency is $4T$, resulting in $f_U = 4f_L$. Thus, the highest attainable frequency is determined by the fastest sampling rate possible.

In the design of this type of DDFS, the following guidelines apply:

- The desired frequency resolution determines the lowest output frequency $f_L$.
- The number of D/A conversions used to generate $f_L$ is $N = 4k = 4f_U/f_L$ provided that four conversions are used to generate $f_U$ ($P = 4$).
- The maximum output frequency $f_U$ is limited by the maximum sampling rate of the DDFS, $\leq 1/4T$. Conversely, $T \leq 1/4f_U$.

The architecture of the complete DDFS is shown in Figure 3.13. To generate $nf_L$, the integer $n$ addresses the register, and each clock cycle $kn$ is added to the content of the accumulator so that the content of the memory address register is increased by $kn$. Each $kn$th point of the memory is addressed, and the content of this memory location is transferred to the D/A converter to produce the output sampled waveform.

To complete the DDFS, the memory size and the length (number of bits) of the memory word must be determined. The word length is determined by system noise requirements. The amplitude of the D/A output is that of an exact sinusoid corrupted with the deterministic noise resulting from truncation caused by the finite length of the digital words (quantization noise). If an ($n$

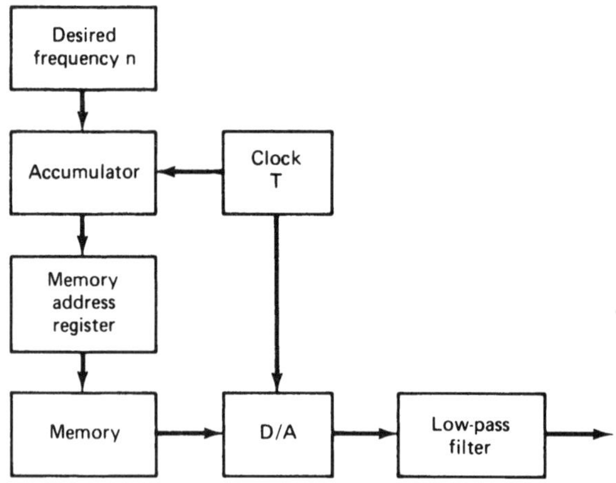


**Figure 3.13** Block diagram of a direct digital frequency synthesizer. (*From* [8]. *Reprinted with permission.*)

+ 1)-bit word length (including one sign bit) is used and the output of the A/D converter varies between 1, the mean noise from the quantization will be

$$\sigma^2 = \frac{1}{12}\left(\frac{1}{2}\right)^{2n} = \frac{1}{3}\left(\frac{1}{2}\right)^{2(n+1)} \tag{3.6}$$

The mean noise is averaged over all possible waveforms. For a worst-case waveform, the noise is a square wave with amplitude $(1/2)(1/2)^n$ and $\sigma = (1/4)(1/2)^{2n}$. For each bit added to the word length, the spectral purity improves by 6 dB.

The main drawback of the DDFS is that it is limited to relatively low frequencies. The upper frequency is directly related to the maximum usable clock frequency. DDFS tends to be noisier than other methods, but adequate spectral purity can be obtained if sufficient low-pass filtering is used at the output. DDFS systems are easily constructed using readily available microprocessors. The combination of DDFS for fine frequency resolution plus other synthesis techniques to obtain higher-frequency output can provide high resolution with very rapid settling time after a frequency change. This is especially valuable for frequency-hopping spread-spectrum systems.

## 3.7 References

1. Tate, Jeffrey P., and Patricia F. Mead: "Crystal Oscillators," in *The Electronics Handbook*, Jerry C. Whitaker (ed.), CRC Press, Boca Raton, Fla., pp. 185–199, 1996.

2. Frerking, M. E.: *Crystal Oscillator Design and Temperature Compensation*, Van Nostrand Reinhold, New York, N. Y., 1978.

3. Dye, D. W.: *Proc. Phys. Soc.*, Vol. 38, pp. 399–457, 1926.

4. Hietala, Alexander W., and Duane C. Rabe: "Latched Accumulator Fractional-*N* Synthesis With Residual Error Reduction," United States Patent, Patent No. 5,093,632, March 3, 1992.

5. Riley, Thomas A. D.: "Frequency Synthesizers Having Dividing Ratio Controlled Sigma-Delta Modulator," United States Patent, Patent No. 4,965,531, October 23, 1990.

6. King, Nigel J. R.: "Phase Locked Loop Variable Frequency Generator," United States Patent, Patent No. 4,204,174, May 20, 1980.

7. Wells, John Norman: "Frequency Synthesizers," European Patent, Patent No. 012579OB2, July 5, 1995.

8. Rohde, Ulrich L.: *Digital PLL Frequency Synthesizers*, Prentice-Hall, Englewood Cliffs, N.J., 1983.

## 3.8 Bibliography

Rohde, Ulrich L.: *Microwave and Wireless Synthesizers: Theory and Design*, John Wiley & Sons, New York, N.Y., pg. 209, 1997.

Chapter

4

# Receiver Characteristics

**Ulrich L. Rohde**

**Jerry C. Whitaker, Editor**

## 4.1 Introduction[1]

The superheterodyne receiver makes use of the heterodyne principle of mixing an incoming signal with a signal generated by a local oscillator (LO) in a nonlinear element (Figure 4.1). However, rather than synchronizing the frequencies, the superheterodyne receiver uses a LO frequency offset by a fixed intermediate frequency (IF) from the desired signal. Because a nonlinear device generates identical difference frequencies if the signal frequency is either above or below the LO frequency (and also a number of other spurious responses), it is necessary to provide sufficient filtering prior to the mixing circuit so that this undesired signal response (and others) is substantially suppressed. The frequency of the undesired signal is referred to as an *image frequency*, and a signal at this frequency is referred to as an *image*. The image frequency is separated from the desired signal frequency by a difference equal to twice the IF. The preselection filtering required at the signal frequency is much broader than if the filtering of adjacent channel signals were required. The channel filtering is accomplished at IF. This is a decided advantage when the receiver must cover a wide frequency band, because it is much more difficult to maintain constant bandwidth in a tunable filter than in a fixed one. Also, for receiving different signal types, the bandwidth can be changed with relative ease at a fixed frequency by switching filters of different bandwidths. Because the IF at which channel selectivity is provided is often lower than the signal band frequencies, it may be easier to provide selectivity at IF, even if wide-band RF tuning is not required.

1. This chapter is based on: Rohde, Ulrich L., and Jerry C. Whitaker: *Communications Receivers: Principles and Design*, 3rd ed., Mcgraw-Hill, New York, N.Y., 2000. Used with permission.

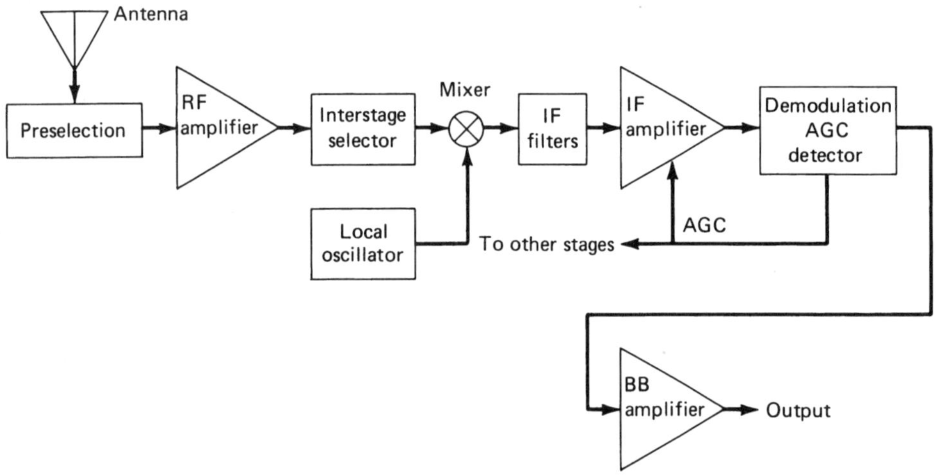


**Figure 4.1** Block diagram of a superheterodyne receiver.

## 4.1a Practical Receivers

Because of the nature of active electronic devices, it is generally easier to provide high stable gain in a fixed-frequency amplifier than in a tunable one, and gain is generally more economical at lower frequencies. Thus, although the superheterodyne receiver does introduce a problem of spurious responses not present in the other receiver types, its advantages are such that it has replaced other types except for special applications.

Referring again to Figure 4.1, the signal is fed from the antenna to a preselector filter and amplifier. The input circuit is aimed at matching the antenna to the first amplifying device so as to achieve the best sensitivity while providing sufficient selectivity to reduce the probability of overload from strong undesired signals in the first amplifier. Losses from the antenna coupling circuit and preselection filters decrease the sensitivity. Because sufficient selectivity must be provided against the image and other principal spurious responses prior to the mixing circuit, the preselection filtering is often broken into two or more parts with intervening amplifiers to minimize the effects of the filter loss on the *noise factor* (NF). The LO provides a strong stable signal at the proper frequency to convert the signal frequency to IF. This conversion occurs in the mixer. (This element has also been referred to as the *first detector*, converter, or *frequency changer*.)

The output from the mixer is applied to the IF amplifier, which amplifies the signal to a suitable power level for the demodulator. This circuit derives from the IF signal the modulation signal, which may be amplified by the baseband amplifier to the level required for output. Normally, the output of an audio amplifier is fed to a headphone or loudspeaker at the radio, or coupled to a transmission line for remote use. A video signal requires development of sweep, intensity, and usually color signals from the amplified video demodulation prior to display. In other cases, the output may be supplied to a data demodulator to produce digital data signals from the baseband signal. The data demodulator may be part of the receiver or may be provided separately as part of a data modem. The data modem may also be fed directly from the receiver at IF. Data demodulation is typically accomplished using digital processing circuits rather than analog demodulators and amplifiers. In this case, the IF amplifier must be designed to provide the appropriate level to an A/D converter so that digital processing can be carried out. Additional IF filtering, data demodulation, and error control coding can all be performed by digital circuits or a microprocessor, either in the receiver or as part of an external modem.

An alternative to IF sampling and A/D conversion is the conversion of the signal to baseband in two separate *coherent demodulators* driven by quadrature LO signals at the IF. The two outputs are then sampled at the appropriate rate for the baseband by two A/D converters or a single multiplexed A/D converter, providing the in-phase and quadrature samples of the baseband signal. Once digitized, these components can be processed digitally to provide filtering, frequency changing, phase and timing recovery, data demodulation, and error control.

## 4.2 The Receiving System

The first essential function of any radio receiver is to effect the transfer of energy picked up by the antenna to the receiver itself through the input circuits. Maximum energy is transferred if the impedance of the input circuit matches that of the antenna (inverse reactance and same resistance) throughout the frequency band of the desired signal. This is not always feasible, and the best energy transfer is not essential in all cases. A receiver may also be connected with other receivers through a *hybrid* or *active multicoupler* to a single antenna. Such arrangements are sometimes very sensitive to mismatches.

There are at least three antenna matching problems in a receiver. The first and, in many cases, most crucial problem is that the receiver may be used from time to time with different antennas whose impedances the potential users cannot specify fully. Second, antennas may be used in mobile applications or in locations subject to changing foliage, buildings, or waves at sea, so that the impedance—even if measured accurately at one time—is subject to change from time to time. Third, at some frequencies, the problems of matching antennas are severely limited by available components, and the losses in a matching network may prove greater than for a simpler lower-loss network with poorer match.

When antenna matching is important over a wide band, it may be necessary to design a network that can be tuned mechanically or electrically under either manual or automatic control in response to a performance measure in the system. In older receivers with a wide tuning range, it was common to have a mechanically tuned preselector that could be adjusted by hand and was generally connected directly to the *variable-frequency oscillator* (VFO) used for the first conversion. At times a trimmer was used in the first circuit to compensate for small antenna mismatches. Thus, tuning of the circuit could be modified to match the effects of the expected antenna impedance range. Modern wide tuning receivers often use one-half-octave switchable filters in the preselector, which may be harder to match, but are much easier to control by computer. Similarly, the first oscillator is generally a microprocessor-controlled synthesizer.

Often the problem of antenna matching design is solved by the user specification that defines one or more "dummy antenna" impedances to be used with a signal generator to test the performance of the receiver for different receiver input circuits. In this case, the user's system is designed to allow for the mismatch losses in performance that result from the use of actual antennas. When it is necessary to measure receiver input impedance accurately, it is best accomplished through a network analyzer.

A number of other receiver input considerations may occur in certain cases. The input circuits may be balanced or unbalanced, or may need to be connectable either way. The input circuits may require grounding, isolation from ground, or either connection. The circuits may need protection from high-voltage discharges or from impulses. They may need to handle, without destruction, high-power nearby signals, both tuned to the receiver frequency and off-tune. Thus, the input circuit can—at times—present significant design challenges.

## 4.2a Gain, Sensitivity, and Noise Figure

Any given receiver is usually required to receive and process a wide range of signal powers, but in most cases it is important that they be capable of receiving distant signals whose power has been attenuated billions of times during transmission. The extent to which such signals can be received usefully is determined by the noise levels received by the antenna (natural and man-made) and those generated within the receiver itself. It is also necessary that the receiver produce a level of output power suitable for the application. Generally the ratio of the output power of a device to the input power is known as the *gain*. The design of a receiver includes gain distribution among the various stages so as to provide adequate receiver gain and an optimum compromise among the other operating characteristics.

While there are limits to the amount of gain that can be achieved practically at one frequency because of feedback, modern receivers need not be gain-limited. When the gain is sufficiently high, the weakest signal power that may be processed satisfactorily is *noise-limited*. This signal level is referred to as the *sensitivity* of the system at a particular time and varies depending on the external noise level. It is possible in some systems for the external noise to fall sufficiently so that the system sensitivity is established by the internal noise of the receiver. A receiver's sensitivity is one of its most important characteristics. There are no universal standards for its measurement, although standards have been adopted for specific applications and by specific user groups. Figure 4.2 shows a block diagram of the test setup and the typical steps involved in determining receiver sensitivity.

### NF

Sensitivity measures depend upon specific signal characteristics. The NF measures the effects of inherent receiver noise in a different manner. Essentially it compares the total receiver noise with the noise that would be present if the receiver generated no noise. This ratio is sometimes called the noise factor $F$, and when expressed in dB, the NF. $F$ is also defined equivalently as the ratio of the S/N of the receiver output to the S/N of the source. The source generally used to test receivers is a signal generator at local room temperature. An antenna, which receives not only signals but noises from the atmosphere, the galaxy, and man-made sources, is unsuitable to provide a measure of receiver NF. However, the NF required of the receiver from a system viewpoint depends on the expected S/N from the antenna. The effects of external noise are sometimes expressed as an equivalent antenna NF.

For the receiver, we are concerned with internal noise sources. Passive devices such as conductors generate noise as a result of the continuous thermal motion of the free electrons. This type of noise is referred to generally as *thermal noise*, and is sometimes called *Johnson noise* after the person who first demonstrated it. Using the statistical theory of thermodynamics, Nyquist showed that the mean-square thermal noise voltage generated by any impedance between two frequencies $f_1$ and $f_2$ can be expressed as

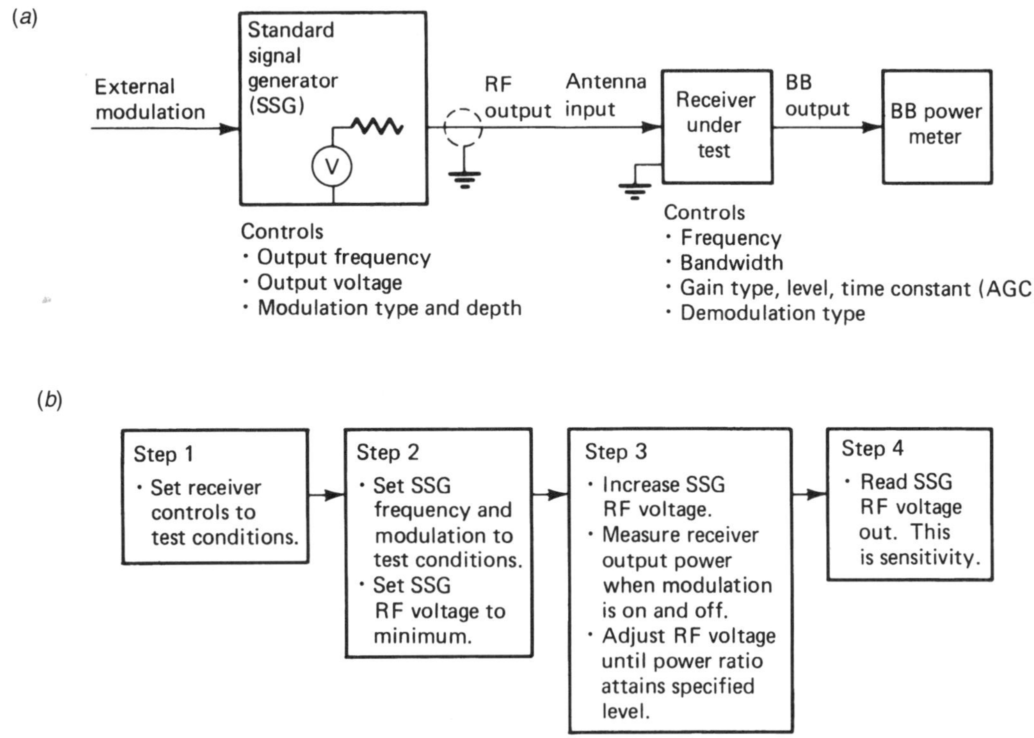


**Figure 4.2** Receiver sensitivity measurement: (*a*) test setup, (*b*) procedure.

$$\overline{V_n^2} = 4kt\int_{f_1}^{f_2} R(f)\,df \tag{4.1}$$

where $R(f)$ is the resistive component of the impedance.

Magnetic substances also produce noise, depending upon the residual magnetization and the applied dc and RF voltages. This is referred to as the *Barkhausen effect*, or *Barkhausen noise*. The greatest source of receiver noise, however, is generally that generated in semiconductors. Like the older thermionic tubes, transistors and diodes also produce characteristic noise. *Shot noise* resulting from the fluctuations in the carrier flow in semiconductor devices produces wide-band noise, similar to thermal noise. Low-frequency noise or $1/f$ noise, also called *flicker effect*, is roughly inversely proportional to frequency and is similar to the contact noise in contact resistors. All of these noise sources contribute to the "excess noise" of the receiver, which causes the NF to exceed 0 dB.

The NF is often measured in a setup similar to that of Figure 4.2, using a specially designed and calibrated white-noise generator as the input. The receiver is tuned to the correct frequency and bandwidth, and the output power meter is driven from a linear demodulator or the final IF amplifier. The signal generator is set to produce no output, and the output power is observed. The generator output is then increased until the output has risen 3 dB. The setting on the generator is the NF in decibels.

The NF of an amplifier can also be calculated as the ratio of input to output S/N, per the equation

$$NF = 10 \log \left[ \frac{(S/N)_1}{(S/N)_0} \right] \tag{4.2}$$

where NF is the noise figure in dB and $(S/N)_1$ and $(S/N)_2$ are the amplifier input and output SNR, respectively.

The NF for a noiseless amplifier or lossless passive device is 0 dB; it is always positive for nonideal devices. The NF of a lossy passive device is numerically equal to the device insertion loss. If the input of a nonideal amplifier of gain $G$ (dB) and noise figure $NF$ (dB) were connected to a matched resistor, the amplifier output noise power $P_{No}$ (dB) would be

$$P_{No} = 10 \log (kT) + 10 \log (B) + G + NF \tag{4.3}$$

where $k$ is Boltzmann's constant (mW/°K), $T$ is the resistor temperature in °K, and $B$ is the noise bandwidth in Hz.

When amplifiers are cascaded, the noise power rises toward the output as noise from succeeding stages is added to the system. Under the assumption that noise powers add noncoherently, the noise figure $NF_T$ of a cascade consisting of two stages of numerical gain $A_1$ and $A_2$ and noise factor $N_1$ and $N_2$, is given by Friis' equation

$$NF_T = 10 \log \left[ \frac{N_1 + (N_2 - 1)}{A_1} \right] \tag{4.4}$$

where the noise factor is $N = 10^{(NF/10)}$ and the numerical gain is $A = 10^{(G/10)}$. The system NF, therefore, is largely determined by the first stage NF when $A_1$ is large enough to make $(N_2 - 1)/\ A_1$ much smaller than $N_1$.

### Minimum Detectable Signal

Another measure of sensitivity is the *minimum detectable signal* (MDS). The measurement procedure is similar to the NF measurement except that a sinusoidal signal generator replaces the noise generator to produce the doubling of output power over noise power alone. This signal power, just equal to the noise power, is defined as the MDS. Because receiver noise varies with bandwidth, so does the MDS, which can be expressed as

$$MDS = k\,T\,B_n\,F \tag{4.5}$$

In dBm, $MDS = -174 + 10\log(B_n + NF)$, where $B_n$ is the noise bandwidth of the receiver. (0 dbm = decibels referenced to 1mW.)

The available thermal noise power per hertz is –174 dBm at 290ºK (63ºF), an arbitrary reference temperature near standard room temperatures. When any two of the quantities in the expression are known, the third may be calculated. As in the case of NF measurements, care is required in measuring MDS, because a large portion of the power being measured is noise, which produces MDS' typical fluctuations.

## 4.3 Selectivity

Selectivity is the property of a receiver that allows it to separate a signal or signals at one frequency from those at all other frequencies. At least two characteristics must be considered simultaneously in establishing the required selectivity of a receiver. The selective circuits must be sufficiently sharp to suppress the interference from adjacent channels and spurious responses. On the other hand, they must be broad enough to pass the highest sideband frequencies with acceptable distortion in amplitude and phase. Each class of signals to be received may require different selectivity to handle the signal sidebands adequately while rejecting interfering transmissions having different channel assignment spacings. However, each class of signal requires about the same selectivity throughout all the frequency bands allocated to that class of service. Older receivers sometimes required broader selectivity at their higher frequencies to compensate for greater oscillator drift. This requirement has been greatly reduced by the introduction of synthesizers for control of LOs and economical high-accuracy and high-stability crystal standards for the reference frequency oscillator. Consequently, except at frequencies above VHF, or in applications where adequate power is not available for temperature-controlled ovens, only the accuracy and stability of the selective circuits themselves may require selectivity allowances today.

Quantitatively the definition of selectivity is the bandwidth for which a test signal $x$ decibels stronger than the minimum acceptable signal at a nominal frequency is reduced to the level of that signal. This measurement is relatively simple for a single selective filter or single-frequency amplifier, and a selectivity curve can be drawn showing the band offset both above and below the nominal frequency as the selected attenuation level is varied. Ranges of 80 to 100 dB of attenuation can be measured readily, and higher ranges—if required—can be achieved with special care. A test setup similar to Figure 4.2 may be employed with the receiver replaced by the selective element under test. Proper care must be taken to achieve proper input and output impedance termination for the particular unit under test. The power output meter need only be sufficiently sensitive, have uniform response over the test bandwidth, and have monotonic response so that the same output level is achieved at each point on the curve. A typical IF selectivity curve is shown in Figure 4.3.

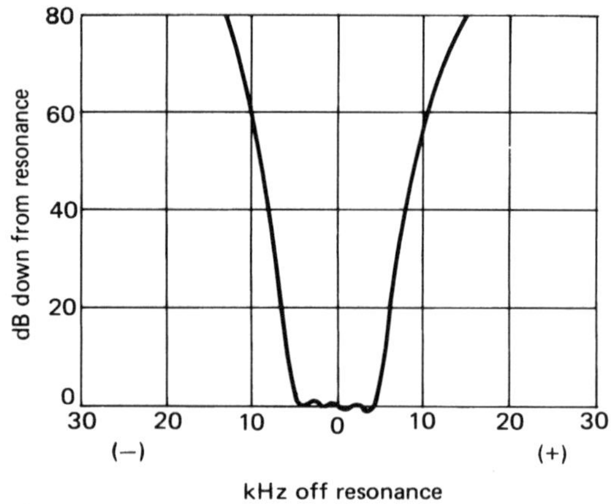


**Figure 4.3** Example of an IF selectivity curve.

The measurement of overall receiver selectivity, using the test setup of Figure 4.2, presents some difficulties. The total selectivity of the receiving system is divided among RF, IF, and baseband selective elements. There are numerous amplifiers and frequency converters, and at least one demodulator intervening between input and output. Hence, there is a high probability of nonlinearities in the nonpassive components affecting the resulting selectivity curves. Some of the effects that occur include overload, modulation distortion, spurious signals, and spurious responses. If there is an AGC, it must be disabled so that it cannot change the amplifier gain in response to the changing signal levels in various stages of the receiver. If there is only an AM or FM demodulator for use in the measurement, distortions occur because of the varying attenuation and phase shift of the circuits across the sidebands.

When measuring complete receiver selectivity, with either a modulated or nonmodulated signal, it is wise to use an output power meter calibrated in decibels. The measurement proceeds as described previously. However, if any unusual changes in attenuation or slope are observed, the generator level may be increased in calibrated steps; it should be noted whether the output changes decibel for decibel. If not, what is being observed at this point is not the selectivity curve, but one of the many nonlinearities or responses of the system.

## 4.4 Dynamic Range

The term *dynamic range*, especially in advertising literature, has been used to mean a variety of things. We must be especially careful in using a common definition when comparing this characteristic of receivers. In some cases, the term has been used to indicate the ratio in decibels between the strongest and weakest signals that a receiver could handle with acceptable noise or distortion. This is the ratio between the signal that is so strong that it causes maximum tolerable distortion and the one that is so weak that it has the minimum acceptable S/N. This measure is of limited value in assessing performance in the normal signal environment where the desired signal may have a range of values, but is surrounded by a dense group of other signals ranging from very weak to very strong. The selective circuits of a receiver can provide protection from many of these signals. However, the stronger ones, because of the nonlinearity of the active devices necessary to provide amplification and frequency conversion, can degrade performance substantially. In modern parlance, dynamic range refers to the ratio of the level of a strong out-of-band signal that in some manner degrades signal performance of the receiver to a very weak signal. The most common weak signal considered is the MDS, and differing strong degrading signal levels may be used. It is, therefore, important to know which definition is meant when evaluating the meaning of the term *dynamic range*.

If the foregoing discussion of dynamic range seems vague, it is because there is not one characteristic but several that is encompassed by the term. Each may have a different numeric value. A receiver is a complex device with many active stages separated by different degrees of selectivity. The response of a receiver to multiple signals of different levels is extremely complex, and the results do not always agree with simple theory. However, such theory provides useful comparative measures. If we think of an amplifier or mixer as a device whose output voltage is a function of the input voltage, we may expand the output voltage in a power series of the input voltage

$$V_o = \Sigma a_n V_i^n \tag{4.6}$$

where $a_1$ is the voltage amplification of the device and the higher-order $a_n$ cause distortion.

Because the desired signal and the undesired interference are generally narrow-band signals, we may represent $V_i$ as a sum of sinusoids of different amplitudes and frequencies. Generally $(A_1 \sin 2\pi f_1 t + A_2 \sin 2\pi f_2 t)^n$ , as a result of trigonometric identities, produces a number of components with different frequencies, $mf_1 \pm (n-m) f_2$ , with $m$ taking on all values from 0 to $n$. These intermodulation (IM) products may have the same frequency as the desired signal for appropriate choices of $f_1$ and $f_2$. When $n$ is even, the minimum difference between the two frequencies for this to happen is the desired frequency itself. This type of *even* IM interference can be reduced substantially by using selective filters.

When $n$ is odd, however, the minimum difference can be very small. Because $m$ and $n - m$ can differ by unity, and each can be close to the signal frequency, if the adjacent interferer is $\xi f$ from the desired signal, the second need be only $(2\xi f)/\ (n-1)\ /$ further away for the product to fall at the desired frequency. Thus, *odd-order* IM products can be caused by strong signals only a few channels removed from the desired signal. Selective filtering capable of reducing such signals substantially is not available in most superheterodyne receivers prior to the final IF. Consequently, odd-order IM products generally limit the dynamic range significantly.

Other effects of odd-order distortion are *desensitization* and *cross modulation*. For the case where $n$ is odd, the presence of the desired signal and a strong interfering signal results in a product of the desired signal with an even order of the interfering signal. One of the resulting components of an even power of a sinusoid is a constant, so the desired signal is multiplied by that constant and an even power of the interferer's signal strength. If the interferer is sufficiently strong, the resulting product will subtract from the desired signal product from the first power term, reducing the effective gain of the device. This is referred to as *desensitization*. If the interferer is amplitude-modulated, the desired signal component will also be amplitude-modulated by the distorted modulation of the interferer. This is known as *cross modulation* of the desired signal by the interfering signal.

This discussion provides a simple theory that can be applied in considering strong signal effects. However, the receiver is far more complicated than the single device, and strong signal performance of single devices by these techniques can become rapidly intractable as higher-order terms must be considered. Another mechanism also limits the dynamic range. LO noise sidebands at low levels can extend substantially from the oscillator frequency. A sufficiently strong off-tune signal can beat with these noise sidebands in a mixer, producing additional noise in the desired signal band. Other characteristics that affect the dynamic range are spurious signals and responses, and blocking.

The effects described here occur in receivers, and tests to measure them are essential to determining the dynamic range. Most of these measurements involving the dynamic range require more than one signal input. They are conducted using two or three signal generators in a test setup such as that indicated in Figure 4.4.

## 4.4a Desensitization

Desensitization measurements are related to the 1-dB compression point and general linearity of the receiver. Two signal generators are used in the setup of Figure 4.4. The controls of the receiver under test are set as specified, usually to one of the narrower bandwidths and with MGC set as in sensitivity measurements so as to avoid effects of the AGC system. The signal in the operating channel is modulated and set to a specified level, usually to produce an output S/N or SINAD measurement of a particular level, for example, 13 dB. The interfering signal is moved off the operating frequency by a predetermined amount so that it does not affect the S/N measurement because of beat notes and is then increased in level until the S/N measurement is reduced by a specified amount, such as 3 dB. More complete information can be obtained by varying the frequency offset and plotting a desensitization selectivity curve. In some cases, limits for this curve are specified. The curve may be carried to a level of input where spurious responses, reciprocal mixing, or other effects prevent an unambiguous measurement. Measurements to 120 dB above sensitivity level can often be achieved.

If the degradation level at which desensitization is measured is set to 1-dB, and the desensitizing signal is well within the passband of the preselector filters, the desensitization level corresponds to the 1 dB *gain compression* (GC), which is experienced by the system up to the first mixer. (See the subsequent discussion of intermodulation and intercept points.) A gain compression (or *blocking*) dynamic range can be defined by comparing the input signal level at 1-dB GC to the MDS, i. e., dynamic range (dB) equals the GC (input dBm) minus the MDS (input dBm). This is sometimes referred to as the *single-tone dynamic range*, because only a single interfering signal is needed to produce GC.

## 4.4b AM Cross Modulation

Although many saturation effects in receivers have been called cross modulation, SSB and FM are not cross-modulated in the same sense as described previously. Cross modulation occurs in AM and VSB signals by a strong modulated signal amplitude-modulating a weak signal through the inherent nonlinearities of the receiver. Cross modulation typically occurs in a band allocated for AM use and requires a much higher interfering signal level than for the generation of IM

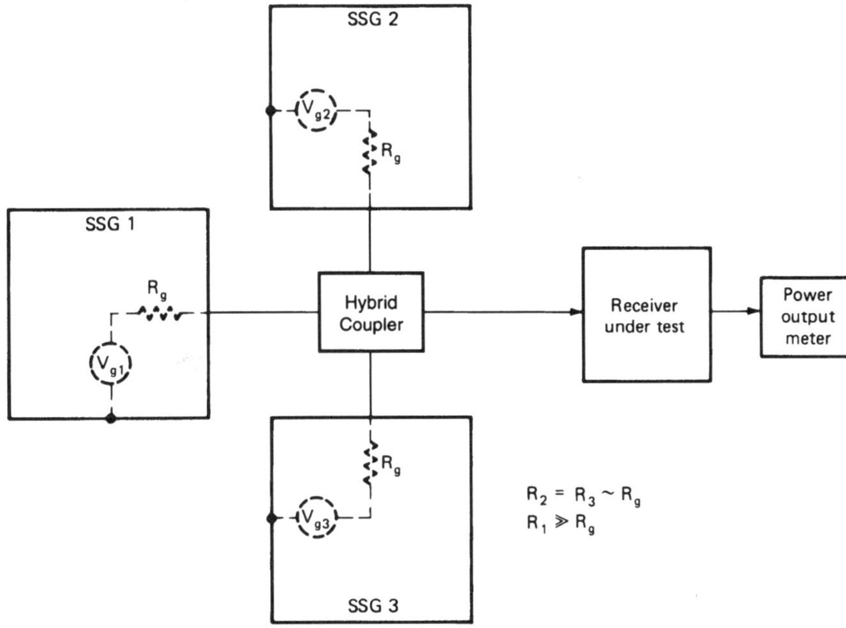


**Figure 4.4** Test setup for measuring the dynamic range properties of a receiver.

products. The typical measurement setup is similar to that for overload measurements, except that the interfering signal is amplitude-modulated, usually at a high level, such as 90 percent. The modulation is at a different frequency than that for the operating channel (if it is modulated), and a band-pass filter is used in the output to assure that the transferred modulation is being measured. The out-of-channel interfering signal is increased in level until the desired signal has a specified level of output at the cross modulation frequency, for example, the equivalent of 10 percent modulation of the desired carrier. One or more specific offsets may be specified for the measurement, or a cross-modulation selectivity curve may be taken by measuring carrier level versus frequency offset to cause the specified degree of cross modulation.

In analog television systems, cross modulation can result in a ghost of an out-of-channel modulation being visible on the operating channel. The so-called three-tone test for television signals is a form of cross-modulation test.

## 4.4c IM

As described in previous sections, IM produces sum and difference frequency products of many orders that manifest themselves as interference. The measurement of the IM distortion performance is one of the most important tests for a communications receiver. No matter how sensitive a receiver may be, if it has poor immunity to strong signals, it will be of little use. Tests for even-order products determine the effectiveness of filtering prior to the channel filter, while odd-order products are negligibly affected by those filters. For this reason, odd-order products are generally much more troublesome than even-order products and are tested for more frequently. The second- and third-order products are generally the strongest and are the ones most frequently tested. A two-signal generator test set is required for testing, depending on the details of the specified test.

For IM measurements, the controls of the receiver under test are set to the specified bandwidths, operating frequency, and other settings as appropriate, and the gain control is set on manual (or AGC disabled). One signal generator is set on the operating frequency, modulated and adjusted to provide a specified S/N (that for sensitivity, for example). The modulation is disabled, and the output level of this signal is measured. This must be done using the IF output, the SSB output with the signal generator offset by a convenient audio frequency, or with the BFO on and offset. Alternatively, the dc level at the AM demodulator can be measured, if accessible. The signal generator is then turned off. It may be left off during the remainder of the test or retuned and used to produce one of the interfering signals.

For second-order IM testing, two signal generators are now set to two frequencies differing from each other by the operating frequency. These frequencies can be equally above and below the carrier frequency at the start, and shifted on successive tests to assure that the preselection filters do not have any weak regions. The signal with the frequency nearest to the operating frequency must be separated far enough to assure adequate channel filter attenuation of the signal (several channels). For third-order IM testing, the frequencies are selected in accordance with the formula given previously so that the one further from the operating frequency is twice as far as the one nearer to the operating frequency. For example, the nearer interferer might be three channels from the desired frequency and the further one, six channels in the same direction.

In either case, the voltage levels of the two interfering signal generators are set equal and are gradually increased until an output equal to the original channel output is measured in the channel. One of several performance requirements may be specified. If the original level is the sensitivity level, the ratio of the interfering generator level to the sensitivity level may have a specified minimum. Alternatively, for any original level, an interfering generator level may be specified that must not produce an output greater than the original level. Finally, an *intercept point* (IP) may be specified.

The IP for the *n*th order of intermodulation occurs because the product is a result of the interfering signal voltages being raised to the *n*th power. With equal voltages, as in the test, the resultant output level of the product increases as

$$V_{dn} = c_n V^n \tag{4.7}$$

where $c_n$ is a proportionality constant and $V$ is the common input level of the two signals. Because the output from a single signal input $V$ at the operating frequency is $G_v V$, there is a theoretical level at which the two outputs would be equal. This value $V_{IPn}$ is the *n*th IP, measured at the input. It is usually specified in dBm. In practice the IPs are not reached because as the amplifiers approach saturation, the voltage at each measured frequency becomes a combination of components from various orders of *n*. Figure 4.5 indicates the input-output power relationships in second- and third-order IPs.

In Equation 4.7 we note that at the IP

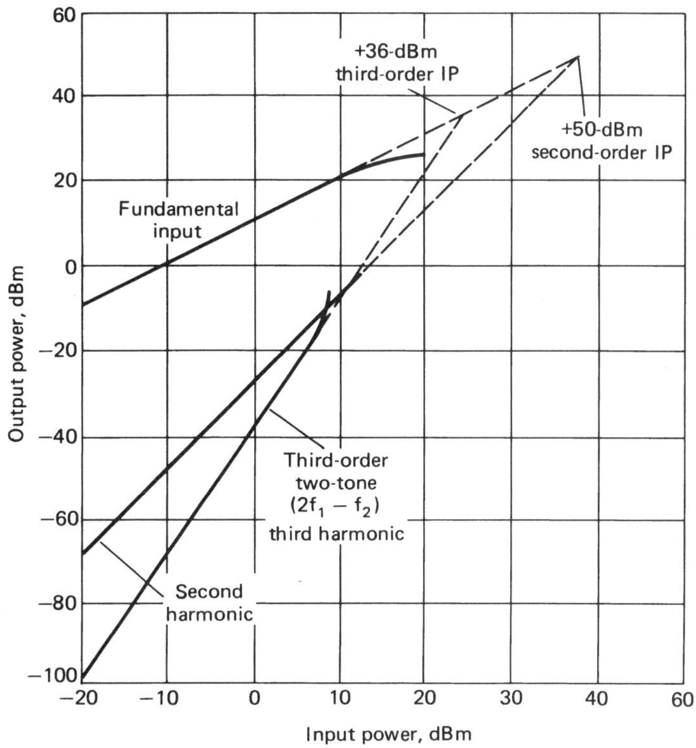


**Figure 4.5** Input/output power relationships for second- and third-order intercept points.

$$V_{dn} = c_n \,(V_{IPn})^n \text{ and } (V_{IPn})_{out} = G_v \,(V_{IPn})_{in} \tag{4.8}$$

This leads to

$$c_n = G_v \,(V_{IPn})^{1-n} \text{ and } V_{dn} = G_v V \left[\frac{V}{V_{IPn}}\right]^{1-n} \tag{4.9}$$

The ratio of signal to distortion becomes $(V_{IPn})/\ V^{n-1}$. In decibels it becomes

$$R_{dn} = 20 \log \left[ \frac{V}{V_{dn}} \right] = (n - 1)\,[20 \log V_{IPn} - 20 \log V] \tag{4.10}$$

If the intercept level is expressed in dBm rather than voltage, then the output power represented by $V$ must be similarly expressed.

The IM products we have been discussing originate in the active devices of the receiver, so that the various voltages or power levels are naturally measured at the device output. The IP is thus naturally referred to the device output and is so specified in most data sheets. In the foregoing discussion, we have referred the IP to the voltage level at the device input. If the input power is required, we subtract from the output intercept level in decibels, the amplifier power gain or loss. The relationship between input and output voltage at the IP is given in Equation 4.8. Reference of the IP to the device input is somewhat unnatural but is technically useful because the receiver system designer must deal with the IP generation in all stages and needs to know at what antenna signal level the receiver will produce the maximum tolerable IM products.

Consider the input power (in each signal) that produces an output IM product equal to the MDS. The ratio of this power to the MDS may be called the *third-order IM dynamic range.* It also is sometimes referred to as the *two-tone dynamic range.* Expressing Equation 4.10 in terms of input power and input IP measured in dBm, we have

$$R_{dn} = (n - 1)\,[IP_{n\,(in)} - P_{(in)}] \tag{4.11}$$

When we substitute MDS for the distortion and $MDS + DR$ for $P_{(in)}$ we obtain

$$\begin{aligned} DR &= (n - 1)\,[IP_{n\,(in)} - MDS - DR], \\ nDR &= (n - 1)\,[IP_{n\,(in)} - MDS] \end{aligned} \tag{4.12}$$

When $n$ is 3, we find the relationship

$$DR = \frac{2\,[IP_{3\,(in)} - MDS]}{3} \tag{4.13}$$

A dynamic range could presumably be defined for other orders of IM, but it is not common to do so. From the three different definitions of dynamic range described in this section, it should be clear why it is important to be careful when comparing receiver specifications for this characteristic.

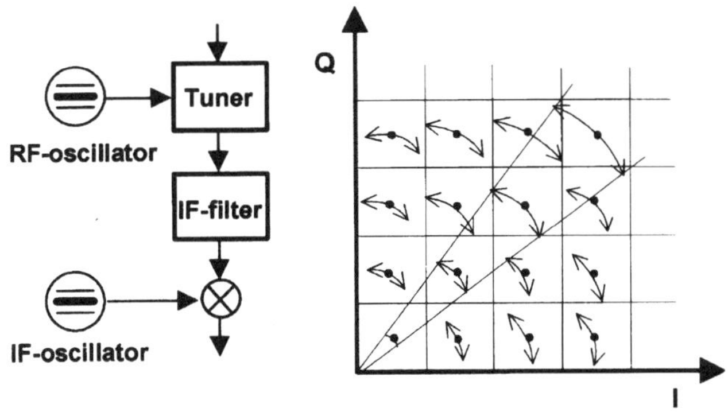


**Figure 4.6** Phase noise is critical to digitally modulated communication systems because of the modulation errors it can introduce. *Inter-symbol interference* (ISI), accompanied by a rise in BER, results when state values become so badly error-blurred that they fall into the regions of adjacent states. This drawing depicts only the results of phase errors introduced by phase noise; in actual systems, thermal noise, AM-to-PM conversion, differential group delay, propagation, and other factors may also contribute to the spreading of state amplitude and phase values. (*Courtesy of Rohde & Schwarz.*)

### 4.4d Error Vector Magnitude

As shown in Figure 4.6, particular sources of amplitude and/or phase error can shift the values of digital emission data states toward decision boundaries, resulting in increased BER because of intersymbol interference. Figures 4.7, 4.8, and 4.9 show three additional sources of such errors.

A figure of merit known as *error vector magnitude* (EVM) has been developed as sensitive indicator of the presence and severity of such errors. The error vector magnitude of an emission is the magnitude of the phasor difference as a function of time between an ideal reference signal and the measured transmitted signal after its timing, amplitude, frequency, phase, and dc offset have been modified by circuitry and/or propagation. Figure 4.10 illustrates the EVM concept.

## 4.5 Gain Control

Receivers must often be capable of handling a signal range of 100 dB or more. Most amplifiers remain linear over only a much smaller range. The later amplifiers in a receiver, which must provide the demodulator with about 1 V on weak signals, would need the capability to handle thousands of volts for strong signals without some form of gain control. Consequently, receivers customarily provide means for changing the gain of the RF or IF amplifiers, or both.

For applications where the received signal is expected to remain always within narrow limits, some form of manually selectable control can be used, which may be set on installation and seldom adjusted. There are few such applications. Most receivers, however, even when an operator is available, must receive signals that vary by tens of decibels over periods of fractions of seconds to minutes. The level also changes when the frequency is reset to receive other signals that

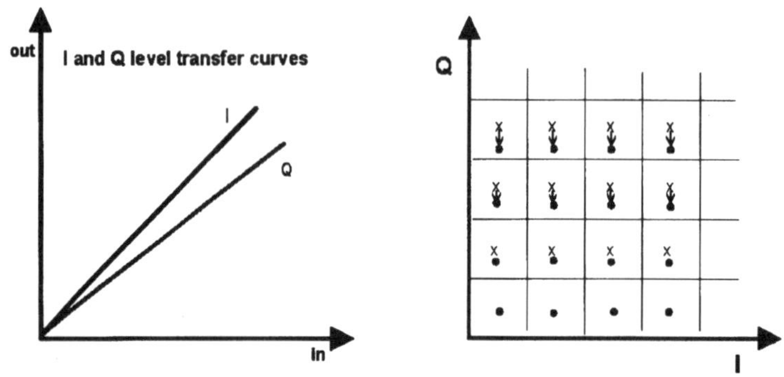


**Figure 4.7** Effect of gain imbalance between *I* and *Q* channels on data signal phase con-stellation. (*Courtesy of Rohde & Schwarz.*)

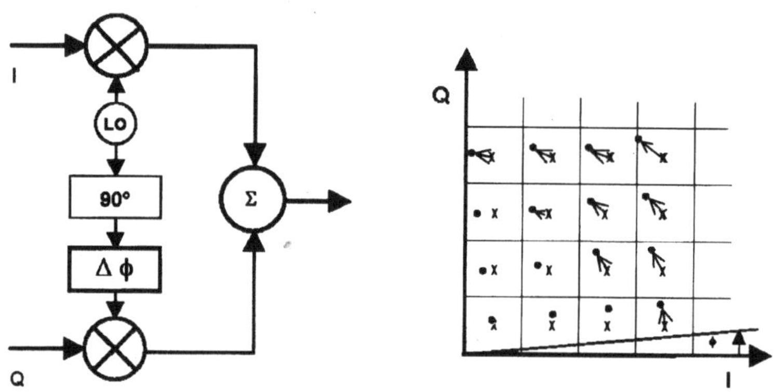


**Figure 4.8** Effect of quadrature offset on data signal phase constellation. (*Courtesy of Rohde & Schwarz.*)

may vary over similar ranges but with substantially different average levels. Consequently, an AGC is very desirable.

Some angle modulation receivers provide gain control by using amplifiers that limit on strong signals. Because the information is in the angle of the carrier, the resulting amplitude distortion is of little consequence. Receivers that must preserve AM or maintain very low angle modulation distortion use amplifiers that can be varied in gain by an external control voltage. In some cases, this has been accomplished by varying the operating points of the amplifying devices, but most modern systems separate solid-state circuits or switched passive elements to obtain variable attenuation between amplifier stages with minimum distortion. For manual control, provision can be made to let an operator set the control voltage for these variable attenuators. For automatic control, the output level from the IF amplifiers or the demodulator is monitored by the AGC cir-

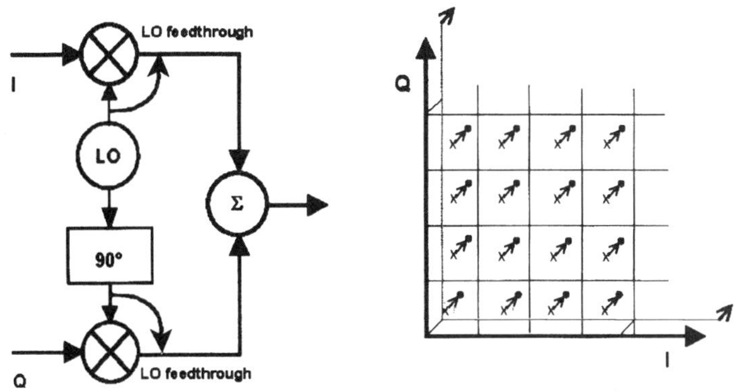


**Figure 4.9** Effect of LO-feedthrough-based *IQ* offset on data signal phase constellation. (*Courtesy of Rohde & Schwarz.*)

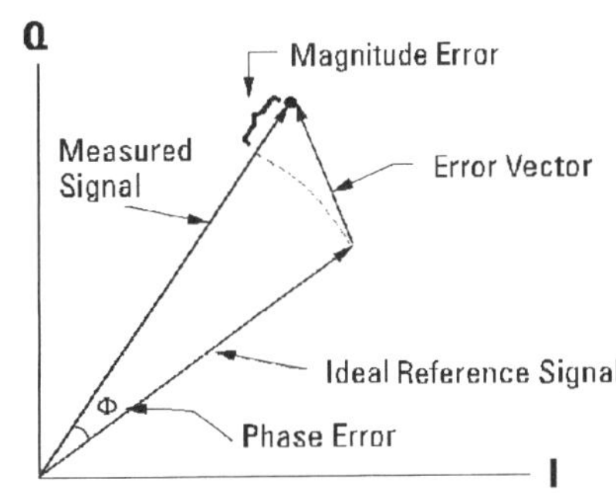


**Figure 4.10** The concept of error vector magnitude. (*Courtesy of Hewlett-Packard.*)

cuit and a low-pass negative-feedback voltage is derived to maintain a relatively constant signal level.

A block diagram of a dual AGC loop system is illustrated in Figure 4.11. One loop is driven by first IF energy that is band-limited, and the other loop is driven by second IF energy that is band-limited by optional second IF filters. The first loop controls a PIN diode pi attenuator ahead of the first mixer. The second loop controls the second IF amplifier stages. In this design, a microprocessor adjusts the time constants of both loops so that time delays introduced by the filters do not cause AGC oscillation.

A number of tests of gain control characteristics are customarily required. MGC may be designed to control gain continuously or in steps. It is important that the steps be small enough that operators do not detect large jumps as they adjust the gain. Because the gain must be controlled over a very wide range, the MGC is easiest to use if it tends to cause a logarithmic variation. Usually, the testing of the MGC is confined to establishing that a specified range of gain control exists and measuring the degree of decibel linearity versus control actuation.

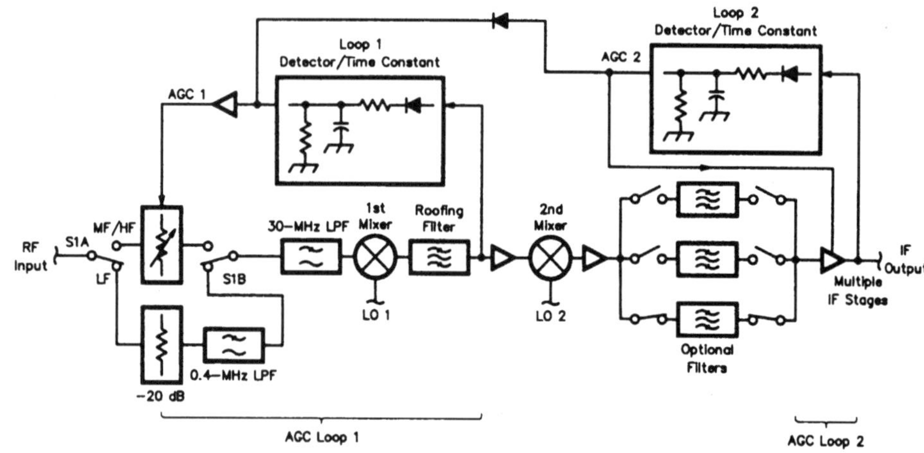


**Figure 4.11** Block diagram of a dual loop AGC system for a communications receiver.

The principal AGC characteristics of importance are the steady-state control range and output-input curve, and the attack and decay times. In a some systems, a variety of time constants are provided for the AGC to allow for different modulation types. For AM voice modulation, the radiated carrier is constant and the lowest sidebands are usually several hundred hertz removed from the carrier. At the receiver, the carrier component can be separated from the demodulated wave by a low-pass filter and can serve as the AGC control voltage. The response time of the filter, which is often just an RC network, need only be fast enough to respond to the fading rate of the medium, which is a maximum of 5 or 10 dB/s in most AM applications. A response time of 0.1 to 0.2 s is required for such a fading rate. For the more common slower rates, responses up to a second or more can be used.

To test for the AGC range and input-output curve, a single signal generator is used (as in Figure 18.8.2) in the AM mode with the receiver's AGC actuated. The signal generator is set to several hundred microvolts, and the baseband output level is adjusted to a convenient level for output power measurement. The signal generator is then tuned to its minimum level and the output level is noted. The signal is gradually increased in amplitude, and the output level is measured for each input level, up to a maximum specified level, such as 2 V. Figure 4.12 shows some typical AGC curves. In most cases, there will be a low-input region where the signal output, rising out of the noise, varies linearly with the input. At some point, the output curve bends over and begins to rise very slowly. At some high level, the output may drop off because of saturation effects in some of the amplifiers. The point at which the linear relationship ends is the AGC *threshold* of the AGC action. The point at which the output starts to decrease, if within a specified range, is considered the upper end of the AGC control range. The difference between these two input levels is the AGC *control range*. If the curve remains monotonic to the maximum input test level, that level is considered the upper limit of the range. A measure of AGC effectiveness is the increase in output from a specified lower input voltage level to an upper input voltage level.

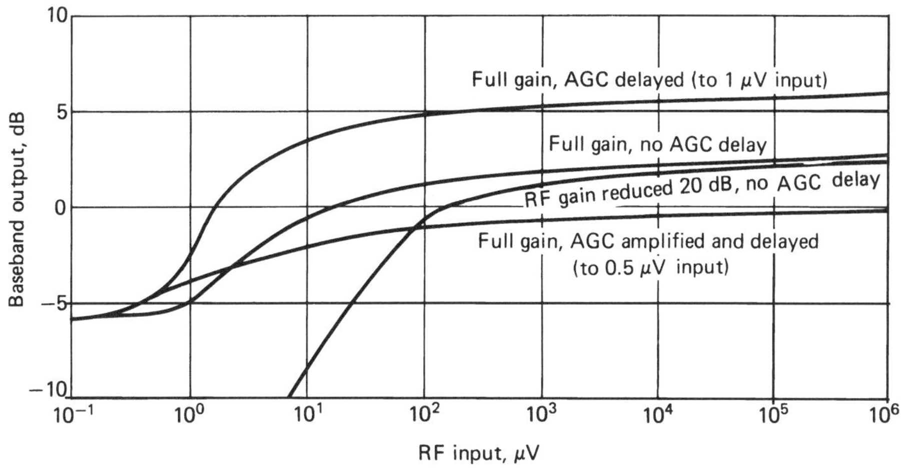


**Figure 4.12** Representative input-output AGC curves.

For example, a good design might have an AGC with a threshold below 1 μV that is monotonic to a level of 1 V and has the 3 dB increase in output between 1 μV and 0.1 V.

The foregoing applies to a purely analog system. The advantage of a DSP-based system include that AGC is handled internally. The receiver still requires a dual-loop AGC of which the input stage AGC will remain analog.

## 4.6 Digital Receiver Characteristics

The foregoing has not exhausted the analog receiver characteristics that may be of interest but has reviewed some of the more significant ones. For example, in FM sets, the *capture ratio* is important. Clearly, an area of increasing interest is the characterization of systems utilizing digital modulation techniques. Because a digital radio system is a hybrid A/D device, many of the test procedures outlined previously for analog receivers are useful and important in characterizing a digital radio system. Additional tests, primarily involving the analysis of bit error rates (BER), must also be run to properly identify any weak points in a receiver design.

### 4.6a BER Testing

The primary method for testing the quality of transmission over a high speed digital communications link is the BER, defined as the number of bit errors divided by the number of bits transmitted. The BER is also used to qualify the sensitivity and noise characteristics of a receiver. The major contributor to BER is *jitter*, which results from noise in the system. This noise causes the output comparator to vary the time of its transition relative to the data clock. If the transition time changes too much, an error will result

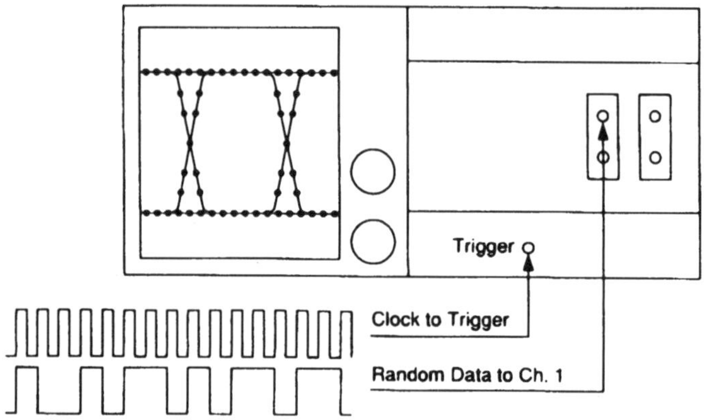


**Figure 4.13** Test setup for eye pattern measurement.

Using a signal analyzer specifically designed for BER testing, jitter can be displayed directly, or the BER is simply tabulated by the analyzer. The format of the testing signal is determined by the application to which the digital radio system is applied. A variety of formats and protocols are used. For end-to-end testing, the analyzer feeds a reference RF signal generator whose characteristics are known, and the output signal is applied to the receiver antenna input.

The noise and jitter on a data waveform provides vital information about the quality of the signal. A typical setup for capturing an *eye pattern* is shown in Figure 4.13. Eye patterns are the traditional method of displaying high-speed digital data (Figure 4.14). Some communications signal analyzers augment this information with a built-in statistical database, which allows additional analysis, including automated noise and jitter measurements on random data. Sophisticated software can also analyze the form of the distribution, providing mean, rms, and standard deviation results.

Proper receiver design involves identifying the areas of the system that are likely to cause problems. LO phase noise is one such area. Phase noise can seriously impair the performance of a digital receiver. Excessive phase noise can increase the BER, especially in systems using phase-based modulation schemes, such as binary PSK and quadrature PSK. For a given statistical phase-error characteristic, BER is degraded according to the percentage of time that the phase error causes the signal position in signal space to cross a decision boundary.

## 4.6b Transmission and Reception Quality

Testing of digital circuits deviates from the typical analog measurements, and yet the analog measurements are still necessary and related. In particular, because of the Doppler effect and the use of digital rather than analog signals, where the phase information is significant, the designer ends up using coding schemes for error-correction—specifically, *forward error correction* (FEC). The S/N as we know it from analog circuits now determines the BER, and its tolerable values depend on the type of modulation used. The actual bit error rate depends on the type of filtering, coding, modulation, and demodulation.

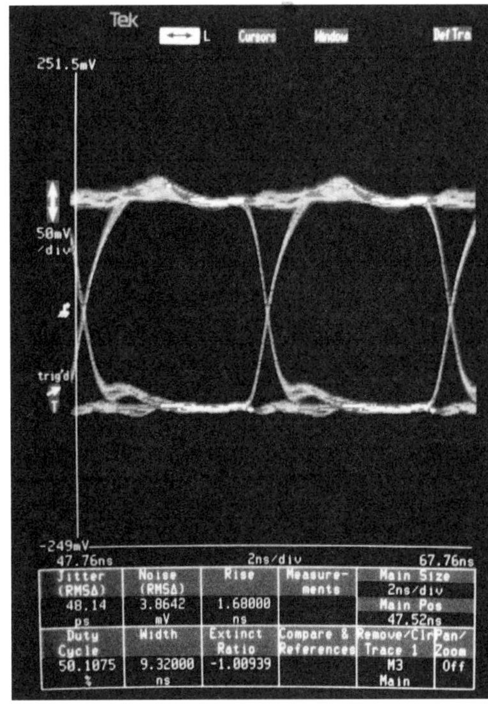


**Figure 4.14** Eye pattern display of BER measurement. (*Courtesy of Tektronix.*)

The *adjacent-channel power ratio* (ACPR), a factor involving the transmitter of a second station, is another problem that receivers must deal with. Given the fact that a transmitter handling digital modulation delivers its power in pulses, its transmissions may affect adjacent channels by producing transient spurious signals similar to what we call *splatter* in analog SSB systems. This is a function of the linearity of the transmitter system all the way out to the antenna, and forces most designers to resort to less-efficient Class A operation. As possible alternatives, some researchers have designed systems using Class D or E modulation.

It is not uncommon to do many linear measurements, and then by using correlation equations, translate these measured results into their digital equivalents. Therefore, the robustness of the signal as a function of antenna signal at the receiver site, constant or known phase relationships, and high adjacent power ratios will provide good transfer characteristics.

## 4.7 Bibliography

Engelson, M., and J. Herbert: “Effective Characterization of CDMA Signals,” *Microwave Journal*, pg. 90, January 1995.

Howald, R.: “Understand the Mathematics of Phase Noise,” *Microwaves & RF*, pg. 97, December 1993.

Johnson, J. B:, “Thermal Agitation of Electricity in Conduction,” *Phys. Rev.*, vol. 32, pg. 97, July 1928.

Nyquist, H.: "Thermal Agitation of Electrical Charge in Conductors," *Phys. Rev.*, vol. 32, pg. 110, July 1928.

Pleasant, D.: "Practical Simulation of Bit Error Rates," *Applied Microwave and Wireless*, pg. 65, Spring 1995.

Rohde, Ulrich L.: "Key Components of Modern Receiver Design—Part 1," *QST*, pg. 29, May 1994.

Rohde, Ulrich L. Rohde and David P. Newkirk: *RF/Microwave Circuit Design for Wireless Applications*, John Wiley & Sons, New York, N.Y., 2000.

"Standards Testing: Bit Error Rate," application note 3SW-8136-2, Tektronix, Beaverton, OR, July 1993.

*Using Vector Modulation Analysis in the Integration, Troubleshooting and Design of Digital RF Communications Systems*, Product Note HP89400-8, Hewlett-Packard, Palo Alto, Calif., 1994.

Watson, R.: "Receiver Dynamic Range; Pt. 1, Guidelines for Receiver Analysis," *Microwaves & RF*, vol. 25, pg. 113, December 1986.

"Waveform Analysis: Noise and Jitter," application note 3SW8142-2, Tektronix, Beaverton, OR, March 1993.

Wilson, E.: "Evaluate the Distortion of Modular Cascades," *Microwaves*, vol. 20, March 1981.

Chapter

# 5

# The Radio Channel

**Ulrich L. Rohde**

**Jerry C. Whitaker, Editor**

## 5.1 Introduction[1]

The transmission of information from a fixed station to one or more mobile stations is considerably influenced by the characteristics of the radio channel [1]. The RF signal arrives at the receiving antenna not only on the direct path but is normally reflected by natural and artificial obstacles in its way. Consequently, the signal arrives at the receiver several times in the form of echoes that are superimposed on the direct signal, as illustrated in Figure 5.1. This superposition may be an advantage as the energy received in this case is greater than in single-path reception. However, this characteristic may be a disadvantage when the different waves cancel each other under unfavorable phase conditions. In conventional car radio reception this effect is known as *fading*. It is particularly annoying when the vehicle stops in an area where the field strength is reduced because of fading (for example, at traffic lights). Additional difficulties arise when digital signals are transmitted. If strong echo signals (compared to the directly received signal) arrive at the receiver with a delay in the order of a symbol period or more, time-adjacent symbols interfere with each other. In addition, the receive frequency may be falsified at high vehicle speeds because of the Doppler effect so that the receiver may have problems estimating the instantaneous phase in the case of angle-modulated carriers. Both effects lead to a high symbol error rate even if the field strength is sufficiently high.

Radio broadcasting systems using conventional frequency modulation are not seriously affected by these interfering effects in most cases. If an analog system is replaced by a digital one that is expected to offer advantages over the previous system, the designer must ensure that the expected advantages—for example, improved audio S/N and the possibility to offer supplementary services to the subscriber—are not achieved at the expense of reception in hilly terrain or at high vehicle speeds because of extreme fading. For this reason, a modulation method combined

1. This chapter is based on: Rohde, Ulrich L., and Jerry C. Whitaker: *Communications Receivers: Principles and Design*, 3rd ed., McGraw-Hill, New York, N.Y., 2000. Used with permission.

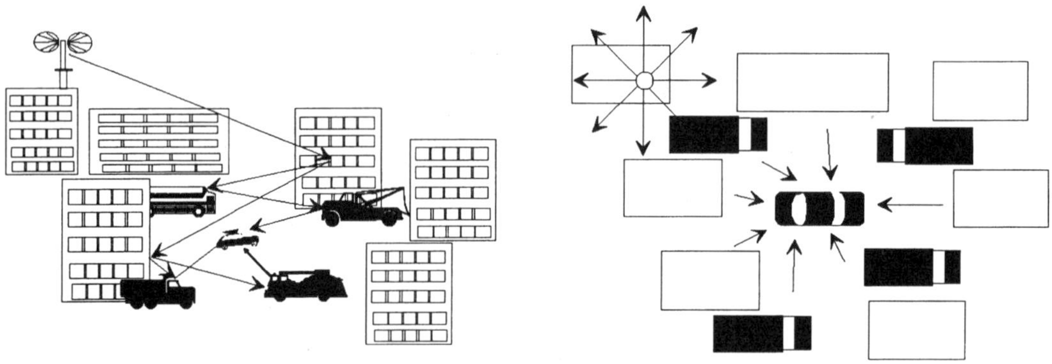

**Figure 5.1** Mobile receiver affected by fading. (*Courtesy of Rohde & Schwarz.*)

with suitable error protection must be found for mobile reception in a typical radio channel that is immune to fading, echo, and Doppler effects.

With a view to this design objective, more detailed information on the radio channel is required. The channel can be described by means of a model. In the worst case, which may be the case for reception in urban areas, it can be assumed that the mobile receives the signal on several indirect paths but not on a direct one. The signals are reflected, for example, by large buildings; the resulting signal delays are relatively long. In the vicinity of the receiver, these paths are split up into a great number of subpaths; the delays of these signals are relatively short. These signals may again be reflected by buildings but also by other vehicles or natural obstacles such as trees. Assuming the subpaths being statistically independent of each other, the superimposed signals at the antenna input cause considerable time- and position-dependent field-strength variations with an amplitude obeying the Rayleigh distribution (Figures 5.2 and 5.3). If a direct path is received in addition to the reflected ones, the distribution changes to the Rice distribution and finally, when the direct path becomes dominant, the distribution follows the Gaussian distribution with the field strength of the direct path being used as the center value.

In a Rayleigh channel the bit error rate increases dramatically compared to the BER in an AWGN (additive white Gaussian noise) channel (Figure 5.4).

## 5.1a Channel Impulse Response

The scenario outlined in the previous section can be demonstrated by means of the *channel impulse response* [1]. Assume that a very short pulse of extremely high amplitude—in the ideal case a *Dirac pulse* $\delta(t)$—is sent by the transmitting antenna at a time $t_0 = 0$. This pulse arrives at the receiving antenna direct and in the form of reflections with different delays $\tau_i$ and different amplitudes because of path losses. The impulse response of the radio channel is the sum of all received pulses (Figure 5.5). Because the mobile receiver and some of the reflecting objects are moving, the channel impulse response is a function of time and of delays $\tau_i$; that is, it corresponds to

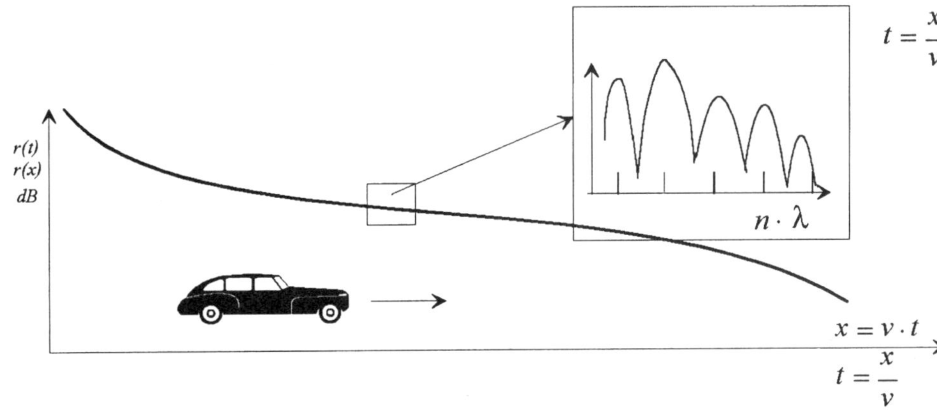


**Figure 5.2** Receive signal as a function of time or position. (*Courtesy of Rohde & Schwarz.*)

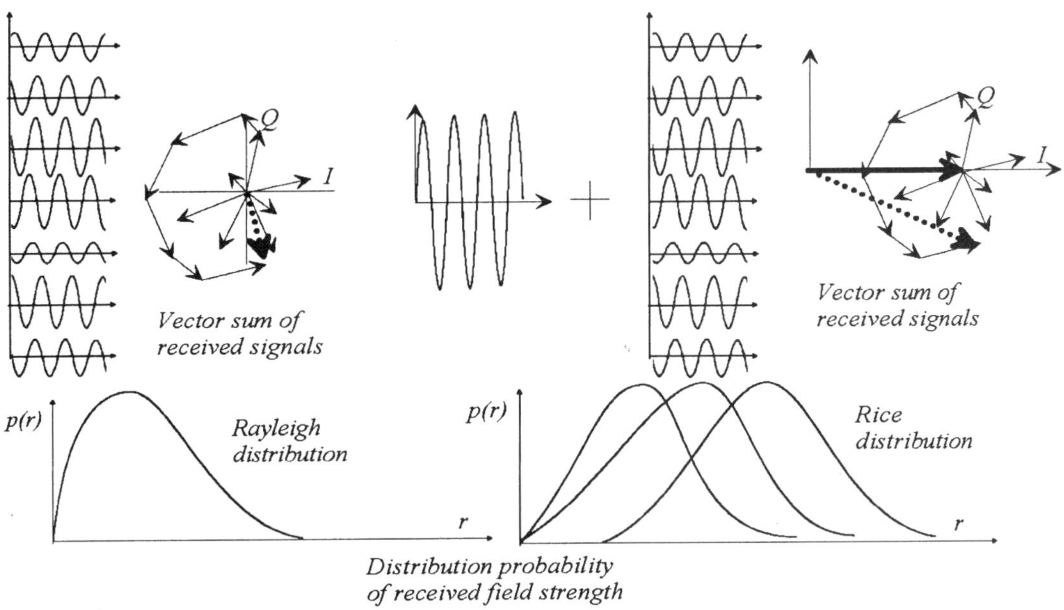


**Figure 5.3** Rayleigh and Rice distribution. (*Courtesy of Rohde & Schwarz.*)

$$h(t, \tau) = \sum_{N} a_i \delta(t - \tau_i) \tag{5.1}$$

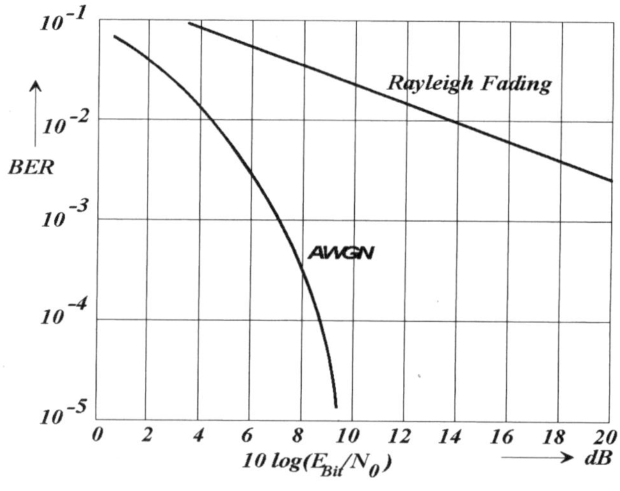


**Figure 5.4** Bit error rate in a Rayleigh channel. (Courtesy of Rohde & Schwarz.)

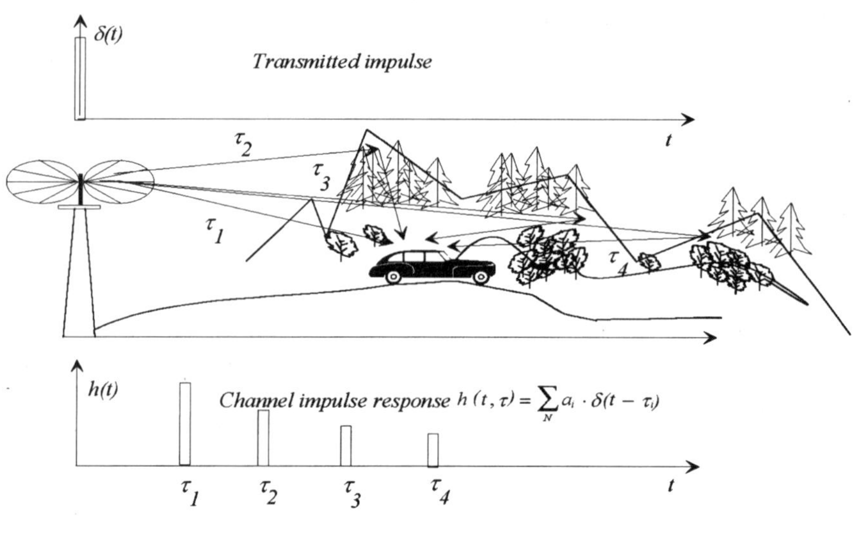


**Figure 5.5** Channel impulse response. (*Courtesy of Rohde & Schwarz*)

This shows that delta functions sent at different times $t$ cause different reactions in the radio channel.

In many experimental investigations, different landscape models with typical echo profiles were created. The most important are:

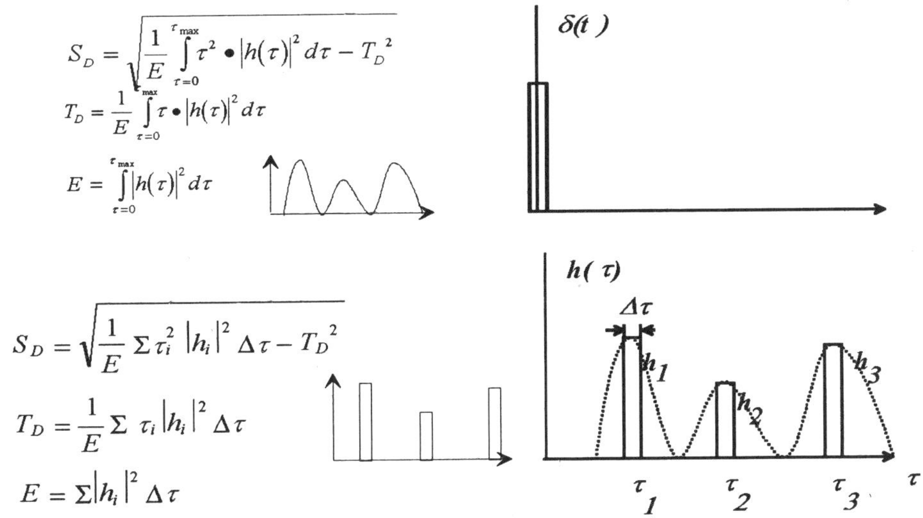


**Figure 5.6** Calculation of delay spread. (*Courtesy of Rohde & Schwarz.*)

- Rural area (RA)
- Typical urban area (TU)
- Bad urban area (BA)
- Hilly terrain (HT)

The channel impulse response tells us how the received power is distributed to the individual echoes. A useful parameter, the *delay spread* can be calculated from the channel impulse response, permitting an approximate description of typical landscape models, as illustrated in Figure 5.6.

The delay spread also roughly characterizes the modulation parameters, carrier frequency, symbol period, and duration of guard interval, which have to be selected in relation to each other. If the receiver is located in an area with a high delay spread (for example, in hilly terrain), echoes of the symbols sent at different times are superimposed when broadband modulation methods with a short symbol period are used. An adjacent transmitter emitting the same information on the same frequency has the effect of an artificial echo (Figure 5.7).

A constructive superposition of echoes is only possible if the symbol period is much greater than the delay spread. The following holds

$$T_s > 10T_d \tag{5.2}$$

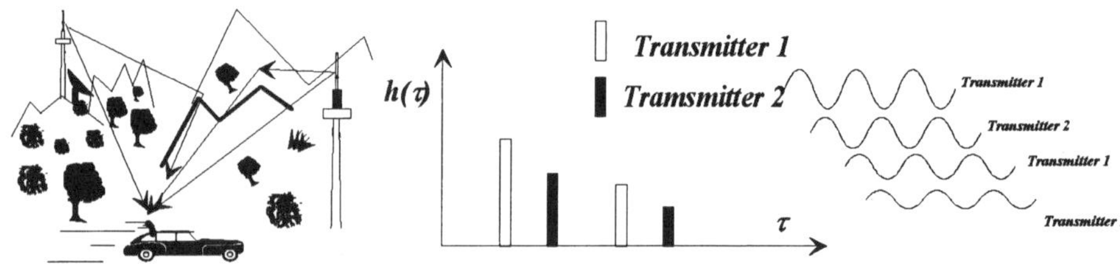


**Figure 5.7** Artificial and natural echoes in the single-frequency network. (*Courtesy of Rohde & Schwarz.*)

This has the consequence that relatively narrowband modulation methods have to be used. If this is not possible, *channel equalizing* is required.

For channel equalizing, a continuous estimation of the radio channel is necessary. The estimation is performed with the aid of a periodic transmission of data known to the receiver. In networks, a midamble consisting of 26 bits—the *training sequence*—can be transmitted with every burst. The training sequence corresponds to a characteristic pattern of *I*/*Q* signals that is held in a memory at the receiver. The baseband signals of every received training sequence are correlated with the stored ones. From this correlation, the channel can be estimated; the properties of the estimated channel will then be fed to the equalizer, as shown in Figure 5.8.

The equalizer uses the *Viterbi algorithm* (maximum sequence likelihood estimation) for the estimation of the phases that most likely have been sent at the sampling times. From these phases the information bits are calculated (Figure 5.9). A well designed equalizer then will superimpose the energies of the single echoes constructively, so that the results in an area where the echoes are moderately delayed (delay times up to 16 μs at the receiver) are better than in an area with no significant echoes (Figure 5.10). The remaining bit errors are eliminated using another Viterbi decoder for the at the transmitter convolutionally encoded data sequences.

The ability of a mobile receiver to work in an hostile environment such as the radio channel with echoes must be proven. The test is performed with the aid of a *fading simulator*, which simulates various scenarios with different delay times and different Doppler profiles. A signal generator produces undistorted *I*/*Q* modulated RF signals that are downconverted into the baseband. Next, the *I*/*Q* signals are digitized and split into different channels where they are delayed and attenuated, and where Doppler effects are superimposed. After combination of these distorted signals at the output of the baseband section of the simulator, the signals modulate the RF carrier, which is the test signal for the receiver under test (Figure 5.11).

## 5.1b Doppler Effect

Because the mobile receiver and some of the reflecting objects are in motion, the receive frequency is shifted as a result of the Doppler effect [1]. In the case of single-path reception, this shift is calculated as follows

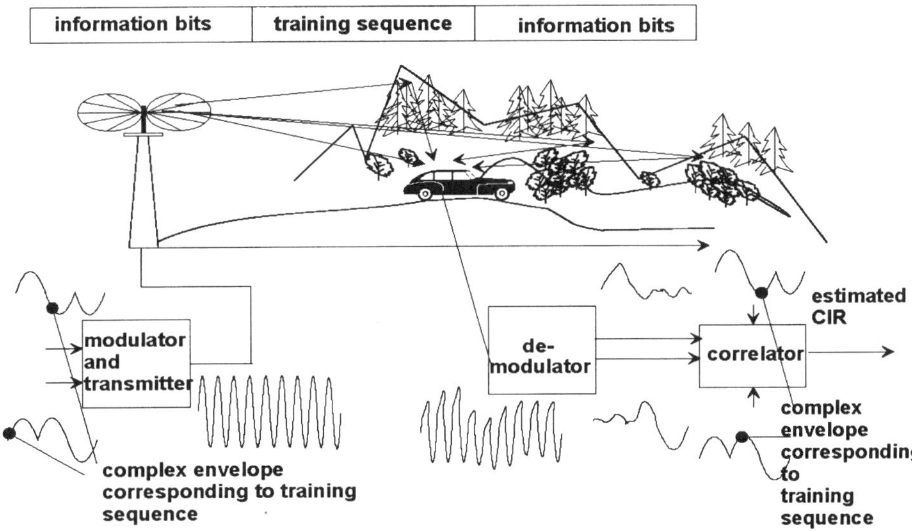


**Figure 5.8** Channel estimation. (*Courtesy of Rohde & Schwarz.*)

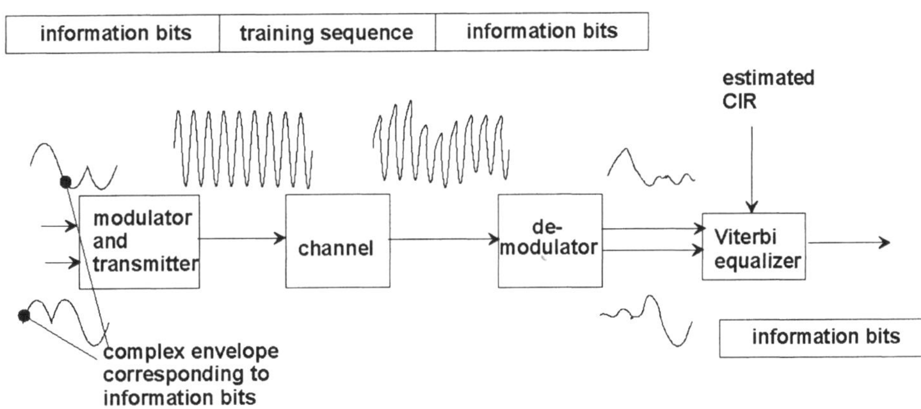


**Figure 5.9** Channel equalization. (*Courtesy of Rohde & Schwarz.*)

$$f_d = \frac{v}{c} f_c \cos \alpha \tag{5.3}$$

Where:

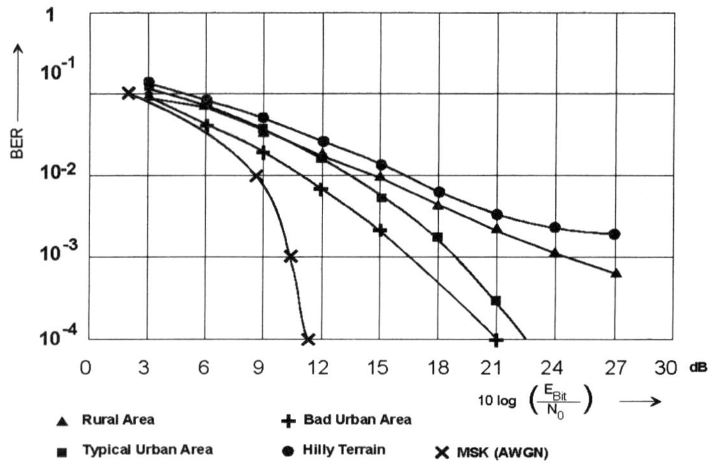


**Figure 5.10** BERs after the channel equalizer in different areas. (*Courtesy of Rohde & Schwarz.*)

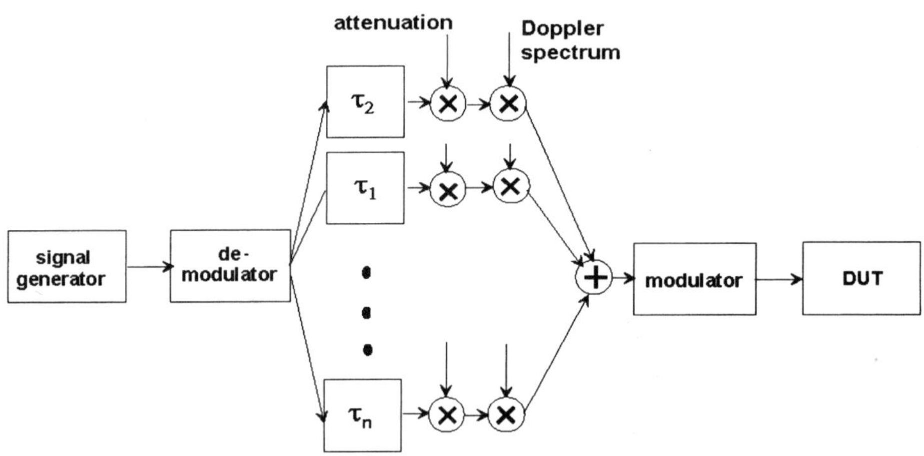


**Figure 5.11** Fading simulator. (*Courtesy of Rohde & Schwarz.*)

$v$ = speed of vehicle
$c$ = speed of light
$f$ = carrier frequency
$\alpha$ = angle between $v$ and the line connecting transmitter and receiver

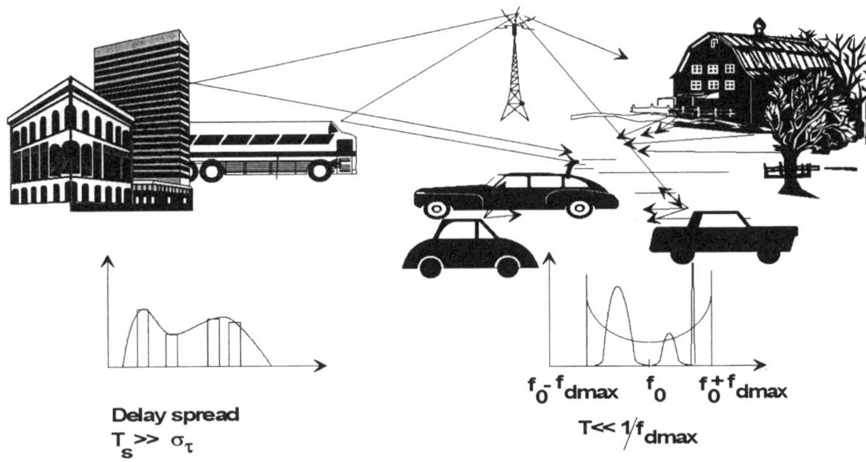


**Figure 5.12** Doppler spread. (*Courtesy of Rohde & Schwarz.*)

In the case of multipath reception, the signals on the individual paths arrive at the receiving antenna with different Doppler shifts because of the different angles $\alpha_i$, and the receive spectrum is spread. Assuming an equal distribution of the angles of incidence, the power density spectrum can be calculated as follows

$$P(f) = \frac{1}{\pi} \frac{1}{\sqrt{f_d^2 - f^2}} \quad \text{for } |f| < |f_d| \tag{5.4}$$

where $f_d$ = maximum Doppler frequency.

Of course, other Doppler spectra are possible in addition to the pure Doppler shift; for example, spectra with a Gaussian distribution using one or several maxima. A Doppler spread can be calculated from the Doppler spectrum analogously to the delay spread shown in Figure 5.12.

## 5.1c Transfer Function

The FFT value of the channel impulse response is the transfer function *H(f,t)* of the radio channel, which is also time-dependent. The transfer function describes the attenuation of frequencies in the transmission channel. When examining the frequency dependence, it will be evident that the influence of the transmission channel on two sine-wave signals of different frequencies becomes greater with increasing frequency difference. This behavior can be adequately described by the *coherence bandwidth*, which is approximately equal to the reciprocal delay spread; that is

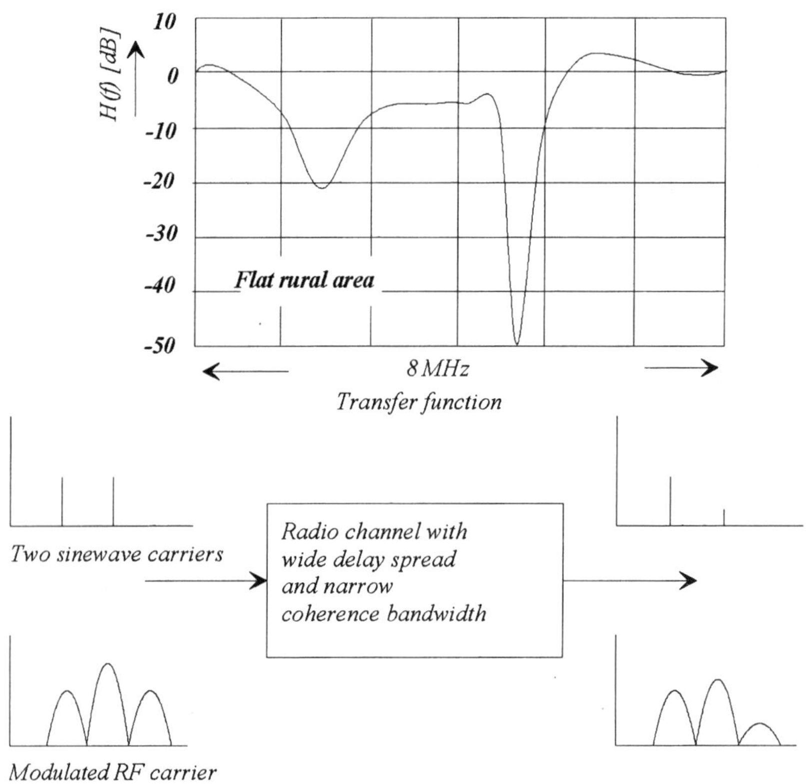


**Figure 5.13** Effect of transfer function on modulated RF signals. (*Courtesy of Rohde & Schwarz.*)

$$(\Delta f)_c = \frac{1}{T_d} \tag{5.5}$$

If the coherence bandwidth is sufficiently wide and—consequently—the associated delay spread is small, the channel is not frequency-selective. This means that all frequencies are subject to the same fading. If the coherence bandwidth is narrow and the associated delay spread wide, even very close adjacent frequencies are attenuated differently by the channel. The effect on a broadband-modulated carrier with respect to the coherence bandwidth is obvious. The sidebands important for the transmitted information are attenuated to a different degree. The result is a considerable distortion of the receive signal combined with a high bit error rate even if the received field strength is high. This characteristic of the radio channel again speaks for the use of narrowband modulation methods. (See Figure 5.13).

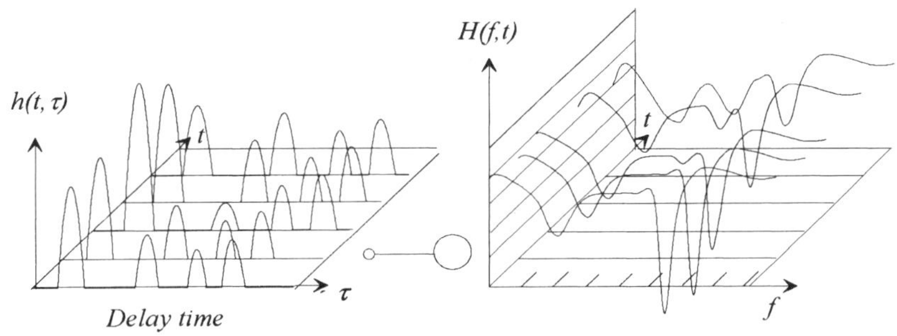


**Figure 5.14** Channel impulse response and transfer function as a function of time. (*Courtesy of Rohde & Schwarz.*)

## 5.1d Time Response of Channel Impulse Response and Transfer Function

The time response of the radio channel can be derived from the Doppler spread. It is assumed that the channel rapidly varies at high vehicle speeds. The time variation of the radio channel can be described by a figure, the *coherence time*, which is analogous to the coherence bandwidth. This calculated value is the reciprocal bandwidth of the Doppler spectrum. A wide Doppler spectrum therefore indicates that the channel impulse response and the transfer function vary rapidly with time, as shown in Figure 5.14. If the Doppler spread is reduced to a single line, the channel is *time-invariant*. In other words, if the vehicle has stopped or moves at a constant speed in a terrain without reflecting objects, the channel impulse response and the transfer function measured at different times are the same.

The effect on information transmission can be illustrated with a simple example. In the case of MPSK modulation using hard keying, the transmitter holds the carrier phase for a certain period of time; that is, for the symbol period $T$. In the case of soft keying with low-pass-filtered baseband signals for limiting the modulated RF carrier, the nominal phase is reached at a specific time—the sampling time. In both cases the phase error $\varphi_f = f_{dTS}$ is superimposed onto the nominal phase angle, which yields a phase uncertainty of $\Delta\varphi = 2\varphi_f$ at the receiver. The longer the symbol period, the greater the angle deviation (Figure 5.15). Considering this characteristic of the transmission channel, a short symbol period of $T_s \ll (\Delta t)_c$ should be used. However, this requires broadband modulation methods.

Figure 5.16 shows the field strength or power arriving at the mobile receiver if the vehicle moves in a Rayleigh distribution channel. Because the phase depends on the vehicle position, the receiver moves through positions of considerably differing field strength at different times (time-dependence of radio channel). In the case of frequency-selective channels, this applies to one frequency only; that is, to a receiver using a narrowband IF filter for narrowband emissions. As Figure 5.16 shows, this effect can be reduced by increasing the bandwidth of the emitted signal and consequently the receiver bandwidth.

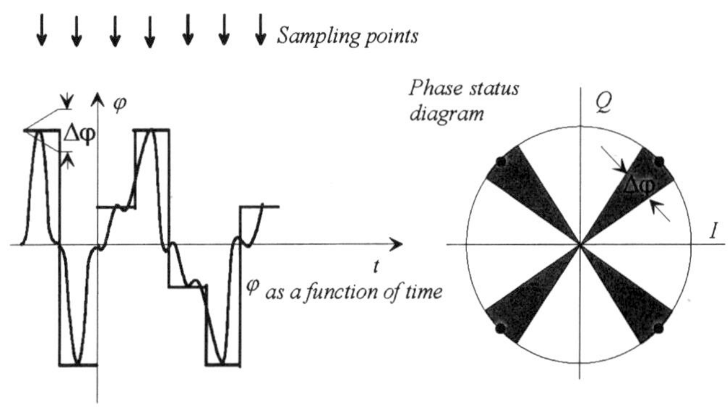


**Figure 5.15** Phase uncertainty caused by Doppler effect. (*Courtesy of Rohde & Schwarz.*)

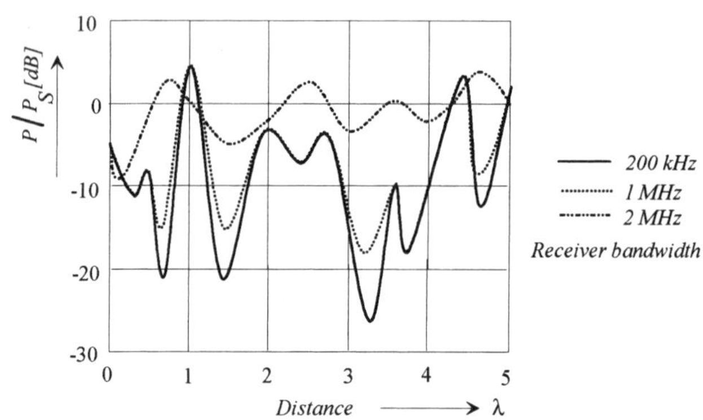


**Figure 5.16** Effect of bandwidth on fading. (*Courtesy of Rohde & Schwarz.*)

## 5.2 References

1. Rohde, Ulrich L., and David P. Newkirk: *RF/Microwave Circuit Design for Wireless Applications*, John Wiley & Sons, New York, N.Y., 2000.

Chapter

# 6 Low Power Amplifiers

Daniel R. von Recklinghausen

## 6.1 Introduction

Amplifiers are the functional building blocks of audio and video systems, and each of those building blocks contains several amplifier stages coupled together. An amplifier may contain its own power supply, while an amplifier stage needs one or more external sources of power. The active component of each amplifier stage is usually a transistor or a field-effect transistor (FET). Other amplifying components, such as vacuum tubes, can also be used in amplifier circuits if the general principles of small- and large-signal voltages and current flows are followed [1–4].

## 6.2 Single-Stage Transistor or FET Amplifier

The single-stage amplifier can best be described as using a single transistor or FET connected as a common emitter or common-source amplifier, using an NPN transistor (Figure 6.1*a*) or an N-channel FET (Figure 6.1*b*) and treating PNP transistors or P-channel FET circuits by simply reversing the current flow and the polarity of the voltages.

### 6.2a DC Conditions

At zero frequency or dc and also at low frequencies, the transistor or FET amplifier stage requires an input voltage $E_1$ equal to the sum of the input voltage of the device (the transistor $V_{be}$ or FET $V_{gs}$) and the voltage across the resistance $R_e$ or $R_s$ between the common node (ground) and the emitter or source terminal. The input current $I_1$ to the amplifier stage is equal to the sum of the current through the external resistor connected between ground and the base or gate and the base current $I_b$ or gate current $I_g$ drawn by the device. In most FET circuits the gate current may be so small that it can be neglected, while in transistor circuits the base current $I_b$ is equal to the collector current $I_c$ divided by the current gain beta of the transistor. The input resistance $R_1$ to the amplifier stage is equal to the ratio of input voltage $E_1$ to input current $I_1$.

The input voltage and the input resistance of an amplifier stage increase as the value of the emitter or source resistor becomes larger.

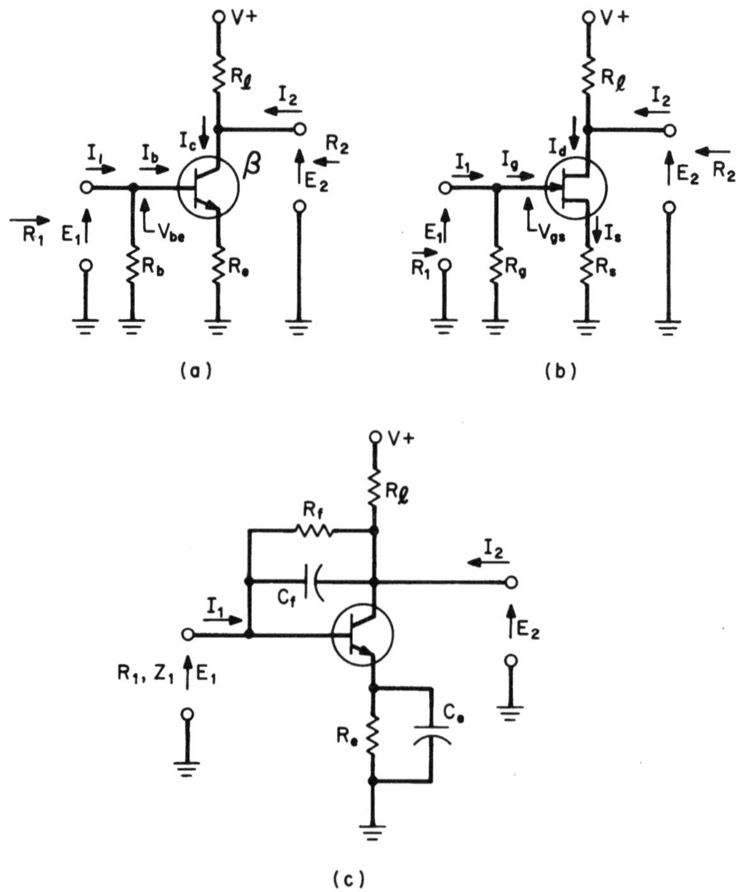


**Figure 6.1** Single-stage amplifier circuits: (*a*) common-emitter NPN, (*b*) common-source N-channel FET, (*c*) single stage with current and voltage feedback.

The output voltage $E_2$ of the amplifier stage, operating without any external load, is equal to the difference of supply voltage V+ and the product of collector or drain load resistor $R_1$ and collector current $I_c$ or drain current $I_d$. An external load will cause the device to draw an additional current $I_2$, which increases the device output current.

As long as the collector-to-emitter voltage is larger than the *saturation voltage* of the transistor, collector current will be nearly independent of supply voltage. Similarly, the drain current of an FET will be nearly independent of drain-to-source voltage as long as this voltage is greater than an equivalent saturation voltage. This saturation voltage is approximately equal to the difference between gate-to-source voltage and *pinch-off voltage*, the latter voltage being the bias voltage which causes nearly zero drain current. In some data sheets for FETs, the pinch-off voltage is given under a different name as *threshold voltage*. At lower supply voltages the collector or drain current will become less until it reaches zero when the drain-to-source voltage is zero or the collector-to-emitter voltage has a very small reverse value.

The output resistance $R_2$ of a transistor or FET amplifier stage is—in effect—the parallel combination of the collector or drain load resistance and the series connection of two resistors, consisting of $R_e$ or $R_s$ and the ratio of collector-to-emitter voltage and collector current or the equivalent drain-to-source voltage and drain current. In actual devices an additional resistor, the relatively large output resistance of the device, is connected in parallel with the output resistance of the amplifier stage.

The collector current of a single-stage transistor amplifier is equal to the base current multiplied by the current gain of the transistor. Because the current gain of a transistor may be specified as tightly as a 2: 1 range at one value of collector current or may have just a minimum value, knowledge of the input current is usually not quite sufficient to specify the output current of a transistor.

## 6.2b Input and Output Impedance, Voltage, and Current Gain

As derived above for a common-emitter or common-source single amplifier stage, input impedance is the ratio of input voltage to input current and output impedance is the ratio of output voltage to output current. As the input current increases, the output current into the external output load resistor will increase by the current-amplification factor of the stage. The output voltage will decrease because the increased current flows from the collector or drain voltage supply source into the collector or drain of the device. Therefore, the voltage amplification is a negative number having the magnitude of the ratio of output-voltage change to input-voltage change.

The magnitude of voltage amplification is often calculated as the product of *transconductance* $G_m$ of the device and load resistance value. This can be done as long as the emitter or source resistor is zero or the resistor is bypassed with a capacitor that effectively acts as a short circuit for all signal changes of interest but allows the desired bias currents to flow through the resistor. In a bipolar transistor the transconductance is approximately equal to the emitter current multiplied by 39, which is the charge of a single electron divided by the product of Boltzmann's constant and absolute temperature in kelvins. In an FET this value will be less and usually is proportional to input bias voltage with reference to the pinch-off voltage.

The power gain of the device is the ratio of output power to input power, often expressed in decibels. Voltage gain or current gain may be stated in decibels but must be so marked.

### AC Gain

The resistor in series with the emitter or source causes negative feedback of most of the output current, which reduces the voltage gain of the single amplifier stage and raises its input impedance (Figures 6.1*a* and *b*). When this resistor $R_e$ is bypassed with a capacitor $C_e$ (Figure 6.1*c*), the amplification factor will be high at high frequencies and will be reduced by approximately 3 dB at the frequency where the impedance of capacitor $C_e$ is equal to the emitter or source input impedance of the device, which in turn is approximately equal to the inverse of the transconductance $G_m$ of the device (Figure 6.2*a*). The gain of the stage will be approximately 3 dB higher than the dc gain at the frequency where the impedance of the capacitor is equal to the emitter or source resistor. These simplifications hold in cases where the product of transconductance and resistance value is much larger than 1.

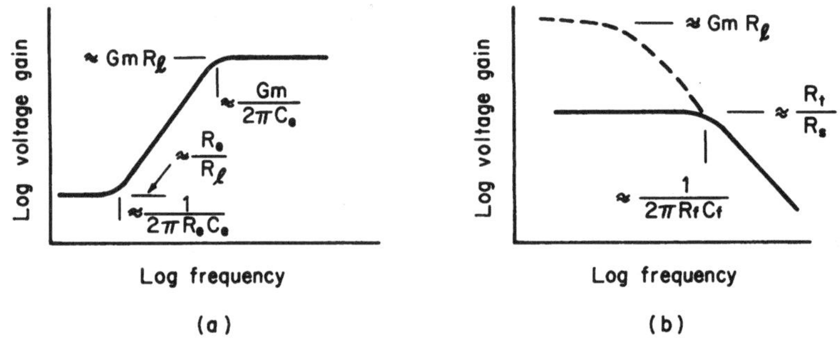


**Figure 6.2** Feedback amplifier voltage gains: (*a*) current feedback, (*b*) voltage feedback.

A portion of the output voltage may also be fed back to the input, which is the base or gate terminal. This resistor $R_f$ will lower the input impedance of the single amplifier stage, reduce current amplification, reduce output impedance of the stage, and act as a supply voltage source for the base or gate. This method is used when the source of input signals and internal resistance $R_S$ is coupled with a capacitor to the base or gate and a group of devices with a spread of current gains, transconductances, or pinch-off voltages must operate with similar amplification in the same circuit. If the feedback element is also a capacitor $C_f$, high-frequency current amplification of the stage will be reduced by approximately 3 dB when the impedance of the capacitor is equal to the feedback resistor $R_f$ and voltage gain of the stage is high (Figure 6.2*b*). At still higher frequencies amplification will decrease at the rate of 6 dB per octave of frequency. It should be noted at this point that the base-collector or gate-drain capacitance of the device has the same effect of limiting high-frequency amplification of the stage, but this capacitor becomes larger as collector-base or drain-gate voltage decreases.

Feedback of the output voltage through an impedance lowers the input impedance of an amplifier stage. Voltage amplification of the stage will be affected only as this lowered input impedance loads the source of input voltage. If the source of input voltage has a finite source impedance and the amplifier stage has very high voltage amplification and reversed phase, the effective amplification for this stage will approach the ratio of feedback impedance to source impedance and also have reversed phase.

## Common-Base or Common-Gate Connection

For this configuration (Figure 6.3*a*), voltage amplification is the same as in the common-emitter or common-source connection, but input impedance is approximately the inverse of the transconductance of the device. As a benefit, high-frequency amplification will be less affected because of the relatively lower emitter-collector or source-drain capacitance and the relatively low input impedance. This is the reason why the cascade connection (Figure 6.3*b*) of a common-emitter amplifier stage driving a common-base amplifier stage exhibits nearly the dc amplification of a common-emitter stage with the wide bandwidth of a common-base stage. The other advantage of a common-base or common-gate amplifier stage is stable amplification at very high frequencies (VHF) and ease of matching to transmission-line impedances, usually 50 to 75Ω

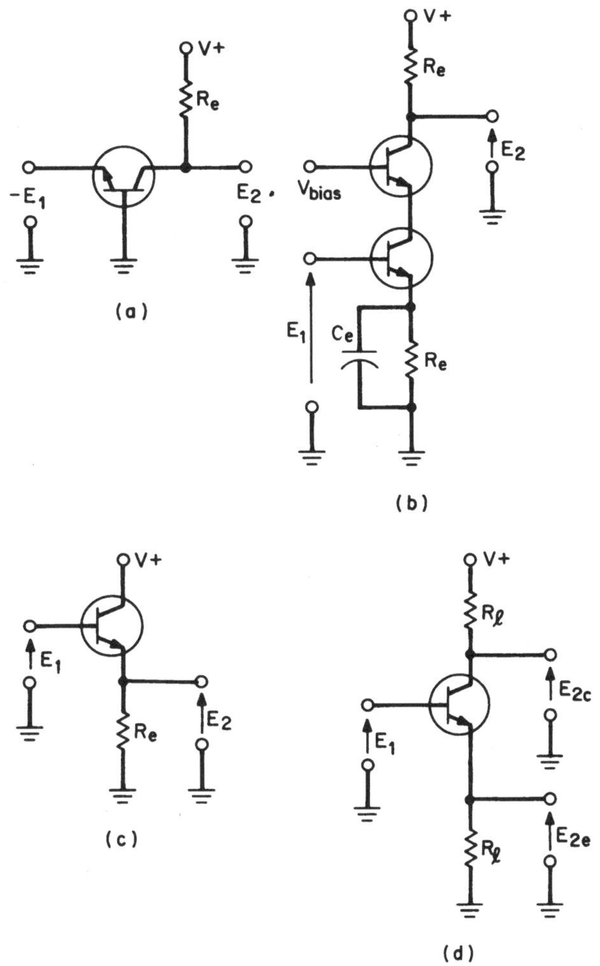


**Figure 6.3** Transistor amplifier circuits: (*a*) common-base NPN, (*b*) cascade NPN, (*c*) common-collector NPN emitter follower, (*d*) split-load phase inverter.

## Common-Collector or Common-Drain Connection of a Transistor or FET

For this case, voltage gain is slightly below 1.000, but the input impedance of a transistor so connected will be equal to the value of the load impedance multiplied by the current gain of the device plus the inverse of the transconductance of the device (Figure 6.3*c*). Similarly, the output impedance of the stage will be the impedance of the source of signals divided by the current gain of the transistor plus the inverse of the transconductance of the device.

When identical resistors are connected between the collector or drain and the supply voltage and the emitter or source and ground, an increase in base or gate voltage will result in an increase of emitter or source voltage that is nearly equal to the decrease in collector or drain voltage. This

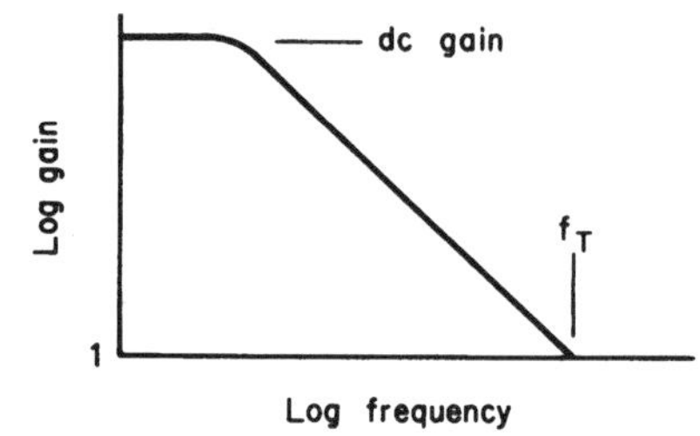


**Figure 6.4** Amplitude-frequency response of a common-emitter or common-source amplifier.

type of connection is known as the *split-load phase inverter*, useful for driving push-pull amplifiers, although the output impedances at the two output terminals are unequal (Figure 6.3*d*).

The current gain of a transistor decreases at high frequencies as the emitter-base capacitance shunts a portion of the transconductance, thereby reducing current gain until it reaches a value of 1 at the transition frequency of the transistor (Figure 6.4). From this it can be seen that the output impedance of an emitter-follower or common-collector stage will increase with frequency, having the effect of an inductive source impedance when the input source to the stage is resistive. If the source impedance is inductive, as it might be with cascaded emitter followers, the output impedance of such a combination can be a negative value at certain high frequencies and be a possible cause of amplifier oscillation. Similar considerations also apply to common-drain FET stages.

## Bias and Large Signals

When large signals have to be handled by a single-stage amplifier, distortion of the signals introduced by the amplifier must be considered. Although feedback can reduce distortion, it is necessary to ensure that each stage of amplification operates in a region where normal signals will not cause the amplifier stage to operate with nearly zero voltage drop across the device or to operate the device with nearly zero current during a portion of the cycle of the signal. Although the amplifier is described primarily with respect to a single-device amplifier stage, the same holds true for any amplifier stage with multiple devices, except that here at least one device must be able to control current flow in the load without being *saturated* (nearly zero voltage drop) or *cut off* (nearly zero current):

If the single-device amplifier load consists of the collector or drain load resistor only, the best operating point should be chosen so that in the absence of a signal, one-half of the supply voltage appears as a quiescent voltage across the load resistor $R_1$. If an additional resistive load $R_1$ is connected to the output through a coupling capacitor $C_c$ (Figure 6.5*a*), the maximum peak load current $I_1$ in one direction is equal to the difference between quiescent current $I_q$ of the stage and the current that would flow if the collector resistor and the external load resistor were connected in series across the supply voltage. In the other direction maximum load current is limited by the quiescent voltage across the device divided by the load resistance. The quiescent current flows in the absence of an alternating signal and is caused by bias voltage or current only. For signals with an equal probability of positive and negative peak excursions, such as audio-frequency waveforms, it is advisable to have the two peak currents equal. This can be accomplished by increasing the quiescent current as the external load resistance decreases.

When several devices contribute current into an external load resistor (Figure 6.5*b*), one useful strategy is to set bias currents so that the sum of all transconductances remains as constant as

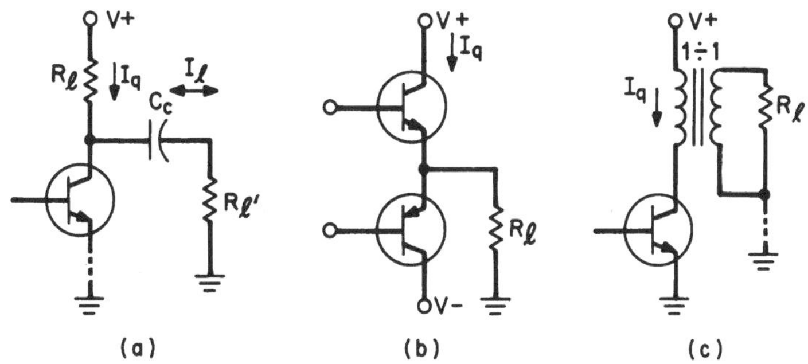


**Figure 6.5** Output load-coupling circuits: (*a*) ac-coupled, (*b*) series-dc parallel-ac push-pull half bridge, (*c*) single-ended, transformer-coupled.

practical, which means a design for minimum distortion. This operating point for one device is near one-fourth of the peak device current for push-pull FET stages and at a lesser value for bipolar push-pull amplifiers.

When the load resistance is coupled to the single-device amplifier stage with a transformer (Figure 6.5*c*), the optimum bias current should be nearly equal to the peak current that would flow through the load impedance at the primary of the transformer with a voltage drop equal to the supply voltage.

## 6.2c Multistage Amplifiers

All practical audio, video, and radio-frequency amplifiers are multistage amplifiers in which cascaded single-stage amplifiers are connected together: Overall feedback then is used to stabilize amplification and quiescent operating points [5].

### DC-Coupled Multistage Amplifiers

Commonly, amplifier stages enclosed in an overall feedback loop are direct-coupled so that the quiescent operating point is determined primarily by the bias of the first stage.

Two cascaded common-emitter amplifier stages can form a gain block useful as a low-cost preamplifier (Figure 6.6*a*). Here, the collector of the first stage is connected to the base of the second stage and to a resistor that supplies collector current to the first stage and base current to the second stage from the supply voltage. Both stages have an emitter resistor connected to the common ground, with the second resistor bypassed with a large capacitor. Base current to the first stage is supplied from the emitter of the second stage through a pair of voltage-divider resistors connected to ground, with audio input signals fed to the base of the first stage through a coupling capacitor. The audio-signal output is taken from the collector of the second stage through a coupling capacitor. The second collector receives its operating current through a resistor and supplies feedback current through a resistor-capacitor network to the emitter of the first stage. When using two NPN transistors and a single positive supply voltage, a low-cost preamplifier with the appropriate feedback network is constructed. The maximum no-feedback voltage ampli-

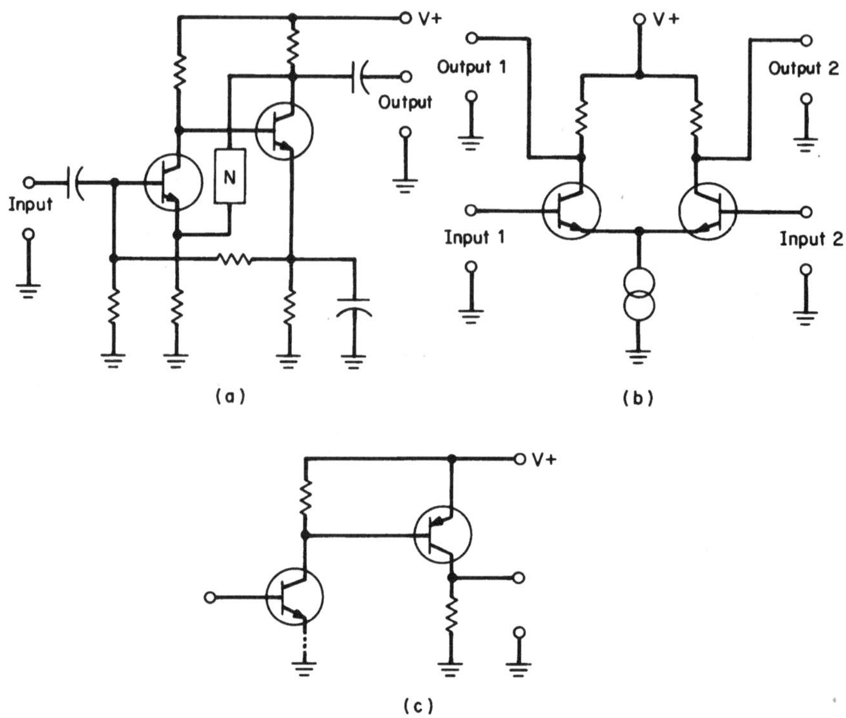


**Figure 6.6** Multistage dc-coupled circuits: (*a*) cascaded two-transistor amplifier, (*b*) emitter-coupled phase inverter, (*c*) cascade NPN and PNP transistors.

fication is approximately equal to the transconductance of the first stage multiplied by the current gain of the second stage and the load resistor value of the second stage.

When two transistors or FETs have their emitters or sources connected together and that junction is supplied with a constant current and output voltages are obtained from identical-value resistors connected between the supply voltage and the two collectors or drains, the long-tailed-pair or emitter-coupled or source-coupled phase-inverter stage is described (Figure 6.6*b*). Here the emitter or source of one device acts as the emitter or source resistor for the other device, and an alternating signal impressed on one input will be amplified with a phase reversal in the same stage while this reversal is not experienced at the other output. This stage, then, is capable of taking one single-ended signal and transforming it into a push-pull signal. The signal applied to the other input now will arrive at the first output without a phase reversal. Thus, the emitter- or source-coupled amplifier is able to amplify the difference between two signals, where one may be an input signal and the other a feedback signal. Its voltage amplification is the same as the normal grounded emitter or grounded source stage, except that the output voltage should now be measured between the two outputs, with the voltage from either output to ground being one-half of that value. This type of input stage is the almost universal input stage of operational amplifiers.

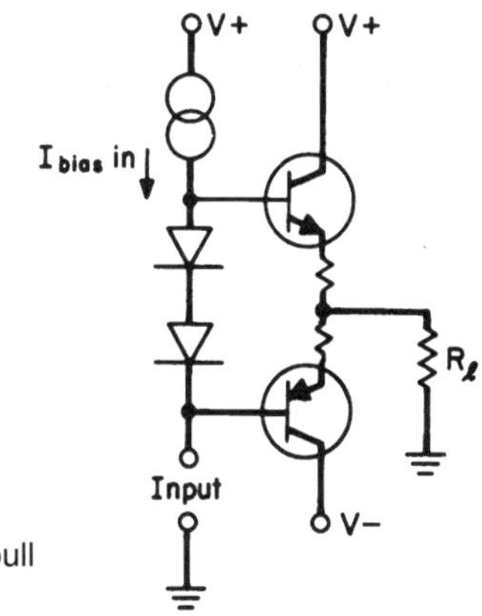


**Figure 6.7** Input signal and bias circuit in a single-ended push-pull amplifier.

Almost the same performance can be expected when the commonly connected emitters or sources are supplied by a resistor instead of a current source or the two input circuits have differing impedances. The long-tailed pair performs best when the two devices are matched and operate at the same temperature. For this reason, matched monolithic transistors and FET provide the best performance, with the additional benefit that, with equally shared current, even-harmonic distortion is substantially reduced when compared with single-ended amplifiers.

Two-device amplifier stages are not restricted to a construction using two devices of the same type (bipolar versus FET) or the same polarity (NPN versus PNP). Using stages of opposite polarity often results in a higher available amplification factor and output voltage when limitations of supply voltage are considered (Figure 6.6*c*).

A common use of transistors or FETs of differing polarity is in series-connected push-pull amplifiers (Figure 6.7), where one stage supplies current of one polarity from one supply voltage to a grounded load and the other stage supplies opposite-polarity current from a supply voltage of opposite polarity. The bias current flows from one supply to the other through the two devices without passing through the load. Here, the two stages may have a relatively low bias current that is often stabilized by a diode per device connected in shunt with the base circuit and kept at the same temperature as the amplifying transistors. As the transistors temperature increases with heat, the base-emitter voltage decreases and the diode forward voltage decreases also, thereby keeping the quiescent current in the transistors within much smaller limits than without diode compensation.

## Cascaded Transistors

Transistors connected in cascade with overall voltage feedback are basic building blocks of amplifiers. When the collectors of two NPN transistors are connected together with the base signal of the second transistor derived entirely from the emitter of the first, an NPN Darlington transistor (Figure 6.8*a*) is described that has only three terminals: the emitter of the second, the base of the first, and the common collector of both. Additional internal base-emitter resistors ensure that leakage currents cannot cause conduction in the last transistor. Such a cascaded transistor may have a current gain nearly equal to the product of the two current gain factors, requires an input voltage which is two base-emitter voltages higher than the emitter voltage, and can reach in

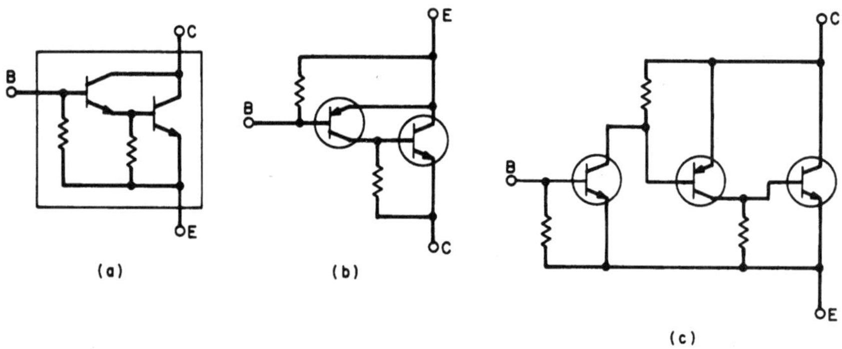


**Figure 6.8** Cascaded circuits: (*a*) compound Darlington or coupled emitter follower, (*b*) compound PNP-NPN transistors, (*c*) triple compound NPN-PNP-NPN transistors.

normal operation a minimum voltage drop between the external collector-emitter terminals equal to one base-emitter voltage. Two PNP transistors can be similarly interconnected.

The collector of a first PNP transistor can be connected to supply the entire base current of a second NPN transistor, with the emitter of the first transistor connected to the collector of the second transistor (Figure 6.8*b*). The entire assembly now functions as a compound PNP transistor having as its base terminal the base of the first PNP transistor, as its collector the emitter of the second NPN transistor, and as its emitter the emitter of the first PNP and the collector of the second NPN transistor. The input voltage now must be one base-emitter voltage lower than the emitter voltage because of the reversed current flow in PNP transistors, and the minimum voltage drop between the collector-emitter terminals is now one base-emitter voltage. The two compound transistors described here are the compound output devices in quasi-complementary push-pull amplifiers.

Three or more transistors of like or mixed polarity may be cascaded, such as PNP-NPN-NPN or NPN-PNP-NPN (Figure 6.8*c*), to form compound PNP or NPN transistors, respectively. Here, the polarity of the input transistor defines the polarity of the compound transistor. The minimum voltage drop and the required input voltage may be different in each connection, becoming highest when only devices of the same polarity are used.

## Parallel-Connected Devices for High Currents

When high currents have to be delivered to a load, several transistors or FETS are often connected in parallel, with each device sharing a portion of the output current (Figure 6.9*a*). Nearly equal current sharing can be achieved when all devices are matched to each other as much as possible. Current sharing can be improved when each of the devices has local current feedback with equal separate emitter or source resistors connected to the common emitter or source connection and all devices share a common heat sink. The emitter resistor for bipolar transistors is typically a fraction of 1 Ω, which allows current sharing to currents as low as a fraction of 1 A.

The circuit layout for a parallel connection must be done carefully to avoid constructing an oscillator circuit at very high frequencies. When using power FETs, it is necessary to connect

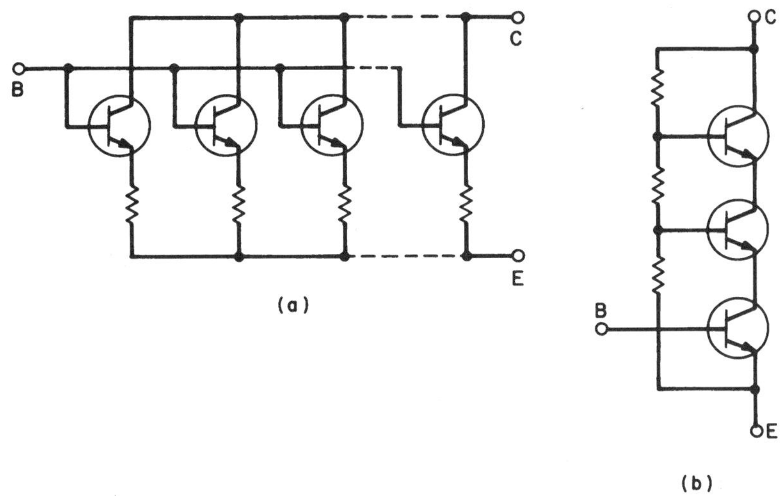


**Figure 6.9** High-voltage and high-current circuits: (*a*) current-sharing parallel transistors, (*b*) totem-pole series-connected transistors.

series resistors of a few ohms in series with each gate lead and, perhaps, to have a ferrite bead in each gate lead to avoid oscillation.

## Series-Connected Devices for High Voltage

When high voltages have to be delivered to a load and single devices are incapable of operating at the maximum peak voltage, several devices can be connected in series to share the voltage while conducting nearly the same current. The resulting *totem-pole connection* for transistors involves connecting the transistors so that the emitter of the second transistor is tied to the collector of the first and the emitter of the third transistor is tied to the collector of the second, and so on (Figure 6.9*b*). A series string of as many equal resistors as there are transistors has its ends connected to the collector of the last transistor and the emitter of the first transistor, and each junction is connected in the same sequence to the base of the same transistor in the sequence, except to the first transistor, whose base receives the input signal. The object of this circuit is to have the first transistor operate as a grounded-emitter device, driving all the others as grounded-base devices. This goal is not perfectly achieved, particularly at high frequencies and at high-output currents where voltage division in the resistor string under load departs from uniformity.

## AC-Coupled Multistage Amplifiers

Amplification of a given signal usually does not include amplification of the dc component of the source of signals. One or more coupling capacitors between stages of amplification reduce the low-frequency response of the system and prevent the dc offset voltages from being propagated to the output (Figure 6.10*a*). When using transistor amplifiers in this fashion, the input impedance of a single-ended input stage after each capacitor may act as a partial rectifier diode for the pulsating signal and produce a low-frequency transient for each pulse. The solution to this

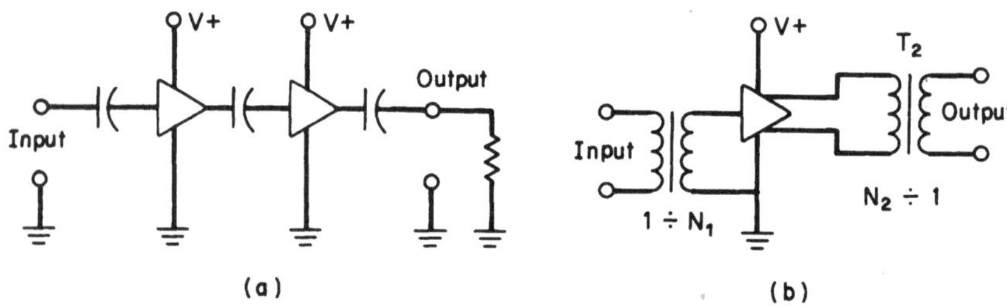


**Figure 6.10** AC amplifier circuits: (*a*) cascaded, capacitor-coupled, (*b*) transformer-coupled.

problem may be, in part, the use of FET circuits, the use of push-pull circuits, and the selection of low-frequency time constants in the amplifier and power supply filtering circuits.

Amplifier stages may be coupled with transformers to the signal source and load and to each other (Figure 6.10*b*). Transformers are excellent devices that can reject common-mode interfering signals which may appear on a program line, and they can also match the source and load impedances to the amplifier circuit. Such an impedance match is needed, for example, in amplifiers operating from a power supply directly connected to the power line where connection of external loudspeakers or headphones would present a shock hazard. Transformers are the only practical components that can match the devices in broadband radio-frequency power amplifiers to source and load.

## 6.2d Power Output Stages

Power output stages of audio or video amplifiers usually are called upon to drive a variety of loads, which may or may not be connected when the signal or the power supply is turned on. Consequently, not only must power amplifiers be stable with any load, but they must be tolerant of excessive signals or loads unless such conditions are prevented from occurring [6].

### Single-Ended Amplifiers

A single-ended amplifier has only one single or compound transistor or FET acting as a variable controlled resistor between power supply and load. The load may be coupled to the output stage through a capacitor or a transformer, which must also return the average direct current to the power supply. Single-ended amplifiers intended for audio-frequency amplification are usually of low power output capability and generally operate from a single power supply voltage. In a single-ended amplifier, transformer- or choke-coupled to the load, the bias current through the device must be at least equal to the peak current through the load and the peak voltage across the load must be less than the supply voltage when the turns ratio of the transformer is 1:1 between the primary and secondary windings.

### Push-Pull Amplifiers

A push-pull amplifier has at least one pair of single or compound output devices that act as variable resistors between supply and load, with the first device *pushing* load current in one direction

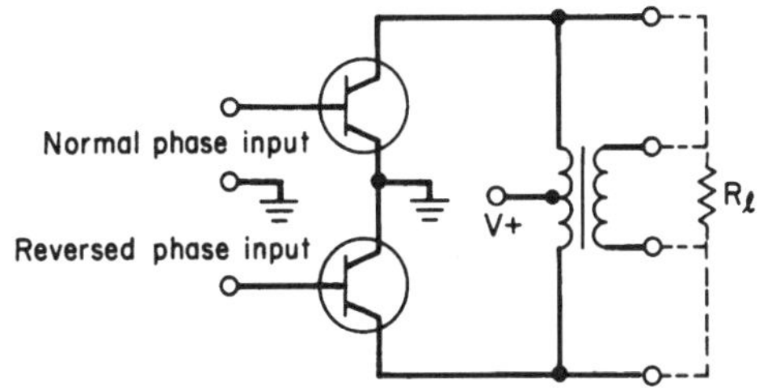


**Figure 6.11** Series-ac, parallel-dc push-pull amplifier.

while the other *pulls* load current in the opposite direction under control of the input signal. When the two devices have the same characteristics and the driving signal is equally balanced, the direct current and all even-harmonic distortion in the load current are canceled [7].

## Parallel-DC, Series-AC Amplifiers

The usual transformer-coupled amplifier (Figure 6.11) has two like devices connected between ground and the end of the primary winding of the transformer, with the supply voltage fed to the center tap of the same winding. Signal voltage is fed to the two devices with opposed phase so that one device increases conduction of current while the other decreases conduction.

The load may be connected to a secondary winding or between the ends of the primary winding. When it is connected in the latter way, maximum conduction of one device, resulting in nearly 0 V at that point, will raise the voltage at the opposite device to almost twice the power supply voltage, which then becomes the peak voltage across the load.

The peak-to-peak voltage across the load then becomes nearly 4 times the power supply voltage, and the peak load current becomes nearly 2 times the power supply voltage divided by the load resistance. The average power supply current is equal to the sum of the average current drawn by each device. Thus, the dc supply load is in parallel, while the ac load signal is in series between the two devices.

The parallel-dc, series-ac push-pull amplifier provides a very high relative power output when supply voltage is low, as in the 12-V automotive electrical system. The transformer- or choke-coupled amplifier makes use of two like devices and is therefore the preferred connection in radio-frequency power amplifiers.

## Series-DC, Parallel-AC Amplifiers

With the availability of complementary transistors as amplifiers, the single-ended or half-bridge amplifier became practical as a transformerless power amplifier in the early 1960s. Prior to that time such amplifiers were constructed by using driver transformers or floating phase inverter amplifiers.

A half-bridge amplifier that is fully balanced has one device connected between the load and one power supply and a second complementary device connected between the load and a second power supply of opposite polarity but the same voltage. The load and the two power supplies are connected to a common ground. The driving voltage is fed to both devices without phase inversion, decreasing conduction in one device while increasing conduction of an opposite current in the other direction.

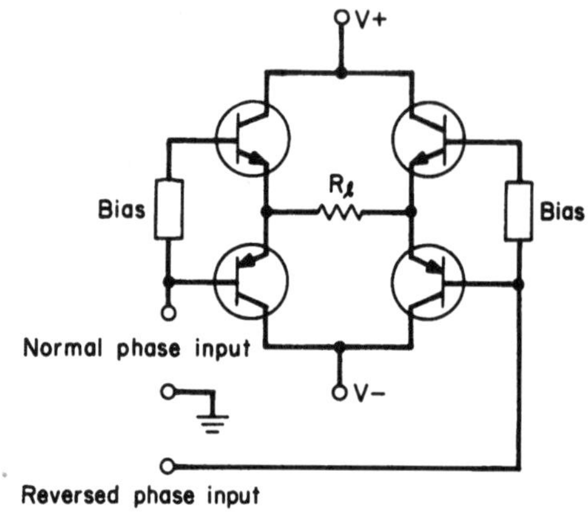


**Figure 6.12** Full-bridge amplifier coupled to a load.

The maximum peak voltage across the load will be slightly less than one supply voltage. The maximum peak-to-peak voltage thus cannot exceed the total of the two supply voltages, which are series-connected. The two devices operate in parallel for ac signals, where one device increases a current of one polarity while the other decreases a current of opposite polarity.

### Full-Bridge Amplifiers

Full-bridge amplifiers are constructed by using two half-bridge amplifiers with the load connected between the two output terminals and the two input terminals driven by signals of opposite polarity from a phase inverter circuit (Figure 6.12). Peak voltage across the load then becomes nearly equal to the total supply voltage, and peak-to-peak load voltage becomes nearly twice the total supply voltage. This type of amplifier connection is preferred over the totem-pole-transistor connection when high-voltage limitations of power devices restrict total available output power into a fixed load resistance without using a transformer.

## 6.3 Classes of Amplifiers

Amplifiers are described as classes depending on the angle of conduction of signal current and voltage-current relationships in the load.

Class A amplifiers (Figure 6.13*a*) conduct signal current throughout the cycle of the signal waveform. They have the lowest distortion before feedback and may be single-ended or push-pull. An ideal Class A amplifier can have a sine-wave output efficiency not exceeding 50 percent at full output.

Class B amplifiers (Figure 6.13*b*) conduct signal current exactly for one-half of the cycle of the input-signal waveform. In a push-pull Class B amplifier, one device conducts for one half-cycle, and the other device conducts for the remaining half-cycle. Linear Class B radio-frequency amplifiers may have only one device, since the second-harmonic components are filtered out in the narrowband matching network. An ideal Class B amplifier can have a maximum sine-wave efficiency not exceeding 78 percent at full power output.

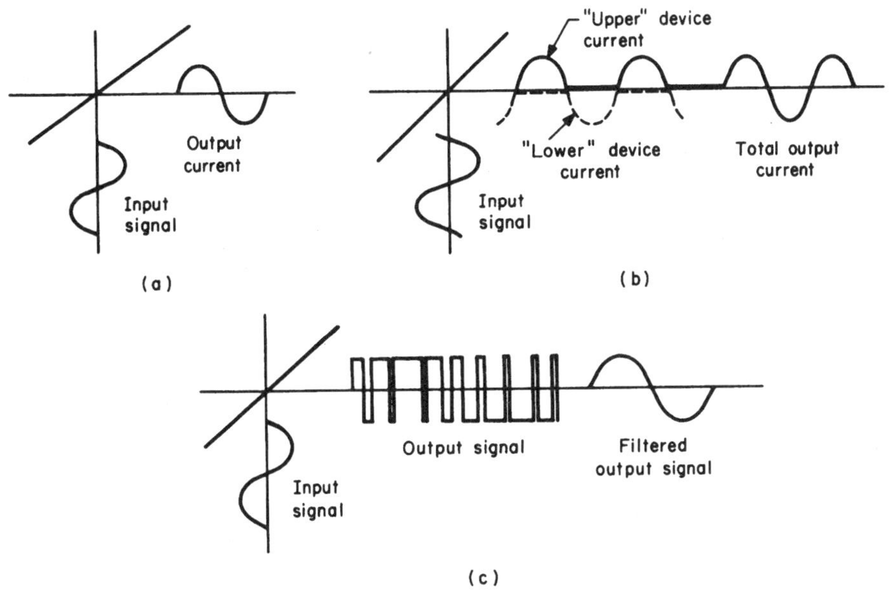


**Figure 6.13** Output current as a function of input signal for amplifiers of various classes: (*a*) Class A, (*b*) Class B, (*c*) Class D.

Class AB amplifiers have a conduction angle between full conduction and half-cycle conduction and efficiencies between Classes A and B. Most conventional audio-frequency amplifiers are adjusted in this way. As lower power outputs are needed with variations in signal amplitude, the efficiency of Class A, AB, and B amplifiers will decrease proportionally to output voltage, decreasing toward zero at very low output. When the load of such an amplifier is a reactive impedance, such as a loudspeaker, efficiency will decrease still further, since any voltampere energy sent to a reactance in one part of a cycle will be returned to the source and the resistance in the circuit in the other part of the cycle.

Class C amplifiers conduct for less than one-half of a complete signal cycle. These amplifiers are used primarily as radio-frequency amplifiers with the load tuned to the signal frequency.

Class D amplifiers (Figure 6.13*c*) are switching amplifiers using a high-frequency carrier signal where the positive pulse on time is proportional to the modulation amplitude. The negative pulse on time completes the rest of the cycle as with Class A amplifiers. In other designs, separate circuits control positive and negative pulses as with Class B amplifiers. The load is isolated from the amplifier output stage with a low-pass filter that does not consume the high-frequency pulse energy. Class D amplifiers have a theoretical efficiency of 100 percent at all signal levels but are difficult to design for wideband low-distortion operation because of the short switching transition times required of the final high-power output stages and the difficulty of design of feedback loops.

Class E amplifiers have as input signals rectangular pulses. The output load is tuned, but the output voltage resembles a damped single pulse.

There is no generally accepted agreement on naming amplifiers of classes above E. Several types of amplifiers have been described with differing letters. These include amplifiers in which several stages are connected in series, drawing power from several power supplies through isolation diodes to provide output signal to a load. For small signals, only the stages connected to the low-voltage supplies conduct current. As these stages saturate, the next stages then conduct current from the next higher power supplies through the saturated stages into the load, and so on.

In a different version, a normal Class B amplifier obtains supply voltage from a high-efficiency switching power supply, the output voltage of which is raised as output voltage demands are increased. In all these amplifier designs attempts are made to improve efficiency with varying signal levels and with program-signal waveforms, which are nonsinusoidal.

## 6.4 Gain Block and the Operational Amplifier

A large number of audio- and video-frequency circuits are constructed by using operational amplifiers because they permit these circuits to be designed with minimum complexity of components. In most applications, the open-circuit voltage gain of an operational amplifier will be much larger than the gain of the amplifier.

An amplifier gain block is *matched* to the output of the previous gain block or other circuit when it is able to extract maximum power from the previous circuit. When the amplifier gain block draws little current from the preceding circuit, the gain block is said to be *bridged* across that circuit and must have a relatively high input impedance.

An operational amplifier connected as a noninverting amplifier can have a high input impedance at its positive input while receiving feedback voltage from a voltage divider connected between output and ground, with the voltage-divider junction connected to the negative input (Figure 6.14*a*). The voltage gain is the voltage ratio of the divider.

An inverting operational amplifier can be used as a gain block, with the input resistor connected between the source and the negative input and matching the desired load of the previous stage. A feedback resistor connected between the output and the same negative input then sets voltage gain equal to the ratio between the two resistors; the positive input is grounded (Figure 6.14*b*). Because there exists only a very low voltage at the negative input, several input resistors can be connected between various input sources and the negative input. In this fashion, these input signals can be mixed together with little danger of feedthrough between signals at each source.

If resistance-capacitance networks are used in place of resistors, equalizer blocks can be designed. With more complex networks, high-pass, low-pass, bandpass, and phase-shifting all-pass blocks result.

These circuits make use of the common-ground mode and thus are unbalanced circuits. An operational amplifier can also amplify the voltage difference between the two wires of a balanced program line while having only little sensitivity to common-mode signals arriving in phase, thereby reducing ground-loop voltages.

The simplest connection involves the use of two identical voltage-divider resistor pairs having their junctions connected to the positive and negative inputs (Figure 6.14*c*). The input terminals of the two networks are then connected to the two wires of the signal line. The return terminal of the network connected to the positive input is grounded, and the other return terminal is connected to the output of the operational amplifier. The *differential voltage gain* of such a circuit is

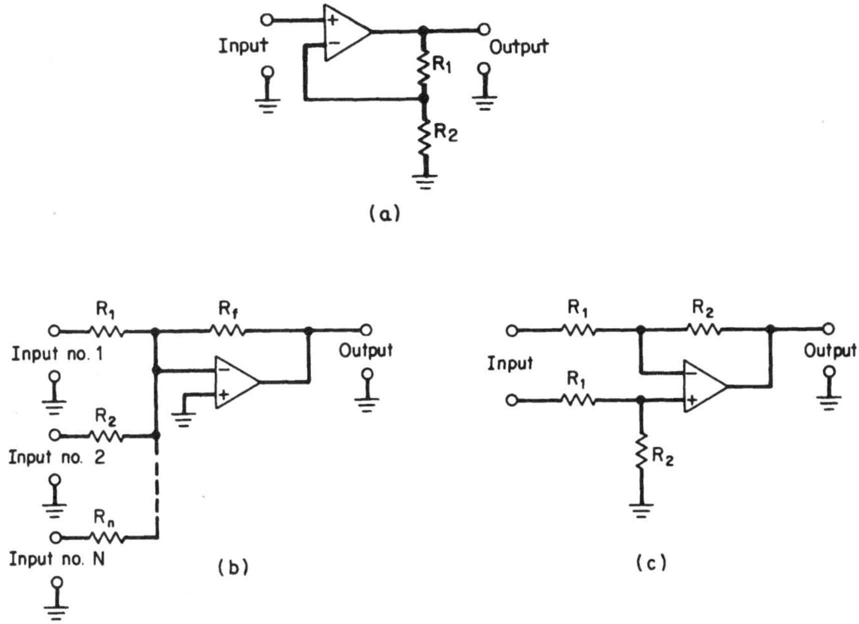


**Figure 6.14** Operational amplifier circuits: (*a*) the gain-block voltage gain = $(1 + R_1/R_2)$ with the same polarity of the input and output signals, (*b*) the gain-block voltage gain = $(R_f/R_1)$, 1 $(R_f/R_2)$ or – $(R_f/R_n)$ with opposite polarity of input and output signals, (*c*) the gain-block differential voltage gain = $(R_2/R_1)$ and low common-mode gain, limited by resistor matching and loop gain.

equal to the ratio of the resistance values, while the *common-mode gain* is limited by resistor accuracy and the residual errors of the operational amplifier, and particularly its common-mode rejection. The input impedances of the positive and negative inputs to the circuit are not equal. A number of gain blocks can be interconnected to become a more complex amplifier system [8].

## 6.4a Feedback and Feed Forward

Feedback is the return of a fraction of the output signal to the input (Figure 6.15). The returned fraction is added to the input signal at the feedback node in the feedback loop of the system. The input signal to the system with feedback for the same output as before feedback is now the vector sum of the original input signal and the feedback signal. Feedback is negative when the new required input signal is larger than the signal without feedback and positive when it is smaller.

Feedback may be acoustic, mechanical, or electronic, depending upon the type of signal amplified. In amplifiers for audio- and video-frequency signals the feedback signal is usually a portion of the output voltage or output current. When the returned fraction is negative and is obtained through a linear network, the reduction ratio of amplifier errors, such as distortion or phase shift, is proportional to the reduction in amplifier gain due to feedback. In the limit, with very high amplifier gain before feedback, the response of the amplifier with respect to frequency

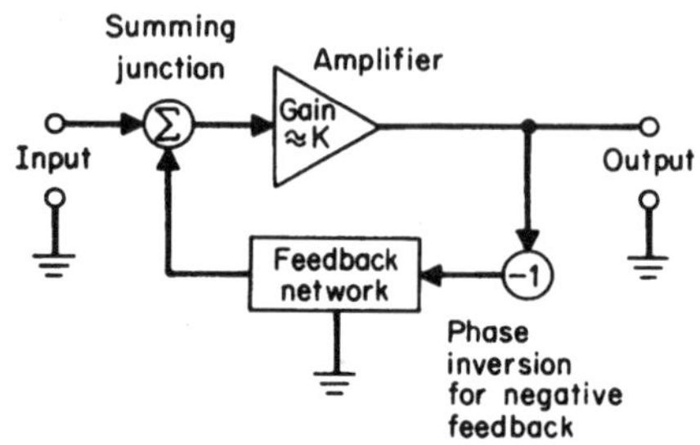


**Figure 6.15** Negative-feedback amplifier. Voltage gain = $K/[1 + (Kb)]$, where $K$ = amplifier voltage gain without feedback and $b$ = gain of the feedback network, usually 1 on less.

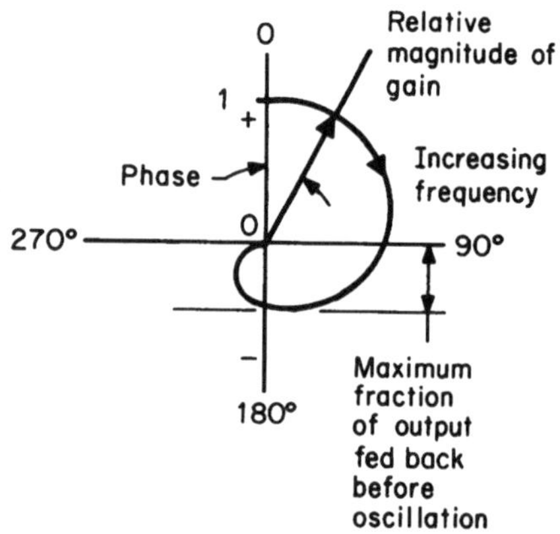


**Figure 6.16** Polar plot of amplifier gain and phase from dc relative to frequency. At higher frequencies, a negative component of the gain characteristics limits the maximum usable feedback before oscillation occurs.

will nearly equal the reciprocal of the loss of the feedback network as measured from output to input [9].

## Linear Feedback

Linear feedback exists when the feedback signal is only a level-independent portion of the output signal. Negative feedback cannot be applied in ever-increasing amounts because all amplifiers have increasing amounts of phase shift as the limits of the frequency range are approached, particularly at high frequencies (Figure 6.16). Whenever phase shift of the feedback signal is between 90 and 270° of phase with respect to the input signal at the feedback node, amplification with feedback will be greater with feedback than without. In the limit, no more positive feedback signal than the original input signal can be returned to the amplifier input before oscillation starts at the frequency where the returned feedback signal is equal to the original input signal in both amplitude and phase. This condition is desirable only in oscillators, not in amplifiers.

Another reason for using little negative feedback is that some intermediate stage of the amplifier may current-limit feeding a capacitor before the output stage is overloaded by input signals. This distortion is known as *slew-rate limiting* and is a cause of transient intermodulation distortion.

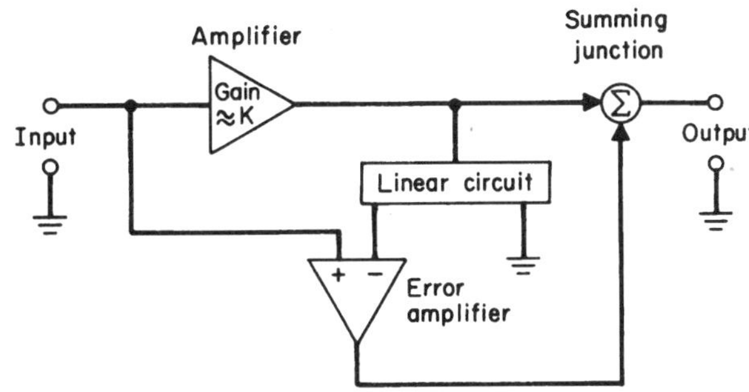


**Figure 6.17** Feed-forward amplifier phase or gain error correction by negative feedback.

## Feed Forward

Feedback is a method of correcting amplifier errors after they have occurred and have been compared with the input signal (Figure 6.17). One feed-forward method is to measure the errors that an amplifier will introduce into the output signal and then to feed these errors, inverted in phase, directly to the output summing junction through a separate path, which may also include an amplifier of lesser output range, since only the error signal will have to be supplied. A portion of the resulting output signal may also be fed back for further error correction.

In feed-forward circuits, the error signal is handled by an amplifier, separate from the amplifier whose errors need correcting. In feedback circuits, the error signal is handled by the amplifier causing the errors.

## Nonlinear Feedback

Precision rectifier circuits make use of nonlinear feedback. Here, one or more diodes in the feedback loop of an operational amplifier result in an output signal which is the half-wave or full-wave rectified signal originally present at the input of the circuit (Figure 6.18). Rectification of signals is a function needed in signal-processing circuits, such as compressors, expanders, meters, and noise reduction circuits for pulse or random noise.

## Voltage Feedback

The output voltage of an amplifier, attenuated in a voltage-divider network, is subtracted from the input voltage, resulting in the amplifier input voltage with feedback. The gain reduction ratio is equal to the reduction ratio of the output impedance of the amplifier, equal to the distortion reduction ratio for signals of the same output voltage, and inversely proportional to the input impedance increase ratio of the amplifier. When the loop gain of the amplifier is very high, as is normal in operational amplifiers, the voltage gain of the amplifier with feedback is nearly equal to the inverse of the loss of the feedback attenuator, and the output impedance of the amplifier becomes very low. Therefore, any variation in amplifier load will have little effect on output voltage until the maximum output-current capability of the amplifier is reached.

In an alternate circuit, the amplifier need not add signal and feedback voltages because only one input terminal is required. Here, the output voltage is converted to a current using an imped-

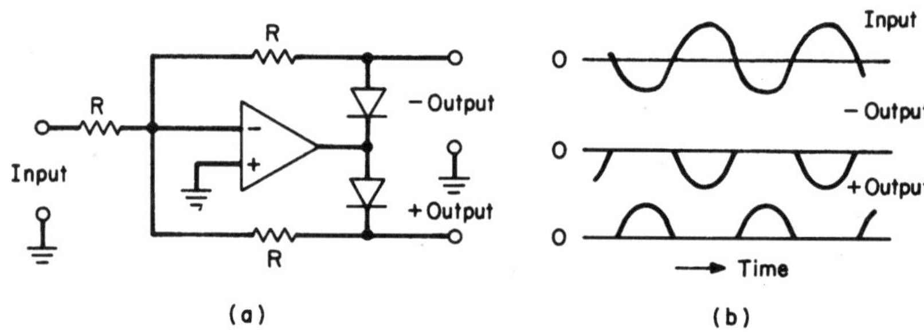


**Figure 6.18** Nonlinear precision-rectifier feedback: (*a*) operational amplifier circuit, (*b*) sine-wave input-output waveforms.

ance and fed to the input terminal, which also receives input current from the signal source, perhaps through a second impedance. The amplifier now amplifies the difference between the two currents. Output impedance and distortion are affected similarly.

Voltage feedback causes the amplifier to become a nearly constant voltage source with a fixed input signal [10].

### Current Feedback

The output current of an amplifier may conveniently be converted to a voltage by passing this current through an impedance connected in series with the load impedance. The resulting voltage is then applied as a feedback voltage to the amplifier. With such a circuit, the internal output impedance ratio of the amplifier will be equal to the inverse of the current gain reduction ratio achieved, and the distortion reduction ratio will be the same as the output-current reduction ratio for the same input signal. Again, the feedback signal may be a feedback voltage, with the amplifier utilizing the difference in feedback and signal voltages, or a feedback current, with the amplifier supplying current gain.

Current feedback causes the amplifier to become a nearly constant current source with a fixed input signal.

## 6.4b Output and Input Impedance

The output impedance of an amplifier usually varies with frequency and is mostly resistive for an amplifier that has a constant fraction of the output fed back to the input or an intermediate stage. The output impedance is sometimes expressed as the *damping factor* of the amplifier, defined as the ratio of nominal load resistance to amplifier internal output impedance. The value of output impedance also includes any impedances connected between the output wiring terminals and the actual output and ground nodes of the circuit.

Another measure of output impedance of an amplifier is *regulation*, usually measured in percentages and defined as the change in output voltage as the nominal load is changed from open-circuit to rated load. Damping factor and regulation are normally rated at mid-frequencies. At the extremes of the frequency range, the output impedance of an amplifier will be different from its midfrequency value because the loop gain of an amplifier decreases, particularly at high frequencies. The output impedance of an amplifier can then be described as a two-terminal network of resistors and reactive elements.

Input impedance of an amplifier may often be set by a physical resistor connected across the input terminals that shunts the input impedance of the circuit. An additional component of the input impedance sometimes rated is the *input capacitance,* partially composed of the wiring capacitance and any capacitors which are part of the radio-frequency filters of the input circuit or are capacitors designed to give the desired termination to certain source signals. In low-noise amplifiers, the input resistance component will be largely determined by feedback to the input circuit.

When a number of amplifiers and other circuits are connected in series so that each amplifier amplifies the output signal from the previous circuit, the connections are often made on a voltage basis, in which each amplifier has a relatively high input impedance and a relatively low output impedance. Here, the amplifier takes very little of the load current that could be provided by the circuit at its input, and its own output voltage changes very little, whether supplying full-load current or not. This type of design is used most often in self-contained equipment or in pieces of equipment operated in close proximity to each other.

Equipment used as part of large distributed systems or with program transmission lines is often designed to present a constant output impedance and input impedance to match the nominal impedance of transmission lines. Audio line impedances of 150 or 600 Ω are common values. At video and radio frequencies, transmission-line impedances of 50, *75,* and 300 Ω are preferred.

Equipment designed to operate in constant-impedance circuits is often rated in decibels with respect to 1 mW (dBm) of output power into a matched load. Gain or loss are given in decibels, and systems are designed on a power gain or loss basis. The advantage of operating circuits at matched transmission-line impedances is that reflections or echoes of signals will not be generated at the receiving end of a traveling signal.

### 6.4c Feed Forward and Correction of Estimated Errors

Some error correction can be accomplished by making a good estimate of the error and then predistorting the signal with opposing distortion. This model may contain amplitude or phase nonlinearities, or both. Such correction can be quite complex and requires considerable accuracy in modeling of distortion components.

### 6.4d Differential Amplifier

Amplifiers that allow the measurement or use of signals generated remote from the point of equipment location are called *instrumentation amplifiers.* These amplifiers have controlled amplification for the difference in voltage between two signals and very low amplification for the sum of the two signals, the measurement of the voltages made with the local ground reference. In general terms, these amplifiers have a controlled differential- or transverse-mode gain and a low common- or longitudinal-mode gain. The common instrumentation amplifier connection uses three operational amplifiers, with the first two circuits amplifying the two signals equally and sharing a common feedback resistor $R_1$ between the two negative inputs. The output signals then pass through two identical resistor attenuators $R_3$ and $R_4$, with the resistor junctions connected to the positive and negative inputs of the third amplifier and resistor $R_4$ completing the connection between positive input and ground and negative input and output. A simplified-version instrumentation amplifier omits the two input stages and connects signal wires directly to the two resistor networks, which now present unequal loads to the two signal wires.

The performance of a differential or instrumentation amplifier is measured by its *common-mode rejection*, which is the ratio of differential- to common-mode gain. This value is largely determined by the accuracy of resistor matching and the ratio of amplifier loop gain to circuit gain. Common signals must not exceed the maximum allowable common-mode input signal for the amplifier.

Differential amplifiers are used in audio, video, and computer equipment when signal sources are widely separated or when ground-loop signals may exist. The two input-signal leads of a differential amplifier are brought directly to the source of signals, often as a pair of twisted wires

## 6.5 References

1. Shockley, W: "The Theory of P-N Junctions in Semiconductors and P-N Junction Transistors," *Proc. JRE*, vol. 41, June 1953.
2. Shockley, W: "A Unipolar Field-Effect Transistor," *Proc. IRE*, vol. 40, November 1952.
3. Kirchner, R. J.: "Properties of Junction Transistors," *Trans. IRE PGA*, AU-3(4), July-August 1955.
4. Trent, R. L.: "Design Principles for Transistor Audio Amplifiers," *Trans. IRE PGA*, AU-3(5), September–October 1955.
5. Garner, L. H.: "High-Power Solid State Amplifiers," *Trans. IRE PGA*, 15(4), December 1967.
6. Fewer, D. R.: "Design Principles for Junction Transistor Audio Power Amplifiers," *Trans. IRE PGA*, AU-3(6), November–December 1955.
7. Petersen, A., and D. B. Sinclair: "A Singled-Ended Push-Pull Audio Amplifier," *Proc. IRE*, vol. 40, January 1952.
8. Widlar, R. J.: "A Unique Current Design for a High Performance Operational Amplifier Especially Suited to Monolithic Construction," *Proc. NEC*, 1965.
9. Black, H. S.: U.S. Patent 2,102,671.
10. Walker, P. J.: "A Current Dumping Audio Power Amplifier," *Wireless World,* December 1975.

## 6.6 Bibliography

Harper, C. A. (ed.): *Handbook of Components for Electronics,* McGraw-Hill, New York, N.Y., 1977.

Lynn, D. K., C. S. Meyer, and D. C. Hamilton (eds.): *Analysis and Design of Integrated Circuits,* McGraw-Hill, New York, N.Y., 1967.

Weinberg, L.: *Network Analysis and Synthesis,* McGraw-Hill, New York, N.Y., 1967.

Chapter

# 7

# Analog/Digital Signal Conversion

Susan A. R. Garrod, K. Blair Benson, Donald G. Fink

Jerry C. Whitaker, Editor

## 7.1 Introduction

Analog-to-digital conversion (A/D) is the process of converting a continuous range of analog signals into specific digital codes. Such conversion is necessary to interface analog pickup elements and systems with digital devices and systems that process, store, interpret, transport, and manipulate the analog values. Analog-to-digital conversion is not an exact process; the comparison between the analog sample and a reference voltage is uncertain by the amount of the difference between one reference voltage and the next [1]. The uncertainty amounts to plus or minus one-half that difference. When words of 8 bits are used, this uncertainty occurs in essentially random fashion, so its effect is equivalent to the introduction of random noise (*quantization noise*). Fortunately, such noise is not prominent in the analog signal derived from the digital version. For example, in 8-bit digitization of the NTSC 4.2 MHz baseband at 13.5 megasamples per second (MS/s), the quantization noise is about 60 dB below the peak-to-peak signal level, far lower than the noise typically present in the analog signal from the camera.

## 7.2 The Nyquist Limit and Aliasing

A critical rule must be observed in sampling an analog signal if it is to be reproduced without spurious effects known as *aliasing*. The rule, first described by Nyquist in 1924 [2], states that the time between samples must be short compared with the rates of change of the analog waveform. In video terms, the sampling rate in megasamples per second must be at least twice the maximum frequency in megahertz of the analog signal. Thus, the 4.2 MHz maximum bandwidth in the luminance spectrum of the NTSC baseband requires that the NTSC signal be sampled at 8.4 MS/s or greater. Conversely, the 13.5 MS/s rate specified in the ITU-R studio digital standard

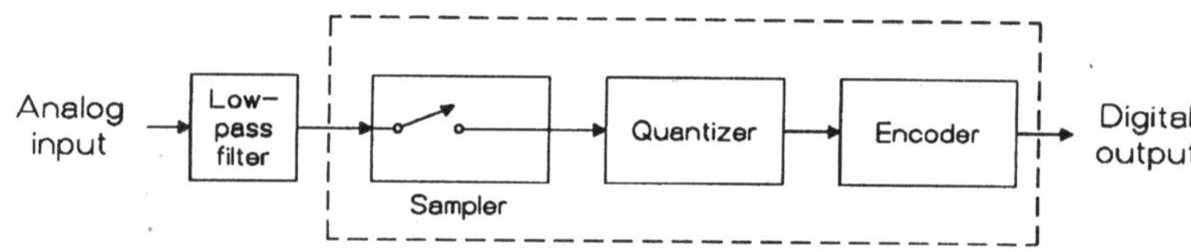


**Figure 7.1** Basic elements of an analog-to-digital converter. (*From* [1]. *Used with permission.*)

can be applied to a signal having no higher frequency components than 6.75 MHz. If studio equipment exceeds this limit—and many cameras and associated amplifiers do—a low-pass filter must be inserted in the signal path before the conversion from analog to digital form takes place. A similar band limit must be met at 3.375 MHz in the chrominance channels before they are digitized in the NTSC system. If the sampling occurs at a rate lower than the Nyquist limit, the spectrum of the output analog signal contains spurious components, which are actually higher-frequency copies of the input spectrum that have been transposed so that they overlap the desired output spectrum. When this output analog signal is displayed, the spurious information shows up in a variety of forms, depending on the subject matter and its motions [1]. Moiré patterns are typical, as are distorted and randomly moving diagonal edges of objects. These aliasing effects often cover large areas and are visible at normal viewing distances.

Aliasing may occur, in fact, not only in digital sampling, but whenever any form of sampling of the image occurs. An example long familiar in motion pictures is that of vehicle wheels (usually wagon wheels) that appear to be rotating backward as the vehicle moves forward. This occurs because the image is sampled by the camera at 24 frames/s. If the rotation of the spokes of the wheel is not precisely synchronous with the film advance, another spoke takes the place of the adjacent one on the next frame, at an earlier time in its rotation. The two spokes are not separately identified by the viewer, so the spoke motion appears reversed. Many other examples of image sampling occur in television. The display similarly offers a series of samples in the vertical dimension, with results that depend not only on the time-vs.-light characteristics of the display device but also, and more important, on the time-vs.-sensation properties of the human eye. These aspects of sampling have a significant bearing on the design of HDTV systems.

## 7.3 The A/D Conversion Process

To convert a signal from the analog domain into a digital form, it is necessary to create a succession of digital words that comprise only two discrete values, 0 and 1 [1]. Figure 7.1 shows the essential elements of the analog-to-digital converter. The input analog signal must be confined to a limited spectrum to prevent spurious components in the reconverted analog output. A low-pass filter, therefore, is placed prior to the converter. The converter proper first samples the analog input, measuring its amplitude at regular, discrete intervals of time. These individual amplitudes then are matched, in the quantizer, against a large number of discrete levels of amplitude (256 levels to convert into 8-bit words). Each one of these discrete levels can be represented by a specific digital word. The process of matching each discrete amplitude with its unique word is car-

**Table 7.1 Binary Values of Amplitude Levels for 8-Bit Words** (*From* [1]. *Used with permission.*)

| Amplitude | Binary Level | Amplitude | Binary Level | Amplitude | Binary Level |
|---|---|---|---|---|---|
| 0 | 00000000 | 120 | 01111000 | 240 | 11110000 |
| 1 | 00000001 | 121 | 01111001 | 241 | 11110001 |
| 2 | 00000010 | 122 | 01111010 | 242 | 11110010 |
| 3 | 00000011 | 123 | 01111011 | 243 | 11110011 |
| 4 | 00000100 | 124 | 01111100 | 244 | 11110100 |
| 5 | 00000101 | 125 | 01111101 | 245 | 11110101 |
| 6 | 00000110 | 126 | 01111110 | 246 | 11110110 |
| 7 | 00000111 | 127 | 01111111 | 247 | 11110111 |
| 8 | 00001000 | 128 | 10000000 | 248 | 11111000 |
| 9 | 00001001 | 129 | 10000001 | 249 | 11111001 |
| 10 | 00001010 | 130 | 10000010 | 250 | 11111010 |
| 11 | 00001011 | 131 | 10000011 | 251 | 11111011 |
| 12 | 00001100 | 132 | 10000100 | 252 | 11111100 |
| 13 | 00001101 | 133 | 10000101 | 253 | 11111101 |
| 14 | 00001110 | 134 | 10000110 | 254 | 11111110 |
| 15 | 00001111 | 135 | 10000111 | 255 | 11111111 |

ried out in the encoder, which, in effect, scans the list of words and picks out the one that matches the amplitude then present. The encoder passes out the series of code words in a sequence corresponding to the sequence in which the analog signal was sampled. This *bit stream* is, consequently, the digital version of the analog input.

The list of digital words corresponding to the sampled amplitudes is known as a *code*. Table 7.1 represents a simple code showing amplitude levels and their 8-bit words in three ranges: 0 to 15, 120 to 135, and 240 to 255. Signals encoded in this way are said to be *pulse-code-modulated*. Although the basic pulse-code modulation (PCM) code sometimes is used, more elaborate codes—with many additional bits per word—generally are applied in circuits where errors may be introduced into the bit stream. Figure 4.2.2 shows a typical video waveform and several quantized amplitude levels based on the PCM coding scheme of Table 7.1.

The sampling rate, even in analog sampling systems, is crucial. Figure 7.3*a* shows the spectral consequence of a sampling rate that is too low for the input bandwidth; Figure 7.3*b* shows the result of a rate equal to the theoretical minimum value, which is impractical; and Figure 7.3*c* shows typical practice. The input spectrum must be limited by a low-pass filter to greatly attenuate frequencies near one-half the sampling rate and above. The higher the sampling rate, the easier and simpler the design of the input filter becomes. An excessively high sampling rate, however, is wasteful of transmission bandwidth and storage capacity, while a low but adequate rate complicates the design and increases the cost of input and output analog filters.

Analog signals can be converted to digital codes using a number of methods, including the following [3]:

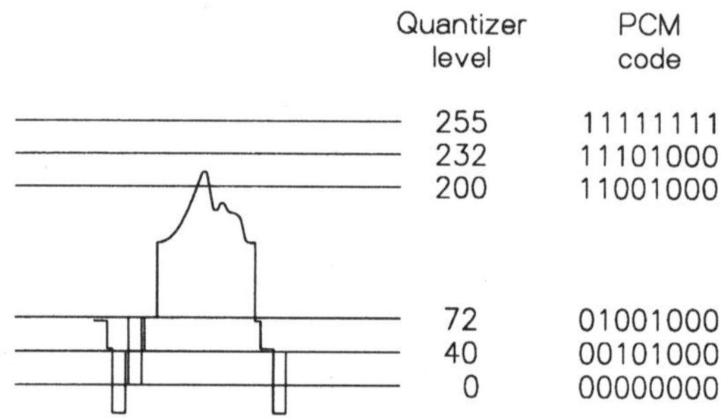


**Figure 7.2** Video waveform quantized into 8-bit words.

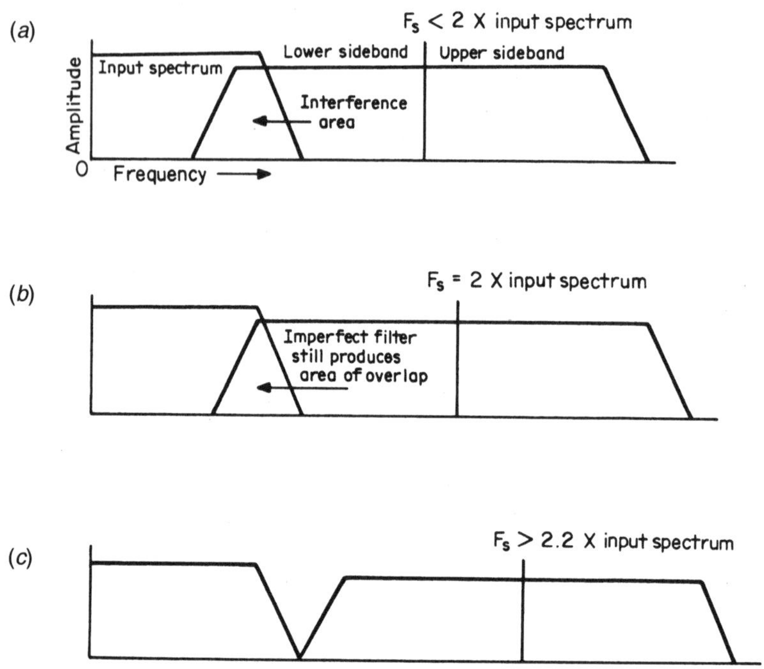


**Figure 7.3** Relationship between sampling rate and bandwidth: (*a*) a sampling rate too low for the input spectrum, (*b*) the theoretical minimum sampling rate ($F_S$), which requires a theoretically perfect filter, (*c*) a practical sampling rate using a practical input filter.

- Integration
- Successive approximation
- Parallel (flash) conversion

- Delta modulation
- Pulse-code modulation
- Sigma-delta conversion

Two of the more common A/D conversion processes are successive approximation and parallel or flash. Very high-resolution digital video systems require specialized A/D techniques that often incorporate one of these general schemes in conjunction with proprietary technology.

## 7.3a Successive Approximation

Successive approximation A/D conversion is a technique commonly used in medium- to high-speed data-acquisition applications. One of the fastest A/D conversion techniques, it requires a minimum amount of circuitry to implement. The conversion times for successive approximation A/D conversion typically range from 10 to 300 μs for 8-bit systems.

The successive approximation A/D converter can approximate the analog signal to form an $n$-bit digital code in $n$ steps. The *successive approximation register* (SAR) individually compares an analog input voltage with the midpoint of one of $n$ ranges to determine the value of 1 bit. This process is repeated a total of $n$ times, using $n$ ranges, to determine the $n$ bits in the code. The comparison is accomplished as follows:

- The SAR determines whether the analog input is above or below the midpoint and sets the bit of the digital code accordingly.
- The SAR assigns the bits beginning with the most significant bit.
- The bit is set to a 1 if the analog input is greater than the midpoint voltage; it is set to a 0 if the input is less than the midpoint voltage.
- The SAR then moves to the next bit and sets it to a 1 or a 0 based on the results of comparing the analog input with the midpoint of the next allowed range.

Because the SAR must perform one approximation for each bit in the digital code, an $n$-bit code requires $n$ approximations. A successive approximation A/D converter consists of four main functional blocks, as shown in Figure 7.4. These blocks are the SAR, the analog comparator, a D/A (digital-to-analog) converter, and a clock.

## 7.3b Parallel/Flash

Parallel or flash A/D conversion is used in high-speed applications such as video signal processing, medical imaging, and radar detection systems. A flash A/D converter simultaneously compares the input analog voltage with $2^n - 1$ threshold voltages to produce an $n$-bit digital code representing the analog voltage. Typical flash A/D converters with 8-bit resolution operate at 100 MHz to 1 GHz.

The functional blocks of a flash A/D converter are shown in Figure 7.5. The circuitry consists of a precision resistor ladder network, $2^n - 1$ analog comparators, and a digital priority encoder. The resistor network establishes threshold voltages for each allowed quantization level. The analog comparators indicate whether the input analog voltage is above or below the threshold at each level. The output of the analog comparators is input to the digital priority encoder. The priority encoder produces the final digital output code, which is stored in an output latch.

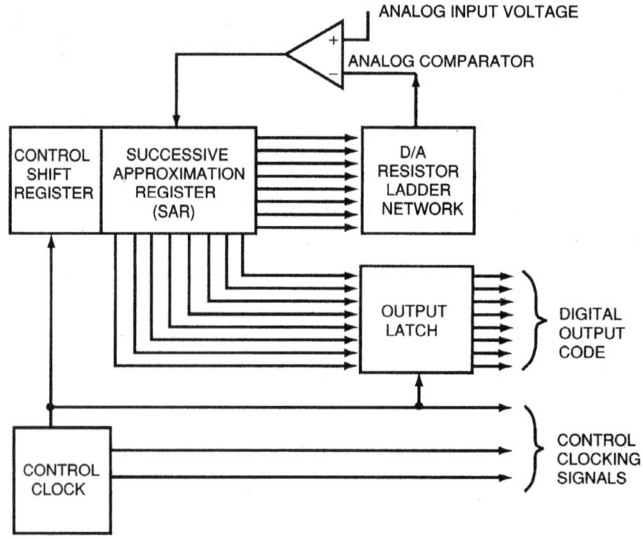


**Figure 7.4** Successive approximation A/D converter block diagram. (*After* [4].)

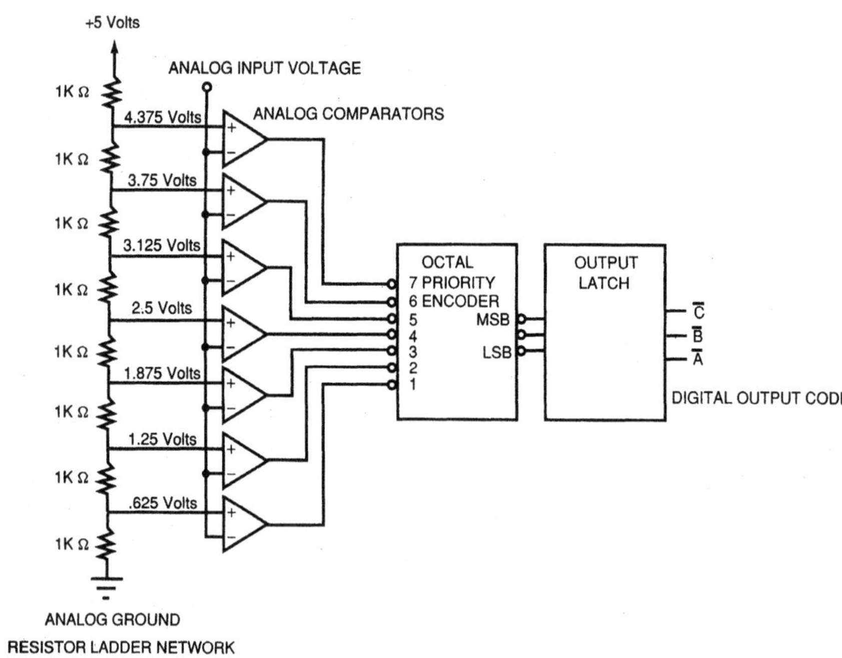


**Figure 7.5** Block diagram of a flash A/D converter. (*After* [5].)

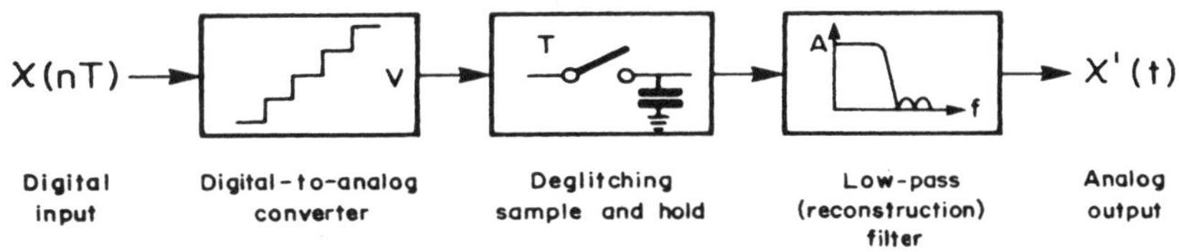


**Figure 7.6** Digital-to-analog converter block diagram.

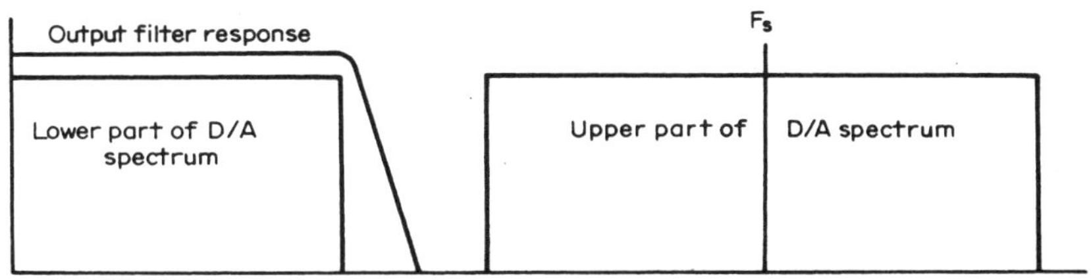


**Figure 7.7** Output filter response requirements for a common D/A converter.

An 8-bit flash A/D converter requires 255 comparators. The cost of high-resolution A/D comparators escalates as the circuit complexity increases and the number of analog converters rises by $2^n - 1$. As a low-cost alternative, some manufacturers produce modified flash converters that perform the A/D conversion in two steps, to reduce the amount of circuitry required. These modified flash converters also are referred to as *half-flash* A/D converters because they perform only half of the conversion simultaneously.

## 7.4 The D/A Conversion Process

The digital-to-analog converter (DAC) is, in principle, quite simple. The digital stream of binary pulses is decoded into discrete, sequentially timed signals corresponding to the original sampling in the A/D. The output is an analog signal of varying levels. The time duration of each level is equal to the width of the sample taken in the A/D conversion process. The analog signal is separated from the sampling components by a low-pass filter. Figure 7.6 shows a simplified block diagram of a D/A. The deglitching sample-and-hold circuits in the center block set up the analog levels from the digital decoding and remove the unwanted high-frequency sampling components. Each digital number is converted to a corresponding voltage and stored until the next number is converted. Figure 7.7 shows the resulting spectrum. The energy surrounding the sampling frequency must be removed, and an output low-pass filter is used to accomplish that task. One cost-effective technique used in a variety of applications is called *oversampling*. A new sampling rate is selected that is a whole multiple of the input sampling rate. The new rate is typically two or four times the old rate. Every second or fourth sample is filled with the input value, while the others are set to zero. The result is passed through a digital filter that distributes the energy in the

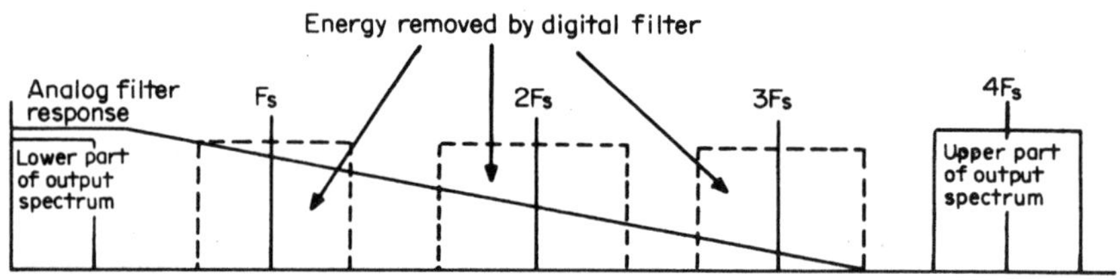


**Figure 7.8** The filtering benefits of oversampling.

real samples among the empty ones and itself. The resulting spectrum (for a 4× oversampling system) is shown in Figure 7.8. The energy around the 4× sample frequency must be removed, which can be done simply because it is so distant from the upper band edge. The response of the output filter is chiefly determined by the digital processing and is therefore very stable with age, in contrast to a strictly analog filter, whose component values are susceptible to drift with age and other variables.

## 7.4a Practical Implementation

To convert digital codes to analog voltages, a voltage weight typically is assigned to each bit in the digital code, and the voltage weights of the entire code are summed [3]. A general-purpose D/A converter consists of a network of precision resistors, input switches, and level shifters to activate the switches to convert the input digital code to an analog current or voltage output. A D/A device that produces an analog current output usually has a faster settling time and better linearity than one that produces a voltage output.

D/A converters commonly have a fixed or variable reference level. The reference level determines the switching threshold of the precision switches that form a controlled impedance network, which in turn controls the value of the output signal. *Fixed-reference* D/A converters produce an output signal that is proportional to the digital input. In contrast, *multiplying* D/A converters produce an output signal that is proportional to the product of a varying reference level times a digital code.

D/A converters can produce bipolar, positive, or negative polarity signals. A *four-quadrant multiplying* D/A converter allows both the reference signal and the value of the binary code to have a positive or negative polarity.

# 7.5 Converter Performance Criteria

The major factors that determine the quality of performance of A/D and D/A converters are *resolution*, *sampling rate*, *speed*, and *linearity* [3]. The resolution of a D/A circuit is the smallest possible change in the output analog signal. In an A/D system, the resolution is the smallest change in voltage that can be detected by the system and produce a change in the digital code. The resolution determines the total number of digital codes, or quantization levels, that will be recognized or produced by the circuit.

**Table 7.2 Representative Sampling of Converter Operating Parameters**

| Converter Type | Sampling Rate | Resolution | S/Nq[1] | Max. Input Frequency | Power Consumption |
|---|---|---|---|---|---|
| A/D | 400 Ms/s[2] | 8 bits | 43 dB | 1 GHz | 3 W |
| | 200 Ms/s | 10 bits | 58 dB | 400 MHz | 2 W |
| | 120 Ms/s | 12 bits | 70 dB | 350 MHz | 1 W |
| | 70 Ms/s | 14 bits | 75 dB | 300 MHz | 1.3 W |
| D/A | **Sampling Rate** | **Resolution** | **Dynamic Range** | | **Power Consumption** |
| | 500 Ms/s[3] | 10 bits | 80 dB | | 250 mW |
| | 300 Ms/s | 12 bits | 85 dB | | 300 mW |
| | 200 Ms/s | 14 bits | 88 dB | | 350 mW |

Notes:
[1] signal-to-quantization noise, [2] megasamples per second, [3] settling rime in megasamples per second

The resolution of a D/A or A/D device usually is specified in terms of the bits in the digital code, or in terms of the *least significant bit* (LSB) of the system. An $n$-bit code allows for $2^n$ quantization levels, or $2^n - 1$ steps between quantization levels. As the number of bits increases, the step size between quantization levels decreases, therefore increasing the accuracy of the system when a conversion is made between an analog and digital signal. The system resolution also can be specified as the voltage step size between quantization levels.

The speed of a D/A or A/D converter is determined by the amount of time it takes to perform the conversion process. For D/A converters, the speed is specified as the *settling time*. For A/D converters, the speed is specified as the *conversion time*. The settling time for a D/A converter varies with supply voltage and transition in the digital code; it is specified in the data sheet with the appropriate conditions stated.

A/D converters have a maximum sampling rate that limits the speed at which they can perform continuous conversions. The sampling rate is the number of times per second that the analog signal can be sampled and converted into a digital code. For proper A/D conversion, the minimum sampling rate must be at least 2 times the highest frequency of the analog signal being sampled to satisfy the Nyquist criterion. The conversion speed and other timing factors must be taken into consideration to determine the maximum sampling rate of an A/D converter. Nyquist A/D converters use a sampling rate that is slightly greater than twice the highest frequency in the analog signal. *Oversampling* A/D converters use sampling rates of $N$ times rate, where $N$ typically ranges from 2 to 64.

Both D/A and A/D converters require a voltage reference to achieve absolute conversion accuracy. Some conversion devices have internal voltage references, whereas others accept external voltage references. For high-performance systems, an external precision reference is required to ensure long-term stability, load regulation, and control over temperature fluctuations.

Measurement accuracy is specified by the converter's linearity. *Integral linearity* is a measure of linearity over the entire conversion range. It often is defined as the deviation from a straight line drawn between the endpoints and through zero (or the *offset value*) of the conversion range.

Integral linearity also is referred to as *relative accuracy.* The offset value is the reference level required to establish the zero or midpoint of the conversion range. *Differential linearity,* the linearity between code transitions, is a measure of the *monotonicity* of the converter. A converter is said to be monotonic if increasing input values result in increasing output values.

The accuracy and linearity values of a converter are specified in units of the LSB of the code. The linearity may vary with temperature, so the values often are specified at +25°C as well as over the entire temperature range of the device.

With each new generation of devices, A/D and D/A converter technology improves, yielding higher sampling rates with greater resolution. Table 7.2 shows some typical values as this book went to press.

## 7.6 References


1. Benson, K. B., and D. G. Fink: "Digital Operations in Video Systems," *HDTV: Advanced Television for the 1990s*, McGraw-Hill, New York, pp. 4.1–4.8, 1990.
2. Nyquist, H.: "Certain Factors Affecting Telegraph Speed," *Bell System Tech. J.*, vol. 3, pp. 324–346, March 1924.
3. Garrod, Susan A. R.: "D/A and A/D Converters," *The Electronics Handbook*, Jerry C. Whitaker (ed.), CRC Press, Boca Raton, Fla., pp. 723–730, 1996.
4. Garrod, Susan, and R. Borns: *Digital Logic: Analysis, Application, and Design*, Saunders College Publishing, Philadelphia, pg. 919, 1991.
5. Garrod, Susan, and R. Borns: *Digital Logic: Analysis, Application, and Design*, Saunders College Publishing, Philadelphia, pg. 928, 1991.

Chapter

# 8

# Digital Filters

**J. A. Chambers, S. Tantaratana, B. W. Bomar**

**Jerry C. Whitaker, Editor**

## 8.1 Introduction

Digital filtering is concerned with the manipulation of discrete data sequences to remove noise, extract information, change the sample rate, and/or modify the input information in some form or context [1]. Although an infinite number of numerical manipulations can be applied to discrete data (e.g., finding the mean value, forming a histogram), the objective of digital filtering is to form a discrete output sequence $y(n)$ from a discrete input sequence $x(n)$. In some manner, each output sample is computed from the input sequence—not just from any one sample, but from many, possibly all, of the input samples. Those filters that compute their output from the present input and a finite number of past inputs are termed *finite impulse response* (FIR) filters; those that use all past inputs are termed *infinite impulse response* (IIR) filters.

## 8.2 FIR Filters

An FIR filter is a linear discrete-time system that forms its output as the weighted sum of the most recent, and a finite number of past, inputs [1]. A *time-invariant* FIR filter has finite memory, and its impulse response (its response to a discrete-time input that is unity at the first sample and otherwise zero) matches the fixed weighting coefficients of the filter. *Time-variant* FIR filters, on the other hand, may operate at various sampling rates and/or have weighting coefficients that adapt in sympathy with some statistical property of the environment in which they are applied.

Perhaps the simplest example of an FIR filter is the *moving average* operation described by the following linear constant-coefficient difference equation:

$$y[n] = \sum_{k=0}^{M} b_k x[n-k] \qquad b_k = \frac{1}{M+1} \tag{8.1}$$

Where:
$y[n]$ = output of the filter at integer sample index $n$
$x[n]$ = input to the filter at integer sample index $n$
$b_k$ = filter weighting coefficients, $k = 0,1,...,M$
$M$ = filter order

In a practical application, the input and output discrete-time signals will be sampled at some regular sampling time interval, $T$ seconds, denoted $x[nT]$ and $y[nT]$, which is related to the sampling frequency by $f_s = 1/T$, samples per second. However, for generality, it is more convenient to assume that $T$ is unity, so that the effective sampling frequency also is unity and the Nyquist frequency is one-half. It is, then, straightforward to scale, by multiplication, this normalized frequency range, i.e. [0, 1/2], to any other sampling frequency.

The output of the simple moving average filter is the average of the $M + 1$ most recent values of $x[n]$. Intuitively, this corresponds to a smoothed version of the input, but its operation is more appropriately described by calculating the frequency response of the filter. First, however, the $z$-domain representation of the filter is introduced in analogy to the $s$- (or *Laplace*) domain representation of analog filters. The $z$-transform of a causal discrete-time signal $x[n]$ is defined by:

$$X(z) = \sum_{n=0}^{\infty} x[n]z^{-n} \tag{8.2}$$

Where:
$X(z)$ = $z$-transform of $x[n]$
$z$ = complex variable

The $z$-transform of a delayed version of $x[n]$, namely $x[n-k]$ with $k$ a positive integer, is found to be given by $z^{-k}X(z)$. This result can be used to relate the $z$-transform of the output, $y[n]$, of the simple moving average filter to its input:

$$Y(z) = \sum_{k=0}^{M} b_k z^{-k} X(z) \qquad b_k = \frac{1}{M+1} \tag{8.3}$$

The $z$-domain transfer function, namely the ratio of the output to input transform, becomes:

$$H(z) = \frac{Y(z)}{X(z)} = \sum_{k=0}^{M} b_k z^{-k} \qquad b_k = \frac{1}{M+1} \tag{8.4}$$

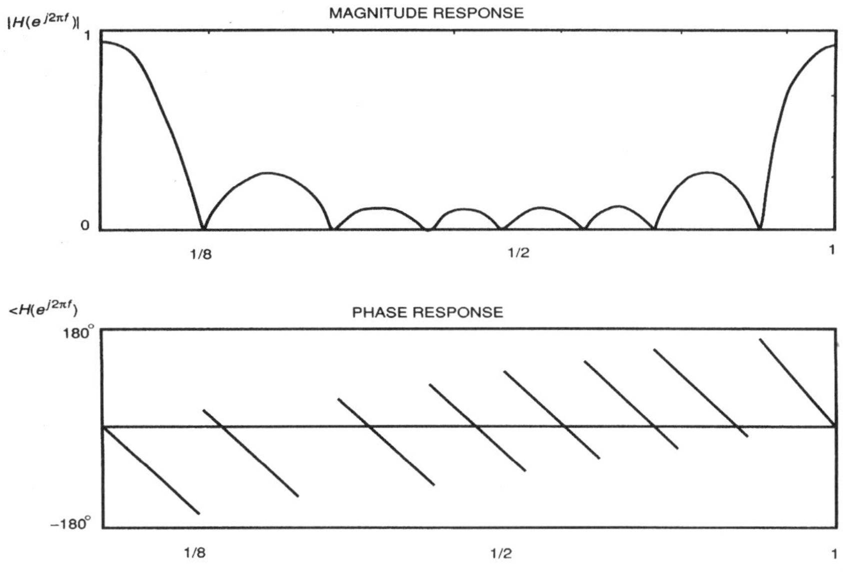


**Figure 8.1** The magnitude and phase response of the simple moving average filter with $M = 7$. (*From* [1]. *Used with permission.*)

Notice the transfer function, $H(z)$, is entirely defined by the values of the weighting coefficients, $b_k$, $k = 0,1,\ldots,M$, which are identical to the discrete impulse response of the filter, and the complex variable $z$. The finite length of the discrete impulse response means that the transient response of the filter will last for only $M + 1$ samples, after which a steady state will be reached. The frequency-domain transfer function for the filter is found by setting

$$z = e^{j2\pi f} \tag{8.5}$$

Where $j = \sqrt{-1}$ and can be written as:

$$H(e^{j2\pi f}) = \frac{1}{M+1}\sum_{k=0}^{M} e^{-j2\pi fk} = \frac{1}{M+1}e^{-j\pi fM}\frac{\sin[\pi f(M+1)]}{\sin(\pi f)} \tag{8.6}$$

The magnitude and phase response of the simple moving average filter, with $M = 7$, are calculated from $H(e^{j2\pi f})$ and shown in Figure 8.1. The filter is seen clearly to act as a crude low-pass smoothing filter with a linear phase response. The sampling frequency periodicity in the magnitude and phase response is a property of discrete-time systems. The linear phase response is due to the $e^{-j\pi fM}$ term in $H(e^{j2\pi f})$ and corresponds to a constant $M/2$ group delay through the filter. A phase discontinuity of ±180° is introduced each time the magnitude term changes sign. FIR filters that have center symmetry in their weighting coefficients have this constant frequency-independent group-delay property that is desirable in applications in which time dispersion is to be avoided, such as in pulse transmission, where it is important to preserve pulse shapes [2].

## 8.2a Design Techniques

Linear-phase FIR filters can be designed to meet various filter specifications, such as low-pass, high-pass, bandpass, and band-stop filtering [1]. For a low-pass filter, two frequencies are required. One is the maximum frequency of the passband below which the magnitude response of the filter is approximately unity, denoted the *passband corner frequency* $f_p$. The other is the minimum frequency of the stop-band above which the magnitude response of the filter must be less than some prescribed level, named the *stop-band corner frequency* $f_s$. The difference between the passband and stop-band corner frequencies is the *transition bandwidth*. Generally, the order of an FIR filter, *M*, required to meet some design specification will increase with a reduction in the width of the transition band. There are three established techniques for coefficient design:

- *Windowing.* A design method that calculates the weighting coefficients by sampling the ideal impulse response of an analog filter and multiplying these values by a smoothing window to improve the overall frequency-domain response of the filter.
- *Frequency sampling.* A technique that samples the ideal frequency-domain specification of the filter and calculates the weighting coefficients by inverse-transforming these values.
- Optimal approximations.

The best results generally can be obtained with the optimal approximations method. With the increasing availability of desktop and portable computers with fast microprocessors, large quantities of memory, and sophisticated software packages, optimal approximations is the preferred method for weighting coefficient design. The impulse response and magnitude response for a $40^{th}$-order optimal half-band FIR low-pass filter designed with the *Parks-McClellan* algorithm [3] are shown in Figure 8.2, together with the ideal frequency-domain design specification. Notice the zeros in the impulse response. This algorithm minimizes the peak deviation of the magnitude response of the design filter from the ideal magnitude response. The magnitude response of the design filter alternates about the desired specification within the passband and above the specification in the stop-band. The maximum deviation from the desired specification is equalized across the passband and stop-band; this is characteristic of an *optimal solution*.

## 8.2b Applications

In general, digitally implemented FIR filters exhibit the following attributes [1]:

- Absence of drift
- Reproducibility
- Multirate realizations
- Ability to adapt to time-varying environments

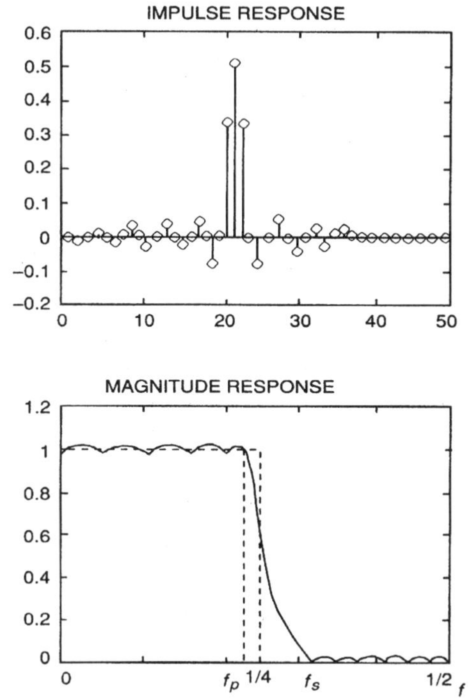


**Figure 8.2** The impulse and magnitude response of an optimal 40th-order half-band FIR filter. (*From* [1]. *Used with permission.*)

These features have led to the widespread use of FIR filters in a variety of applications, particularly in telecommunications. The primary advantage of the fixed-coefficient FIR filter is its unconditional stability because of the lack of feedback within its structure and its exact linear phase characteristics. Nonetheless, for applications that require sharp, selective, filtering—in standard form—they do require relatively large orders. For some applications, this may be prohibitive; therefore, recursive IIR filters are a valuable alternative.

## 8.2c Finite Wordlength Effects

Practical digital filters must be implemented with finite precision numbers and arithmetic [1]. As a result, both the filter coefficients and the filter input and output signals are in discrete form. This leads to four types of finite wordlength effects:

- *Discretization* (quantization) of the filter coefficients has the effect of perturbing the location of the filter poles and zeroes. As a result, the actual filter response differs slightly from the ideal response. This deterministic frequency response error is referred to as *coefficient quantization error.*
- The use of finite precision arithmetic makes it necessary to quantize filter calculations by rounding or truncation. *Roundoff noise* is that error in the filter output that results from rounding or truncating calculations within the filter. As the name implies, this error looks like low-level noise at the filter output.

- Quantization of the filter calculations also renders the filter slightly nonlinear. For large signals this nonlinearity is negligible, and roundoff noise is the major concern. However, for recursive filters with a zero or constant input, this nonlinearity can cause spurious oscillations called *limit cycles*.
- With fixed-point arithmetic it is possible for filter calculations to overflow. The term *overflow oscillation* refers to a high-level oscillation that can exist in an otherwise stable filter because of the nonlinearity associated with the overflow of internal filter calculations. Another term for this is *adder overflow limit cycle*.

## 8.3 Infinite Impulse Response Filters

A digital filter with impulse response having infinite length is known as an *infinite impulse response* filter [1]. Compared to an FIR filter, an IIR filter requires a much lower order to achieve the same requirement of the magnitude response. However, whereas an FIR filter is always stable, an IIR filter may be unstable if the coefficients are not chosen properly. Because the phase of a stable causal IIR filter cannot be made linear, FIR filters are preferable to IIR filters in applications for which linear phase is essential.

Practical *direct form* realizations of IIR filters are shown in Figure 8.3. The realization shown in Figure 8.3*a* is known as *direct form I*. Rearranging the structure results in *direct form II*, as shown in Figure 8.3*b*. The results of transposition are *transposed direct form I* and *transposed direct form II*, as shown in Figures 8.3*c* and 8.3*d*, respectively. Other realizations for IIR filters are *state-space structure*, *wave structure*, and *lattice structure*. In some situations, it is more convenient or suitable to use software realizations that are implemented by programming a general-purpose microprocessor or a digital signal processor. (See [1] for details on IIR filter implementations.)

Designing an IIR filter involves choosing the coefficients to satisfy a given specification, usually a magnitude response parameter. There are various IIR filter design techniques, including:

- Design using an analog prototype filter, in which an analog filter is designed to meet the (analog) specification and the analog filter transfer function is transformed to a digital system function.
- Design using digital frequency transformation, which assumes that a given digital low-pass filter is available, and the desired digital filter is then obtained from the digital low-pass filter by a digital frequency transformation.
- Computer-aided design (CAD), which involves the execution of algorithms that choose the coefficients so that the response is as close as possible to the desired filter.

The first two methods are easily accomplished; they are suitable for designing standard filters (low-pass, high-pass, bandpass, and band-stop). The CAD approach, however, can be used to design both standard and nonstandard filters.

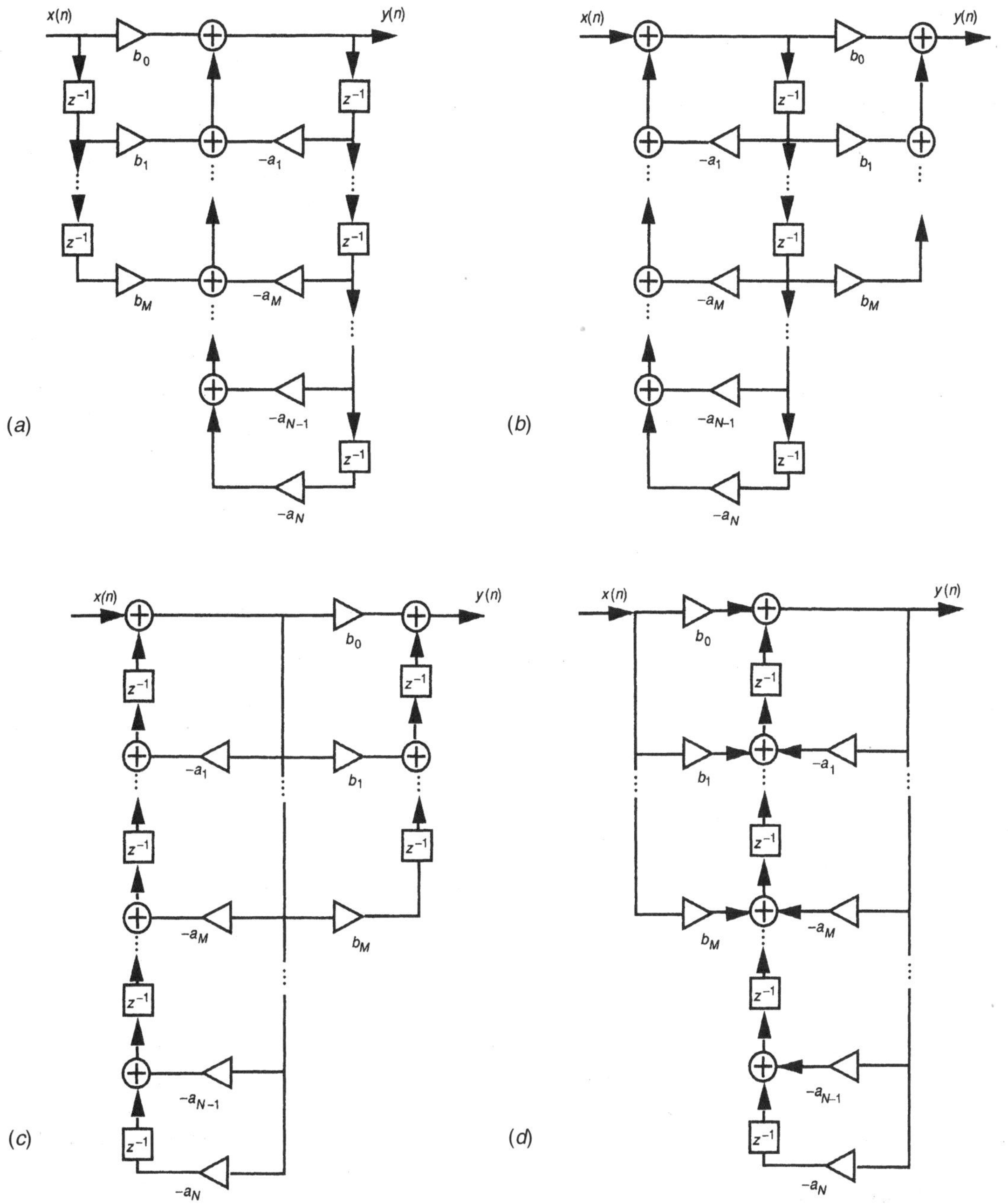


**Figure 8.3** Direct form realizations of IIR filters: (*a*) direct form I, (*b*) direct form II, (*c*) transposed direct form I, (*d*) transposed direct form II. (*From* [1]. *Used with permission.*)

## 8.4 Reference


1. Chambers, J. A., S. Tantaratana, and B. W. Bomar: "Digital Filters," *The Electronics Handbook*, Jerry C. Whitaker (ed.), CRC Press, Boca Raton, Fla., pp. 749–772, 1996.

2. Lee, E. A., and D. G. Messerschmitt: *Digital Communications*, $2^{nd}$ ed., Kluwer, Norell, Mass., 1994.

3. Parks, T. W., and J. H. McClellan: "A Program for the Design of Linear Phase Infinite Impulse Response Filters," *IEEE Trans. Audio Electroacoustics*, AU-20(3), pp. 195–199, 1972.

Chapter

# 9

# Digital Modulation

**Rodger E. Ziemer, Oktay Alkin**

**Jerry C. Whitaker, Editor**

## 9.1 Introduction

Digital modulation is necessary before digital data can be transmitted through a channel, be it a satellite link or HDTV. *Modulation* is the process of varying some attribute of a carrier waveform as a function of the input intelligence to be transmitted. Attributes that can be varied include amplitude, frequency, and phase.

## 9.2 Digital Modulaton Techniques

With digital modulation, the message sequence is a stream of digits, typically of binary value [1]. In the simplest case, parameter variation is on a symbol-by-symbol basis; no memory is involved. Carrier parameters that can be varied under this scenario include the following:

- Amplitude, resulting in *amplitude-shift keying* (ASK)
- Frequency, resulting in *frequency-shift keying* (FSK)
- Phase, resulting in *phase-shift keying* (PSK)

So-called higher-order modulation schemes impose memory over several symbol periods. Such modulation techniques can be classified as *binary* or *M*-ary, depending on whether one of two possible signals or $M > 2$ signals per signaling interval can be sent. (Binary signaling may be defined as any signaling scheme in which the number of possible signals sent during any given signaling interval is two. *M*-ary signaling, on the other hand, is a signaling system in which the number of possible signals sent during any given signaling interval is *M*.) For the case of *M*-ary modulation when the source digits are binary, it is clear that several bits must be grouped to make up an *M*-ary word.

Another classification for digital modulation is *coherent* vs. *noncoherent*, depending upon whether a reference carrier at the receiver coherent with the received carrier is required for

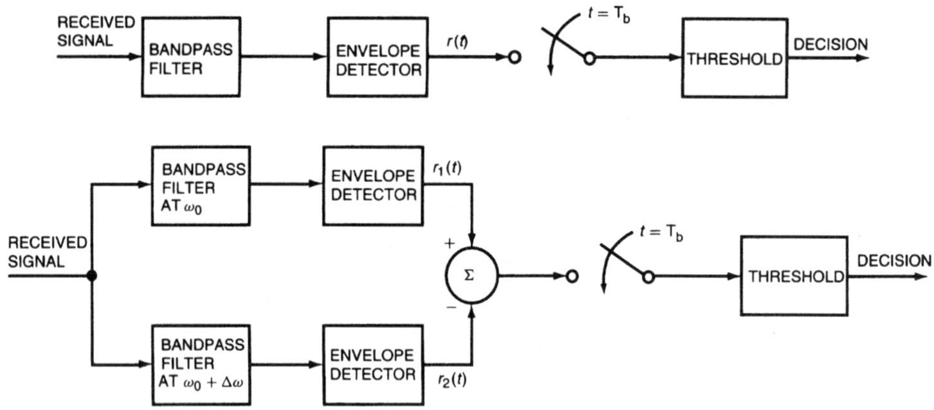


**Figure 9.1** Receiver systems for noncoherent detection of binary signals: (*a*) ASK, (*b*) FSK. (*From* [1]. *Used with permission.*)

demodulation (the coherent case) or not (the noncoherent case). For situations in which it is difficult to maintain phase stability—for example, in channels subject to fading—it is useful to employ a modulation technique that does not require the acquisition of a reference signal at the receiver that is phase-coherent with the received carrier. ASK and FSK are two modulation techniques that lend themselves well to noncoherent detection. Receivers for detection of ASK and FSK noncoherently are shown in Figure 9.1.

One other binary modulation technique is, in a sense, noncoherent: *differentially coherent PSK* (DPSK). With DPSK, the phase of the preceding bit interval is used as a reference for the current bit interval. This technique depends on the channel being sufficiently stable so that phase changes resulting from channel perturbations from a given bit interval to the succeeding one are inconsequential. It also depends on there being a known phase relationship from one bit interval to the next. This requirement is ensured by differentially encoding the bits before phase modulation at the transmitter. Differential encoding is illustrated in Table 9.1. An arbitrary reference bit is chosen to start the process. In the table a *1* has been chosen. For each bit of the encoded sequence, the present bit is used as the reference for the following bit in the sequence. A *0* in the message sequence is encoded as a transition from the state of the reference bit to the opposite state in the encoded message sequence. A *1* is encoded as no change of state. Using these rules, the result is the encoded sequence shown in the table.

## 9.2a QPSK

Consider the case of an MPSK signal where $M = 4$, commonly referred to as *quadriphase-shift keying* (QPSK) [1]. This common modulation technique utilizes four signals in the signal set distinguished by four phases 90° apart. For the case of an MASK signal where $M = 4$, a *quadrature-amplitude-shift keying* (QASK) condition results. With QASK, both the phase and amplitude of

**Table 9.1 Example of the differential encoding process** (*After* [1].)

| **Message Sequence** | | 1 | 0 | 0 | 1 | 1 | 1 | 0 |
|---|---|---|---|---|---|---|---|---|
| **Encoded Sequence** | 1 | 1 | 0 | 1 | 1 | 1 | 1 | 0 |
| **Transmitted Phase Radians** | 0 | 0 | p | 0 | 0 | 0 | 0 | p |

the carrier take on a set of values in one-to-one correspondence with the digital data to be transmitted.

Several variations of QPSK have been developed to meet specific operational requirements. One such scheme is referred to as *offset* QPSK (OQPSK) [2]. This format is produced by allowing only ±90° phase changes in a QPSK format. Furthermore, the phase changes can take place at multiples of a half-symbol interval, or a bit period. The reason for limiting phase changes to ±90° is to prevent the large envelope deviations that occur when QPSK is filtered to restrict sidelobe power, and regrowth of the sidelobes after amplitude limiting is used to produce a constant-envelope signal. This condition often is encountered in satellite systems in which, because of power-efficiency considerations, hard limiting repeaters are used in the communications system.

Another modulation technique closely related to QPSK and OQPSK is *minimum shift keying* (MSK) [2]. MSK is produced from OQPSK by weighting the in-phase and quadrature components of the baseband OQPSK signal with half-sinusoids. The phase changes linearly over a bit interval. As with OQPSK, the goal of MSK is to produce a modulated signal with a spectrum of reduced sidelobe power, one that behaves well when filtered and limited. Many different forms of MSK have been proposed and investigated over the years.

A modulation scheme related to 8-PSK is *π/4-differential QPSK* (π/4-DQPSK) [3]. This technique is essentially an 8-PSK format with differential encoding where, from a given phase state, only specified phase shifts of ± π/4 or ± 3π/4 are allowed.

*Continuous phase modulation* (CPM) [4] comprises a host of modulation schemes. These formats employ continuous phase trajectories over one or more symbols to get from one phase to the next in response to input changes. CPM schemes are employed in an attempt to simultaneously improve power and bandwidth efficiency.

## 9.2b Signal Analysis

The ideal choice of a modulation technique depends on many factors. Two of the most basic are the *bandwidth efficiency* and *power efficiency*. These parameters are defined as follows:

- Bandwidth efficiency is the ratio of the bit rate to the bandwidth occupied for a digital modulation scheme. Technically, it is dimensionless, but for convenience it is usually given the dimensions of bits/second/hertz.

- Power efficiency is the energy per bit over the noise power spectral density ($E_b/N_o$) required to provide a given probability of bit error for a digital modulation scheme.

Computation of these parameters is beyond the scope of this chapter. Interested readers are directed to [1] for a detailed discussion of performance parameters.

## 9.3 Digital Coding

Two issues are fundamental in assessing the performance of a digital communication system [5]:

- The reliability of the system in terms of accurately transmitting information from one point to another
- The rate at which information can be transmitted with an acceptable level of reliability

In an ideal communication system, information would be transmitted at an infinite rate with an infinite level of reliability. In reality, however, fundamental limitations affect the performance of any communication system. No physical system is capable of instantaneous response to changes, and the range of frequencies that a system can reliably handle is limited. These real-world considerations lead to the concept of *bandwidth.* In addition, random noise affects any signal being transmitted through any communication medium. Finite bandwidth and additive random noise are two fundamental limitations that prevent designers from achieving an infinite rate of transmission with infinite reliability. Clearly, a compromise is necessary. What makes the situation even more challenging is that the reliability and the rate of information transmission usually work against each other. For a given system, a higher rate of transmission normally means a lower degree of reliability, and vice versa. To favorably affect this balance, it is necessary to improve the efficiency and the robustness of the communication system. *Source coding* and *channel coding* are the means for accomplishing this task.

### 9.3a Source Coding

Most information sources generate signals that contain redundancies [5]. For example, consider a picture that is made up of pixels, each of which represents one of 256 grayness levels. If a fixed coding scheme is used that assigns 8 binary digits to each pixel, a 100 × 100 picture of random patterns and a 100 × 100 picture that consists of only white pixels would both be coded into the same number of binary digits, although the white-pixel version would have significantly less information than the random-pattern version.

One simple method of source encoding is the *Huffman* coding technique, which is based on the idea of assigning a code word to each symbol of the source alphabet such that the length of each code word is approximately equal to the amount of information conveyed by that symbol. As a result, symbols with lower probabilities get longer code words. Huffman coding is achieved through the following process:

- List the source symbols in descending order of probabilities.
- Assign a binary 0 and a binary 1, respectively, to the last two symbols in the list.
- Combine the last two symbols in the list into a new symbol with its probability equal to the sum of two symbol probabilities.
- Reorder the list, and continue in this manner until only one symbol is left.
- Trace the binary assignments in reverse order to obtain the code word for each symbol.

A tree diagram for decoding a coded sequence of symbols is shown in Figure 9.2. It can easily be verified that the *entropy* of the source under consideration is 2.3382 bits/symbol, and the average code-word length using Huffman coding is 2.37 bits/symbol.

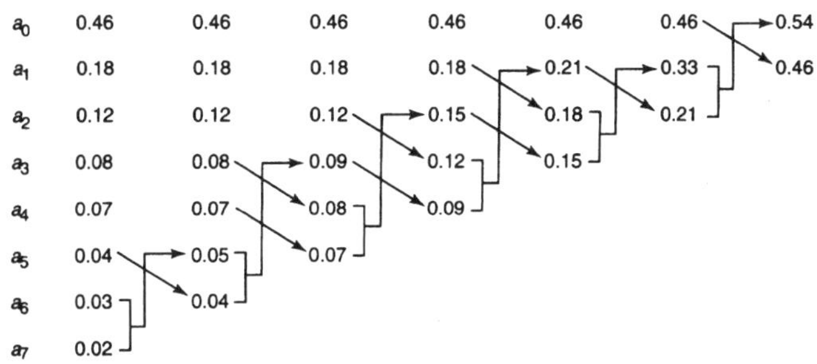


**Figure 9.2** The Huffman coding algorithm. (*From* [5]. *Used with permission.*)

At this point it is appropriate to define *entropy*. In a general sense, entropy is a measure of the disorder or randomness in a closed system. With regard to digital communications, it is defined as a measure of the number of bits necessary to transmit a message as a function of the probability that the message will consist of a specific set of symbols.

## 9.3b Channel Coding

The previous section identified the need for removing redundancies from the message signal to increase efficiency in transmission [5]. From an efficiency point of view, the ideal scenario would be to obtain an average word length that is numerically equal to the entropy of the source. From a practical perspective, however, this would make it impossible to detect or correct errors that may occur during transmission. Some redundancy must be added to the signal in a controlled manner to facilitate detection and correction of transmission errors. This process is referred to as channel coding.

A variety of techniques exist for detection and correction of errors. For the purposes of this chapter, however, it is sufficient to understand that error-correction coding is important to reliable digital transmission and that it adds to the total bit rate of a given information stream. For closed systems, where retransmission of garbled data is possible, a minimum of error-correction overhead is practical. The error-checking *parity* system is a familiar technique. However, for transmission channels where 2-way communication is not possible, or the channel restrictions do not permit retransmission of specific packets of data, robust error correction is a requirement. More information on the basic principles of error correction can be found in [5].

## 9.3c Error-Correction Coding

Digital modulation schemes in their basic form have dependency between signaling elements over only one signaling division [1]. There are advantages, however, to providing memory over several signaling elements from the standpoint of error correction. Historically, this has been accomplished by adding redundant symbols for error correction to the encoded data, and then using the encoded symbol stream to modulate the carrier. The ratio of information symbols to total encoded symbols is referred to as the *code rate*. At the receiver, demodulation is accomplished followed by decoding.

The drawback to this approach is that redundant symbols are added, requiring a larger transmission bandwidth, assuming the same data throughput. However, the resulting signal is more immune to channel-induced errors resulting from, among other things, a marginal S/N for the channel. The end result for the system is a *coding gain*, defined as the ratio of the signal-to-noise ratios without and with coding.

There are two widely used coding methods:

- *Block coding*, a scheme that encodes the information symbols block-by-block by adding a fixed number of error-correction symbols to a fixed block length of information symbols.
- *Convolutional coding*, a scheme that encodes a sliding window of information symbols by means of a shift register and two or more modulo-2 adders for the bits in the shift register that are sampled to produce the encoded output.

Although an examination of these coding methods is beyond the scope of this chapter, note that coding used in conjunction with modulation always expands the required transmission bandwidth by the inverse of the code rate, assuming the overall bit rate is held constant. In other words, the power efficiency goes up, but the bandwidth efficiency goes down with the use of a well-designed code. Certain techniques have been developed to overcome this limitation, including *trellis-coded modulation* (TCM), which is designed to simultaneously conserve power and bandwidth [6].

## 9.4 Reference


1. Ziemer, Rodger E.: "Digital Modulation," *The Electronics Handbook*, Jerry C. Whitaker (ed.), CRC Press, Boca Raton, Fla., pp. 1213–1236, 1996.
2. Ziemer, R., and W. Tranter: *Principles of Communications: Systems, Modulation, and Noise*, 4th ed., Wiley, New York, 1995.
3. Peterson, R., R. Ziemer, and D. Borth: *Introduction to Spread Spectrum Communications*, Prentice-Hall, Englewood Cliffs, N. J., 1995.
4. Sklar, B.: *Digital Communications: Fundamentals and Applications*, Prentice-Hall, Englewood Cliffs, N. J., 1988.
5. Alkin, Oktay: "Digital Coding Schemes," *The Electronics Handbook*, Jerry C. Whitaker (ed.), CRC Press, Boca Raton, Fla., pp. 1252–1258, 1996.
6. Ungerboeck, G.: "Trellis-Coded Modulation with Redundant Signal Sets," parts I and II, *IEEE Comm. Mag.*, vol. 25 (Feb.), pp. 5-11 and 12-21, 1987.

Chapter

# 10

# DSP Devices and Systems

**Ken Polhmann**

## 10.1 Introduction

To efficiently process digital signals, considerable computational power is required. The impressive advancements in the performance of microprocessors intended for personal computer applications have enabled a host of new devices intended for communications systems. For receivers, the most important of these is the *digital signal processor* (DSP), which is a class of processor intended for a specific application or range of applications. The DSP is, in essence, a microprocessor that sacrifices flexibility (or *instruction set*) for speed. There are a number of tradeoffs in DSP design, however, with each new generation of devices, those constraints are minimized while performance is improved.

## 10.2 Fundamentals of Digital Signal Processing[1]

Digital signal processing is used to generate, analyze, or otherwise manipulate signals in the digital domain [1]. Digital processing of acquired waveforms offers several advantages over processing of continuous-time signals. Fundamentally, the use of unambiguous discrete samples promotes:

- Use of components with lower tolerances.
- Predetermined system accuracy.
- Identically reproducible circuits.
- Theoretically unlimited number of successive operations on a sample.
- Reduced sensitivity to external effects such as noise, temperature, and aging.

1. Portions of this chapter were adapted from: Pohlmann, Ken: *Principles of Digital Audio*, McGraw-Hill, New York, N.Y., 2000. Used with permission.

The programmable nature of discrete-time signals permits changes in function without changes in hardware. Some operations implemented with digital processing are difficult or impossible using analog means. On the other hand, DSP also has certain disadvantages, including:

- The technology always requires power; there is no passive form of DSP circuitry.
- Digital representation of a signal requires a larger bandwidth than the corresponding analog signal.
- DSP technology can be expensive to develop.
- Circuits capable of performing fast computation are required.
- When used for analog applications, A/D and D/A conversion are required.

DSP presents rich possibilities for professional video and audio applications. Error correction, multiplexing, sample rate conversion, speech and music synthesis, data reduction and data compression, filtering, adaptive equalization, dynamic compression and expansion, reverberation, ambience processing, time alignment, mixing and editing, encryption and watermarking, and acoustical analysis can all be performed with digital signal processing.

## 10.2a Discrete Systems

A discrete system is any system that accepts one or more discrete input signals $x(n)$ and produces one or more discrete output signals $y(n)$ in accordance with a set of operating rules [1]. The input and output discrete time signals are represented by a sequence of numbers. If an analog signal $x(t)$ is sampled every $T$ seconds, the discrete time signal is $x(nT)$, where $n$ is an integer. Time can be *normalized* so that the signal is written as $x(n)$.

Linearity and time–invariance are two important criteria for discrete systems. A *linear system* exhibits the property of *superposition:* the response of a linear system to a sum of signals is the sum of the responses to each individual input. That is, the input $x_1(n) + x_2(n)$ yields the output $y_1(n) + y_2(n)$. A linear system exhibits the property of *homogeneity*: the amplitude of the output of a linear system is proportional to that of the input. That is, an input $ax(n)$ yields the output $ay(n)$. Combining these properties, a linear discrete system with the input signal $ax_1(n) + bx_2(n)$ produces an output signal $ay_1(n) + by_2(n)$ where $a$ and $b$ are constants. The input signals are treated independently, output amplitude is proportional to that of the input, and no new signal components are introduced. As described in the following paragraphs, all $z$-transforms and Fourier transforms are linear.

A discrete time system is time-invariant if the input signal $x(n - k)$ produces an output signal $y(n - k)$ where $k$ is an integer. In other words, a *linear time-invariant discrete* (LTD) system behaves the same way at all times. For example, an input delayed by $k$ samples generates an output delayed by $k$ samples.

A discrete system is *causal* if at any instant the output signal corresponding to any input signal is independent of the values of the input signal after that instant. In other words, there are no output values before there has been an input signal. The output does not depend on future inputs.

## 10.2b Impulse Response and Convolution

The impulse response $h(t)$ gives a full description of a linear time-invariant discrete system in the time domain [1]. A LTD system, like any discrete system, converts an input signal into an output

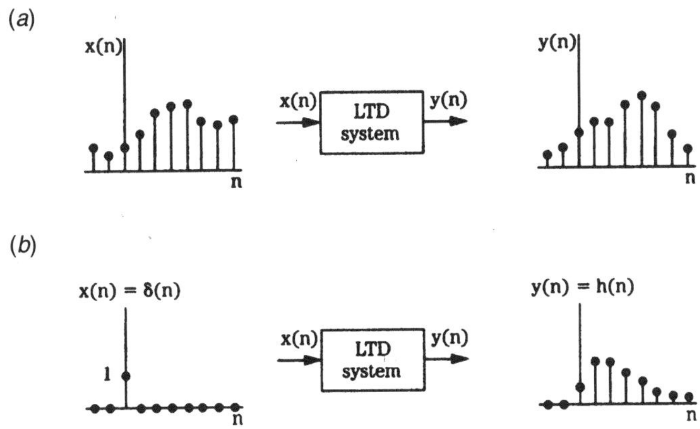


**Figure 10.1** Two properties of linear-time-invariant discrete (LTD) systems: (*a*) LDT systems produce an output signal based on the input, (*b*) an LTD system can be characterized by its impulse response. (*van den Enden and Verhoeckx. From* [1]. *Used with permission.*)

signal, as shown in Figure 10.1*a*. However, an LTD system has a special property such that when an impulse (a *delta function*) is applied to an LTD system, the output is the system's impulse response, as shown in Figure 10.1*b*. The impulse response describes the system in the time domain, and can be used to reveal the frequency response of the system in the frequency domain. Practically speaking, most digital filters are LTD systems, and yield this property. A system is stable if any input signal of finite amplitude produces an output signal of finite amplitude. In other words, the sum of the absolute value of every input and the impulse response must yield a finite number. Useful discrete systems are stable.

Furthermore, the impulse response can be sampled and used to filter a signal. Audio and video samples themselves are impulses, represented as numbers. The signal could be filtered, for example, by using the samples as *scaling values*: all of the values of a filter's impulse response are multiplied by each signal value. This yields a series of filter impulse responses scaled to each signal sample. To obtain the result, each scaled filter impulse response is substituted for its multiplying signal sample. The filter response can extend over many samples; thus, several scaled values might overlap. When these are added together, the series of sums forms the new filtered signal values.

This is the process of *convolution*. The output of a linear system is the convolution of the input and the system's impulse response. Convolution is a time-domain process that is equivalent to the multiplication of the frequency responses of two networks. Convolution in one domain (such as the time domain) is equivalent to multiplication in the conjugate domain (such as frequency). Furthermore, the duality exists such that multiplication in the time domain is equivalent to convolution in the frequency domain.

Because convolution is not an intuitive phenomenon, a graphical illustration of its nature is useful. Consider the waveform in Figure 10.2*a*. It can be divided into discrete pieces such that $x(t) = x_1(t) + x_2(t) + x_3(t) + ..$

In other words

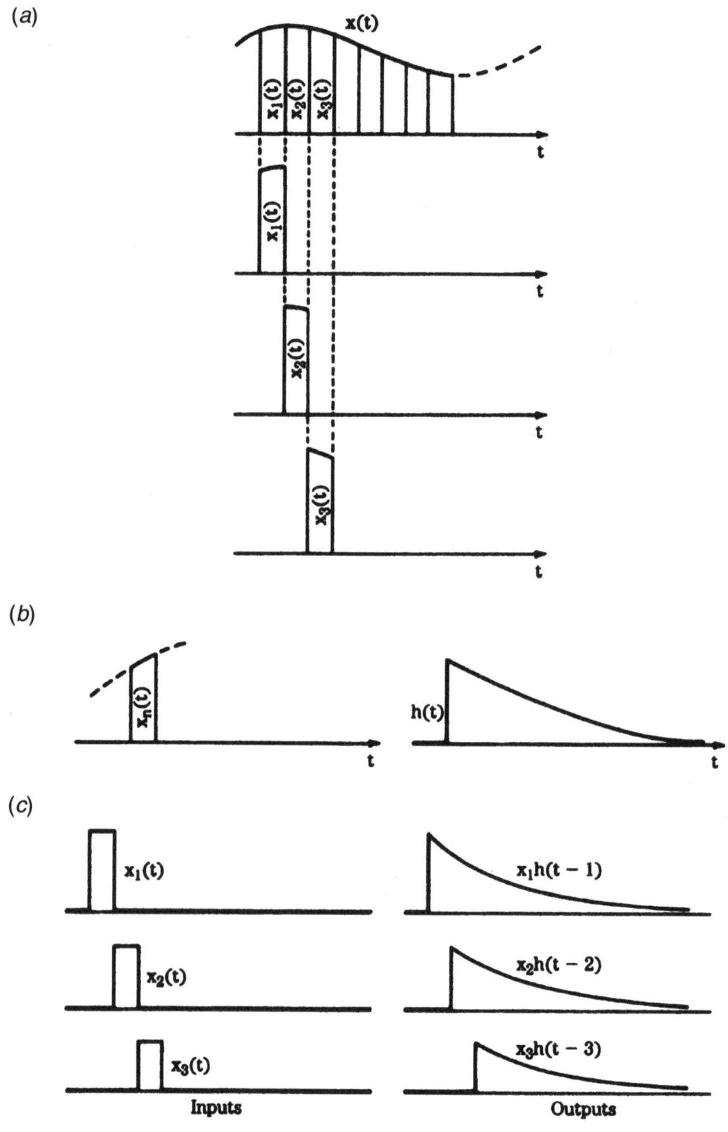


**Figure 10.2** A graphical representation of convolution: (*a*) the samples comprising a discrete signal may be considered singly, (*b*) when applied to a discrete processing system such as a digital filter, each sample produces an output response, (*c*) the overall response is the summation of the individual responses. (*Blesser. From* [1]. *Used with permission.*)

$$x(t) = \sum_{k=-x}^{x} x_k(t) \tag{10.1}$$

where $k = 1, 2, 3, \ldots$

Consider a network that produces an output $h(t)$ when a single piece of the waveform is input, as shown in Figure 10.2*b*. The output $h(t)$ defines the network; from this single response we can find the network's response to any input. Similarly, the inputs that follow produce outputs that are scaled and delayed by the delay of the input, as shown in Figure 10.3*c*. The sum of the individual responses is the full response to the input waveform:

$$y(n) = \sum_{k=1}^{x} h(k)x(n-k) \tag{10.2}$$

Equivalently,

$$y(n) = \sum_{k=1}^{x} x(k)h(n-k) \tag{10.3}$$

This is *convolution*, mathematically expressed as:

$$y(n) = h(n)*x(n) \quad \text{where * denotes convolution} \tag{10.4}$$

To view convolution in action, consider a series of snapshots of the terms present at five consecutive sample times

$$\begin{bmatrix} t = 0\text{T} & t = 1\text{T} & t = 2\text{T} & t = 3\text{T} & t = 4\text{T} \\ x_0h_0 & x_0h_1 & x_0h_2 & x_0h_3 & x_0h_4 \\ & x_1h_0 & x_1h_1 & x_1h_2 & x_1h_3 \\ & & x_2h_0 & x_2h_1 & x_2h_2 \\ & & & x_3h_0 & x_3h_1 \\ & & & & x_4h_0 \end{bmatrix} \tag{10.5}$$

The response is the sum of the terms in each column:

$$\begin{bmatrix} y_o = x_0h_0 \\ y_1 = x_0h_1 + x_1h_0 \\ y_2 = x_0h_2 + x_1h_1 + x_2h_0 \\ y_3 = x_0h_3 + x_2h_2 + x_2h_1 + x_3h_0 \\ y_4 = x_0h_4 + x_1h_3 + x_2h_2 + x_3h_1 + x_4h_0 \end{bmatrix} \quad (10.6)$$

The convolved response is found by reversing the impulse response, and aligning $h_0$ with the current $x_k$ sample to generate the ordered weighted product. The rest of the sequence is obtained by moving the reversed impulse response until it has passed through the duration of the samples of interest, be it finite or infinite in length.

More generally, when two waveforms are multiplied together, their spectra are convolved, and if two spectra are multiplied, their determining waveforms are multiplied. The response to any input waveform can be determined from the impulse response of the network, and its response to any part of the input waveform. As noted, the convolution of two signals in the time domain corresponds to multiplication of their Fourier transforms in the frequency domain (as well as the dual correspondence). It is apparent, then, that any signal can be considered to be a sum of impulses.

## 10.2c Complex Numbers

Analog and digital networks share a common mathematical basis [1]. Fundamentally, whether the discussion is one of resistors, capacitors, and inductors, or scaling, delay, and addition (all linear, time-invariant elements), processors can be understood through complex numbers. A complex number $z$ is any number that can be written in the form $z = x + jy$ where $x$ and $y$ are real numbers, and where $x$ is the real part, and $jy$ is the imaginary part of the complex number. An imaginary number is any real number multiplied by $j$, where $j$ is the square root of –1. The form $x + jy$ is the *rectangular form* of a complex number, and represents the two-dimensional aspects of numbers. For example, the real part can denote distance, and the imaginary part can denote direction. A vector can thus be constructed, showing the indicated location.

A waveform can also be described by a complex number. This is often expressed in polar form, with two parameters: $r$ and $\theta$. The form $re^{j\theta}$ also can be used. If a dot is placed on a circle and rotated, perhaps representing a waveform changing over time, the dot's location can be expressed by a complex number. A location of 450 would be expressed as 0.707 + 0.707$j$. A location of 900 would be 0 + 1$j$, 135° would be –0.707 + 0.707$j$, and 180° would be –1 + 0$j$. The size of the circle could be used to indicate the magnitude of the number.

The $j$ operator can also be used to convert between imaginary and real numbers. A real number multiplied by an imaginary number becomes complex, and an imaginary number multiplied by an imaginary number becomes real. Multiplication by a complex number is analogous to phase shifting. For example, multiplication by $j$ represents a 90° phase shift, and multiplication by 0.707 + 0.707$j$ represents a 45° phase shift. In the digital domain, phase shift is performed by time delay. A digital network comprised of delays can be analyzed by changing each delay to a phase shift. For example, a delay of 10° corresponds to the complex number 0.984 – 0.174$j$. If the input signal is multiplied by this complex number, the output result would be a signal of the same magnitude, but delayed by 10°.

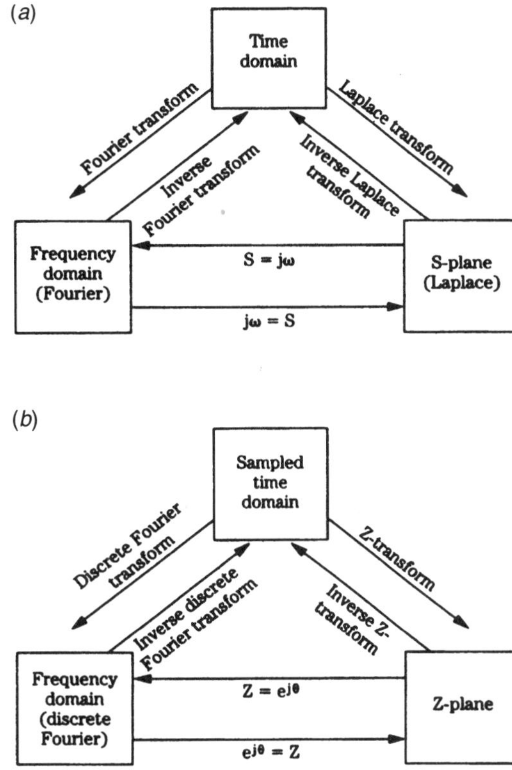


**Figure 10.3** Transforms are used to mathematically convert a signal from one domain to another: (*a*) analog signals can be expressed in the time, frequency, and *s*-plane domains, (*b*) discrete signals can be expressed in the sampled-time, frequency, and *z*-plane domains. (*From* [1]. *Used with permission.*)

## 10.2d Mathematical Transforms

Signal processing, either analog or digital, can be considered in either of two domains [1]. Together, they offer two perspectives on a unified theory. For analog signals, the domains are time and frequency. For sampled signals, they are discrete time and discrete frequency. A transform is a mathematical tool used to move between the time and frequency domains. Continuous transforms are used with signals continuous in time and frequency; series transforms are applied to continuous time, discrete frequency signals; and discrete transforms are applied to discrete time and frequency signals.

The analog relationships between a continuous signal, its Fourier transform, and Laplace transform are shown in Figure 10.3*a*. The discrete-time relationships between a discrete signal, its discrete Fourier transform, and *z*-transform are shown in Figure 10.3*b*.

The Laplace transform is used to analyze continuous time and frequency signals; it maps a time domain function $x(t)$ into a frequency domain, complex frequency function $X(s)$. The Laplace transform takes the form

$$X(s) = \int_{-x}^{x} x(t)e^{-st}dt \qquad (10.7)$$

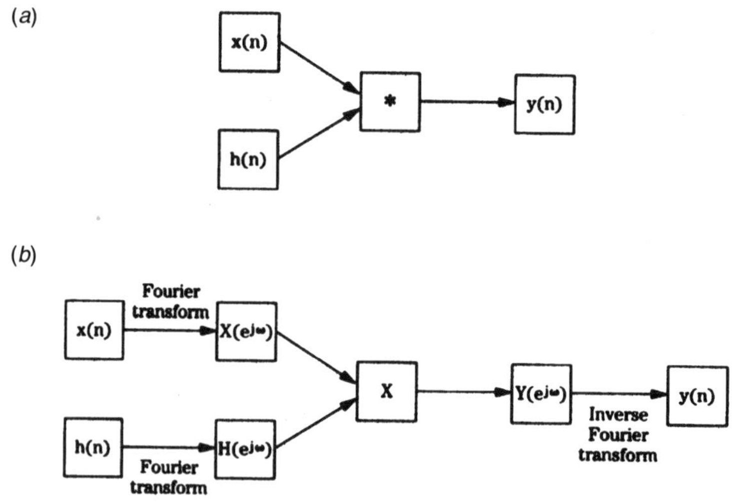


**Figure 10.4** Given an input signal $x(n)$ and impulse response $h(n)$ the output signal $y(n)$ can be calculated through: (*a*) direct convolution, or (*b*) Fourier transformation, multiplication, and inverse Fourier transformation. In practice, the latter method is often an easier calculation. (*From* [1]. *Used with permission.*)

The inverse Laplace transform performs the reverse mapping. Laplace transforms are useful for analog design.

The Fourier transform is a special kind of Laplace transform; it maps a time domain function $x(t)$ into a frequency domain function $X(jw)$, where $X(jw)$ describes the spectrum (frequency response) of the signal $x(t)$. The Fourier transform takes the form

$$X(j\omega) = \int_{-x}^{x} x(t)e^{-j\omega t}dt \tag{10.8}$$

This equation (and the inverse Fourier transform), are identical to the Laplace transforms when $s = jw$; the Laplace transform equals the Fourier transform when the real part of $s$ is zero. The Fourier series is a special case of the Fourier transform and results when a signal contains only discrete frequencies, and the signal is periodic in the time domain.

Figure 10.4 shows how transforms are used. Specifically, two methods can be used to compute an output signal: convolution in the time domain, and multiplication in the frequency domain. Although convolution is conceptually concise, in practice, the second method using transforms and multiplication in the frequency domain is usually preferable. Transforms also are invaluable in analyzing a signal, to determine its spectral characteristics. In either case, the effect of filtering a discrete signal can be predictably known.

The Fourier transform for discrete signals generates a continuous spectrum but is difficult to compute. Thus, a sampled spectrum for discrete time signals of finite duration is implemented as the discrete Fourier transform (DFT). Just as the Fourier transform generates the spectrum of a continuous signal, the DFT generates the spectrum of a discrete signal, expressed as a set of harmonically related sinusoids with unique amplitude and phase. The DFT takes samples of a waveform and operates on them as if they were an infinitely long waveform comprised of sinusoids, harmonically related to a fundamental frequency corresponding to the original sample period. An inverse DFT can recover the original sampled signal.

The DFT is the Fourier transform of a sampled signal. When a finite number of samples ($N$) are considered, the $N$-point DFT transform is expressed as

$$X(m) = \sum_{n=0}^{N-1} x(n)e^{(-j(2\pi/N))mn} \tag{10.9}$$

The $X(m)$ term is often called bin $m$ and describes the amplitude of the frequencies in signal $x(n)$, computed at $N$ equally spaced frequencies. The $m = 0$, or bin 0 term describes the dc content of the signal, and all other frequencies are all harmonically related to the fundamental frequency corresponding to $m = 1$, or bin 1. Bin numbers thus specify the harmonics that comprise the signal, and the amplitude in each bin describes the power spectrum (square of the amplitude). The DFT thus describes all the frequencies contained in signal $x(n)$. There are identical positive and negative frequencies; usually only the positive half is shown, and multiplied by 2 to obtain the actual amplitudes.

An example of DFT operations is shown in Figure 10.5. The input signal to be analyzed is a simple periodic function $x(n) = \cos(2\pi n/6)$. The function is periodic over six samples because $x(n) = x(n + 6)$. Three $N$-point DFTs are used, with $N = 6$, 12, and 16. In the first two cases, $N$ is equal to 6 or is an integer multiple of 6; a larger $N$ yields greater spectral resolution. In the third case, $N = 16$, the discrete spectrum positions cannot exactly represent the input signal; spectral leakage occurs in all bins. In all cases, the spectrum is symmetrical.

The DFT is computation-intensive, requiring $N^2$ complex multiplications and $N(N - 1)$ complex additions. The DFT is often generated with the fast Fourier transform (FFT), a collection of fast and efficient algorithms for spectral computation that takes advantage of computational symmetries and redundancies in the DFT; it requires $N\log_2 N$ computations, 100 times fewer than DFT. The FFT can only be used when $N$ is an integral power of 2; zero samples can be padded to satisfy this requirement. The FFT is not another type of transformation, but rather an efficient method of calculating the DFT. In general, a number of short length DFTs are calculated, then the results are combined. The FFT can be applied to various calculation methods and strategies, including analysis of signals and filter design.

The FFT will transform a time series, such as the impulse response of a network, into the real and imaginary parts of the impulse response in the frequency domain. In this way, the magnitude and phase of the network's transfer function can be obtained. An inverse FFT can produce a time domain signal. FFT filtering is accomplished through multiplication of spectra. The impulse response of the filter is transformed to the frequency domain. Real and imaginary arrays, obtained by FFT transformation of overlapping segments of the signal, are multiplied by filter arrays, and an inverse FFT produces a filtered signal. Because the FFT can be efficiently computed, it can be used as an alternative to time domain convolution if the overall number of multiplications is fewer.

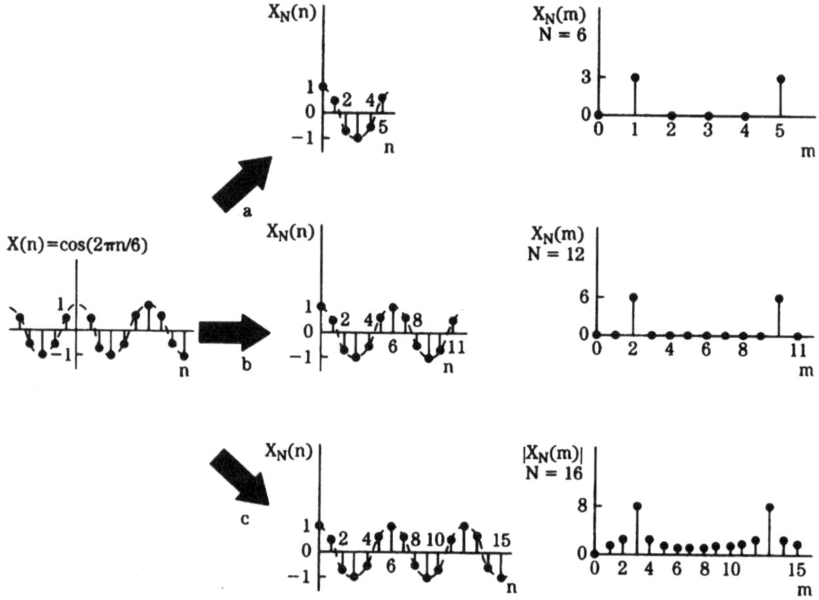


**Figure 10.5** Examples of a periodic signal applied to an *N*-point DFT for three different values of *N*. Greater resolution is obtained as *N* is increased. When *N* is not equal to an integral number of waveform periods, spectral leakage occurs. (*van den Enden and Verhoeckx. From* [1]. *Used with permission.*)

The *z*-transform operates on discrete signals in the same way that the Laplace transform operates on continuous signals. In the same way that the Laplace transform is a generalization of the Fourier transform, the *z*-transform is a generalization of the DFT. Whereas the Fourier transform operates on a particular complex value $e^{-jw}$ the *z*-transform operates with any complex value. When $z = e^{jw}$, the *z*-transform is identical to the Fourier transform. The DFT is thus a special case of the *z*-transform. The *z*-transform of a sequence $x(n)$ is defined as

$$X(z) = \sum_{n=-x}^{x} x(n)z^{-n} \tag{10.10}$$

where $z$ is a complex variable and $z^{-1}$ represents a unit delay element. The *z*-transform has an inverse transform, often obtained through partial fraction expansion.

Whereas the DFT is used for literal operations, the *z*-transform is a mathematical tool used in digital signal processing theory. Convolution in the time domain is equivalent to multiplication in the *z*-domain. For example, we could take the *z*-transform of the convolution equation, such that the *z*-transform of an input multiplied by the *z*-transform of a filter's impulse response is equal to the *z*-transform of the filter output. In other words, the ratio of the filter output transform to the filter input transform—that is, the *transfer function* $H(z)$—is the *z*-transform of the impulse response. Furthermore, this ratio, the transfer function $H(z)$, is a fixed function determined by the filter. In the $z$-domain, given an impulse input, the transfer function equals the output.

## 10.2e Unit Circle and Region of Convergence

The Fourier transform of a discrete signal corresponds to the $z$-transform on the unit circle in the $z$-plane. The equation $z = e^{jw}$ defines the *unit circle* in the complex plane. The evaluation of the $z$-transform along the unit circle yields the frequency response of the function.

The variable $z$ is complex, and $X(z)$ is the function of the complex variable. The set of $z$ in the complex plane for which the magnitude of $X(z)$ is finite is said to be in the *region of convergence*. The set of $z$ in the complex plane for which the magnitude of $X(z)$ is infinite is said to diverge, and is outside the region of convergence. The function $X(z)$ is defined over the entire $z$-plane but is only valid in the region of convergence. The complex variable $s$ is used to describe complex frequency; this is a function of the Laplace transform. The $s$ variable lies on the complex $s$-plane. The $s$-plane can be mapped to the $z$-plane; vertical lines on the $s$-plane map as circles in the $z$-plane.

Because there is a finite number of samples, practical systems must be designed within the region of convergence. The unit circle is the smallest region in the $z$-plane that falls within the region of convergence for all finite stable sequences. Poles must be placed inside the unit circle on the $z$-plane for proper stability. Improper placement of the poles constitutes an instability.

Mapping from the $s$-plane to the $z$-plane is an important process. Theoretically, this function allows the designer to choose an analog transfer function and find the $z$-transform of that function. Unfortunately, the $s$-plane generally does not map into the unit circle of the $z$-plane. Stable analog filters, for example, do not always map into stable digital filters. This is avoided by multiplying by a transform constant, used to match analog and digital frequency response. There also is a nonlinear relationship between analog and digital break frequencies that must be accounted for. The nonlinearities are known as *warping effects* and the use of the constant is known as *prewarping* the transfer function.

Often, a digital implementation can be derived from an existing analog representation. For example, a stable analog filter can be described by the system function $H(s)$. Its frequency response is found by evaluating $H(s)$ at points on the imaginary axis of the $s$-plane. In the function $H(s)$, $s$ can be replaced by a rational function of $z$, which will map the imaginary axis of the $s$-plane onto the unit circle of the $z$-plane. The resulting system function $H(z)$ is evaluated along the unit circle and will take on the same values of $H(s)$ evaluated along its imaginary axis.

## 10.2f Poles and Zeros

Summarizing, the transfer function $H(z)$ of a linear, time-invariant discrete-time filter is defined to be the $z$-transform of the impulse response $h(n)$. The spectrum of a function is equal to the $z$-transform evaluated on the unit circle. The transfer function of a digital filter can be written in terms of its $z$ transform; this permits analysis in terms of the filter's poles and zeros. The zeros are the roots of the numerator's polynomial of the transfer function of the filter, and the poles are the denominator's roots. Mathematically, zeros make $H(z) = 0$, and poles make $H(z)$ nonanalytic. When the magnitude of $H(z)$ is plotted as a function of $z$, poles appear at a distance above the $z$-plane and zeros touch the $z$-plane. One might imagine the flat $z$-plane and above it a flexible contour—the *magnitude transfer function*—passing through the poles and zeros, with peaks on top

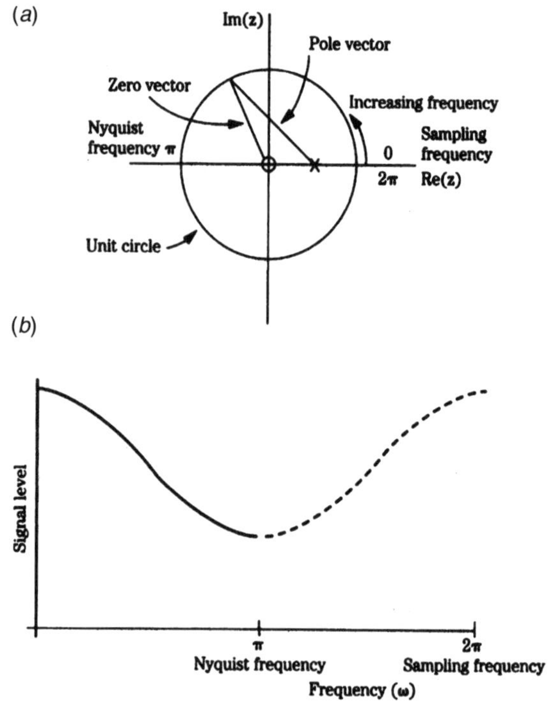


**Figure 10.6** The frequency response of a filter can be obtained by dividing the magnitude of the zero vector by that of the pole vector: (*a*) an example of a *z*-plane plot of a lowpass filter showing the pole and zero locations, (*b*) examination of the plot reveals the filter frequency response. (*From* [1]. *Used with permission.*)

of poles, and valleys centered on zeros. Tracing the rising and falling of the contour around the unit circle yields the frequency response. For example, the gain of a filter at any frequency can be measured by the magnitude of the contour. The phase shift at any frequency is the angle of the complex number that represents the system response at that frequency.

If we plot $|z| = 1$ on the complex plane, we get the unit circle; $|z| > 1$ specifies all points on the complex plane that lie outside the unit circle; and $|z| < 1$ specifies all points inside it. The *z*-transform of a sequence can be represented by plotting the locations of the poles and zeros on the complex plane.

Figure 10.6*a* shows an example of a *z*-plane plot. Among other approaches, the response can be analyzed by examining the relationships between the pole and zero vectors. In the *z*-plane, angular frequency is represented as an angle, with a rotation of 360° corresponding to the sampling frequency. The Nyquist frequency is thus located at π in the figure. The example shows a single pole (*X*) and zero (0). The amplitude of the frequency response can be determined by dividing the magnitude of the zero vector by that of the pole vector. The frequency response from 0 to the Nyquist frequency is seen to be that of a lowpass filter, as shown in Figure 10.6*b*. Similarly, the phase response can be determined by subtracting the argument of the pole vector from that of the zero vector. As the positions of the pole and zero are varied, the response of the filter changes. For example, if the pole is moved along the negative real axis, the filter's response changes to that of a highpass filter.

Zeros are created by summing input samples, and poles are created by feedback. The order of a filter is equal to the number of poles or zeros it exhibits, whichever is greater. A filter is stable only if all its poles are inside the unit circle of the z-plane. Zeros can lie anywhere. When all

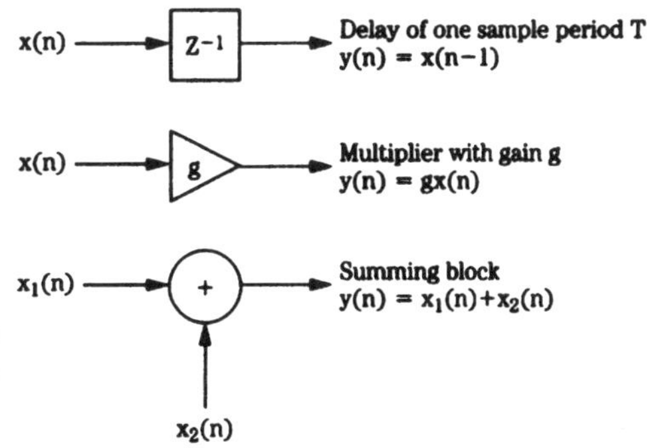


**Figure 10.7** The three basic elements in any DSP system: delay, multiplication, and summation. (*From* [1]. *Used with permission.*)

zeros lie inside the unit circle, the system is called a *minimum-phase network*. If all poles are inside the unit circle and all zeros are outside, and if poles and zeros are always reflections of one another in the unit circle, the system is a *constant-amplitude*, or *all-pass network*. If a system has zeros only, except for the origin, and they are reflected in pairs in the unit circle, the system is phase linear. No real function can have more zeros than poles. When the coefficients are real, poles and zeros occur in complex conjugate pairs; their plot is symmetrical across the real $z$-axis. The closer its location to the unit circle, the greater the effect of each pole and zero on frequency response.

## 10.3 DSP Elements

Successful DSP applications require sophisticated hardware and software [1]. However, all DSP processing can be considered in three basic processing operations (Figure 10.7):

- **Summing**, where multiple digital values are added to produce a single result.
- **Multiplication**, where a gain change is accomplished by multiplying the sample value by a coefficient.
- **Time delay**, where a digital value is stored for one sample period ($n - 1$).

The delay element (realized with shift registers or memory locations) is alternatively notated as $z^{-1}$ because a delay of one sampling period in the time domain corresponds to multiplication by $z^{-1}$ in the $z$-domain; thus $z^{-1}x(n) = x(n - 1)$. Delays can be cascaded, for example, a $z^{-2}$ term describes a two-sample ($n$ - 2) delay. Although it is usually most convenient to operate with sample numbers, the time of a delay can be obtained by taking $nT$, where $T$ is the sampling interval. Figure 10.8 shows two examples of simple networks and their impulse responses. LTD systems such as these are completely described by the impulse response.

In practice, these elemental operations are performed many times for each sample, in specific configurations depending on the desired result. In this way, algorithms can be devised to perform operations useful to signal processing, such as reverberation, equalization, data compression, limiting, and noise removal. Of course, for real-time operation, all processing for each sample must be completed within one sampling period.

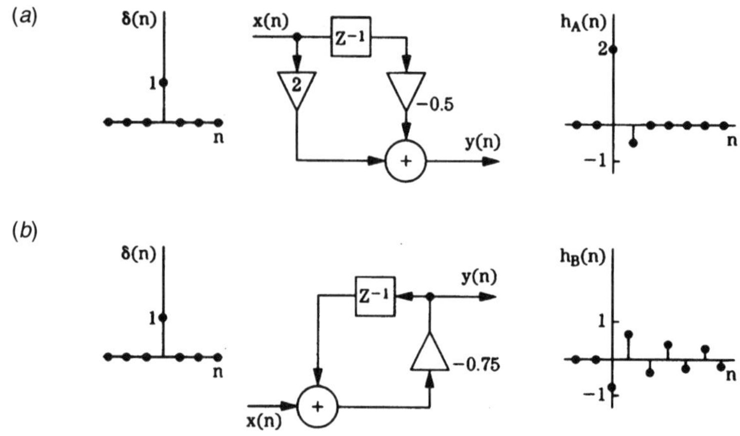


**Figure 10.8** LTD systems can be characterized by their impulse responses: (*a*) simple nonrecursive system and its impulse response, (*b*) simple recursive system and its impulse response. (*van den Enden and Verhoeckx. From* [1]. *Used with permission.*)

## 10.3a Sources of Errors

In general, errors in digital processors can be classified as coefficient errors, limit cycle errors, overflow, truncation, and round-off errors [1]. Coefficient errors occur when a coefficient is not specified with sufficient accuracy. Limit cycle error might occur when a signal is removed from a filter, leaving a decaying sum. This decay might become zero or might oscillate at a constant amplitude, known as *limit cycle oscillation*. This effect can be eliminated, for example, by offsetting the filter output so that truncation always produces a zero output. Overflow occurs when a register length is exceeded, resulting in a computational error. In the case of wrap-around, when a 1 is added to the maximum value positive two's complement number, the result is the maximum value negative number. In short, the information has overflowed into a nonexistent bit. To prevent this, *saturating arithmetic* can be used so that when the addition of two positive numbers would result in a negative number, the maximum positive sum is substituted instead.

Truncation and round-off errors occur whenever the word length of a computed result is limited. Errors accumulate both inside the processor during calculation, and when word length is reduced for output through a D/A converter. However, A/D conversion always results in quantization error, and computation error can appear in different guises. For example, when two $n$-bit numbers are multiplied, the number of output bits will be $2n - 1$. Thus, multiplication almost doubles the number of bits required to represent the output. Although many hardware multipliers can perform double precision computation, a finite word length must be maintained following multiplication, thus limiting precision. Discarded data results in an error that is analogous to A/D quantization. To be properly modeled, multiplication must be followed by quantization. Multiplication does not introduce error, but inability to keep the extra bits does.

Rather than truncate a word—for example, following multiplication—the value can be *rounded*; that is, the word is taken to the nearest available value. This results in a peak error of 1/2 LSB, and a RMS value of $1/(12)^{1/2}$, or 0.288 LSB. This round-off error will accumulate over successive calculations. In general, the number of calculations must be large for significant error.

However, in addition, dither information can be lost during computation. For example, when a properly dithered 16-bit word is input to a 32-bit processor, even though computation is of high precision, the output signal can be truncated to 16 bits for conversion through the output D/A converter. For example, a 16-bit signal that is delayed and scaled by a 12-dB attenuation would result in a 12-bit undithered signal. To overcome this, digital dithering to the resolution of the next processing step should be used in a computation.

## 10.3b DSP Integrated Circuits

A DSP chip is a specialized hardware device that performs digital signal processing under the control of software algorithms [1]. DSP chips are stand-alone processors, often independent of host CPUs (central processing units), and are specially designed for operations used in spectral and numerical applications. For example, large numbers of multiplications are possible, as well as special addressing modes such as bit-reverse and circular addressing. When the memory and input/output circuits are added, the result is an integrated digital signal processor. Such a general-purpose DSP chip is software programmable, and thus can be used for a variety of signal-processing applications. Alternatively, a custom signal processor can be designed to accomplish a specific task.

DSP chips are designed according to two arithmetic types, fixed integer and floating point, which define the format of the data they operate on. A fixed integer chip uses two's complement, binary integer data. Floating point chips use integer and floating point numbers (represented as a mantissa and exponent). The dynamic range of a fixed integer chip is based on its word length; data must be scaled to prevent overflow; this can increase programming complexity. The scientific notation used in a floating point chip allows larger dynamic range, without overflow problems. However, the resolution of floating point representation is limited by the word length of the exponent.

DSP chips often use a pipelining architecture so that several instructions can be paralleled. For example, a fetch (fetch instruction from memory and update program counter), decode (decode instruction and generate operand address), read (read operand from memory), and execute (perform necessary operations) can be effectively executed in one clock cycle with pipelining. A pipeline manager, aided by proficient user programming, helps ensure rapid processing.

DSP chips, like all computers, are comprised of input and output devices, an arithmetic unit, a control unit, and memory, interconnected by buses. All computers originally used a single sequential bus (*von Neumann architecture*), shared by data, memory addresses, and instructions. However, in a DSP chip a particularly large number of operations must be performed quickly for real-time operation. Thus, parallel bus structures are used (such as the *Harvard architecture*) that store data and instructions in separate memories and transfers them via separate buses. For example, a chip can have separate buses for program, data, and DMA, providing parallel program fetches, data reads, as well as DMA operations with slower peripherals.

General purpose DSP chips tend to follow a similar architecture; typical elements include the following:

- Multiply-accumulate unit
- Data address generator
- Data RAM
- Coefficient RAM

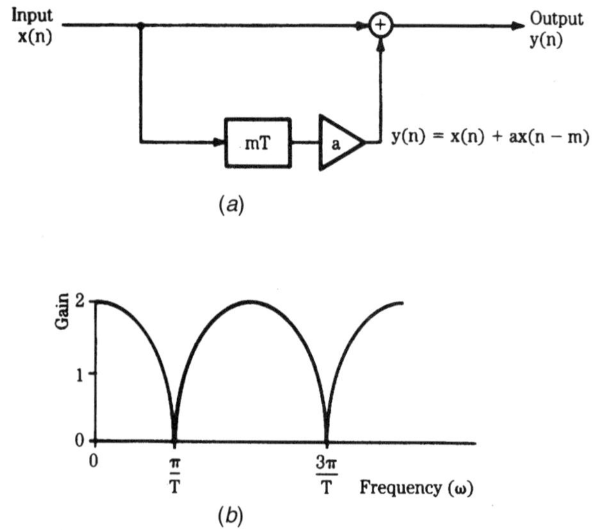


**Figure 10.9** A delay block can be used to create an echo circuit: (*a*) the circuit contains an *mT* delay and gain stage, (*b*) with shorter delay times, a comb filter response will result. (*Bloom, Berkhout, and Eggermont. From* [1]. *Used with permission.*)

- Coefficient address generator
- Program control unit
- Program ROM

These components are connected by the following buses:

- The data bus
- Coefficient bus
- Control bus

## 10.3c DSP Applications

In addition to digital filtering, some of the most powerful and creative applications of digital signal processing come in the form of specialized processing of audio and video signals. Building on the basic operations of multiplication and delay, sophisticated operations can be developed. Reverberation perhaps epitomizes the degree of time manipulation possible in the digital domain; it is possible to synthesize reverberation to both simulate natural acoustical environments and to create acoustical environments that could not physically exist.

### Digital Delay

A delay block is a simple storage unit, such as a memory location [1]. A sample is placed in memory, stored, then recalled some time later and output. A delay unit can be described by the equation $y(n) = x(n - m)$ where $m$ is the delay in samples. Generally, when the delay is small, the frequency response of the signal is altered; when the delay is longer, an echo results. Just as in filtering, a simple delay can be used to create sophisticated effects. For example, Figure 10.9*a*

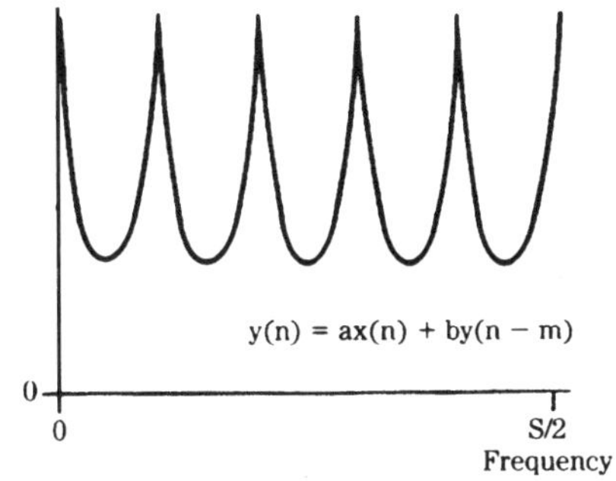


**Figure 10.10** A recursive comb filter creates a delay with feedback, yielding a toothed frequency response. (*From* [1]. *Used with permission.*)

shows an echo circuit using a delay block. Delay $mT$ is of duration $m$ samples, and samples are multiplied by a gain coefficient (a scaling factor) less than unity. If the delay time is set between 10 and 50 ms, an echo results; with shorter fixed delays, a comb filter response results, as shown in Figure 10.9*b*. Peaks and dips are equally spaced through the frequency response from 0 Hz to the Nyquist frequency. The number of peaks depends on the delay time; the longer the delay, the more peaks.

A comb filter can be either recursive or nonrecursive. It cascades a series of delay elements, creating a new response. Mathematically, a nonrecursive comb filter can be designed by adding the input sample to the same sample but delayed

$$y(n) = x(n) + ax(n-m) \tag{10.11}$$

where $m$ is the delay time in samples. A recursive comb filter creates a delay with feedback. The delayed signal is attenuated and fed back into the delay

$$y(n) = ax(n) + by(n-m) \tag{10.12}$$

This yields a response as shown in Figure 10.10. The number of peaks depends on the duration of the delay; the longer the delay, the greater the number of peaks.

An all-pass filter is one that has a flat frequency response from 0 Hz to the Nyquist frequency. However, its phase response causes different frequencies to be delayed by different amounts. An all-pass filter can be described as

$$y(n) = -ax(n) + x(n-1) + by(n-1) \tag{10.13}$$

If the delay in the foregoing circuits is replaced by a digital all-pass filter or a cascade of all-pass filters, a *phasing effect* is achieved

$$y(n) = -ax(n) + x(n-m) + by(n-m) \tag{10.14}$$

The effect becomes more pronounced as the delay increases. The system exhibits nonuniformly spaced notches in its frequency response, varying independently in time.

**Table 10.1 Characteristics of the TMS320C55x Processors** (*After* [2]. *Courtesy of Texas Instruments.*)

| Parameter | VC5510 |
|---|---|
| **Memory** | |
| On-chip SARAM | 32 K words (64 K bytes) |
| On-chip DARAM | 128 K words (256 K bytes) |
| On-chip ROM | 16 K words (32 K bytes) |
| Total addressable memory (internal/external) | 8M words (16 M bytes) |
| On-chip bootloader (in ROM) | Yes |
| **Peripherals** | |
| McBSPs | 3 |
| DMA controller | Yes |
| EHPI (16-bit) | Yes |
| Configurable instruction cache | 24K bytes |
| Timers | 2 |
| Programmable DPLL clock generator | Yes |
| **General Purpose I/O Pins** | |
| Dedicated input/output | Yes |
| XF—dedicated output | 1 |
| Multiplexed with McBSP (input/output) | 21 |
| Multiplexed with timer (output only) | 2 |
| **CPU Cycle Time/Speed** | |
| 160 MHz (6.25 ns) | Yes |
| 200 MHz (5 ns) | Yes |
| Package Type | 240-pin BGA |

## 10.3d Example DSP Device

The functional description of a representative DSP device will help to illustrate the concepts outline previously in this chapter. Table 10.1 lists the overall characteristics for members of the TMS320C55x generation of fixed-point digital signal processors (Texas Instruments). Features for the high performance, low-power C55x CPU include the following [2]:

- Advanced multiple-bus architecture with one internal program memory bus and five internal data buses (three dedicated to *reads* and two dedicated to *writes*)
- Unified program/data memory architecture

- Dual 17-bit × 17-bit multipliers coupled to 40-bit dedicated adders for non-pipelined single-cycle *multiply accumulate* (MAC) operations
- *Compare, select, and store unit* (CSSU) for the add/compare section of the Viterbi operator
- Exponent encoder to compute an exponent value of a 40-bit accumulator value in a single cycle
- Two address generators with eight auxiliary registers and two auxiliary register arithmetic units
- Data buses with bus holders
- 8 M × 16-bit (16 Mbyte) total addressable memory space
- Single-instruction repeat or block repeat operations for program code
- Conditional execution
- Seven-stage pipeline for high instruction throughput
- Instruction buffer unit that loads, parses, queues, and decodes instructions to decouple the program fetch function from the pipeline
- Program flow unit that coordinates program actions among multiple parallel CPU functional units
- Address data flow unit that provides data address generation and includes a 16-bit arithmetic unit capable of performing arithmetic, logical, shift, and saturation operations
- Data computation unit containing the primary computation units of the CPU, including a 40-bit arithmetic logic unit, two multiply-accumulate units, and a shifter

Because many DSP chips are used in portable systems, power consumption is a key operating parameter. As a result, power control features are an important DSP specification. Hardware and software functions designed to conserve power include:

- Software-programmable *idle domains* that provide configurable low-power modes
- Automatic power management
- Advanced low-power CMOS process

### Functional Overview

The C55x architecture achieves power-efficient performance through increased parallelism and a focus on reduction in power dissipation. The CPU supports an internal bus structure composed of the following elements [2]:

- One program bus
- Three data read buses
- Two data write buses
- Additional buses dedicated to peripheral and DMA activity

These buses provide the ability to perform up to three data reads and two data writes in a single cycle. In parallel, the DMA controller can perform up to two data transfers per cycle independent

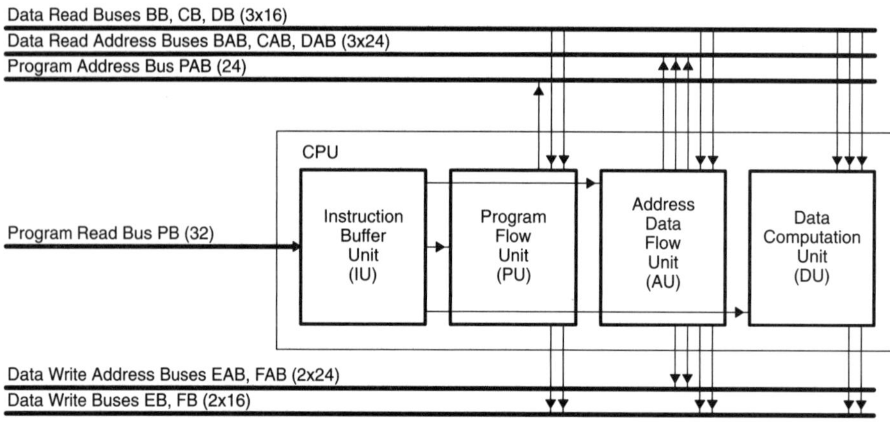


**Figure 10.11** Functional Block Diagram of the TMS320C55x DSP series. (*From* [2]. *Courtesy of Texas Instruments.*)

of CPU activity. The C55x CPU provides two multiply-accumulate units each capable of 17-bit × 17-bit multiplication in a single cycle. A central 40-bit arithmetic/logic unit (ALU) is supported by an additional 16-bit ALU. Use of ALUs is subject to instruction set control. This programmability provides the capacity to optimize parallel activity and power consumption. These resources are managed in the address data flow unit (AU) and data computation unit (DU) of the C55x CPU.

The C55x architecture supports a variable byte width instruction set for improved code density. The instruction buffer unit (IU) performs 32-bit program fetches from internal or external memory and queues instructions for the program unit (PU). The program unit decodes the instructions, directs tasks to AU and DU resources, and manages the fully-protected pipeline.

A configurable instruction cache is also available to minimize external memory accesses, improving data throughput and conserving system power.

The C55x architecture is built around four primary blocks:

- The instruction buffer unit
- Program flow unit
- Address data flow unit
- Data computation unit

These functional units exchange program and data information with each other and with memory through multiple dedicated internal buses. Figure 10.11 shows the principal blocks and bus structure in the C55x devices.

Program fetches are performed using the 24-bit program address bus (PAB) and the 32-bit program read bus (PB). The functional units read data from memory via three 16-bit data read buses named B-bus (BB), C-bus (CB), and D-bus (DB). Each data read bus also has an associ-

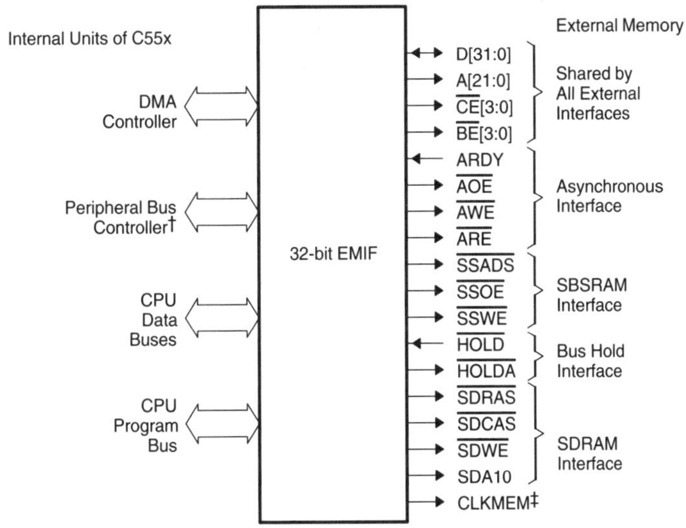


† This connection allows the CPU to access the EMIF registers.
‡ The CLKMEM signal is shared by the SDRAM and SBSRAM interfaces.

**Figure 10.12** Block diagram of EMIF for the TMS320C55x DSP. (*From* [2]. *Courtesy of Texas Instruments.*)

ated 24-bit data read address bus (BAB, CAB, and DAB). Single operand reads are performed on the D-bus. Dual-operand reads use C-bus and D-bus. B-bus provides a third read path and can be used to provide coefficients for dual-multiply operations.

Program and data writes are performed on two 16-bit data write buses called E-bus (EB) and F-bus (FB). The write buses also have associated 24-bit data write address buses (EAB and FAB). Additional buses are present on the C55x devices to provide dedicated service to the DMA controller and the peripheral controller.

All C55x DSP devices use the same CPU structure but are capable of supporting different on-chip peripherals and memory configurations. The on-chip peripherals include:

- Digital phase-locked loop (DPLL) clock generation
- Instruction cache
- External memory interface (EMIF)
- Direct memory access (DMA) controller
- 16-bit enhanced host port interface (EHPI)
- Multichannel serial ports (McBSPs)
- 16-bit timers with 4-bit prescalers
- General-purpose I/O (GPIO) pins
- Trace FIFO (for emulation purposes only)

Peripheral control registers are mapped to an I/O space separate from the main memory space. The peripheral bus controller handles exchange of data between peripherals and the CPU via dedicated peripheral buses.

The EMIF supports a "glueless" interface from the C55x to a variety of external memory devices. For each memory type, the EMIF supports 8-bit, 16-bit, and 32-bit accesses for both reads and writes. For writes, the EMIF controls the byte enable signals and the data bus to perform 8-bit transfers or 16-bit transfers. For reads, the entire 32-bit bus is read. Then, it is internally parsed by the EMIF. The EMIF block diagram, Figure 10.12, shows the interface between external memory and the internal resources of the C55x.

## 10.4 References

1. Pohlmann, Ken: *Principles of Digital Audio*, McGraw-Hill, New York, N.Y., 2000.
2. *TMS320C55x DSP Functional Overview*, Texas Instruments, Dallas, TX, literature No. SRPU312, June 2000.

Chapter

# 11

# Adaptive Receiver Processing

**Ulrich L. Rohde**

**Jerry C. Whitaker, Editor**

## 11.1 Introduction[1]

The term *adaptive processing* is applied to techniques intended to modify the receiver characteristics with changing signal environment so as to achieve improved performance. The use of diversity reception may be considered a simple form of adaptation that operates on several samples of the signal and interference to produce a reduced outage fraction or reduced error rate under certain conditions of fading. HF frequency management, changing the frequency of a radio link in response to the diurnal, seasonal, and sunspot cycles, is another simple, manual form of adaptation that has been used for many years.

In recent years, two additional forms of adaptivity have appeared—*adaptive antenna processing* (sometimes referred to as *adaptive null steering*) and *adaptive equalization*. With the availability of low-cost low-power microprocessors, the implementation of automatic adaptive frequency management has become attractive, as well as other adaptive processes. Adaptive antenna processing arose out of the antijam requirements of radar systems, but is equally applicable to communications use, when several antenna elements are available. In adaptive antenna processing, each of the inputs are modified in amplitude and delay (or phase) prior to their combination to achieve a desired signal improvement. The adaptive control process by which this is accomplished and the techniques for input modification must be designed for the specific application. Adaptive equalization came about because of developments intended to counteract intersymbol interference generated in data transmission over the telephone network by channel distortion. The concepts are now being used to combat multipath and other distortion in radio communications.

---

1. This chapter is based on: Rohde, Ulrich L., and Jerry C. Whitaker: *Communications Receivers: Principles and Design*, 3rd. ed., McGraw-Hill, New York, N.Y., 2000. Used with permission.

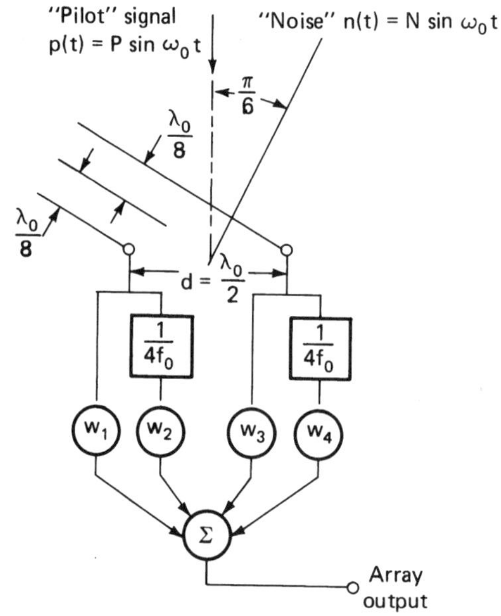


**Figure 11.1** Simple array configuration. (*From* [1]. *Reprinted with permission of IEEE.*)

## 11.2 Adaptive Antenna Processing

Much of the original work on adaptive antenna processing was related to military applications (References [1] to [4]). The techniques described in [2] to [4] are based on the fact that objectionable interference generally does not come from the same direction as the signal, so that separated antenna elements receive the two sets of radiation with different delays. By appropriate amplitude and/or delay variations in the two channels, a single interferer may be nulled, while the signal remains available. Because it is the differences in delays that are primarily responsible for the signal differences, at least when the antenna elements are identical, broadband processing requires that the amplitude and delay be adjusted in each channel. However, for narrowband signals, carrier phase difference may be substituted for delay.

To illustrate how this occurs, a simple two-element example from Widrow et al. [1] will be used before the more complex general expressions are given. Figure 11.1 shows the configuration. The antennas are assumed to be simple omnidirectional elements, with the signal arriving at angle $\theta$ and the interferer at angle $\Phi$ The signal amplitude in each antenna element is the same, but the phase differs because of the arrival angle $\phi$. Analogously, the interference amplitudes are the same, and phases differ because of angle $\theta$. Each received channel is split into $I$ and $Q$ components, which are separately multiplied by different values or *weights*. This results in an effective amplitude and phase change in each channel. If we use the usual complex notation for the narrowband signals, we have for the signal and interferer components at the output

$$S_{out} = W_1 s \exp\left[j\left(d \sin \Phi\right)/2\right] + W_2 s \exp\left[-j\left(d \sin \Phi\right)/2\right]$$

$$N_{out} = W_1 n \exp[\,j\,(d \sin \theta) \,/\, 2] + W_2 n \exp[-j\,(d \sin \theta) \,/\, 2] \tag{11.1}$$

where $s$ and $n$ are the amplitudes of the two waves, and the phases are referred to the phase at the center of the array. We would like to eliminate the interferer output $N_{out}$ while keeping the signal output $S_{out}$ equal to $s_i$, for example. This requires four real equations to be satisfied

$$\begin{aligned}
&w_{1i} \cos\left(\frac{d \sin \theta}{2}\right) - w_{1q} \sin\left(\frac{d \sin \theta}{2}\right) \\
&\quad + w_{2i} \cos\left(\frac{d \sin \theta}{2}\right) + w_{2q} \sin\left(\frac{d \sin \theta}{2}\right) = 0 \\
&w_{1q} \cos\left(\frac{d \sin \theta}{2}\right) + w_{1i} \sin\left(\frac{d \sin \theta}{2}\right) \\
&\quad + w_{2q} \cos\left(\frac{d \sin \theta}{2}\right) - w_{2i} \sin\left(\frac{d \sin \theta}{2}\right) = 0 \\
&w_{1i} \cos\left(\frac{d \sin \Phi}{2}\right) - w_{1q} \sin\left(\frac{d \sin \Phi}{2}\right) \\
&\quad + w_{2i} \cos\left(\frac{d \sin \Phi}{2}\right) + w_{2q} \sin\left(\frac{d \sin \Phi}{2}\right) = 1 \\
&w_{1q} \cos\left(\frac{d \sin \Phi}{2}\right) - w_{1i} \sin\left(\frac{d \sin \Phi}{2}\right) \\
&\quad + w_{2q} \cos\left(\frac{d \sin \Phi}{2}\right) + w_{2i} \sin\left(\frac{d \sin \Phi}{2}\right) = 0
\end{aligned} \tag{11.2}$$

These are four linear equations with constant coefficients, and may be solved in the usual way to produce

$$W_1 \equiv w_{1i} + j\, w_{1q}$$

$$W_1 = -\frac{\sin[(d \sin\theta)/2] + j\cos[(d\sin\theta)/2]}{2\ \sin[d(\sin\Phi - \sin\theta)/2]}$$

(11.3)

$$W_1 = -\frac{j\exp[-j(d\sin\theta)/2]}{2\ \sin[d(\sin\Phi - \sin\theta)/2]}$$

$$W_2 \equiv w_{2i} + j\, w_{2q}$$

$$W_2 = -\frac{\sin[(d \sin\theta)/2] - j\cos[(d\sin\theta)/2]}{2\ \sin[d(\sin\Phi - \sin\theta)/2]}$$

$$W_2 = \frac{j\exp[j(d\sin\theta)/2]}{2\ \sin[d(\sin\Phi - \sin\theta)/2]}$$

(11.4)

The final forms on the right also arise from a direct solution of Equation (11.1) by letting

$$W_1 = W\exp[-j(d\sin\theta)/2] \text{ and } W_2 = -W\exp[j(d\sin\theta)/2]$$

(11.5)

which satisfy Equation (11.1) with $N_{out} = 0$, and then solving for $W$. The problem of automatically controlling the coefficients to find these values remains and will be discussed after the forms of equation and solution have been indicated for the more general network of Figure 11.2.

In Figure 11.2 there are now $n$ array elements providing separate inputs. The bandwidth is assumed narrow, so there are also $n$ complex weights that must be adjusted to produce an optimum setting in some sense. In principle, using the approach given previously, $n$ inputs should allow $n - 1$ interfering signals to be nulled, except at some angles where this may not be possible, while still retaining adequate signal strength. More generally, however, there may be more or fewer interferers, so that some criterion other than nulling may be desirable. To solve the problem, which becomes a solution of $n$-linear equations subject to some constraints, matrix algebra is convenient. The equations analogous to Equation (11.1) are

$$S_{out} = \mathbf{W}_T\, \mathbf{S}_{in} \qquad (11.6)$$

$$N_{out} = \mathbf{W}_T\, \mathbf{N}_{in} \qquad (11.7)$$

Boldface characters are used to represent matrices, in this case one-dimensional matrices (vectors); the subscript $T$ represents the matrix transpose and the superscript $-1$ represents the

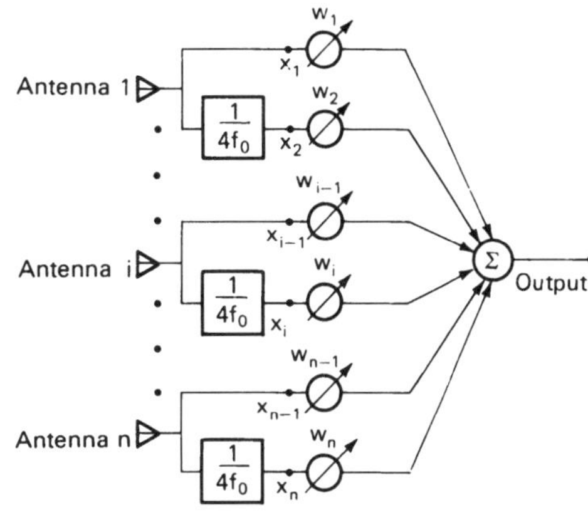


**Figure 11.2** Generalized form of an adaptive array. (*From* [1]. *Reprinted with permission of IEEE.*)

inverse. The inner product of two vectors may be written $A_T\mathbf{B}$ or $B_T\mathbf{A}$, since both forms produce the same scalar value. The vectors **W**, $N_{in}$, and $S_{in}$ are defined as

$$\mathbf{W} = \begin{Vmatrix} W_1 \\ W_2 \\ W_3 \\ \cdot \\ \cdot \\ \cdot \\ W_n \end{Vmatrix} \quad N_{in} = \begin{Vmatrix} \sum N_{1\ in} \\ \sum N_{2\ in} \\ \sum N_{3\ in} \\ \cdot \\ \cdot \\ \cdot \\ \sum N_{n\ in} \end{Vmatrix} \quad S_{in} = \begin{Vmatrix} S_{1\ in} \\ S_{2\ in} \\ S_{3\ in} \\ \cdot \\ \cdot \\ \cdot \\ S_{n\ in} \end{Vmatrix} \tag{11.8}$$

In order to select $W_1$, it is necessary to place some further constraints on the system. In the simple example, it was assumed that $N_{out}$ should be zero. This is not generally possible, especially when the random noise is added to the interferers or if the interferers happen to be more in number than the number of $W_1$. The two principal constraints that have been used in the past are the *minimum* S/N (MSN), also called the *minimum signal-to-interference ratio* (MSIR), and the *least-mean-square* (LMS) error. The former requires that it be possible to identify the signal from the interferers, the latter requires a local reference that has a higher correlation with the signal than with any of the interfering sources. Both criteria lead to similar solutions.

In the MSN case, we have

$$P_m = S_{out} * S_{out} = (\mathrm{W_T}\,\mathbf{S}_{\mathrm{in}})* (\mathrm{S}_{in\ T}\,\mathbf{W}) \tag{11.9}$$

$$P_n = E\{N_{out} * N_{out}\} = E\{(\mathrm{W}_T\,\mathrm{N}_{in})* (\mathrm{N}_{in\ T}\,\mathbf{W})\} \tag{11.10}$$

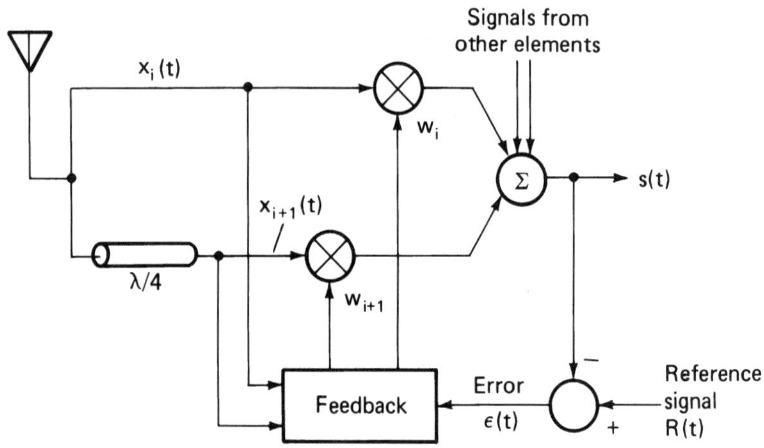


**Figure 11.3** Functional block diagram of a control circuit for an adaptive array. (*From* [5]. *Reprinted with permission of IEEE.*)

$$P_n = W_{T*} E\{\mathrm{N}_{in} * \mathbf{N}_{in\ T}\}\ \mathbf{W} \equiv \mathrm{W}_{T*}\ \mathbf{MW} \qquad (11.11)$$

where $E\{\ ...\}$ is the expectation of a random variable and $*$ indicates the complex conjugate. It has been shown [1] that to maximize the S/N

$$\mathbf{MW} = \mathbf{S} \qquad (11.12)$$

$$\mathbf{M} = E\{\mathrm{N}_{in} * \mathrm{N}_{in\ T}\} \qquad (11.13)$$

where μ is an arbitrary constant. The weights **W** can be solved for by multiplying Equation (11.12) by $\mathrm{M}^{-1}$.

The drawback in this equation is that, in practice, the statistics of the interference are not necessarily known, so that the value of **M** cannot be determined a priori. Moreover, the interferers will change from time to time, so that no fixed values of **W** could be used. It is therefore necessary to estimate the various interferer expectations from previously received signal samples. A compromise must be reached between the number of samples required to approximate the expectations and the time in which the values of the statistics might change. The functional circuit suitable for implementing this MSN algorithm is shown in Figure 11.3.

Identical circuits are used for weighting each element. The circuit may be implemented using either digital or analog circuits, and applying the weights directly at RF or in the IF stages of the receiver. While it might seem that a more complex receiver is required to implement at IF, since *n* frequency converters and IF amplifiers are required, it must be realized that as a practical matter, the levels input to the control unit from both individual and sum channels must have sufficient level to overcome circuit noise and thresholds and provide adequate output power to perform the weighting operation. Alternatively, in a digital implementation they must have suffi-

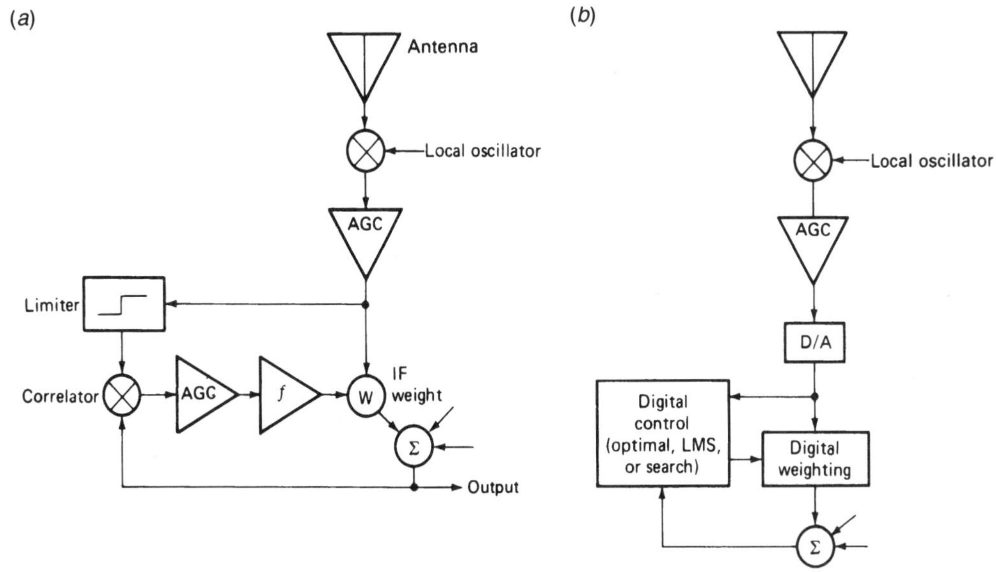


**Figure 11.4** Block diagrams of analog and digital implementations of control loops for an MSN algorithm: (*a*) analog correlation control (analog IF weights), (*b*) all-digital control and weighting. (*From* [6]. *Reprinted with permission of IEEE.*)

cient level to drive the A/D converters. Figure 11.4 gives diagrams indicating in somewhat more detail the analog and digital implementation of the control circuits.

In Figure 11.3, a fixed reference voltage is inserted in the control loop. In the absence of a reference matrix, such as this represents, the MSN algorithm attempts to null signals coming from all directions, including the desired signal. Suitable reference levels can prevent nulling in a preselected direction. This is important if there are only a few interferers and the signal has a comparable energy. Where the signal is comparatively weak, this is less important. This might be true, for example, in spread-spectrum cases where the spread has caused the general noise level and other users to provide a background noise level greater than the signal level (prior to despreading to achieve the receiver's processing gain). Without the reference (or in other directions when there is a reference), the adaptation operates to invert the stronger signals' powers about the background noise level, the strongest first [3]. The speed with which the algorithm converges in the presence of strong interferers is also a problem with which the designer must be concerned.

The LMS algorithm differs from the MSN in that it assumes a capability to distinguish between signal and noise. To achieve this, Widrow et al. [1] have compared the summed signal with a local reference signal, which in its simplest form is a replica of the received signal. Figure 11.5 is a block diagram of this circuit, including diagrams of the control circuits. It will be noted that the form of these circuits is similar to those for the MSN algorithm. The solution for the LMS algorithm is given in [1] as

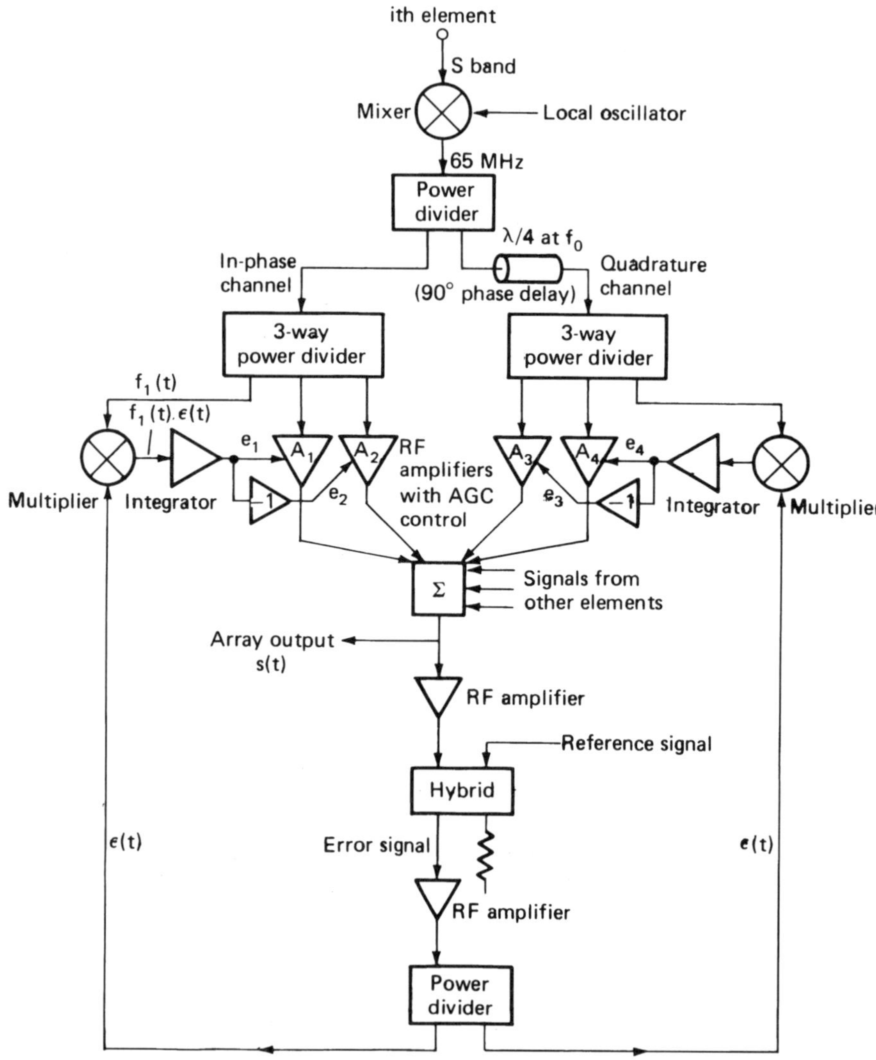


**Figure 11.5** Block diagram of an LMS algorithm. (*From* [6]. *Reprinted with permission of IEEE.*)

$$\boldsymbol{\emptyset}\,\mathbf{W} = \emptyset_{xd} \tag{11.14}$$

where **Ø** resembles **M**, except that it includes the received signal terms as well as the noise. $\emptyset_{xd}$ is the expectancy of the product of the input vector with reference signal *d*. When $d = 0$, the control algorithm reduces to that of the MSN. In [1], this case is shown with an added pilot signal, generated in a direction where it is desired not to reduce the output. This is equivalent to the MSN with an offset vector. Because of uncertainties in timing at a receiver, in many cases it is not possible to produce a good reference signal unless local demodulation is being accomplished. Demodulation may not be possible in the face of interference. Therefore, the reference is not available at the start of processing. In this case, a two-mode system may be used with an MSN while the interference is high, switching to an LMS when the reduction in the stronger interferers allows recovery of the signal. This depends strongly on the transient behavior of the adaptive algorithm.

## 11.3 Adaptive Equalization

Adaptive equalization of radio communications channels, while developed to improve the performance of a channel with a single input, has much in common with adaptive antenna processing. In adaptive equalization, the interferers are symbols sent at earlier or later times. As a result of multipath and other channel distortions, tails of prior symbols or precursors of symbols yet to come cause *intersymbol interference* that reduces or eliminates tolerance to noise that would exist in a channel free of these problems. One of the earliest adaptive techniques used to combat radio multipath was the RAKE system [7]. Subsequently, the use of an inverse ionosphere for combating multipath was suggested [8], and various other techniques were proposed for automatic equalization (References [9] and [10]). Meanwhile the problems of adapting the switched telephone system to transmit higher data rates in the face of unpredictable distortion had also led to the development of adaptive linear transversal filter equalizers [11]. The rapid development of digital processing technology in recent years has led to ever-growing efforts in the area. Several survey treatments (References [12] to [14]) have appeared, providing extensive reference and background material for further study. Here, we attempt to present the general current approaches.

For channels with small distortion from phase or amplitude dispersion, or small amplitude multipath, the *linear transversal equalizer* is useful (Figure 11.6). As with adaptive antenna processing, a reference may be used in the control system. It is common to provide a reference transmission prior to data transmission to allow initial setting of the weighting vector and subsequently update the values using the differences between the output and the expected output of the equalizer (Figure 11.7). This type of equalizer proved useful in reducing error rates in telephone channels and thus allowed higher data rates with acceptable performance. After the line is connected, a training signal, whose characteristic is known at the receiver, is sent. This allows for the initial setting of the weighting coefficients, using an LMS-type algorithm. After training, the signal outputs are compared with the stored expected signal values, and an MSN or other algorithm is used to correct for slow changes. The same sort of algorithm can be useful to correct dispersion in propagation on a single-path channel or a multipath channel of the sort that has a strong main path and a number of weak multipaths.

This type of equalizer is not effective when there is substantial multipath (Figure 11.8). In such a case, a nonlinear decision feedback configuration added to the linear circuit is necessary. The block diagram for such a circuit is illustrated in Figure 11.9. In this case, the already decided output symbols are passed to a transversal filter, the outputs of which are weighted and summed at the input to the decision process along with the outputs from the weighted and the summed outputs from the linear segment from the symbols yet to be decided. This nonlinear process rep-

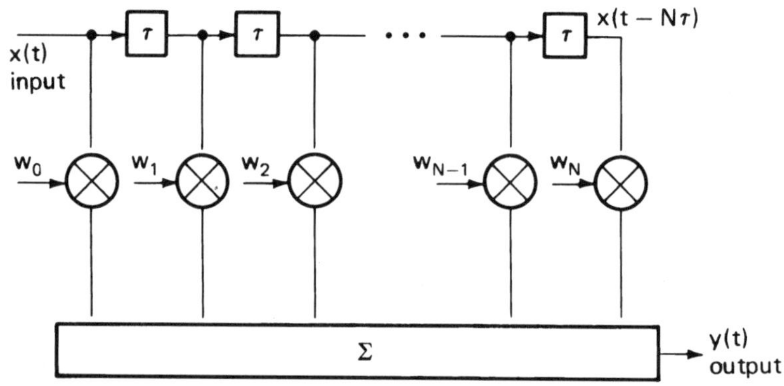


**Figure 11.6** Block diagram of a linear transversal equalizer. (*From* [13]. *Reprinted with permission of IEEE.*)

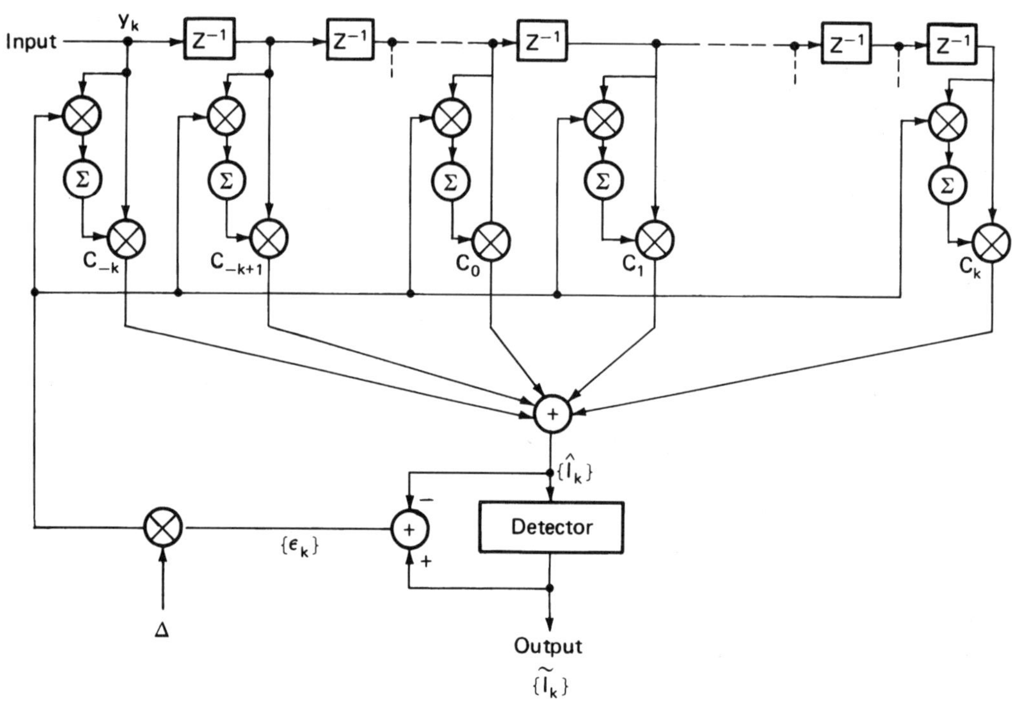


**Figure 11.7** Block diagram of control functions for a linear transversal equalizer. (*From* [12]. *Courtesy of J. G. Proakis and Academic Press.*)

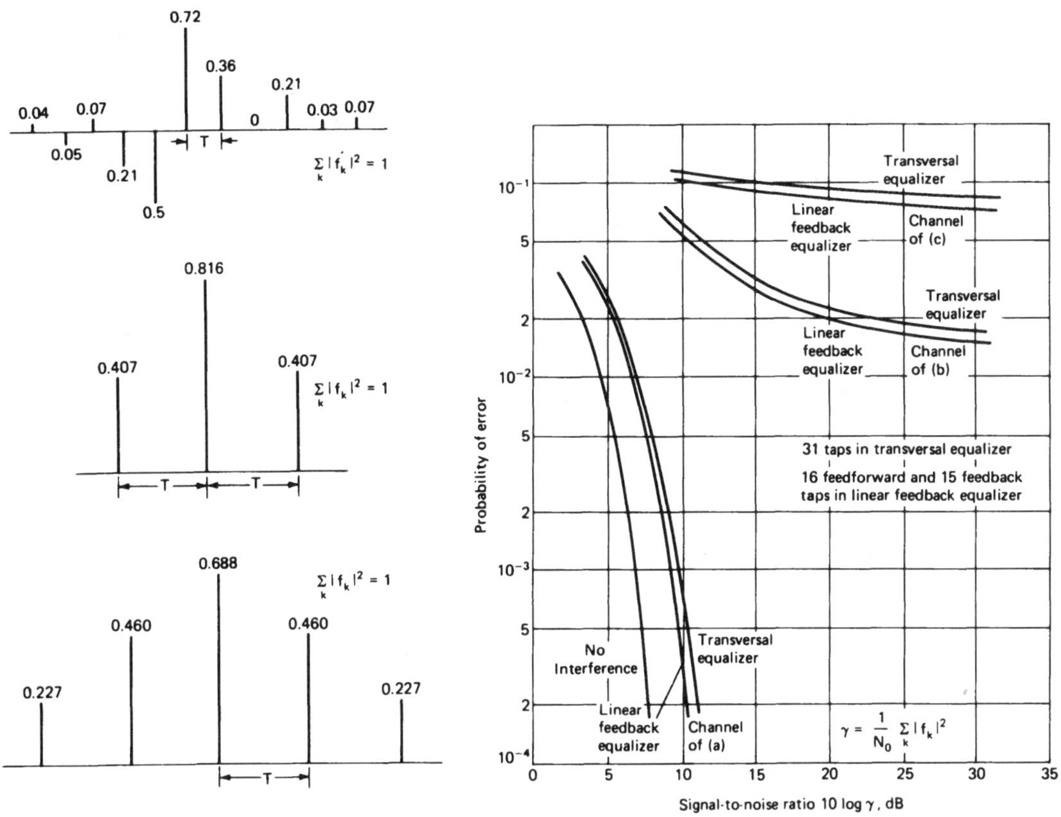


**Figure 11.8** Effect of different degrees of channel distortion. (*From* [12]. *Courtesy of J. G. Proakis and Academic Press.*)

resents a difference from the adaptive antenna structure, but results in much improved performance when channels with significant levels of much delayed multipath are present.

Referring now back to Figure 11.5, the similarity of the adaptive equalizer and the adaptive antenna processor is easy to see if we consider the $n$ outputs of the transversal filter, the equivalent of the $n$ separate antenna signals. If we adopt the notations of Widrow [4], the sum output $S$ has the form

$$S = \mathrm{W}_T \mathbf{X} \tag{11.15}$$

where the $X_i$ are the signal variations at the various taps of the equalizer. If the stored reference signal is $D$, then when it is known that a training preamble is being sent, the error becomes

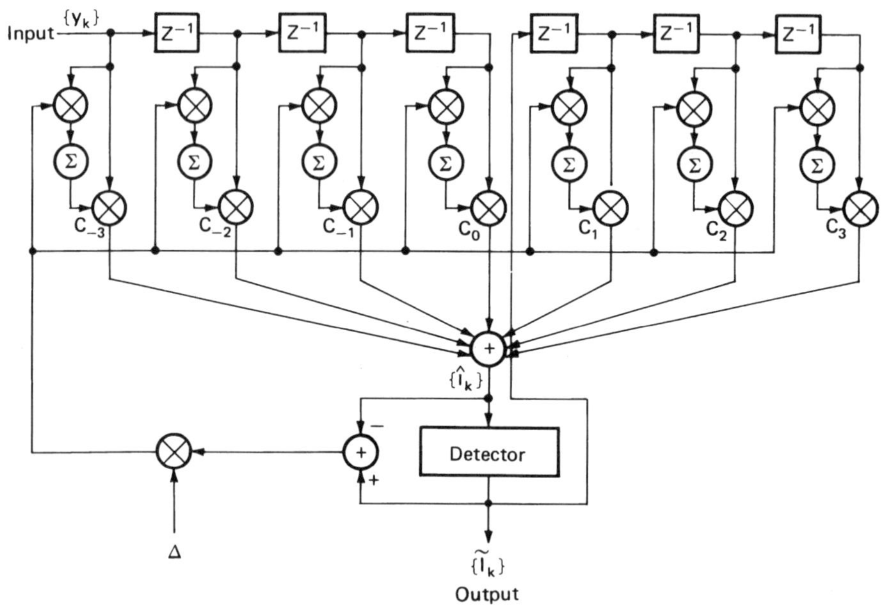


**Figure 11.9** Nonlinear adaptive equalizer configuration. (*From* [12]. *Courtesy of J. G. Proakis and Academic Press.*)

$$\varepsilon = S - D \tag{11.16}$$

The LMS criterion applied to this situation gives rise to the same expressions and diagrams as shown previously. If an analog delay line is used, ε is a time function whose square must be integrated over the period of the delay line to obtain the equivalent of the expectation for the successive digital samples, in the case of a digital delay line.

Another algorithm that has been used [11] replaces the LMS criterion with a peak distortion criterion. The peak criterion is defined as the sum of the absolute values of the error at sample times other than that at which the maximum value of $S$ occurs. This gives rise to a *zero forcing algorithm*, where the outputs at all sample times other than the maximum are forced to zero. The maximum value may be set to unity. This is a satisfactory criterion when the peak distortion is less than the maximum sample and when the noise is small, as in telephone channels or high-grade radio channels. However, it is not likely to be of use where distortion and noise are high. As with adaptive antenna processing, various modifications of the control algorithms have been tried to improve the convergence period so that more of the available transmission period can be used for information transmission.

The use of *linear feedback equalization* (i. e., the placement of the tap corresponding to the decision sample time within the equalizer, rather than at the right-hand side) proved of little value in good or bad channels [12]. Linear equalization, instead, is most useful in the feedforward mode. These observations led to the development of the nonlinear feedback equalizer (Figure 11.9). The fact that the feedback delay line is used for the already decided symbols leads to

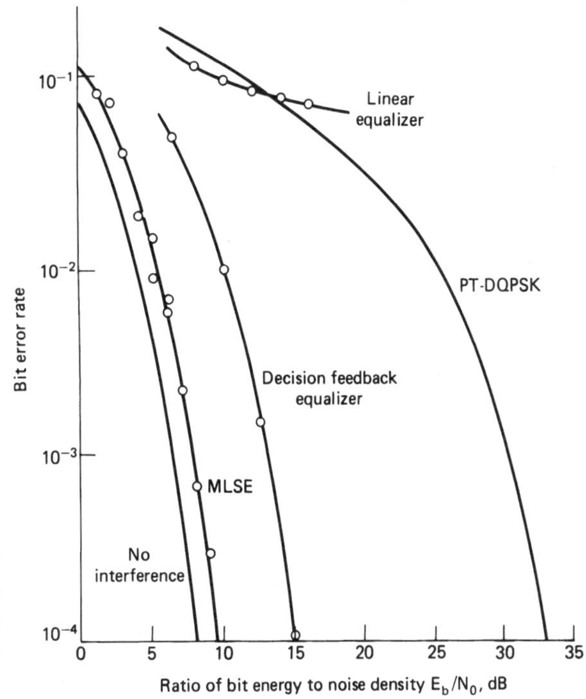


**Figure 11.10** Simulation results of binary signaling through a time-invariant multipath channel. (*From* [14]. *Courtesy of Naval Ocean Systems Center.*)

the setting of coefficients in that section on a different basis. With the LMS criterion, it can be shown that the circuit will completely eliminate the interference from the already decided symbols as long as the decision is correct. Because the time averaging or summing is generally slow enough that the weighting coefficients do not change significantly during a period of many output symbols, the effect of occasional errors is of little importance. The feedback equalization technique has been used to produce a high-speed digital transmission through HF multipath channels.

The adaptive equalizers discussed here use recursive methods of setting the weighting coefficients, so that every decision is affected by the $n$ samples of the waveform in the delay line and a number of previous $n$ or more sets of prior samples. This leads to the possibility of algorithms that accumulate $kn^2$ samples and make each decision on the basis of the entire statistics, using maximum-likelihood techniques [12]. As each new sample enters the system, the oldest sample is removed and the process is repeated. Properly designed, such a program should produce the best possible equalization for the number of points selected. Some experimental work of this type has been done using a *maximum-likelihood sequence estimation* (MLSE) or a Viterbi algorithm to reduce the computations. Figure 11.10 [15] indicates some comparisons of different equalization techniques in a simulated time-invariant channel.

In the previous discussions, the delay is the inverse of the symbol frequency. This is the maximum delay that can be used, and is used in most adaptive equalizer designs. In order for the equalizer to perform satisfactorily, symbol timing must be recovered very accurately. If the delay is made a fraction of a symbol, more timing error can be tolerated, but the processing load is increased. For binary symbols this is probably of minor significance, but if the symbols are higher-order, it may be necessary to use a *fractionally spaced* equalizer. A delay of one-half or one-third may prove more practical than improving symbol timing recovery.

## 11.4 References


1. Widrow, B., P. E. Mantey, L. J. Griffiths, and B. B. Goode: "Adaptive Antenna Systems," *Proc. IEEE*, vol. 55, pg. 2143, December 1967.
2. Applebaum. S. P.: "Adaptive Arrays," *IEEE Trans.*, vol. AP-24, pg. 585, September 1976.
3. Compton, R. T., Jr., R. J. Huff, W. G. Swarner, and A. A. Ksienski: "Adaptive Arrays for Communication Systems: An Overview of Research at the Ohio State University," *IEEE Trans.*, vol. AP-24, pg. 599, September 1976.
4. Widrow, B., and J. M. McCool: "A Comparison of Adaptive Algorithms Based on the Methods of Steepest Descent and Random Search," *IEEE Trans.*, vol. AP-24, pg. 615, September 1976.
5. Riegler, R. L., and R. T. Compton, Jr.: "An Adaptive Array for Interference Rejection," *Proc. IEEE*, vol. 61, June 1973.
6. Rossweiler, G. C., F. Wallace, and C. Ottenhoff: "Analog versus Digital Null-Steering Controllers," *ICC '77 Conf. Rec.*, 1977.
7. Price, R., and P. E. Green, Jr.: "A Communication Technique for Multipath Channels," *Proc. IRE*, vol. 46, ph. 555, March 1958.
8. Hulst, G. D.: "Inverse Ionosphere," *IRE Trans.*, vol. CS-8, pg. 3, March 1960.
9. Gibson, E. D.: "Automatic Equalization Using Time-Domain Equalizers," *Proc. IEEE*, vol. 53, pg. 1140, August 1965.
10. Di Toro, M. J.: "Communications in Time-Frequency Spread Media," *Proc. IEEE*, vol. 56, October 1968.
11. Lucky, R. W.: "Automatic Equalization for Digital Communication," *Bell Sys. Tech. J.*, vol. 44, pg. 547, April 1965.
12. Proakis, J. G.: "Advances in Equalization for Intersymbol Interference," in *Advances in Communication Systems: Theory & Application*, A. V. Balakrishnan and A. J. Viterbi (eds.), Academic Press, New York, N.Y., 1975.
13. Monsen, P.: "Fading Channel Communications," *IEEE Commun.*, vol. 18, pg. 16, January 1980.
14. Qureshi, S.: "Adaptive Equalization," *IEEE Commun.*, vol. 20, pg. 9, March 1982.
15. Hoff, L. E., and A. R. King: "Skywave Communication Techniques," Tech. Rep. 709, Naval Ocean Systems Center, San Diego, Calif, March 30,1981.

Chapter

# 12

# Television Reception Principles

**K. Blair Benson**

**L. H. Hoke, Jr., L. E. Donovan, J. D. Knox, D. E. Manners, W. G. Miller, R. J. Peffer, J. G. Zahnen**

## 12.1 Introduction

Television receivers provide black-and-white or color reproduction of pictures and the accompanying monaural or stereophonic sound from signals broadcast through the air or via cable distribution systems. The broadcast channels in the U.S. are 6 MHz wide for transmission on conventional 525-line standards.

## 12.2 Basic Operating Principles

The minimum signal level at which television receivers provide usable pictures and sound, called the *sensitivity level*, generally is on the order of 10 to 20 μV. The maximum level encountered in locations near transmitters may be as high as several hundred millivolts. The FCC has set up two standard signal level classifications, Grades A and B, for the purpose of licensing television stations and allocating coverage areas. Grade A is to be used in urban areas relatively near the transmitting tower, and Grade B use ranges from suburban to rural and fringe areas a number of miles from the transmitting antenna. The FCC values are expressed in microvolts per meter (μV/m), where meter is the signal wavelength [1].

The standard transmitter field-strength values for the outer edges of these services for Channels 2 through 69 are listed in Table 12.1. Included for reference in the table are the signal levels for what may be considered "city grade" in order to give an indication of the wide range in signal level that a receiver may be required to handle. The actual antenna terminal voltage into a matched receiver load, listed in the second column, is calculated from the following equation:

$$e = E\frac{96.68}{\sqrt{f_1 \times f_2}} \quad (12.1)$$

**Table 12.1 Television Service Operating Parameters**

| Band and Channels | Frequency (MHz) | City grade | | Grade A | | Grade B | |
|---|---|---|---|---|---|---|---|
| | | μV/m | μV | μV/m | μV | μV/m | μV |
| VHF 2–6 | 54–88 MHz | 5,010 | 7030 | 2510 | 3520 | 224 | 314 |
| VHF 7–13 | 174–216 MHz | 7,080 | 3550 | 3550 | 1770 | 631 | 315 |
| UHF 14–69 | 470–806 MHz | 10,000 | 1570 | 5010 | 787 | 1580 | 248 |
| UHF 70–83[1] | 806–890 MHz | 10,000 | 1570 | 5010 | 571 | 1580 | 180 |

1. Receiver coverage of Channels 70 to 83 has been on a voluntary basis since July 1982. This frequency band was reallocated by the FCC to land mobile use in 1975 with the provision that existing transmitters could continue indefinitely.

Where
$e$ = terminal voltage, μV, 300 Ω
$E$ = field, μV/m
$f_1$ and $f_2$ = band-edge frequencies, MHz

Many sizes and form factors of receivers are manufactured. Portable personal types include pocket-sized or hand-held models with picture sizes of 2 to 4 in (5 to 10 cm) diagonal for monochrome and 5 to 6 in (13 to 15 cm) for color powered by either batteries or ac. Conventional cathode ray tubes (CRTS) for picture displays in portable sets have essentially been supplanted by flat CRTs and liquid crystal displays.

Larger screen sizes are available in monochrome where low cost and light weight are prime requirements. However, except where extreme portability is important, the vast majority of television program viewing is in color. The 19-in (48-cm) and 27-in (69-cm) sizes now dominate the market, although the smaller 13-in (33-cm) size is popular as a second or semiportable set.

The television receiver functions can be broken down into several interconnected blocks. With the rapidly increasing use of large-scale integrated circuits, the isolation of functions has become more evident in the design and service of receivers, while at the same time the actual parts count has dropped dramatically. The typical functional configuration of a receiver using a tri-gun picture tube, shown in Figure 12.1, will serve as a guide for the following description of receiver design and operation. The discussions of each major block, in turn, are accompanied with more detailed subblock diagrams.

## 12.2a Tuner Principles

The purpose of the tuner, and the following *intermediate amplifier* (IF), is to select the desired radio frequency (RF) signals in a 6 MHz channel, to the exclusion of all other signals, available from the antenna or cable system and to amplify the signals to a level adequate for demodulation. Channel selection is accomplished with either mechanically switched and manually tuned circuits, or in varactor tuners with electrically switched and controlled circuit components. A mechanical tuner consists of two units, one for the VHF band from 54 to 88 and 174 to 216

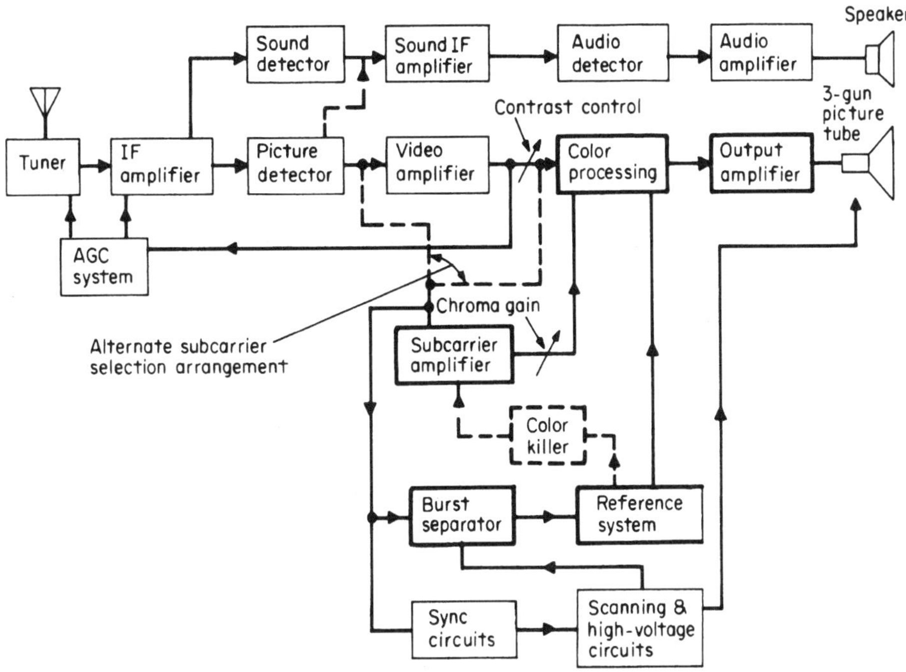


**Figure 12.1** Fundamental block diagram of a color receiver with a tri-gun picture tube display.

MHz, and the other for the UHF band from 470 to 806 MHz. Two separate antenna connections are provided for the VHF and UHF sections of the tuner.

Varactor tuners, on the other hand, have no moving parts or mechanisms and consequently are less than a third the volume of their mechanical equivalent. Part of this smaller size is the result of combining the VHF and UHF circuits on a single printed circuit board in the same shielded box with a common antenna connection, thus eliminating the need for an outrigger coupling unit.

## Selectivity

The tuner bandpass generally is 10 MHz in order to ensure that the picture and sound signals of the full 6 MHz television channel are amplified with no significant imbalance in levels or phase distortion by the skirts of the bandpass filters. This bandpass characteristic usually is provided by three tuned circuits:

- A single-tuned preselector between the antenna input and the RF amplifier stage
- A double-tuned interstage network between the RF and mixer stages
- A single-tuned coupling circuit at the mixer output.

**Table 12.2 Potential VHF Interference Problems**

| Desired Channel | Interfering Signals | Mechanism |
|---|---|---|
| 5 | Channel. 11 picture | 2 × ch. 5 osc. – ch. 11 pix = IF |
| 6 | Channel 13 picture | 2 × ch. 6 osc. – ch. 13 pix = IF |
| 7 and 8 | Channel 5, FM (98–108 MHz) | Ch. 5 pix + FM = ch. 7 and 8 |
| 2–6 | Channel 5, FM (97–99 MHz) | 2 × (FM – ch. 5) = IF |
| 7–13 | FM (88–108 MHz) | 2 × FM = ch. 7–13 |
| 6 | FM (89–92 MHz) | Ch. 6 pix + FM – ch. 6 osc. = IF |
| 2 | 6 m amateur (52–54 MHz) | 2 × ch. 2 pix – 6 m = ch. 2 |
| 2 | CB (27 MHz) | 2 × CB = ch. 2 |
| 5 and 6 | CB (27 MHz) | 3 × CB = ch. 5 and 6 |

The first two circuits are frequency-selective to the desired channel by varying either or both the inductance and capacitance. The mixer output is tuned to approximately 44 MHz, the center frequency of the IF channel.

The purpose of the RF selectivity function is to reduce all signals that are outside of the selected television channel. For example, the input section of VHF tuners usually contains a high-pass filter and trap section to reject signals lower than Channel 2 (54 MHz), such as standard broadcast, amateur, and citizen's band (CB) emissions. In addition, a trap is provided to reduce FM broadcast signal in the 88 to 108 MHz band. A list of the major interference problems is tabulated in Table 12.2 for VHF channels. In Table 12.3 for UHF channels, the formula for calculation of the interfering channels is given in the second column, and the calculation for a receiver tuned to Channel 30 is given in the third column.

## VHF Tuner

A block diagram of a typical mechanical tuner is shown in Figure 12.2. The antenna is coupled to a tunable RF stage through a bandpass filter to reduce spurious interference signals in the IF band, or from FM broadcast stations and CB transmitters. Another bandpass filter is provided in the UHF section for the same purpose. The typical responses of these filters are shown in Figures 12.3.*a* and *b*.

The RF stage provides a gain of 15 to 20 dB (approximately a 10:1 voltage gain) with a bandpass selectivity of about 10 MHz between the –3 dB points on the response curve. The response between these points is relatively flat with a dip of only a decibel or so at the midpoint. Therefore, the response over the narrower 6 MHz television channel bandwidth is essentially flat.

VHF tuners have a rotary shaft that switches a different set of three or four coils or coil taps into the circuit at each VHF channel position (2 to 13). The circuits with these switched coils are the following:

- RF input preselection
- RF input coupling (single-tuned for monochrome, double-tuned for color)
- RF-to-mixer interstage

**Table 12.3 Potential UHF Interference Problems**

| Interference Type | Interfering Channels | Channel 30 Example |
|---|---|---|
| IF beat | $N \pm 7, \pm 8$ | 22, 23, 37, 38 |
| Intermodulation | $N \pm 2, \pm 3, \pm 4, \pm 5$ | 25–28, 32–35 |
| Adjacent channel | $N + 1, -1$ | 29, 31 |
| Local oscillator | $N \pm 7 \times$ | 23, 37 |
| Sound image | $N + 1/6\ (2 \times 41.25)$ | 44 |
| Picture image | $N + 1/6\ (2 \times 45.75)$ | 45 |

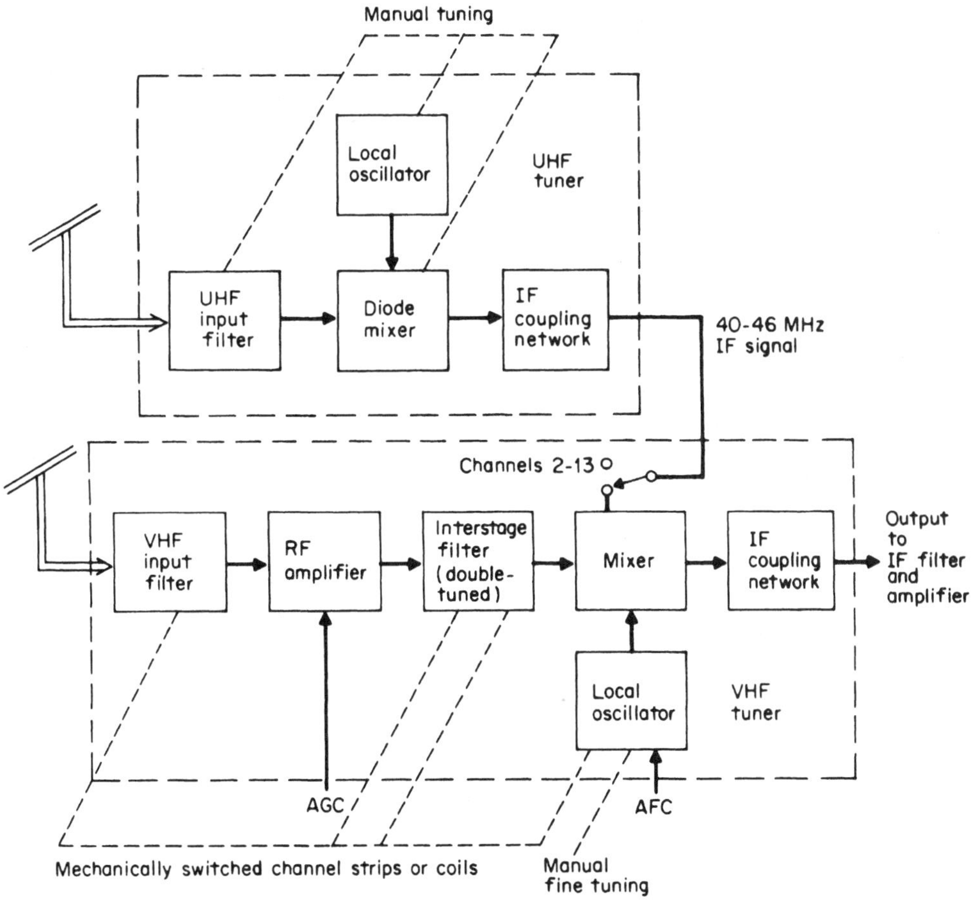


**Figure 12.2** Typical mechanical-tuner configuration.

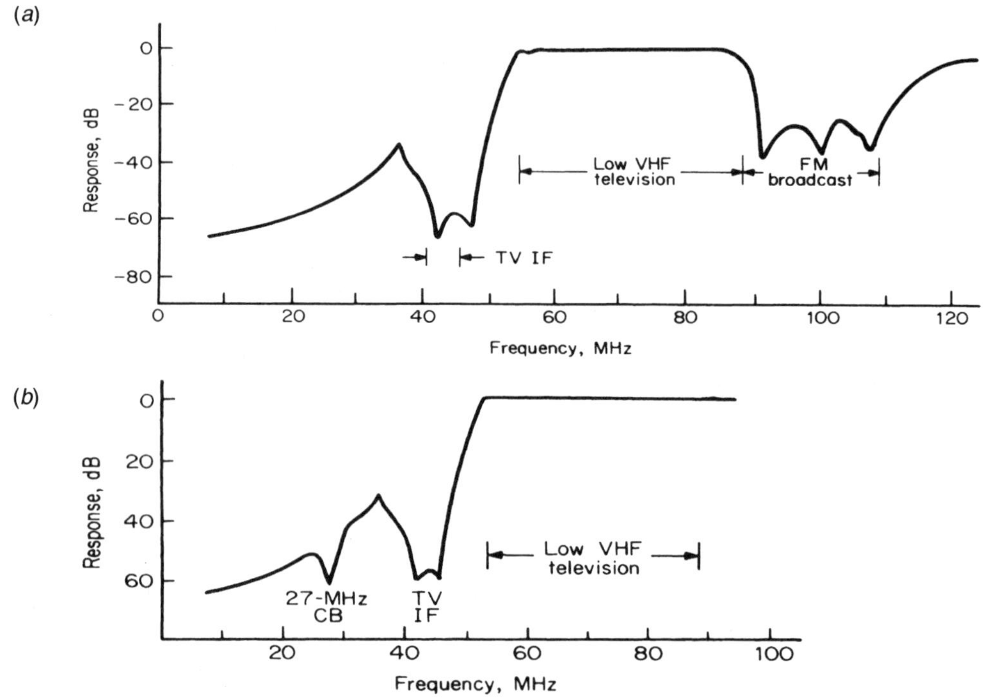


**Figure 12.3** Filter response characteristics: (*a*) response of a tuner input FM bandstop filter, (*b*) response of a tuner input with CB and IF traps.

In the first switch position (Channel 1), the RF stage is disabled and the mixer stage becomes an IF amplifier stage, centered on 44 MHz for the UHF tuner.

The mixer stage combines the RF signal with the output of a tunable local oscillator to produce an IF of 43.75 MHz for the picture carrier signal and 42.25 MHz for the sound carrier signal. The local oscillator signal thus is always 45.75 MHz above that of the selected incoming picture signal. For example, the frequencies for Channel 2 are listed in Table 12.4.

These frequencies were chosen to minimize interference from one television receiver into another by always having the local-oscillator signal above the VHF channels. Note that the oscillator frequencies for the low VHF channels (2 to 6) are between Channels 6 (low VHF) and 7 (high VHF), and the oscillator frequencies for the high VHF fall above these channels.

The picture and sound signals of the full 6 MHz television channel are amplified with no significant imbalance in levels or phase distortion by the skirts of the bandpass filters. This bandpass characteristic usually is provided by three tuned circuits:

- A single-tuned preselector between the antenna input and the RF amplifier stage
- Double-tuned interstage network between the RF and mixer stages
- Single-tuned coupling circuit at the mixer output

**Table 12.4 Local Oscillator and IF Frequencies**

| Parameter | Channel 2, MHz | Channel 6, MHz | Channel 7, MHz |
|---|---|---|---|
| Channel width | 54–60 | 82–88 | 174–180 |
| Local oscillator | 101.00 | 129.00 | 221.00 |
| Less picture signal | 55.25 | 83.25 | 175.25 |
| IF picture signal | 45.75 | 45.75 | 45.75 |

The first two are frequency-selective to the desired channel by varying either or both the inductance and capacitance. The mixer output is tuned to approximately 44 MHz, the center frequency of the IF channel.

## UHF Tuner

The UHF tuner contains a tunable input circuit to select the desired channel, followed by a diode mixer. As in a VHF tuner, the local oscillator is operated at 45.75 MHz above the selected input channel signal. The output of the UHF mixer is fed to the mixer of the accompanying VHF tuner, which functions as an IF amplifier. Selection between UHF and VHF is made by applying power to the appropriate tuner RF stage.

Mechanical UHF tuners have a shaft that when rotated moves one set of plates of variable air-dielectric capacitors in three resonant circuits. The first two are a double-tuned preselector in the amplifier-mixer coupling circuit, and the third is the tank circuit of the local oscillator. In order to meet the discrete selection requirement of the FCC, a mechanical detent on the rotation of the shaft and a channel-selector indicator are provided, as illustrated in Figure 12.4.

The inductor for each tuned circuit is a rigid metal strip, grounded at one end to the tuner shield and connected at the other end to the fixed plate of a three-section variable capacitor with the rotary plates grounded. The three tuned circuits are separated by two internal shields that divide the tuner box into three compartments.

## Tuner-IF Link Coupling

With mechanically switched tuners, it has usually been necessary to place the tuner behind the viewer control panel and connect it to the IF section, located on the chassis, with a foot or so length of shielded 50 or 75 Ω coaxial cable. Because the output of the tuner and the input of the IF amplifier are high-impedance-tuned circuits, for maximum signal transfer, it is necessary to couple these to the cable with impedance-matching networks.

Two common resonant circuit arrangements are shown in Figure 12.5. The low-side capacitive system has a low-pass characteristic that attenuates the local oscillator and mixer harmonic currents ahead of the IF amplifier. This can be an advantage in controlling local-oscillator radiation and in reducing the generation of spurious signals in the IF section. On the other hand, the low-side inductance gives a better termination to the link cable and therefore reduces interstage cable loss. The necessary bandpass characteristics can be obtained either by undercoupled stagger tuning or by overcoupled synchronous tuning, as illustrated in Figure 12.5.

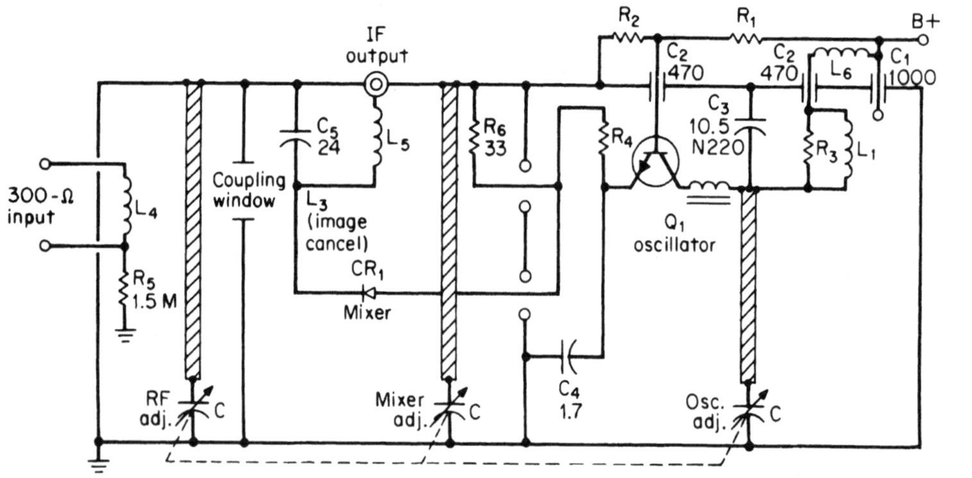


**Figure 12.4** UHF rotary turner. (*Courtesy of Philips Consumer Electronics.*)

## 12.2b Advanced Tuner Systems

Mechanically tuned television receivers were the mainstay of consumer sets since the beginning of TV broadcasting. It was not until the 1970s and 1980s that electronically tuned systems became practical. More recent technological trends include microprocessor-based control of the tuning functions. Many consumer models have, in fact, completely dispensed with the conventional tuner controls, in favor of using the remote control as the primary user interface. This being the case, considerable integration of functions can be gained, resulting in performance improvements and cost savings.

The preceding sections on mechanically tuned receivers are, however, still important today because they form the foundations for the all-electronic tuning systems that are prevalent today.

### Varactor Tuner

The varactor diode forms the basis for electronic tuning, which is accomplished by a change in capacitance with the applied dc voltage to the device. One diode is used in each tuned circuit. Unlike variable air-dielectric capacitors, *varicaps* have a resistive component in addition to their capacitance that lowers the $Q$ and results in a degraded noise figure. Therefore, varactor UHF tuners usually include an RF amplifier stage, making it functionally similar to a VHF tuner. (See Figure 12.6.)

The full UHF band can be covered by a single varicap in a tuned circuit because the ratio of highest and lowest frequencies in the UHF bands is less than 2:1 (1.7). However, the ratio of the highest to lowest frequencies in the two VHF bands is over twice (4.07) that of the UHF band. This is beyond the range that typically can be covered by a tuned circuit using varicaps. This problem is solved by the use of band switching between the low and high VHF channels. This is accomplished rather simply by short-circuiting a part of the tuning coil in each resonant tank cir-

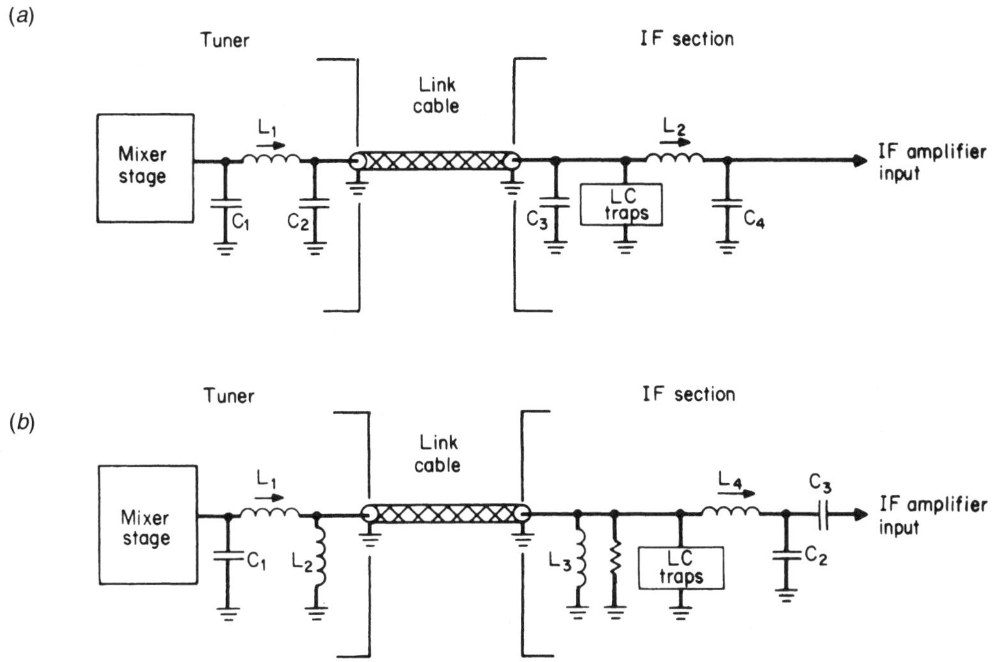


**Figure 12.5** Tuner-to-IF section link coupling: (*a*) low-side capacitive coupling, (*b*) low-side inductive coupling.

cuit to reduce its inductance. The short circuit is provided by a diode that has a low resistance in the forward-biased condition and a low capacitance in the reverse-biased condition. A typical RF and oscillator circuit arrangement is shown in Figure 12.7. Applying a positive voltage to $V_s$ switches the tuner to high VHF by causing the diodes to conduct and lower the inductance of the tuning circuits.

## Tuning Systems

The purpose of the tuning system is to set the tuner, VHF or UHF, to the desired channel and to fine-tune the local oscillator for the video carrier from the mixer to be set at the proper IF frequency of 46.75 MHz. In mechanical tuners, this obviously involves an adjustment of the rotary selector switch and the capacitor knob on the switch shaft. In electronically tuned systems, the dc tuning voltage can be supplied from the wiper arm of a potentiometer control connected to a fixed voltage source as shown in Figure 12.8*a*.

Alternatively, multiples of this circuit, as shown in Figure 12.8*b*, can provide preset fine-tuning for each channel. This arrangement most commonly is found in cable-channel selector boxes supplied with an external cable processor.

In digital systems, such as that shown in Figure 12.8*c*, the tuning voltage can be read as a digital word from the memory of a keyboard and display station (or remote control circuit). After

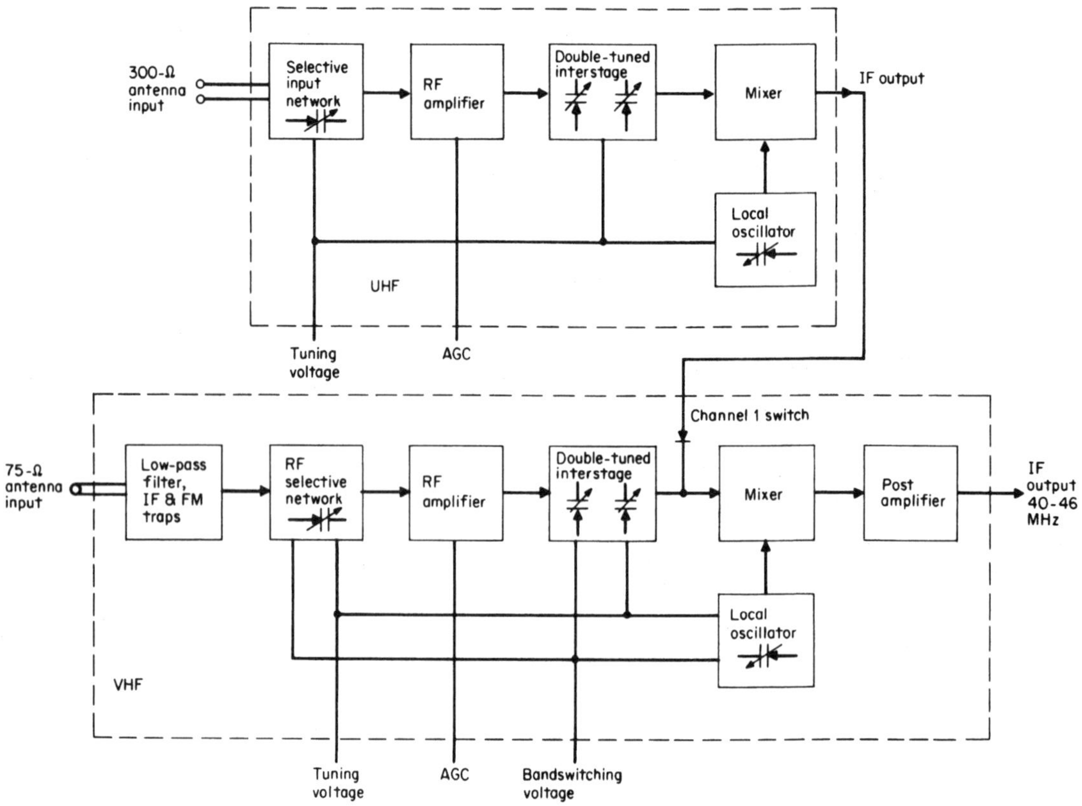


**Figure 12.6** Typical varactor-tuned block configuration.

conversion from a digital code to an analog voltage, the tuning control voltage is sent to the tuner.

Figure 12.8*d* shows a microprocessor system using a phase-locked loop to compare a medium-frequency square-wave signal from the channel selector keyboard, corresponding to a specific channel, with a signal divided down by 256 from the local oscillator. The error signal generated by the difference in these two frequencies is filtered and used to correct the tuning voltage being supplied to the tuner.

## 12.2c Intermediate-Frequency (IF) Amplifier Requirements

The IF picture and sound carrier frequencies standardized for conventional television receivers were chosen with prime consideration of possible degradation of the picture from interfering signals. The picture-carrier frequency is 45.75 MHz and, with the local oscillator above the received RF signal, the sound carrier frequency is 41.25 MHz.

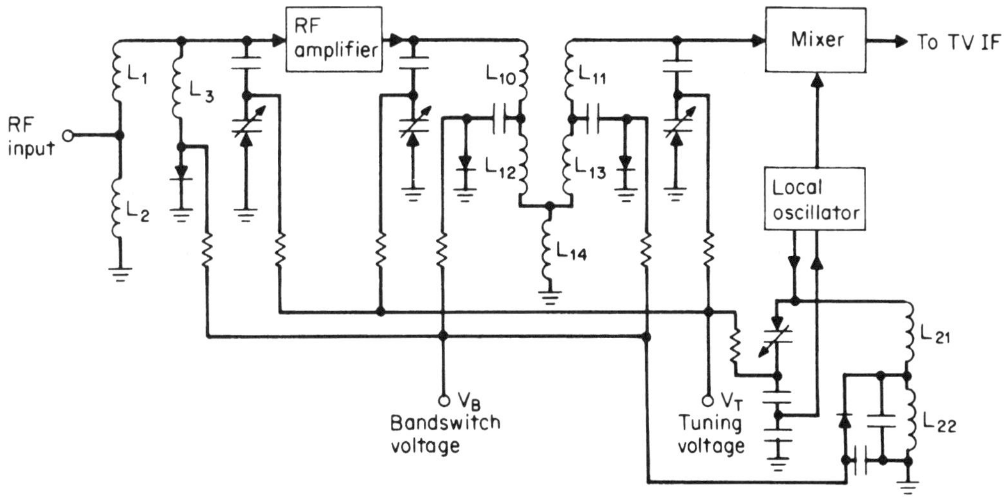


**Figure 12.7** VHF tuner band-switching and tuning circuits. $+V_B$ = high VHF (active tuning inductors = $L_3$ in parallel with $L_1$ and $L_2$, $L_{10}$, $L_{11}$, $L_{21}$). $-V_B$ = low VHF (active tuning inductors = $L_1 + L_2$, $L_{12}$, $L_{11} + L_{13}$, $L_{14}$, $L_{21} + L_{22}$).

The three factors given greatest emphasis in the choice were the following:

- Interference from other nontelevision services
- Interference from the fundamental and harmonics of local oscillators in other television receivers
- Spurious responses from the image signal in the mixer conversion and from harmonics of the IF signals

Analysis of the relationships indicates the soundness of the choice of 45.75 MHz for the IF picture carrier. The important advantages include the following:

- No images from the mixer conversion process fall within the VHF band selected by the tuner except for a negligible interference on the edge of the Channel 7 passband from the image from another receiver tuned to Channel 6.
- All channels are clear of picture harmonics except that the fourth harmonic of the IF picture carrier falls near the Channel 8 picture carrier. This can cause a noticeable beat pattern in another receiver if the offending receiver has not been designed with adequate shielding.
- Local oscillator radiation does not interfere with another receiver on any channel or on any channel image.
- No UHF signal falls on the image frequency of another station.

On the other hand, it should be pointed out that channels for certain public safety communications are allocated in the standard IF band. Because these transmitters radiate high power levels,

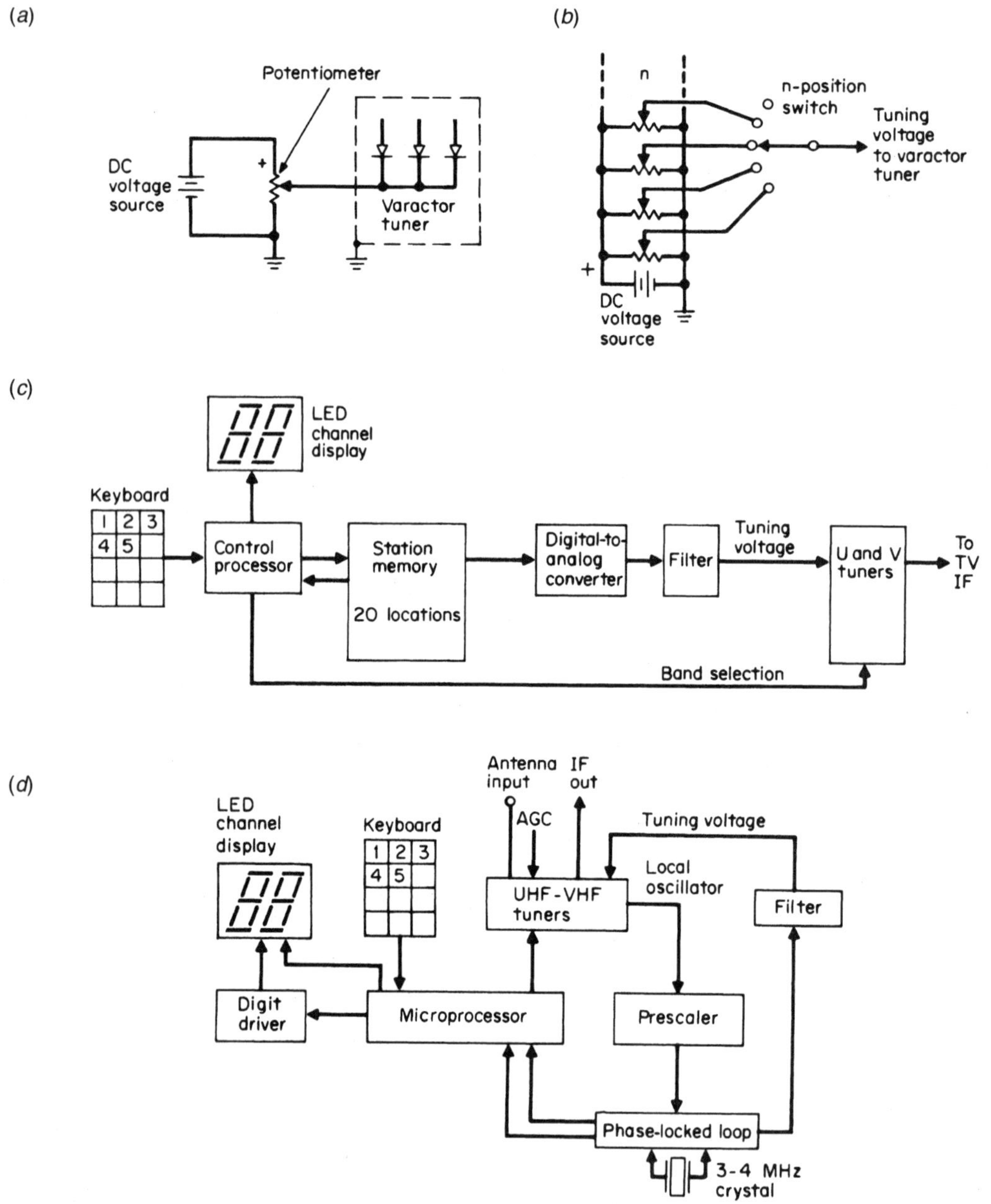


**Figure 12.8** Varactor-tuned systems: (*a*) simple potentiometer controlled varactor-tuned system, (*b*) multiple potentiometers providing *n*-channel selection, (*c*) simplified memory tuning, (*d*) microprocessor-based PLL tuning system.

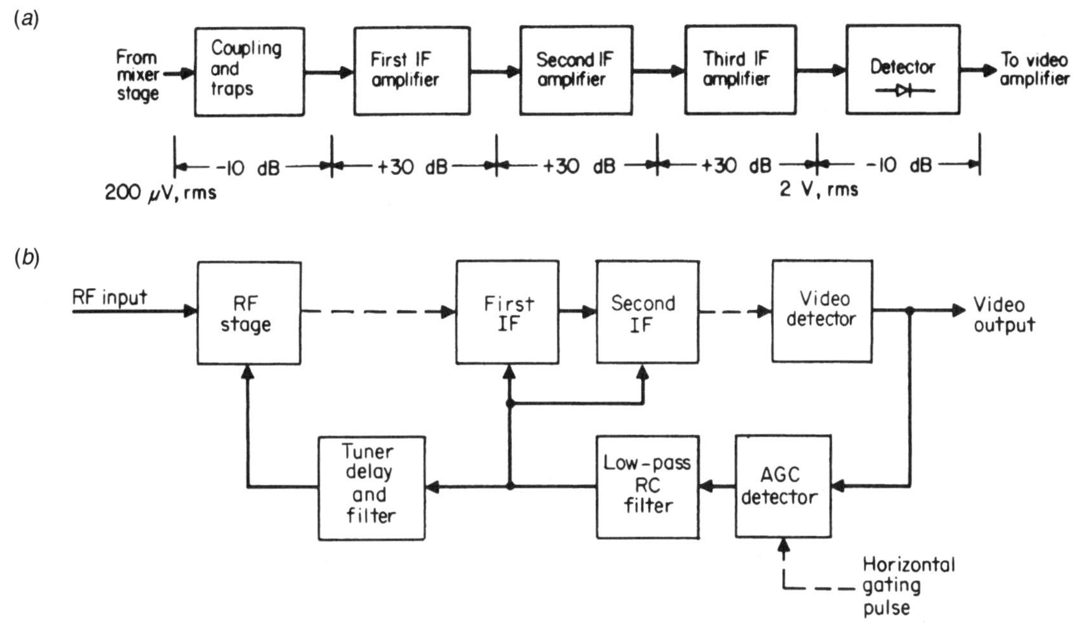


**Figure 12.9** IF amplifier system: (*a*) typical IF amplifier strip block diagram and gain distribution, (*b*) receiver AGC system.

receivers require thorough shielding of the IF amplifier and IF signal rejection traps in the tuner ahead of the mixer. In locations where severe cases of interference are encountered, the addition of a rejection filter in the antenna input may be necessary.

## Gain Characteristics

The output level of the picture and sound carriers from the mixer in the tuner is about 200 μV. The IF section provides amplification to boost this level to about 2 V, which is required for linear operation of the detector or demodulator stage. This is an overall gain of 80 dB. The gain distribution in a typical IF amplifier using discrete gain stages is shown in Figure 12.9*a*.

Automatic gain control (AGC) in a closed feedback loop is used to prevent overload in the IF, and the mixer stage as well, from strong signals (see Figure 12.9*b*). Input-signal levels may range from a few microvolts to several hundred microvolts, thus emphasizing the need for AGC. The AGC voltage is applied only to the IF for moderate signal levels so that the low-noise RF amplifier stage in the tuner will operate at maximum gain for relatively weak tuner input signals. A "delay" bias is applied to the tuner gain control to block application of the AGC voltage except at very high antenna signal levels. As the antenna signal level increases, the AGC voltage is applied first to the first and second IF stages. When the input signal reaches about 1 mV, the AGC voltage is applied to the tuner, as well.

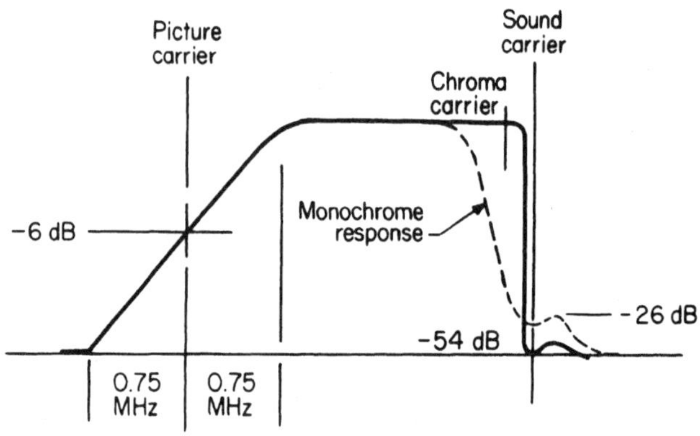


**Figure 12.10** Ideal IF amplitude response for color and monochrome reception.

## Response Characteristics

The bandpass of the IF amplifier must be wide enough to cover the picture and sound carriers of the channel selected in the tuner while providing sharp rejection of adjacent channel signals. Specifically the upper adjacent-picture carrier and lower adjacent-sound channel must be attenuated 40 and 50 dB, respectively, to eliminate visible interference patterns in the picture. The sound carrier at 4.5 MHz below the picture carrier must be of adequate level to feed either a separate sound IF channel or a 4.5 MHz intercarrier sound channel. Furthermore, because in the vestigial sideband system of transmission the video carrier lower sideband is missing, the response characteristic is modified from flat response to attenuate the picture carrier by 50 percent (6 dB).

In addition, in color receivers the color subcarrier at 3.58 MHz below the picture carrier must be amplified without time delay relative to the picture carrier or distortion. Ideally, this requires the response shown in Figure 12.10. Notice that the color IF is wider and has greater attenuation of the channel sound carrier in order to reduce the visibility of the 920 kHz beat between the color subcarrier and the sound carrier.

These and other more stringent requirements for color reception are illustrated in Figure 12.11. Specifically:

- IF bandwidth must be extended on the high-frequency video side to accommodate the color subcarrier modulation sidebands that extend to 41.25 MHz (as shown in Figure 12.11). The response must be stable and, except in sets with *automatic chroma control* (ACC), the response must not change with the input signal level (AGC), in order to maintain a constant level of color saturation.

- More accurate tuning of the received signal must be accomplished in order to avoid shifting the carriers on the tuner IF passband response. Deviation from their prescribed positions will alter the ratio of luminance to chrominance (saturation). While this is corrected in receivers with *automatic fine tuning* (AFT) and ACC, it can change the time relationship between color and luminance that is apparent in the color picture as chroma being misplaced horizontally.

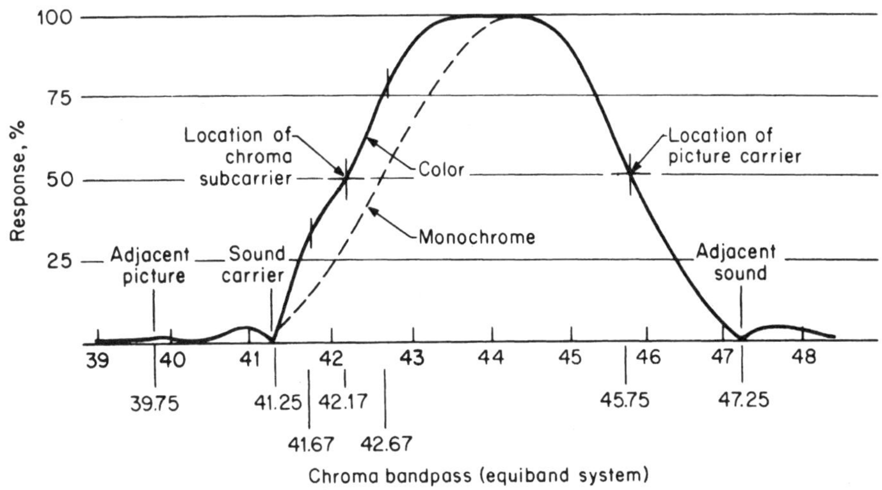


**Figure 12.11** Overall IF bandwidth for color reception.

- Color subcarrier presence as a second signal dictates greater freedom from overload of amplifier and detector circuits, which can result in spurious intermodulation signals visible as beat patterns. These cannot be removed by subsequent filtering.
- Envelope delay (time delay) of the narrow-band chroma and wide-band luminance signals must be equalized so that the horizontal position of the two signals match in the color picture.

## Surface Acoustic Wave (SAW) Filter

A SAW filter can provide the entire passband shape and adjacent-channel attenuation required for a television receiver. A typical amplitude response and group delay characteristic are shown in Figure 12.12. The sound carrier (41.25 MHz) attenuation of the SAW filter has been designed to operate with a synchronous detector, hence the lesser attenuation than in a conventional LC bandpass filter. The response of an LC discrete stage configuration shows a 60 dB attenuation of the sound carrier, necessary for suppression of the 920 kHz sound-chroma beat when used with a diode detector. In addition, with SAW technology it is possible to make wider adjacent-channel traps, which improve their performance and—in part—makes allowance for the temperature coefficient of the substrate materials used in SAW filters. This drift may be as great as 59 kHz per 10°C.

The schematic diagram of a SAW filter IF circuit is shown in Figure 12.13. The filter typically has an insertion loss of 15 to 20 dB and therefore requires a preamplifier to maintain a satisfactory overall receiver SNR.

The SAW filter consists of a piezoelectric substrate measuring 4 to 8 mm by 0.4 mm thick, upon which has been deposited a pattern of two sets of interleaved aluminum fingers. The width of the fingers may be on the order of 50 to 500 mm thick and 10 to 20 μm wide. (See Figure 12.14.

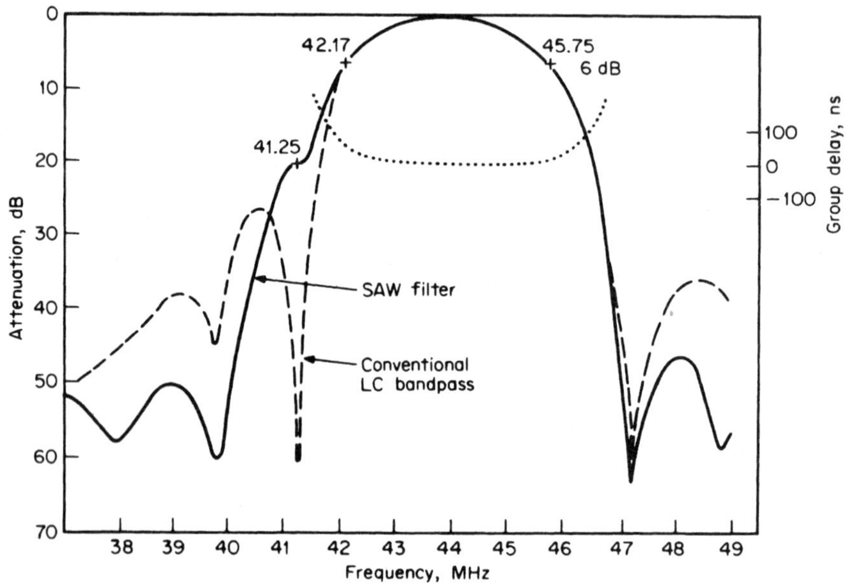


**Figure 12.12** Surface acoustic wave (SAW) filter response.

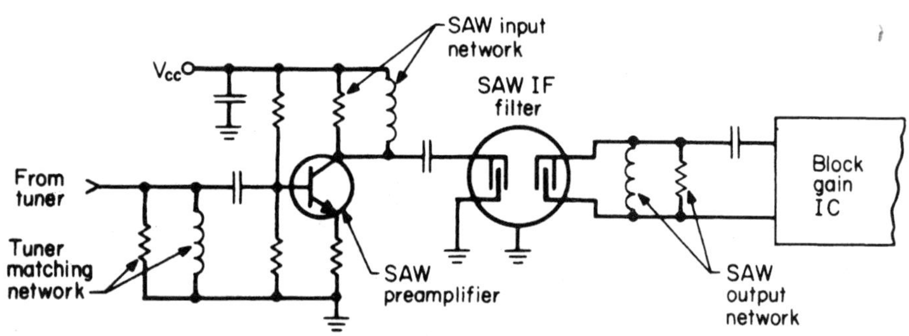


**Figure 12.13** Schematic of a block filter/block gain configuration.

Although quartz and other materials have been researched for use as SAW filter substrates, for television applications lithium niobate and lithium tantalate are typical. When one set is driven by an electric signal voltage, an acoustic wave moves across the surface to the other set of fingers that are connected to the load. The transfer amplitude-frequency response appears as a (sin $x$)/$x$ (Figure 12.15).

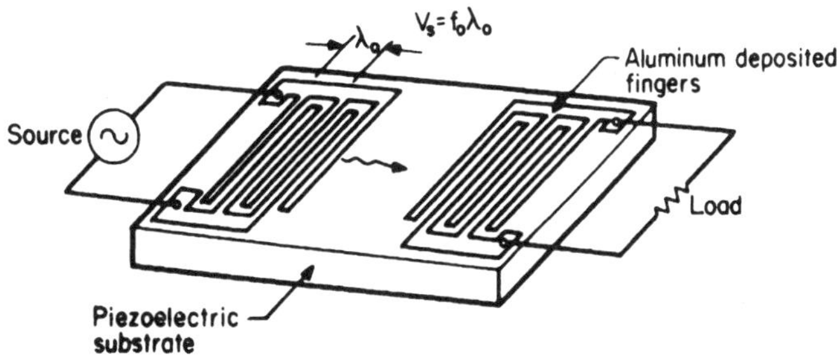


**Figure 12.14** Physical implementation of a SAW filter.

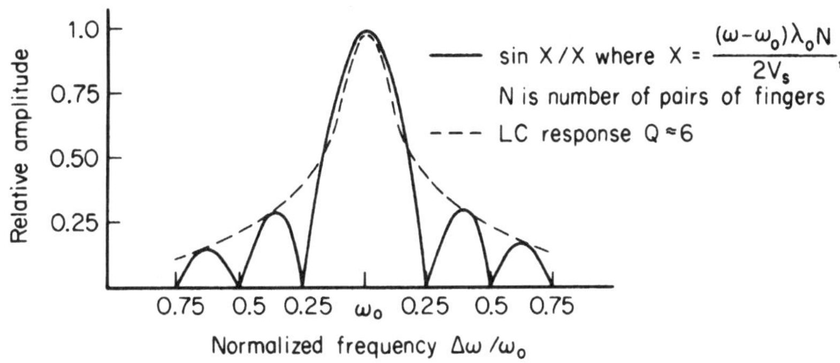


**Figure 12.15** Response of a uniform interdigitated SAW filter. (After [2].)

A modification to the design in Figure 12.14 that gives more optimum television bandpass and trap response consists of varying the length of the fingers to form a diamond configuration, illustrated in Figure 12.16. This is equivalent to connecting several transducers with slightly different resonant frequencies and bandwidths in parallel. Other modification consists of varying the aperture spacings, distance between the transducers, and the passive coupler strip patterns in the space between the transducers.

## Gain Block Requirements

Integrated-circuit gain blocks have the same basic requirements as discrete stage IF amplifiers (see Figure 12.17). These are: high gain, low distortion, and a large linear gain-control range under all operating conditions. One typical differential amplifier configuration used in IC gain blocks that meets these objectives has a gain of nearly 20 dB and a gain-control range of 24 dB. A direct-coupled cascade of three stages yields an overall gain of 57 dB and a gain-control range of 64 dB. The gain-control system internal to this IC begins to gain-reduce the third stage at an IC input level of 100 μV of IF carrier. With increasing input signal level, the third stage gain

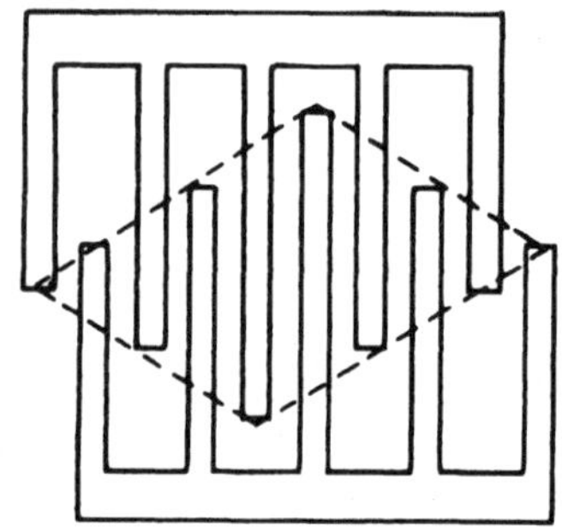

**Figure 12.16** SAW apodized IDT pattern.

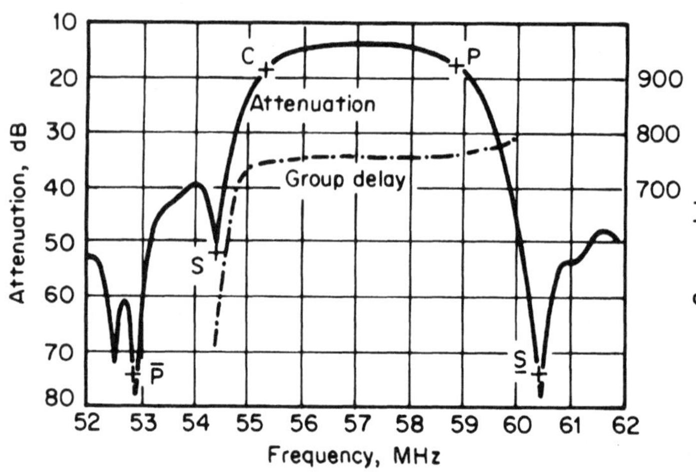


**Figure 12.17** Frequency response of SAW filter picture-carrier output. (After [3].)

reduces to 0 dB and then is followed by the second stage to a similar level, followed by the first in the same manner. By this means, a noise figure of 7 dB is held constant over an IF input signal range of 40 dB. The need for a preamplifier ahead of the SAW IF becomes less important when the IF amplifier noise figure is maintained constant by this cascaded control system.

The high gain and small size of an integrated-circuit IF amplifier places greater importance on PC layout techniques and ground paths if stability is to be achieved under a wide range of operating conditions. These considerations also carry over to the external circuits and components.

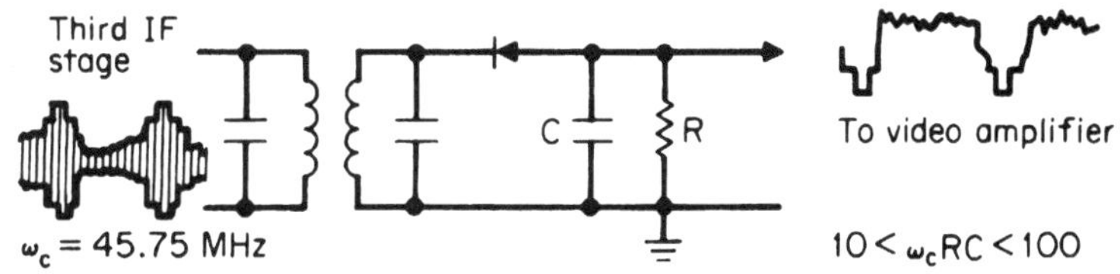


**Figure 12.18** Television receiver envelope detector.

## 12.2d Video Signal Demodulation

The function of the video demodulator is to extract the picture signal information that has been placed on the RF carrier as amplitude modulation. The demodulator receives the modulated carrier signal from the IF amplifier that has boosted the peak-to-peak level to 1 or 2 V. The modulation components extend from dc to 4.5 MHz. The output of the demodulator is fed directly to the video amplifier.

There are four types of demodulators commonly used in television receivers:

- Envelope detector
- Transistor detector
- Synchronous detector
- Feedback balanced diode

### Envelope Detector

Of the several types of demodulators, the envelope detector is the simplest. It consists of a diode rectifier feeding a parallel load of a resistor and a capacitor (Figure 12.18). In other words, it is a half-wave rectifier that charges the capacitor to the peak value of the modulation envelope.

Because of the large loss in the diode, a high level of IF voltage is required to recover 1 or 2 volts of demodulated video. In addition, unless the circuit is operated at a high signal level, the curvature of the diode impedance curve near cutoff results in undesirable compression of peak-white signals. The requirement for large signal levels and the nonlinearity of detection result in design problems and certain performance deficiencies, including the following:

- Beat signal products will occur between the color subcarrier (42.17 MHz), the sound carrier (41.25 MHz), and high-amplitude components in the video signal. The most serious is a 920 Hz (color to sound) beat and 60 Hz buzz in sound from vertical sync and peak-white video modulation.
- Distortion of luminance toward black of as much as 10 percent and asymmetric transient response. Referred to as *quadrature distortion*, this characteristic of the vestigial sideband is aggravated by nonlinearity of the diode. (See Figure 12.19.)
- Radiation of the fourth harmonic of the video IF produced by the detection action directly from the chassis, which can interfere with reception of VHF Channel 8 (180 to 186 MHz).

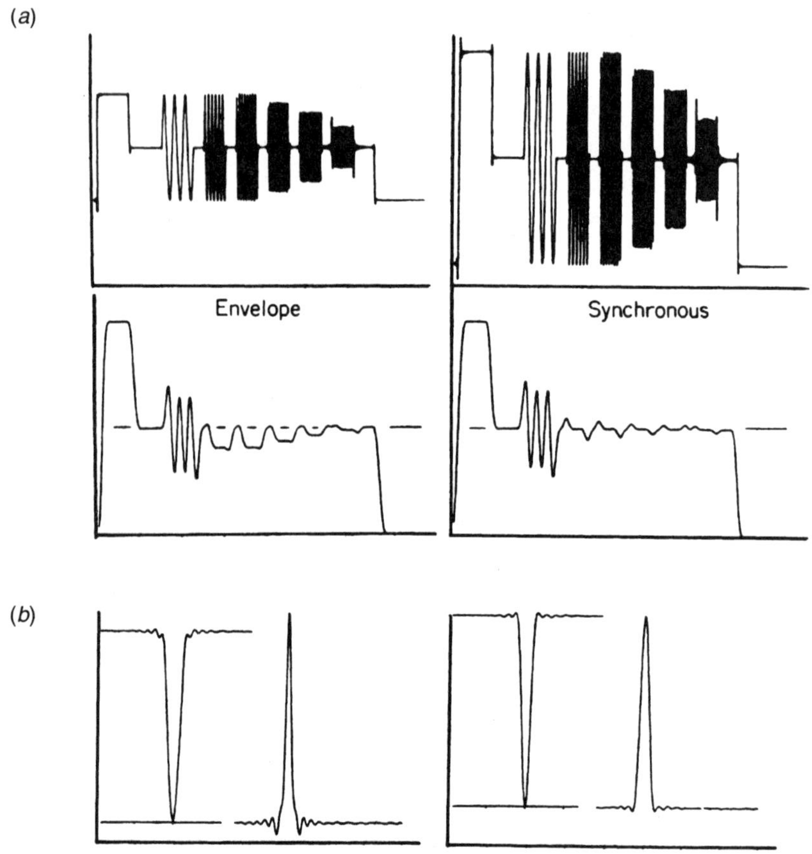


**Figure 12.19** Comparison of quadrature distortion in envelope and synchronous detectors: (*a*) axis shift, (*b*) inverted and normal 2T-pulse response. (*After* [4].)

Even with these deficiencies, the diode envelope detector was used in the majority of the monochrome and color television receivers dating back to vacuum tube designs up to the era of discrete transistors.

## Transistor Detector

A transistor biased near collector cutoff and driven with a modulated carrier at an amplitude greater than the bias level (see Figure 12.20) provides a demodulator that can have a gain of 15 or 20 dB over that of a diode. Consequently, less gain is required in the IF amplifier; in fact, in some receiver designs, the third IF stage has been eliminated. Unfortunately, this is offset by the same deficiencies in signal detection as the diode envelope detector.

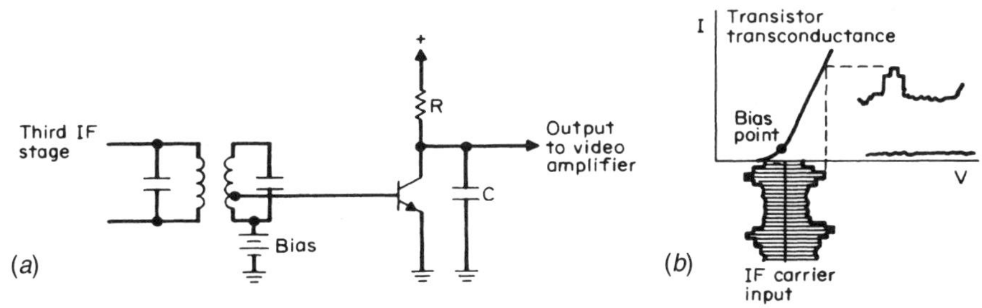


**Figure 12.20** Transistor detector: (*a*) schematic diagram, (*b*) detection characteristic.

### Synchronous Detector

The synchronous detector is basically a balanced rectifier in which the carrier is sampled by an unmodulated carrier at the same frequency as the modulated carrier. The unmodulated reference signal is generated in a separate high-*Q* limiting circuit that removes the modulation. An alternative system for generating the reference waveform is by means of a local oscillator phase-locked to the IF signal carrier.

The advantages of synchronous demodulation are:

- Higher gain than a diode detector
- Low level input, which considerably reduces beat-signal generation
- Low level detection operation, which reduces IF harmonics by more than 20 dB
- Little or no quadrature distortion (see Figure 12.19), depending upon the lack of residual phase modulation (purity) of the reference carrier used for detection
- Circuit easily formatted in an IC

## 12.2e Automatic Gain Control (AGC)

Each amplifier, mixer, and detector stage in a television receiver has a number of operating conditions that must be met in order to achieve optimum performance. Specifically, these conditions include that:

- The input level is greater than the internally generated noise by a factor in excess of the minimum acceptable SNR.
- Input level does not overload the amplifier stages, thus causing a bias shift.
- Bias operates each functional component of the system at its optimum linearity point, that is, the lowest third-order product for amplifiers, and highest practical second-order for mixers and detectors.
- Spurious responses in the output are 50 dB below the desired signal.

The function of the automatic gain control system is to maintain signal levels in these stages at the optimum value over a large range of receiver input levels. The control voltage to operate the system usually is derived from the video detector or the first video amplifier stage. Common implementation techniques include the following:

- Average AGC, which operates on the principle of keeping the carrier level constant in the IF. Changes in modulation of the carrier will affect the gain control, and therefore it is used only in low-cost receivers.
- Peak or sync-clamp AGC, which compares the video sync-tip level with a fixed dc level. If the sync-tip amplitude exceeds the reference level, a control voltage is applied to the RF and IF stages to reduce their gain and thus restore the sync-tip level to the reference level.
- Keyed or gated AGC, which is similar to sync-clamp AGC. The stage where the comparison of sync-tip and reference signals takes place is activated only during the sync-pulse interval by a horizontal flyback pulse. Because the AGC circuit is insensitive to spurious noise signals between sync pulses, the noise immunity is considerably improved over the other two systems.

## AGC Delay

For best receiver SNR, the tuner RF stage is operated at maximum gain for RF signals up to a threshold level of 1 mV. In discrete amplifier chains, the AGC system begins to reduce the gain of the second IF stage proportionately as the RF signal level increases from just above the sensitivity level to the second-stage limit of gain reduction (20 to 25 dB). For increasing signals, the first IF stage gain is reduced. Finally, above the control delay point of 1mV, the tuner gain is reduced. A plot of the relationships between receiver input RF level and the gain characteristics of the tuner and IF are shown in Figure 12.21*a* and the noise figure is shown in Figure 12.21*b*.

## System Analysis

The interconnection of the amplifier stages (RF, mixer, and IF), the detector, and the lowpass-filtered control voltage is—in effect—a feedback system. The loop gain of the feedback system is nonlinear, increasing with increasing signal level. There are two principal constraints that designers must cope with:

- First, the loop gain should be large to maintain good regulation of the detector output over a wide range of input signal levels. As the loop gain increases, the stability of the system will decrease and oscillation can occur.
- Second, the ability of a receiver to reject impulse noise is inversely proportional to the loop gain. Excessive impulse noise can saturate the detector and reduce the RF-IF gain, thereby causing either a loss in picture contrast or a complete loss of picture. This problem can be alleviated by bandwidth-limiting the video signal fed to the AGC detector, or the use of keyed or gated AGC to block false input signals (except during the sync pulse time).

A good compromise between regulation of video level and noise immunity is realized with a loop-gain factor of 20 to 30 dB.

The filter network and filter time constants play an important part in the effectiveness of AGC operation. The filter removes the 15.750 kHz horizontal sync pulses and equalizing pulses, the

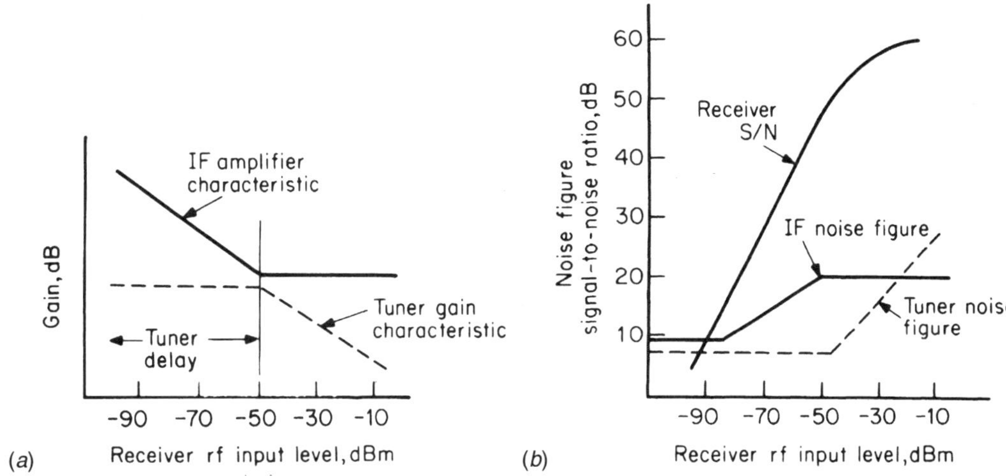


**Figure 12.21** Automatic gain control principles: (*a*) gain control as a function of input level, (*b*) noise figure of RF and IF stages with gain control and the resulting receiver SNR.

latter in blocks in the 60 Hz vertical sync interval. The filter time constants must be chosen to eliminate or minimize the following problems:

- *Airplane flutter*, a fluctuation in signal level caused by alternate cancellation and reinforcement of the received signal by reflections from an airplane flying overhead in the path between the transmitting and receiving antennas. The amplitude may vary as much as 2 to 1 at rates from 50 to 150 Hz. If the time constants are too long, especially that of the control voltage to the RF stage, the gain will not change rapidly enough to track the fluctuating level of the signal. The result will be a flutter in contrast and brightness of the picture.
- *Vertical sync pulse sag*, resulting from the AGC system speed of response being so fast that it will follow changes in sync pulse energy during the vertical sync interval. Gain increases during the initial half-width equalizing pulses, then decreases during the slotted vertical sync pulse, increases again during the final equalizing pulses, and then returns to normal during the end of vertical blanking and the next field of picture. Excessive sag can cause loss of proper interlace and vertical jitter or bounce. Sag can be reduced by limiting the response of the AGC control loop, or through the use of keyed AGC.
- Lock-out of the received signal during channel switching, caused by excessive speed of the AGC system. This can result in as much as a 2:1 decrease in pull-in range for the AGC system. In keyed AGC systems, if the timing of the horizontal gating pulses is incorrect, excessive or insufficient sync can result at the sync separator, which—in turn—will upset the operation of the AGC loop.

**Table 12.5 Typical AFC Closed-Loop Characteristics**

| Parameter | Value |
|---|---|
| Pull-in range | ±750 kHz |
| Hold-in range | ±1.5 MHz |
| Frequency error for ±500 kHz offset | < 50 kHz |

## 12.2f Automatic Frequency Control (AFC)

Also called *automatic fine-tuning* (AFT), the AFC circuit senses the frequency of the picture carrier in the IF section and sends a correction voltage to the local oscillator in the tuner section if the picture carrier is not on the standard frequency of 45.75 MHz.

Typical AFC systems consist of a frequency discriminator prior to the video detector, a low-pass filter, and a varactor diode controlling the local oscillator. The frequency discriminator in discrete transistor IF systems has typically been the familiar balanced-diode type used for FM radio receivers with the components adjusted for wide-band operation centering on 45.75 MHz. A small amount of unbalance is designed into the circuit to compensate for the unbalanced sideband components of vestigial sideband signal characteristics. The characteristics of AFC closed loops are shown in Table 12.5.

In early solid-state designs, the AFC block was a single IC with a few external components. Current designs have included the AFC circuit in the form of a synchronous demodulator on the same IC die as the other functions of the IF section.

## 12.2g Sound Carrier Separation Systems

Television sound is transmitted as a frequency-modulated signal with a maximum deviation of ±25 kHz (100 percent modulation) capable of providing an audio bandwidth of 50 to 15,000 Hz. The frequency of the sound carrier is 4.5 MHz above the RF picture carrier. The basic system is monaural with dual-channel stereophonic transmission at the option of the broadcaster.

The *intercarrier* sound system passes the IF picture and sound carriers (45.75 and 41.25 MHz, respectively) through a detector (nonlinear stage) to create the intermodulation beat of 4.5 MHz. The intercarrier sound signal is then amplified, limited, and FM-demodulated to recover the audio. Block diagrams of intercarrier sound systems are shown in Figures 12.22 and 12.23. In a discrete component implementation, the intercarrier detector is typically a simple diode detector feeding a 4.5 MHz resonant network. If an IC IF system is used, the sound and picture IF signals are carried all the way to the video detector, where one output port of the balanced synchronous demodulator supplies both the 4.5 MHz sound carrier and the composite baseband video.

The coupling network between the intercarrier detector and the sound IF amplifier usually has the form of a half-section high-pass filter that is resonant at 4.5 MHz. This form gives greater attenuation to the video and sync pulses in the frequency range from 4.5 MHz to dc (carrier), thereby reducing buzz in the recovered audio, especially under the conditions of low picture carrier. An alternative implementation uses a piezoelectric ceramic filter that is designed to have a bandpass characteristic at 4.5 MHz and needs no in-circuit adjustment.

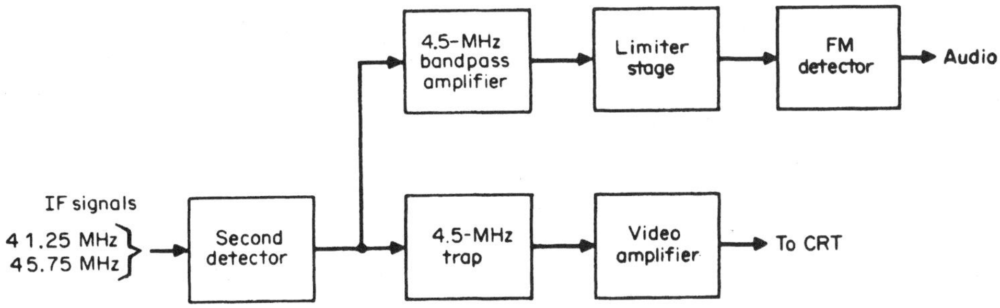


**Figure 12.22** A typical monochrome intercarrier sound system.

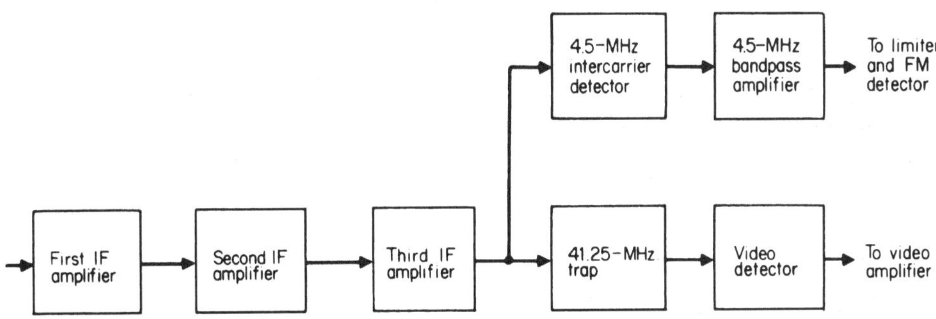


**Figure 12.23** An intercarrier sound takeoff ahead of the video detector.

Audio buzz results when video-related phase-modulated components of the visual carrier are transferred to the sound channel. The generation of *incidental carrier phase modulation* (ICPM) can be transmitter-related or receiver-related; however, the transfer to the sound channel takes place in the receiver. This can occur in the mixer or the detection circuit. Here, a synchronous detector represents little improvement over an envelope diode detector unless a narrow-band filter is used in the reference channel [5].

Split-carrier sound processes the IF picture and sound carriers as shown in Figure 12.24.

*Quasi-parallel* sound utilizes a special filter such as the SAW filter of Figure 12.14 to eliminate the Nyquist slope in the sound detection channel, thereby eliminating a major source of ICPM generation in the receiver. The block diagram of this system is shown in Figure 12.25.

Nearly all sound channels in present-day television receivers are designed as a one- or two-IC configuration. The single IC contains the functions of sound IF amplifier-limiter, FM detector, volume control, and audio output. Two-chip systems usually incorporate stereo functionality or audio power amplification.

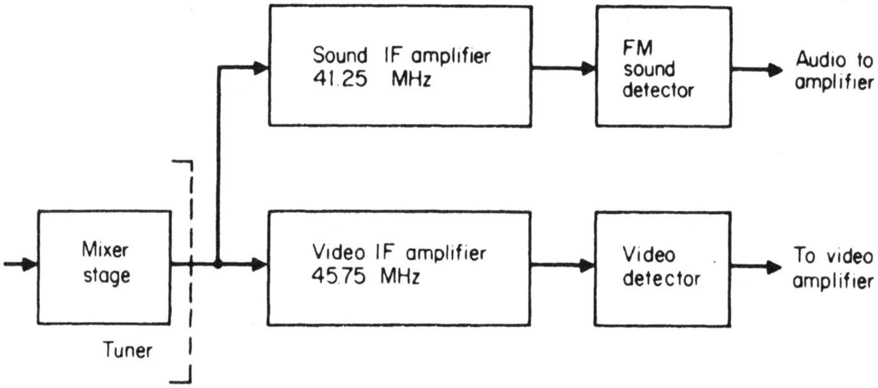


**Figure 12.24** Split-carrier sound system.

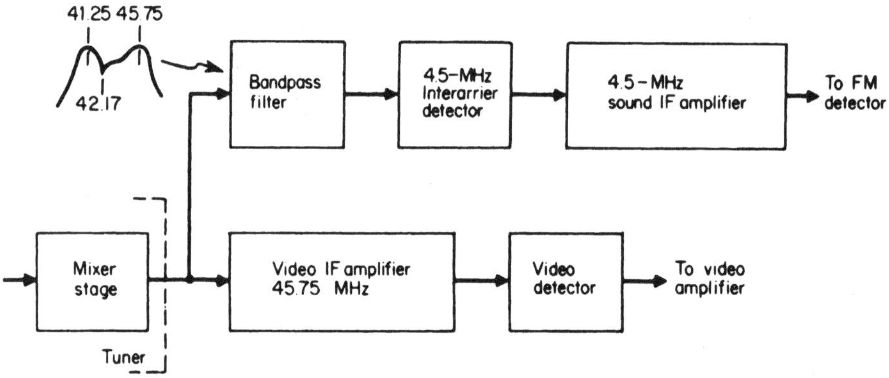


**Figure 12.25** Quasi-parallel sound system.

Four types of detector circuits typically are used in ICs for demodulation of the FM sound carrier:

- The *quadrature detector*, also known as the *gated coincidence detector* and *analog multiplier*, measures the instantaneous phase shift across a reactive circuit as the carrier frequency shifts. At center frequency (zero deviation) the LC phase network gives a 90'° phase shift to $V_2$ compared with $V_1$. As the carrier deviates, the phase shift changes proportionately to the amount of carrier deviation and direction.
- The *balanced peak detector*, which utilizes two peak or envelope detectors, a differential amplifier, and a frequency-selective circuit or piezoceramic discriminator.

- The *differential peak detector*, which operates at a low voltage level and does not require square-wave switching pulses. Therefore, it creates less harmonic radiation than the quadrature detector. In some designs, a low-pass filter is placed between the limiter and peak detector to further reduce harmonic radiation and increase AM rejection.
- The *phase-locked-loop detector*, which requires no frequency-selective LC network to accomplish demodulation. In this system, the voltage-controlled oscillator (VCO) is phase-locked by the feedback loop into following the deviation of the incoming FM signal. The low-frequency error voltage that forces the VCO to track is—in fact—the demodulated output.

## 12.2h Video Amplifiers

A range of video signals of 1 to 3 V at the second detector has become standard for many practical reasons, including optimum gain distribution between RF, IF, and video sections and distribution of signal levels so that video detection and sync separation may be effectively performed. The video amplifier gain and output level are designed to drive the picture tube with this input level. Sufficient reserve is provided to allow for low percentage modulation and signal strengths below the AGC threshold.

### Picture Controls

A video gain or contrast control and a brightness or background control are provided to allow the viewer to select the contrast ratio and overall brightness level that produce the most pleasing picture for a variety of scene material, transmission characteristics, and ambient lighting conditions. The contrast control usually provides a 4:1 gain change. This is accomplished either by attenuator action between the output of the video stage and the CRT or by changing the ac gain of the video stage by means of an ac-coupled variable resistor in the emitter circuit. The brightness control shifts the dc bias level on the CRT to raise or lower the video signal with respect to the CRT beam cutoff voltage level. (See Figure 12.26.)

### AC and dc Coupling

For perfect picture transmission and reproduction, it is necessary that all shades of gray are demodulated and reproduced accurately by the display device. This implies that the dc level developed by the video demodulator, in response to the various levels of video carrier, must be carried to the picture tube. Direct coupling or dc restoration is often used, especially in color receivers where color saturation is directly dependent upon luminance level. (See Figure 12.27.)

Many low cost monochrome designs utilize only ac coupling with no regard for the dc information. This eases the high-voltage power supply design as well as simplifying the video circuitry. These sets will produce a picture in which the average value of luminance remains nearly constant. For example, a night scene having a level of 15 to 20 IRE units and no peak-white excursions will tend to brighten toward the luminance level of the typical daytime scene (50 IRE units). Likewise a full-raster white scene with few black excursions will tend to darken to the average luminance level condition by use of partial dc coupling in which a high-resistance path exists between the second detector and the CRT. This path usually has a gain of one-half to one-fourth that of the ac signal path.

The transient response of the video amplifier is controlled by its amplitude and phase characteristics. The low-frequency transient response, including the effects of dc restoration, if used, is

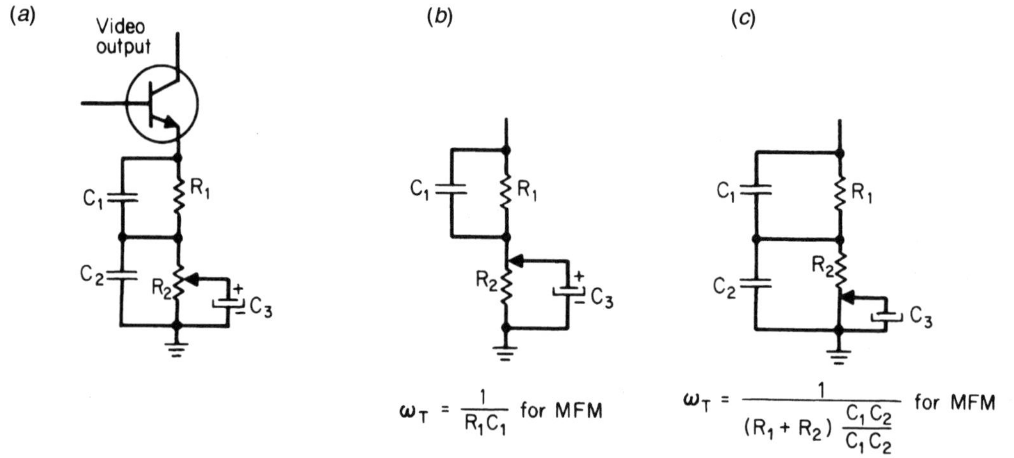


**Figure 12.26** Contrast control circuits: (*a*) contrast control network in the emitter circuit, (*b*) equivalent circuit at maximum contrast (maximum gain), (*c*) minimum contrast.

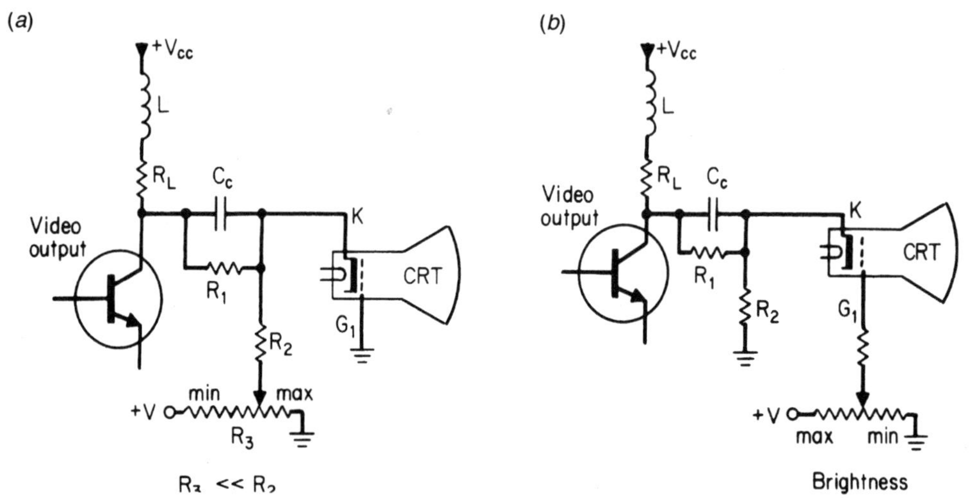


**Figure 12.27** CRT luminance drive circuit: (*a*) brightness control in CRT cathode circuit, (*b*) brightness control in CRT grid circuit.

measured in terms of distortion to the vertical blanking pulse. Faithful reproduction requires that the change in amplitude over the pulse duration, usually a decrease from initial value called *sag* or *tilt*, be less than 5 percent. In general, there is no direct and simple relationship between the sag and the lower 3 dB cutoff frequency. However, lowering the 3 dB cutoff frequency will reduce the tilt, as illustrated in Figure 12.28.

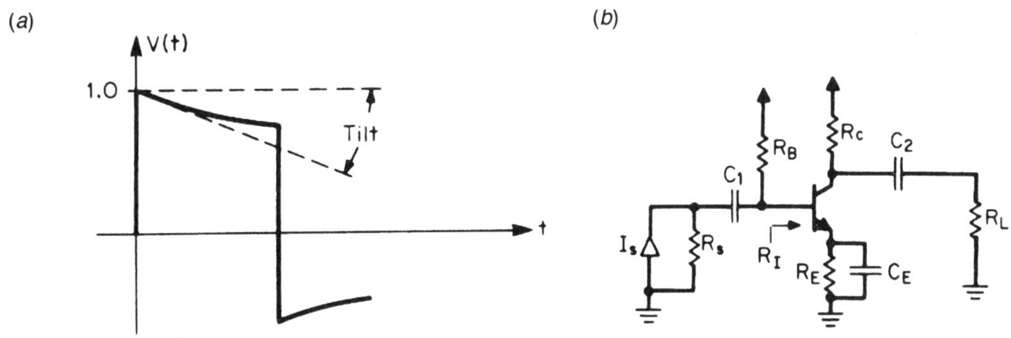


**Figure 12.28** Video stage low-frequency response: (*a*) square-wave output showing tilt, (*b*) RC time constant circuits in the common-emitter stage that affect low-frequency response.

## Low-Frequency Response Requirements

The effect of inadequate low-frequency response appears in the picture as vertical shading. If the response is so poor as to cause a substantial droop of the top of the vertical blanking pulse, then incomplete blanking of retrace lines can occur.

It is not necessary or desirable to extend the low-frequency response to achieve essentially perfect LF square-wave reproduction. First, the effect of tilt produced by imperfect LF response is modified if dc restoration is employed. Direct-current restorers, particularly the fast-acting variety, substantially reduce tilt, and their effect must be considered in specifying the overall response. Second, extended LF response makes the system more susceptible to instability and low-frequency interference. Current coupling through a common power supply impedance can produce the low-frequency oscillation known as "motorboating." Motorboating is not usually a problem in television receiver video amplifiers because they seldom employ the number of stages required to produce regenerative feedback, but in multistage amplifiers the tendency toward motorboating is reduced as the LF response is reduced.

A more commonly encountered problem is the effect of airplane flutter and *line bounce*. Although a fast-acting AGC can substantially reduce the effects of low-frequency amplitude variations produced by airplane reflections, the effect is so annoying visually as to warrant a sacrifice in LF response to bring about further reduction. A transient in-line voltage amplitude, commonly called a line bounce, also can produce an annoying brightness transient that can similarly be reduced through a sacrifice of LF response. Special circuit precautions against line bounce include the longest possible power supply time constant, bypassing the picture tube electrodes to the supply instead of ground, and the use of coupling networks to attenuate the response sharply below the LF cutoff frequency. The overall receiver response is usually an empirically determined compromise.

The high-frequency transient characteristic is usually expressed as the amplifier response to an ideal input voltage or current step. This response is shown in Figure 12.29 and described in the following terms:

- Rise time $\tau_R$ is the time required for the output pulse to rise from 10 to 90 percent of its final (steady-state) value.

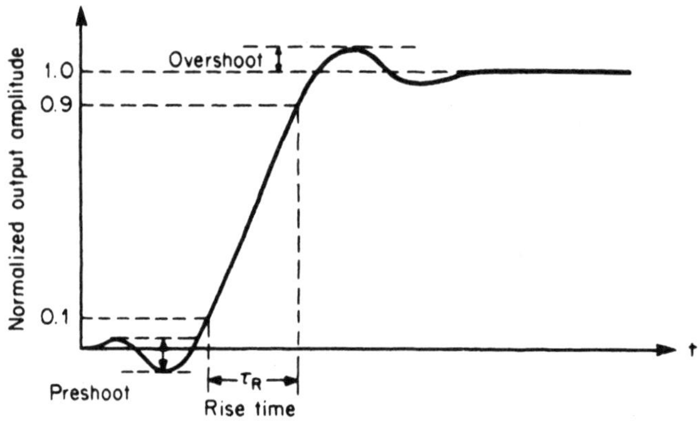


**Figure 12.29** Video amplifier response to a step input.

- Overshoot is the amplitude by which the transient rise exceeds its final value, expressed as a percentage of the final value.
- Preshoot is the amplitude by which the transient oscillatory output waveform exceeds its initial value.
- Smear is an abnormally slow rise as the output wave approaches its final value.
- Ringing is an oscillatory approach to the final value.

In practice, rise times of 0.1 to 0.2 µs are typical. Overshoot, smear, and ringing amplitude are usually held to 5 percent of the final value, and ringing is restricted to one complete cycle.

## 12.2i Color Receiver Luminance Channel

Suppression of the chroma subcarrier is necessary to reduce objectionable dot crawl in and around colored parts of the picture, as well as reduce the distortion of luminance levels resulting from the nonlinear transfer characteristic of CRT electron guns. Traditionally, a simple high-*Q* LC trap, centered around the color subcarrier, has been used for rejection, but this necessitates a trade off between luminance channel bandwidth and the stop band for the chroma sidebands. Luminance channel comb filtering largely avoids this compromise and is one reason why it is commonly used.

The luminance channel also provides the time delay required to correct the time delay registration with the color difference signals, which normally incur delays in the range of from 300 to 1000 ns in their relatively narrow-bandwidth filters.

While delay circuits having substantially flat amplitude and group delay out to the highest baseband frequency of interest can and have been used, this is not necessarily required nor desirable for cost-effective overall design. Because the other links in the chain (i.e., tuner, IF, traps at 4.5 and 3.58 MHz, and CRT driver stage) may all contribute significant linear distortion individually, it is frequently advantageous to allow these distortions to occur and use the delay block as an overall group delay and/or amplitude equalizer.

Although it is well known that, for "distortionless" transmission, a linear system must possess both uniform amplitude and group delay responses over the frequency band of interest, the limitations of a finite bandwidth lead to noticeably slower rise and fall times, rendering edges less sharp or distinct. By intentionally distorting the receiver amplitude response and boosting the relative response to the mid and upper baseband frequencies to varying degrees, both faster rise and fall times can be developed along with enhanced fine detail. If carried too far, however, objectionable outlining can occur, especially to those transients in the white direction. Furthermore, the visibility of background noise is increased.

For several reasons—including possible variations in transient response of the transmitted signal, distortion due to multipath, antennas, receiver tolerances, SNR, and viewer preference—it is difficult to define a fixed response at the receiver that is optimum under all conditions. Therefore, it is useful to make the amplitude response variable so it can be controlled to best suit the individual situation. Over the range of adjustment, it is assumed that the overall group delay shall remain reasonably flat across the video band. The exact shape of the amplitude response is directly related to the desired time domain response (height and width of preshoot, overshoot, and ringing) and chroma subcarrier sideband suppression.

Because the peaked signal later operates on the nonlinear CRT gun characteristics, large white preshoots and overshoots can contribute to excessive beam currents, which can cause CRT spot defocusing. To alleviate this, circuits have been developed that compress large excursions of the peaking component in the white direction. For best operation, it is desirable that the signals being processed have equal preshoot and overshoot.

Low level, high frequency noise in the luminance channel can be removed by a technique called *coring*. One coring technique involves nonlinearly operating on the peaking or edge-enhancement signal, discussed earlier in this section. The peaking signal is passed through an amplifier having low or essentially no gain in the mid-amplitude range. When this modified peaking signal is added to the direct video, the large transitions will be enhanced, but the small ones (noise) will not be, giving the illusion that the picture sharpness has been increased while the noise has been decreased.

## 12.2j Chroma Subcarrier Processing

In the *equiband chroma* system, typical of practically all consumer receivers, the chroma amplifier must be preceded by a bandpass filter network that complements the chroma sideband response produced by the tuner and IF. Frequencies below 3 MHz also must be attenuated to reduce not only possible video cross-color disturbances but also crosstalk caused by the lower-frequency *I* channel chroma information.

Another requirement is that the filter have a gentle transition from passband to stop band in order to impart a minimum amount of group delay in the chroma signal, which then must be compensated by additional group delay circuitry in the luminance channel. The fourth-order high-pass filter is a practical realization of these requirements.

As described previously, this stage also serves as the chroma gain control circuit. The usual implementation in an IC consists of a differential amplifier having the chroma signal applied to the current source. The gain-control dc voltage is applied to one side of the differential pair to divert the signal current away from the output side.

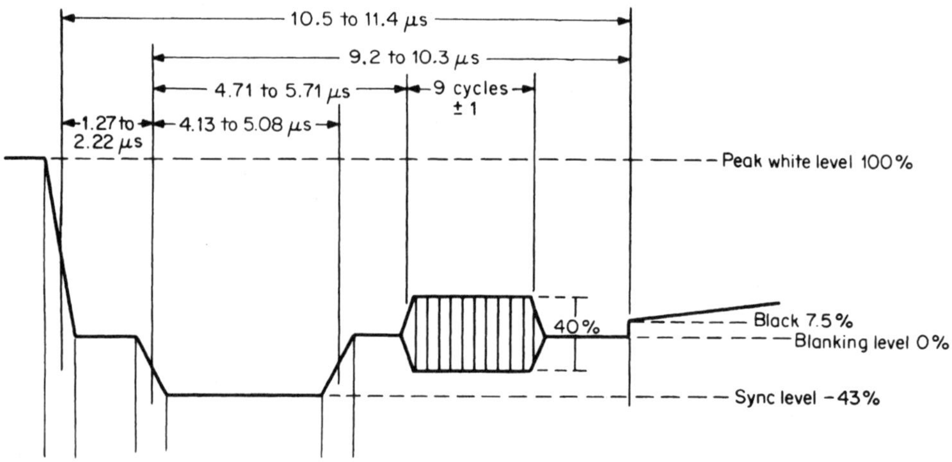


**Figure 12.30** Horizontal blanking interval specification.

## Burst Separation

Complete separation of the color synchronizing burst from video requires *time gating*. The gate requirements are largely determined by the horizontal sync and burst specifications, illustrated in Figure 12.30. It is essential that all video information be excluded. It is also desirable that both the leading and trailing edges of burst be passed so that the complementary phase errors introduced at these points by quadrature distortion average to zero. Widening the gate pulse to minimize the required timing accuracy has a negligible effect on the noise performance of the reference system and may be beneficial in the presence of echoes. The ≈ 2 μs spacing between the trailing edges of burst and horizontal blanking determines the total permissible timing variation. Noise modulation of the gate timing should not permit noise excursions to encroach upon the burst because the resulting cross modulation will have the effect of increasing the noise power delivered to the reference system.

## Burst Gating Signal Generation

The gate pulse generator must provide both steady-state phase accuracy and reasonable noise immunity. The horizontal flyback pulse has been widely used for burst gating because it is derived from the horizontal scan oscillator system, which meets the noise immunity requirements and, with appropriate design, can approximate the steady-state requirements. A further improvement in steady-state phase accuracy can be achieved by deriving the gating pulse directly from the trailing edge of the horizontal sync pulse. This technique is utilized in several chroma system ICs.

The burst gate in conventional discrete component circuits has the form of a conventional amplifier that is biased into linear conduction only during the presence of the gating pulse. In the IC implementation, the complete chroma signal is usually made available at one input of the *automatic phase control* (APC) burst-reference phase detector. The gating pulse then enables the phase detector to function only during the presence of burst.

## Color Subcarrier Reference Separation

The color subcarrier reference system converts the synchronizing bursts to a continuous carrier of identical frequency and close phase tolerance. Theoretically, the long-term and repetitive phase inaccuracies should be restricted to the same value, approximately ±5°. Practically, if transmission variations considerably in excess of this value are encountered, and if operator control of phase ("hue control") is provided, the long-term accuracy need not be so great. Somewhat greater instantaneous inaccuracies can be tolerated in the presence of thermal noise so that an rms phase error specification of 5 to 10° at an S/N of unity may be regarded as typical.

## Reference Systems

Three types of reference synchronization systems have been used:

- Automatic phase control of a VCO
- Injection lock of a crystal
- Ringing of a crystal filter

Best performance can be achieved by the APC loop. In typical applications, the figure of merit can be made much smaller (better) for the APC loop than for the other systems by making the factor $(1/y) + m$ have a value considerably less than 1, even as small as 0.1. The parts count for each type system, at one time much higher for the APC system, is no longer a consideration because of IC implementations where the oscillator and phase detector are integrated and only the resistors and capacitors of the filter network and oscillator crystal are external.

The APC circuit is a phase-actuated feedback system consisting of three functional components:

- A phase detector
- Low-pass filter
- DC voltage-controlled oscillator

The overall system is illustrated in Figure 12.31 The characteristics of these three units define both the dynamic and static loop characteristics and hence the overall system performance.

The phase detector generates a dc output $E$ whose polarity and amplitude are proportional to the direction and magnitude of the relative phase difference $d\phi$ between the oscillator and synchronizing (burst) signals.

The VCO is an IC implementation that requires only an external crystal and simple phase-shift network. The oscillator can be shifted ±45° by varying the phase-control voltage. This leads to symmetrical pull-in and hold-in ranges.

## Chroma Demodulation

The chroma signal can be considered to be made up of two amplitude-modulated carriers having a quadrature phase relationship. Each of these carriers can be individually recovered through the

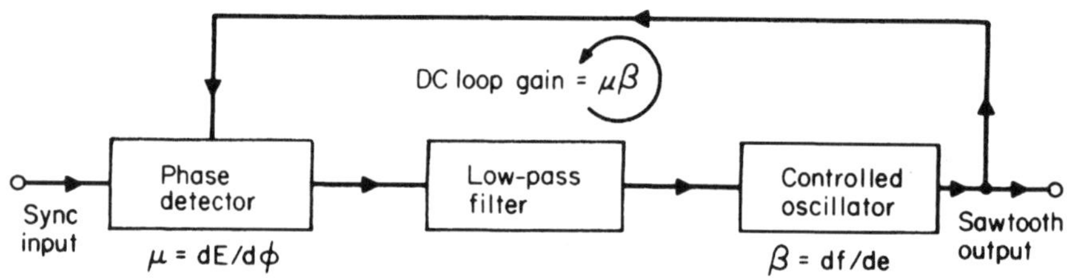


**Figure 12.31** Automatic phase control (APC) system block diagram.

use of a synchronous detector. The reference input to the demodulator is that phase which will demodulate only the *I* signal. The output contains no *Q* signal.

Demodulation products of 7.16 MHz in the output signal can contribute to an optical interference moiré pattern in the picture. This is related to the line geometry of the shadow mask. The 7.16 MHz output also can result in excessively high line-terminal radiation from the receiver. A first-order low-pass filter with cutoff of 1 to 2 MHz usually provides sufficient attenuation. In extreme cases, an LC trap may be required.

### Demodulation Axis

Over the double sideband region (±500 kHz around the subcarrier frequency) the chrominance signal can be demodulated as pure quadrature AM signals along either the *I* and *Q* axis or R–Y and B–Y axis. The latter signal leads to a simpler matrix for obtaining the color drive signals R, G, and B.

Current practice has moved away from the classic demodulation angles for two main reasons:

- Receiver picture tube phosphors have been modified to yield greater light output and can no longer produce the original NTSC primary colors. The chromaticity coordinates of the primary colors, as well as the RGB current ratios to produce white balance, vary from one CRT manufacturer to another.
- The white-point setup has, over the years, moved from the cold 9300 K of monochrome tubes to the warmer illuminant C and D65, which produce more vivid colors, more representative of those that can be seen in natural sunlight.

## 12.2k RGB Matrixing and CRT Drive

Because RGB primary color signals driving the display are required as the end output in a television receiver, it is necessary to combine or "matrix" the demodulated color-difference signals with the luminance signal. Several circuit configurations can be used to accomplish this task.

In the color-difference drive matrixing technique, R–Y, G–Y, and B–Y signals are applied to respective control grids of the CRT while luminance is applied to all three cathodes; the CRT thereby matrixes the primary colors. This approach has the advantage that gray scale is not a function of linearity matching among the three channels because at any level of gray, the color-difference driver stages are at the same dc level. Also, because the luminance driver is common, any dc drift shows up only as a brightness shift. Luminance channel frequency response uniformity is ensured by the common driver.

RGB drive, wherein RGB signals are applied to respective cathodes and $G_1$ is dc biased, requires less drive and has none of the potential color fidelity errors of the color-difference system. RGB drive, however, places higher demands on the drive amplifiers for linearity, frequency response, and dc stability, plus requiring a matrixing network in the amplifier chain.

Low-level RGB matrixing and CRT drive are commonly used, especially with CRT devices that have unitized guns in which the common $G_1$ and $G_2$ elements require differential cathode bias adjustments and drive adjustments to yield gray-scale tracking. In this technique, RGB signals are matrixed at a level of a few volts and then amplified to a higher level (100 to 200 V) suitable for CRT cathode drive.

Direct current stability, frequency response, and linearity of the three stages, even if somewhat less than ideal, should be reasonably well matched to ensure overall gray-scale tracking. Bias and gain adjustments should be independent in their operation, rather than interdependent, and should minimally affect those characteristics listed previously in this section.

Figure 12.32 illustrates a simple example of one of the three CRT drivers. If the amplifier black-level bias voltage equals the black level from the RGB decoder, drive adjustment will not change the amplifier black-level output voltage level or affect CRT cutoff. Furthermore, if $R_B >> R_E$, drive level will be independent of bias setting. Note also that frequency response-determining networks are configured to be unaffected by adjustments.

Frequently, the shunt peaking coil can be made common to all three channels, because differences between the channels are predominantly color-difference signals of relatively narrow bandwidth. Although the frequency responses could be compensated to provide the widest possible bandwidth, this is usually not necessary when the frequency response of preceding low-level luminance processing (especially the peaking stage) is factored in. One exception in which output stage bandwidth must be increased to its maximum is in an application, television receiver or video monitor, where direct RGB inputs are provided for auxiliary services, such as computers, teletext, and S-VHS wideband VCRs.

## Comb Filter Processing

The frequency spectrum of a typical NTSC composite video signal is shown in Figure 12.33*a*. A comb filter, characterized by 100 percent transmission of desired frequencies of a given channel and substantially zero transmission for the undesired interleaved signal spectrum, can effectively separate chroma and luminance components from the composite signal. Such a filter can be easily made, in principle, by delaying the composite video signal one horizontal scan period (63.555 μs in NTSC-M) and adding or subtracting to the undelayed composite video signal (Figure 12.33*b* and *c*).

The output of the sum channel will have frequencies at *f* (horizontal), and all integral multiples thereof reinforce in phase, while those interleaved frequencies will be out of phase and will cancel. This can be used as the luminance path. The difference channel will have integral frequency multiples cancel while the interleaved ones will reinforce. This channel can serve as the chrominance channel. The filter characteristic and interleaving are shown in Figure 12.33*c*.

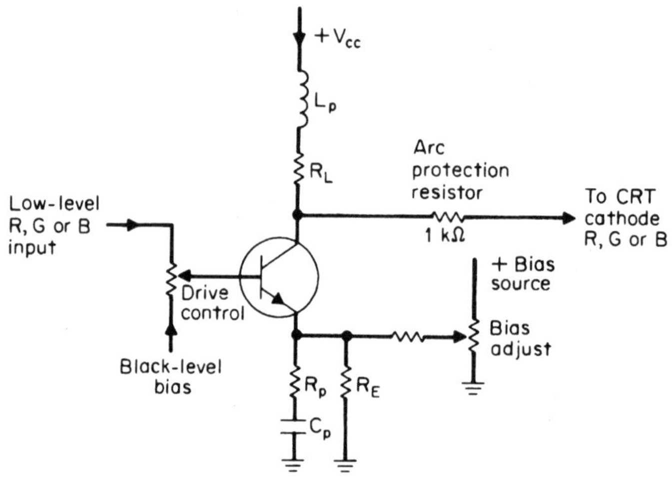


**Figure 12.32** Simplified R, G, and B output stage.

## Automatic Circuits

The relative level of the chroma subcarrier in the incoming signal is highly sensitive to transmission path disorders, thereby introducing objectionable variations in saturation. These can be observed between one received signal and another or over a period of time on the same channel unless some adaptive correction is built into the system. The color burst reference, transmitted at 40 IRE units peak-to-peak, is representative of the same path distortions and is normally used as a reference for automatic gain controlling the chroma channel. A balanced peak detector or synchronous detector, having good noise rejection characteristics, detects the burst level and provides the control signal to the chroma gain-controlled stage.

Allowing the receiver chroma channel to operate during reception of a monochrome signal will result in unnecessary cross color and colored noise, made worse by the ACC increasing the chroma amplifier gain to the maximum. Most receivers, therefore, cut off the chroma channel transmission when the received burst level goes below approximately 5 to 7 percent. Hysteresis has been used to minimize the flutter or threshold problem with varying signal levels.

*Burst-referenced* ACC systems perform adequately when receiving correctly modulated signals with the appropriate burst-to-chroma ratio. Occasionally, however, burst level may not bear a correct relation to its accompanying chroma signal, leading to incorrectly saturated color levels. It has been determined that most viewers are more critical of excessive color levels than insufficient ones. Experience has shown that when peak chroma amplitude exceeds burst by greater than 2/1, limiting of the chroma signal is helpful. This threshold nearly corresponds to the amplitude of a 75 percent modulated color bar chart (2.2/1). At this level, negligible distortion is introduced into correctly modulated signals. Only those that are nonstandard are affected significantly. The output of the peak chroma detector also is sent to the chroma gain-controlled stage.

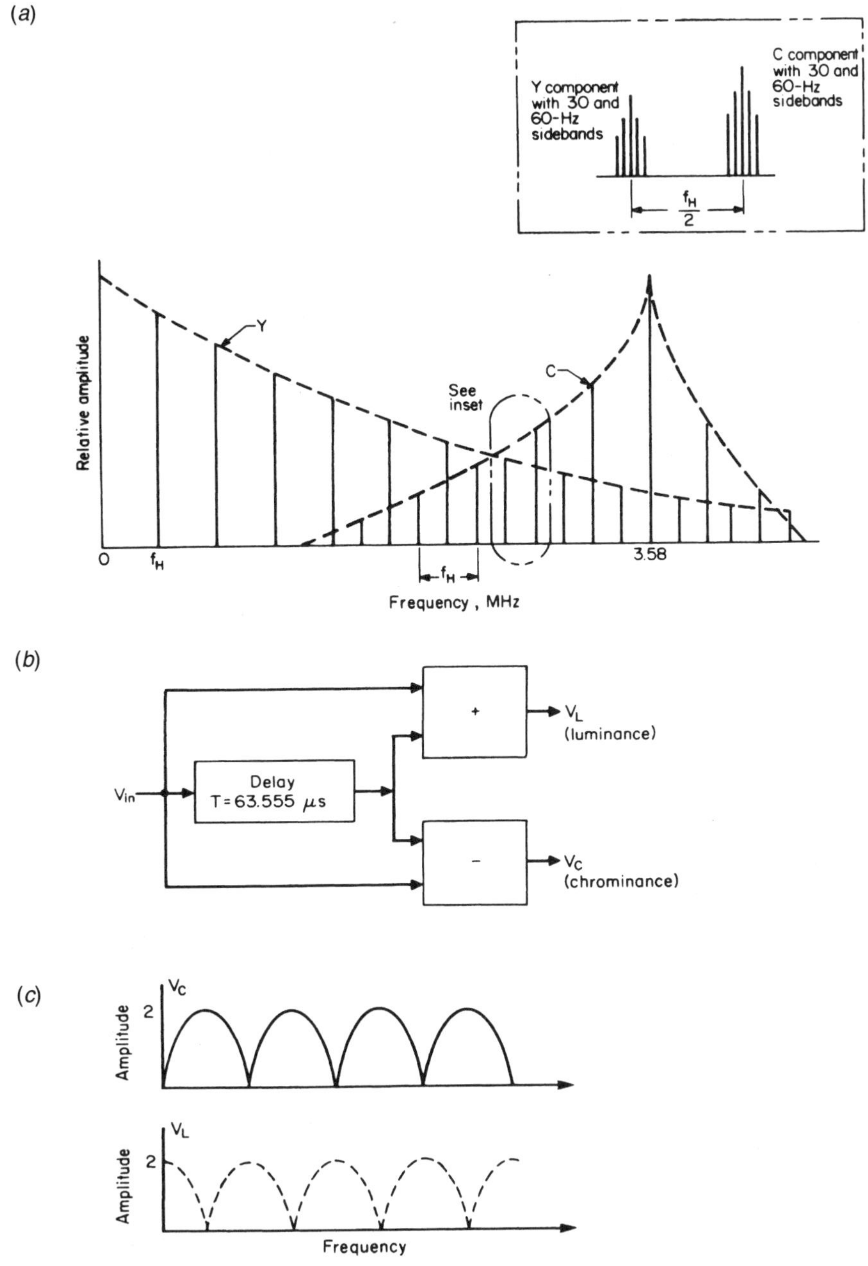


**Figure 12.33** Color system filtering: (*a*) frequency spectrum of NTSC color system, showing interleaving of signals, (*b*) simplified 1H delay comb filter block diagram, (*c*) chrominance $V_C$ and luminance $V_L$ outputs of the comb filter.

### Tint

One major objective in color television receiver design is to minimize the incidence of flesh-tone reproduction with incorrect hue. Automatic hue-correcting systems can be categorized into two classes:

- Static flesh-tone correction, achieved by selecting the chroma demodulating angles and gain ratios to desensitize the resultant color-difference vector in the flesh-tone region (+*I* axis). The demodulation parameters remain fixed, but the effective *Q* axis gain is reduced. This has the disadvantage of distorting hues in all four quadrants.
- Dynamic flesh-tone corrective systems, which can adaptively confine correction to the region within several degrees of the positive *I* axis, leaving all other hues relatively unaffected. This is typically accomplished by detecting the phase of the incoming chroma signal and modulating the phase angle of the demodulator reference signal to result in an effective phase shift of 10 to 15° toward the *I* axis for a chroma vector that lies within 30° of the I axis. This approach produces no amplitude change in the chroma. In fact, for chroma saturation greater than 70 percent, the system is defeated on the theory that the color is not a flesh tone.

A simplification in circuitry can be achieved if the effective correction area is increased to the entire positive-*I* 180° sector. A conventional phase detector can be utilized and the maximum correction of approximately 20° will occur for chroma signals having phase of ±45° from the *I* axis. Signals with phase greater or less than 45° will have increasingly lower correction values, as illustrated in Figure 12.34.

The *vertical interval reference* (VIR) signal, as shown in Figure 12.34, provides references for black level, luminance, and—in addition to burst—a reference for chroma amplitude and phase. While originally developed to aid broadcasters, it has been employed in television receivers to correct for saturation and hue errors resulting from transmitter or path distortion errors.

## 12.21 Scanning Synchronization

The scan-synchronizing signal, consisting of the horizontal and vertical sync pulses, is removed from the composite video signal by means of a sync-separator circuit. The classic approach has been to ac-couple the video signal to an overdriven transistor amplifier (illustrated in Figure 12.35) that is biased off for signal levels smaller than $V_{CO}$ and saturates for signal levels greater than $V_{sat}$, a range of approximately 0.1 V.

Vertical synchronizing information can be recovered from the output of the sync separator by the technique of integration. The classic two-section RC integrator provides a smooth, ramp waveform that corresponds to the vertical sync block as shown in Figure 12.36. The ramp is then sent to a *relaxation* or *blocking* oscillator, which is operating with a period slightly longer than the vertical frame period. The upper part of the ramp will trigger the oscillator into conduction to achieve vertical synchronization. The oscillator will then reset and wait for the threshold level of the next ramp. The vertical-hold potentiometer controls the free-running frequency or period of the vertical oscillator.

One modification to the integrator design is to reduce the integration (speed up the time constants) and provide for some differentiation of the waveform prior to applying it to the vertical oscillator (shown in Figure 12.37). Although degrading the noise performance of the system, this technique provides a more certain and repeatable trigger level than the full integrator, thereby leading to an improvement in interlace over a larger portion of the hold-in range. A second bene-

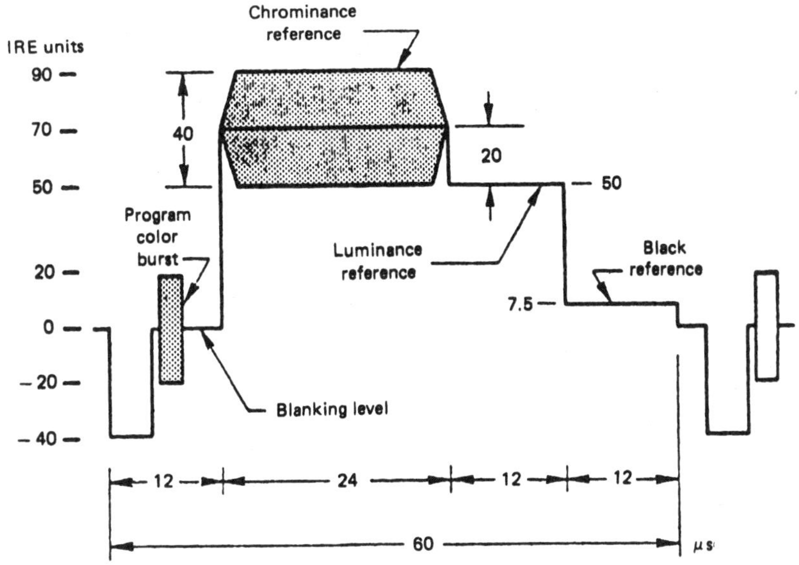


**Figure 12.34** Vertical interval reference (VIR) signal. Note that the chrominance and the program color burst are in phase.

fit is to provide a more stable vertical lock when receiving signals that have distorted vertical sync waveforms. These can be generated by video cassette recorders in nonstandard playback modes, such as fast/slow forward, still, and reverse. Prerecorded tapes having antipiracy nonstandard sync waveforms also contribute to the problem.

## Vertical Countdown Synchronizing Systems

A vertical scan system using digital logic elements can be based upon the frequency relationship of 525/2 that exists between the horizontal and vertical scans. Such a system can be considered to be phase-locked for both horizontal scan and vertical scan, which will result in improved noise immunity and picture stability. Vertical sync is derived from a pulse train having a frequency of twice the horizontal rate. The sync is therefore precisely timed for both even and odd fields, resulting in excellent interlace. This system requires no hold control.

The block diagram of one design is shown in Figure 12.38. The 31.5 kHz clock input is converted to horizontal drive pulses by a divide-by-2 flip-flop. A two-mode counter, set for 525 counts for standard interlaced signals and 541 for noninterlaced signals, produces the vertical output pulse. Noninterlaced signals are produced by a variety of VCR cameras, games, and picture test generators. The choice of 541 allows the counter to continue until the arrival of vertical sync from the composite video waveform. The actual vertical sync then resets the counter.

Other systems make use of clock frequencies of 16 and 32 times the horizontal frequency. Low-cost crystal or ceramic resonator-controlled oscillators can be built to operate at these fre-

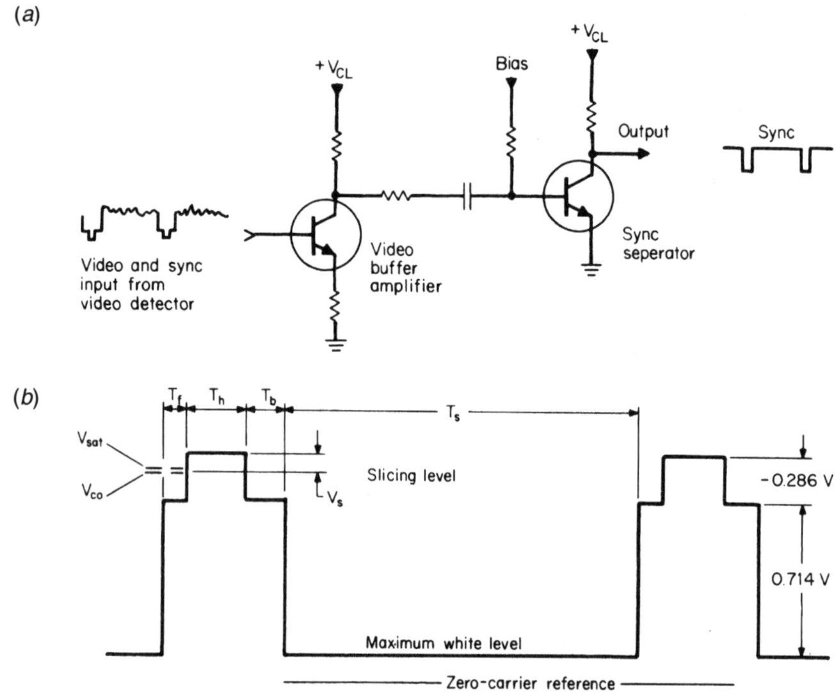


**Figure 12.35** Sync separator: (*a*) typical circuit, (*b*) sync waveform.

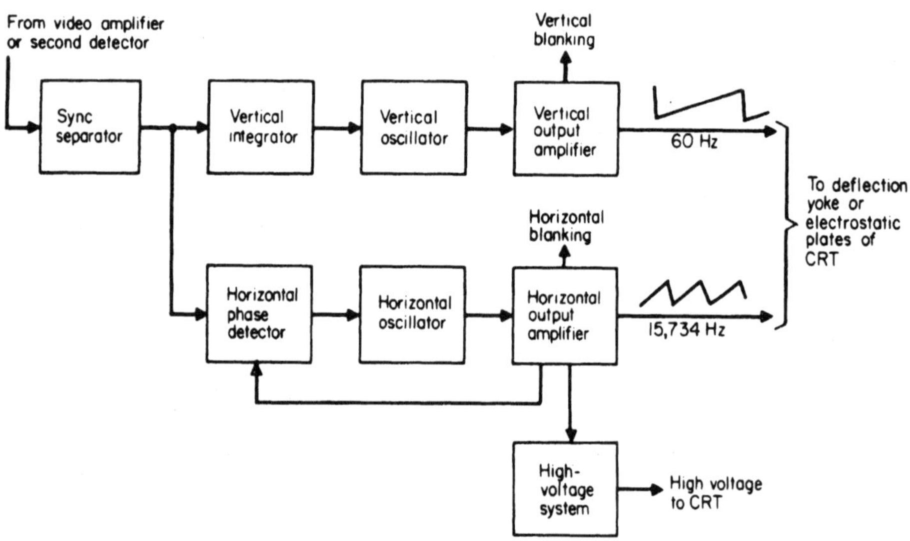


**Figure 12.36** Picture scanning section functional block diagram.

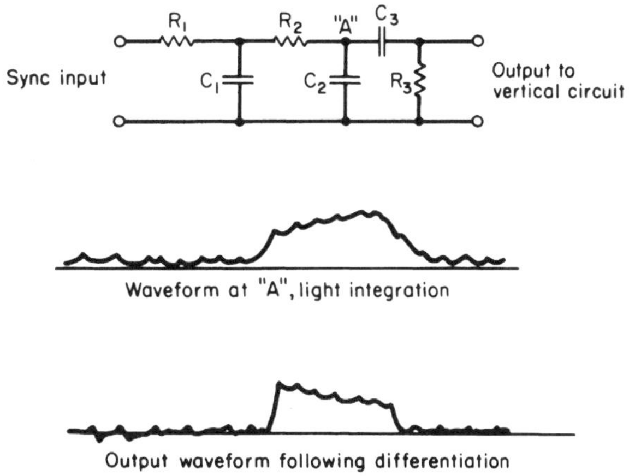


**Figure 12.37** Modified integrator followed by a differentiator.

quencies. As in the first system, dual-mode operation is necessary in order to handle standard interlaced and nonstandard sync waveforms. An exact 525/2 countdown is used with interlaced signals. For noninterlaced signals, the systems usually operate in a free-running mode with injection of the video-derived sync pulse causing lock.

A critical characteristic, and probably the most complex portion of any countdown system, is the circuitry that properly adjusts for nonstandard sync waveforms. These can be simple 525 noninterlaced fields, distorted vertical sync blocks, blocks having no horizontal serrations, fields with excessive or insufficient lines, and a combination of the above.

## Impulse Noise Suppression

The simplest type suppression that will improve the basic circuit in a noise environment of human origin consists of a parallel resistor and capacitor in series with the sync charging capacitor. Variations of this simple circuit, which include diode clamps and switches, have been developed to speed up the noise suppression performance while not permitting excessive tilt in the sync output during the vertical block.

More complex solutions to the noise problem include the noise canceler and noise gate. The canceler monitors the video signal between the second detector and the sync separator. A noise spike, which exceeds the sync or pedestal level, is inverted and added to the video after the isolation resistor. This action cancels that part of the noise pulse which would otherwise produce an output from the sync separator.

For proper operation, the canceler circuit must track the sync-tip amplitude from a strong RF signal to the fringe level. The noise gate, in a similar manner, recognizes a noise pulse of large amplitude and prevents the sync separator from conducting, either by applying reverse bias or by opening a transistor switch in the emitter circuit.

## Horizontal Automatic Phase Control

Horizontal scan synchronization is accomplished by means of an APC loop, with theory and characteristics similar to those used in the chroma reference system. Input signals to the horizon-

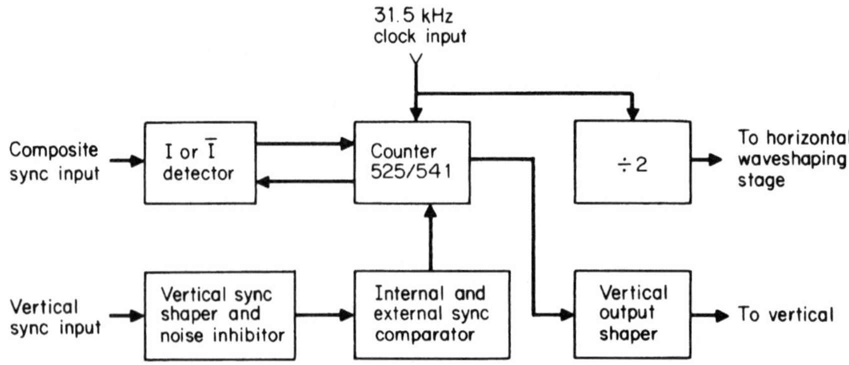


**Figure 12.38** Vertical countdown sync processor, using a $2f_H$ clock.

tal APC loop are sync pulses from the sync separator and horizontal flyback pulses, which are integrated to form a sawtooth waveform. The phase detector compares the phase (time coincidence) of these two waveforms and sends a dc-coupled low-pass filtered error signal to the voltage-controlled oscillator to cause the frequency to shift in the direction of minimal phase error between the two input signals.

The recovery of the APC loop to a step transient input involves the parameters of the natural resonant frequency, as well as the amount of overshoot permitted after the correct phase has been reached. Both of these characteristics can be evaluated by use of a jitter generator, which creates a time base error between alternate fields. The resultant sync error and system dynamics can be seen in the picture display of the receiver shown in Figure 12.39. This type of disturbance occurs when the two or three playback heads of a consumer helical-scan VCR switch tracks. It will also occur when the receiver is operated from a signal that does not have horizontal slices in the vertical sync block.

Phase detector circuitry in the form of classic double-diode bridge detectors, in either the balanced or unbalanced configuration, can sometimes be found in low-cost receivers. Integrated circuits commonly utilize a gated differential amplifier (Figure 12.40.) in which the common feed from the current source is driven by the sync pulse, and the sawtooth waveform derived from the flyback pulse is applied to one side of the differential pair.

## Vertical Scanning

Class-B vertical circuits consist essentially of an audio frequency amplifier with current feedback. This approach maintains linearity without the need for an adjustable linearity control. Yoke impedance changes caused by temperature variations will not affect the yoke current, thus, a thermistor is not required. The current-sensing resistor must, of course, be temperature stable. An amplifier that uses a single NPN and a single PNP transistor in the form of a complementary output stage is given in Figure 12.41. Quasi-complementary, Darlington outputs and other common audio output stage configurations also can be used.

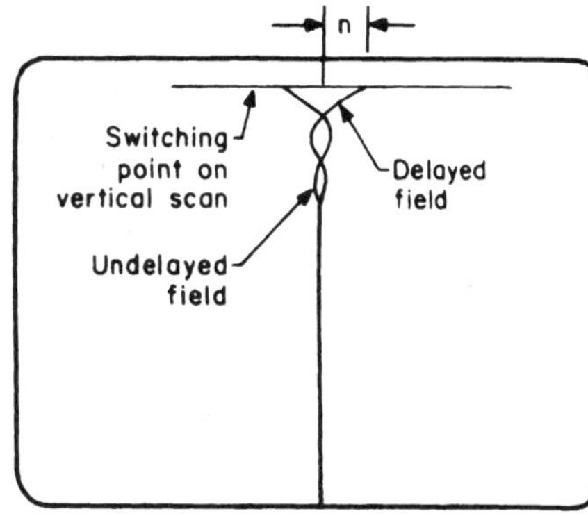


**Figure 12.39** The pattern on a receiver screen displaying a vertical line with the even and odd fields offset by *n* ms.

Establishing proper dc bias for the output stages is quite critical. Too little quiescent current will result in crossover distortion that will impose a faint horizontal white line in the center of the picture even though the distortion may not be detectable on an oscilloscope presentation of the yoke current waveform. Too much quiescent current results in excessive power dissipation in the output transistors.

## Retrace Power Reduction

The voltage required to accomplish retrace results in a substantial portion of the power dissipation in the output devices. The supply voltage to the amplifier and corresponding power dissipation can be reduced by using a *retrace switch* or *flyback generator* circuit to provide additional supply voltage during retrace. One version of a retrace switch is given in Figure 12.42. During the trace time, the capacitor is charged to a voltage near the supply voltage. As retrace begins, the voltage across the yoke goes positive, thus forcing $Q_R$ into saturation. This places the cathode of $C_R$ at the supply potential and the anode at a level of 1.5 to 2 times the supply, depending upon the values of $R_R$ and $C_R$.

## Horizontal Scanning

The horizontal scan system has two primary functions:

- It provides a modified sawtooth-shaped current to the horizontal yoke coils to cause the electron beam to travel horizontally across the face of the CRT.
- It provides drive to the high-voltage or flyback transformer to create the voltage needed for the CRT anode.

Frequently, low-voltage supplies also are derived from the flyback transformer. The major components of the horizontal-scan section consist of a driver stage, horizontal output device (either bipolar transistor or SCR), yoke current damper diode, retrace capacitor, yoke coil, and flyback transformer, as illustrated in Figure 12.43.

During the scan or retrace interval, the deflection yoke may be considered a pure inductance with a dc voltage impressed across it. This creates a sawtooth waveform of current (see Figure 12.44). This current flows through the damper diode during the first half scan. It then reverses

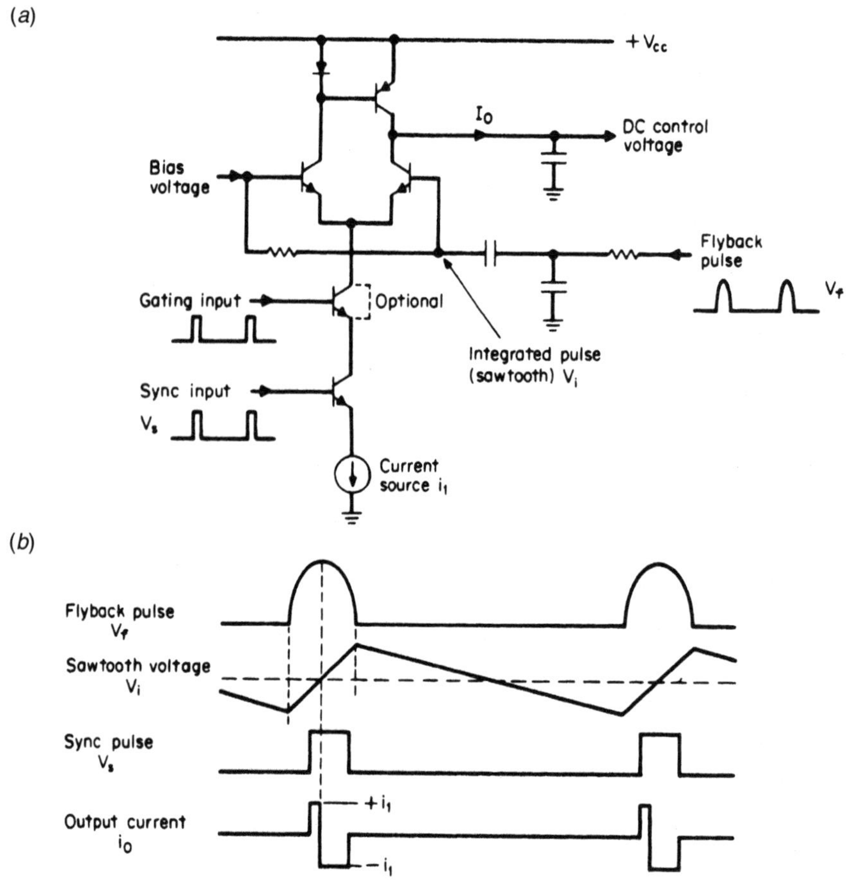


**Figure 12.40** Gated phase detector in an IC implementation: (*a*) circuit diagram, (*b*) pulse-timing waveforms.

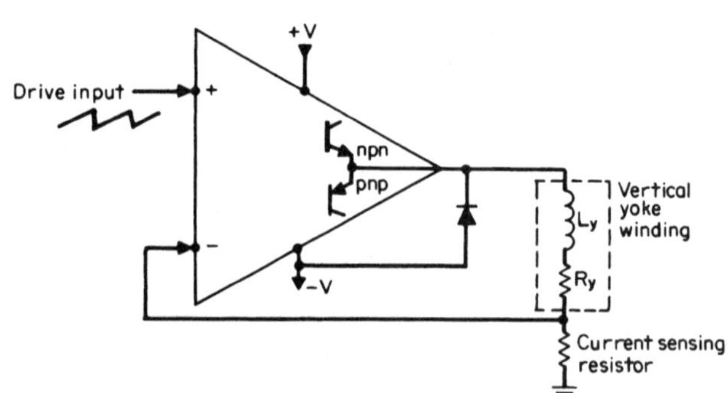


**Figure 12.41** A class-B vertical output stage.

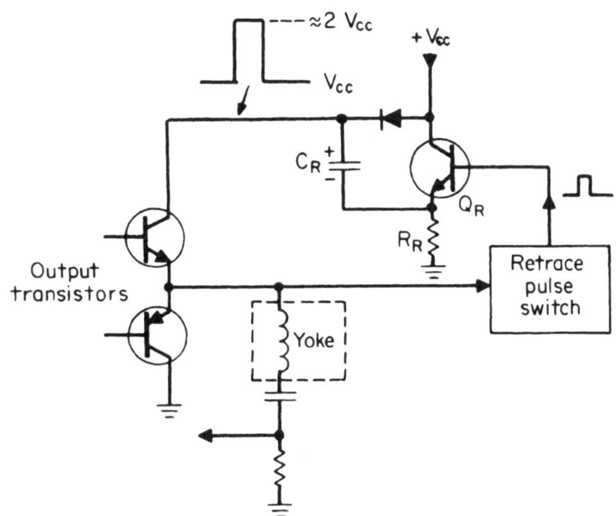


**Figure 12.42** Simplified schematic of a retrace switch.

direction and flows through the horizontal output transistor collector. This sawtooth-current waveform deflects the electron beam across the face of the picture tube. A similarly shaped current flows through the primary winding of the high-voltage output transformer.

At the beginning of the retrace interval, the transformer and yoke inductances transfer energy to the retrace-tuning capacitor and the accompanying stray capacitances, thereby causing a half sine wave of voltage to be generated. This high-energy pulse appears on the transistor collector and is stepped up, via the flyback transformer, to become the high voltage for the picture tube anode. Finally, at the end of the cycle, the damper diode conducts, and another horizontal scan is started.

## 12.2m High Voltage Generation

High voltage in the range 8 to 16 kV is required to supply the anode of monochrome picture tubes. Color tubes have anode requirements in the range 20 to 30 kV and focus voltage requirements of 3 to 12 kV. The horizontal flyback transformer is the common element in the generation of these high voltages. There are three common variations of this design:

- The flyback with a half-wave rectifier
- Flyback driving a voltage multiplier
- Voltage-adding integrated flyback

Consider the simplified horizontal-scan circuit shown in Figure 12.43. The voltage at the top of the high voltage (HV) winding consists of a series of pulses delivered during retrace from the stored energy in the yoke field. The yoke voltage pulses are then multiplied by the turns ratio of

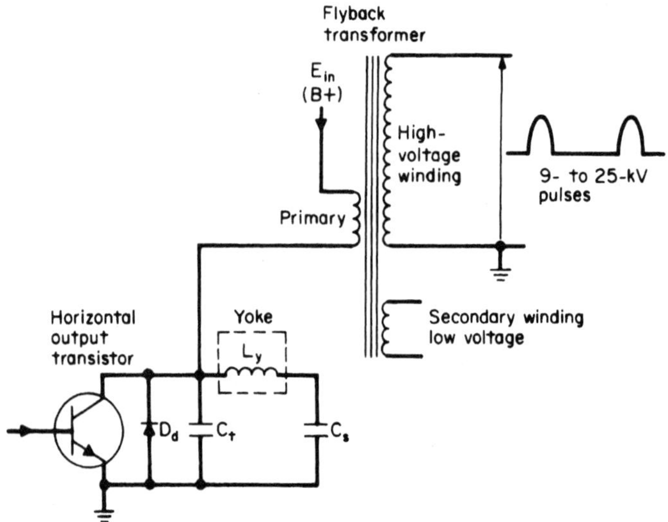


**Figure 12.43** Simplified horizontal scan circuit.

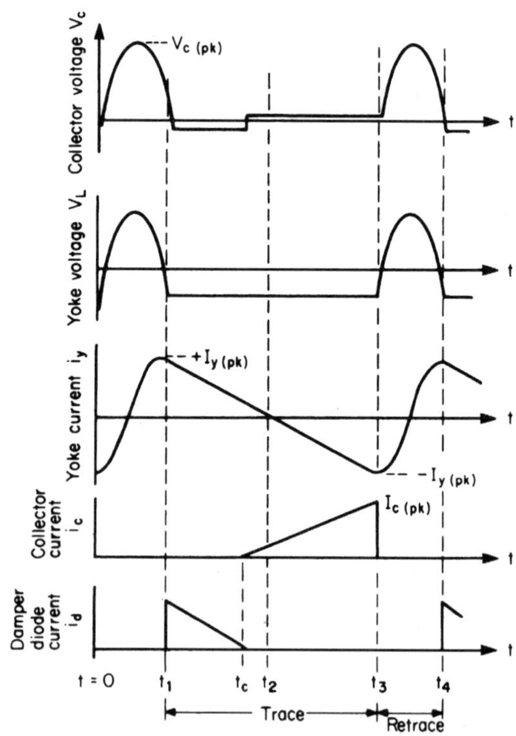


**Figure 12.44** Horizontal scan waveform timing diagrams.

the HV winding to primary winding. The peak voltage across the primary during retrace is given by:

$$V_{p(pk)} = E_{in} 0.8\left(1.79 + 1.57\frac{T_{trace}}{T_{retrace}}\right) \qquad (12.2)$$

Where:
$E_{in}$ = supply voltage (B+)
0.8 = accounts for the pulse shape factor with third harmonic tuning
$T_{trace}$ = trace period, ≈ 52.0 μs
$T_{retrace}$ = retrace period, ≈ 11.5 μs

## Flyback with Half-Wave Rectifier

The most common HV supply for small-screen monochrome and color television receivers uses a direct half-wave rectifier circuit. The pulses at the top of the HV winding are rectified by the single diode, or composite of several diodes in series. The charge is then stored in the capacitance of the anode region of the picture tube. Considerable voltage step-up is required from the primary to the HV winding. This results in a large value of leakage inductance for the HV winding, which decreases its efficiency as a step-up transformer.

Harmonic tuning of the HV winding improves the efficiency by making the total inductance and the distributed capacitance of the winding plus the CRT anode capacitance resonate at an odd harmonic of the flyback pulse frequency. For the single-rectifier circuit, usually the third harmonic resonance is most easily implemented by proper choice of winding configuration, which results in appropriate leakage inductance and distributed capacitance values. This will result in HV pulse waveforms, which will give an improvement in the HV supply regulation (internal impedance) as well as a reduction in the amplitude of ringing in the current and voltage waveforms at the start of scan.

## Voltage Multiplier Circuits

A voltage multiplier circuit consists of a combination of diodes and capacitors connected in such a way that the dc output voltage is greater than the peak amplitude of the input pulse. The tripler was a common HV system for color television receivers requiring anode voltages of 25 to 30 kV dating back to the 1970s. The considerably reduced pulse voltage required from the HV winding resulted in a flyback transformer with a more tightly coupled HV winding.

## Integrated Flybacks

Most medium- and large-screen color receivers utilize an integrated flyback transformer in which the HV winding is segmented into three or four parallel-wound sections. These sections are series-connected with a diode between adjacent segments. These diodes are physically mounted as part of the HV section. The transformer is then encapsulated in high-voltage polyester or epoxy.

Two HV winding construction configurations also have been used. One, the layer or solenoid-wound type, has very tight coupling to the primary and operates well with no deliberate harmonic tuning. Each winding (layer) must be designed to have balanced voltage and capacitance

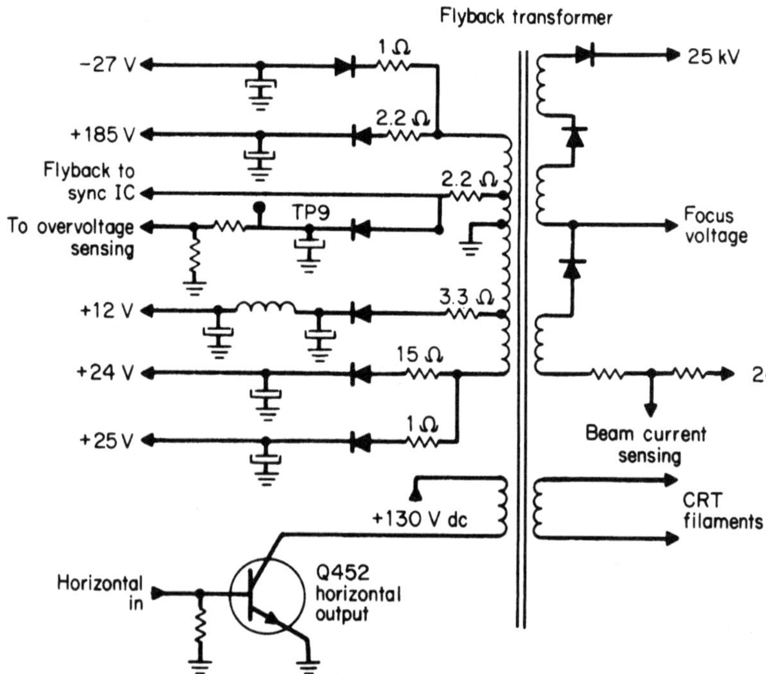


**Figure 12.45** Auxiliary power sources derived from the horizontal output transformer. (*Courtesy of Philips.*)

with respect to the primary. The second, a bobbin or segmented-winding design, has high leakage inductance and usually requires tuning to an odd harmonic (e.g., the ninth). Regulation of this construction is not quite as good as the solenoid-wound primary winding at a horizontal-frequency rate. The +12 V, +24 V, +25 V, and −27 V supplies are scan-rectified. The +185 V, overvoltage sensing, focus voltage, and 25 kV anode voltage are derived by retrace-rectified supplies. The CRT filament is used directly in its ac mode.

Flyback-generated supplies provide a convenient means for isolation between different ground systems, as required for an iso-hot chassis.

## Power Supplies

Most receivers use the flyback pulse from the horizontal transformer as the source of power for the various dc voltages required by the set. Using the pulse waveform at a duty cycle of 10 or 15 percent, by proper winding direction and grounding of either end of the winding, several different voltage sources can be created.

Scan rectified supplies are operated at a duty cycle of approximately 80 percent and are thus better able to furnish higher current loads. Also, the diodes used in such supplies must be capable of blocking voltages that are nine to ten times larger than the level they are producing. Diodes

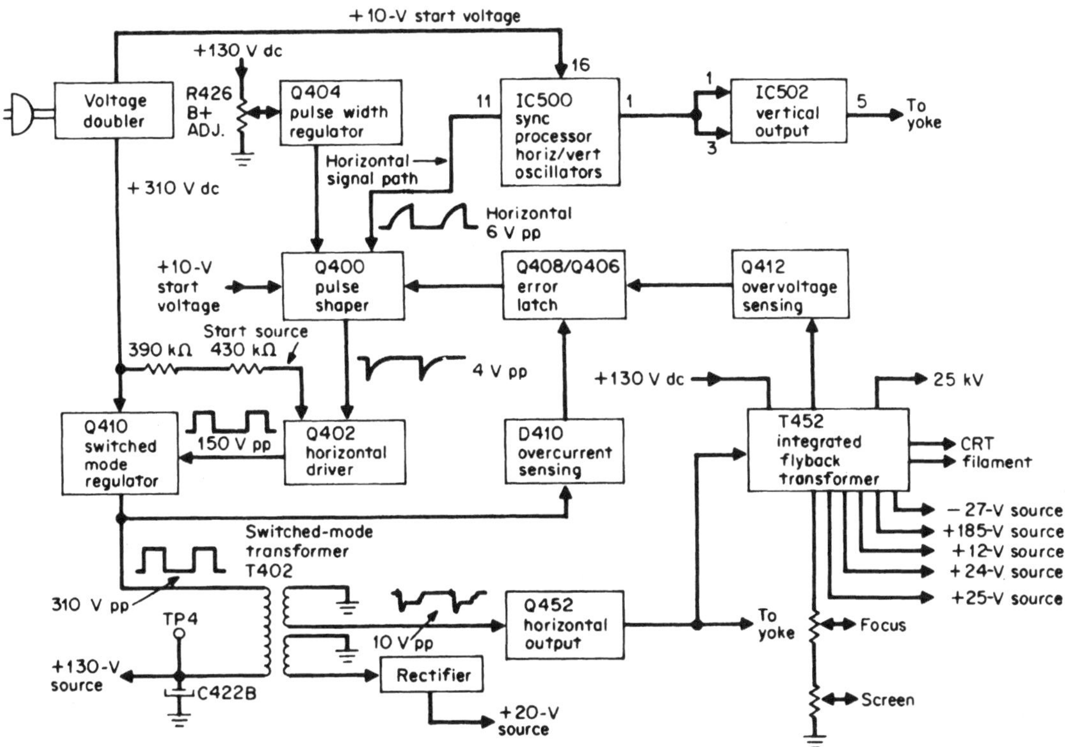


**Figure 12.46** Color receiver power supply using a switched-mode regulator system. (*Courtesy of Philips.*)

having fast-recovery characteristics are used to keep the power dissipation at a minimum during the turn-off interval because of the presence of this high reverse voltage.

A typical receiver system containing the various auxiliary power supplies derived from flyback transformer windings is shown in Figure 12.45. Transistor *Q* 452 switches the primary winding at a horizontal frequency rate. The +12 V, +24 V, +25 V, and –27 V supplies are scan-rectified. The +185 V, overvoltage sensing, focus voltage, and 25 kV anode voltage are derived by retrace-rectified supplies. The CRT filament is used directly in its ac mode.

As noted in the previous section, flyback-generated supplies provide a convenient means for isolation between different ground systems. Figure 12.46 shows the block diagram of such a television receiver power supply system [6, 7].

## 12.3 References


1. FCC Regulations, 47 CFR, 15.65, Washington, D.C.
2. A. DeVries et al, "Characteristics of Surface-Wave Integratable Filters (SWIFS)," *IEEE Trans.*, vol. BTR-17, no. 1, p. 16.
3. Yamada and Uematsu, "New Color TV with Composite SAW IF Filter Separating Sound and Picture Signals," *IEEE Trans.*, vol. CE-28, no. 3, p. 193. Copyright 1982 IEEE.
4. Neal, C. B., and S. Goyal, "Frequency and Amplitude Phase Effects in Television Broadcast Systems," *IEEE Trans.*, vol. CE-23, no. 3, pg. 241, August 1977.
5. Fockens, P., and C. G. Eilers: "Intercarrier Buzz Phenomena Analysis and Cures," *IEEE Trans. Consumer Electronics*, vol. CE-27, no. 3, pg. 381, August 1981.
6. *IEEE Guide for Surge Withstand Capability,* (SWC) Tests, ANSI C37.90a-1974/IEEE Std. 472-1974, IEEE, New York, 1974.
7. "Television Receivers and Video Products," UL 1410, Sec. 71, Underwriters Laboratories, Inc., New York, 1981.

Chapter

# 13

# ATSC DTV Receiver Systems

**Jerry C. Whitaker, Editor**

## 13.1 Introduction

The introduction of a new television system must be viewed as a chain of elements that begins with image and sound pickup and ends with image display and sound reproduction. The DTV receiver is a vital link in the chain. By necessity, the ATSC system places considerable requirements upon the television receiver. The level of complexity of a DTV-compliant receiver is unprecedented, and that complexity is made possible only through advancements in large-scale integrated circuit design and fabrication.

The goal of any one-way broadcasting system, such as television, is to concentrate the hardware requirements at the source as much as possible and to make the receivers—which greatly outnumber the transmitters—as simple and inexpensive as possible. Despite the significant complexity of a DTV receiver, this principal has been an important design objective from the start.

### 13.1a Noise Figure

One of the more important specifications that receiver designers must consider is the *noise figure* (NF). A number of factors enter into the ultimate carrier-to-noise ratio within a given receiver. For example, the receiver planning factors applicable to UHF DTV service are shown in Table 13.1 [1]. A consumer can enhance the noise performance of the installation by improving any contributing factor; examples include installation of a *low-noise amplifier* (LNA) or a high-gain antenna.

The assumptions made in the DTV system planning process were based on *threshold-of-visibility* (TOV) measurements taken during specified tests and procedures [2]. These TOV numbers were correlated to *bit error rate* (BER) results from the same tests. When characterizing the Grand Alliance VSB modem hardware in the lab and in field tests, only BER measurements were taken. In Table 13.2, the results of these tests are expressed in equivalent TOV numbers derived from the BER measurements. It should be noted that the exact amount of adjacent-channel or co-channel interference entering the receiver terminals is a function of the overall antenna gain pattern used (not just the front/back ratio), which is also a function of frequency.

**Table 13.1 Receiver Planning Factors Used by the FCC** (*After* [1].)

| Planning Factors | Low VHF | High VHF | UHF |
|---|---|---|---|
| Antenna impedance (ohms) | 75 | 75 | 75 |
| Bandwidth (MHz) | 6 | 6 | 6 |
| Thermal noise (dBm) | –106.2 | –106.2 | –106.2 |
| Noise figure (dB) | 10 | 10 | 7 |
| Frequency (MHz) | 69 | 194 | 615 |
| Antenna factor (dBm/dBμ) | –111.7 | –120.7 | –130.7[1] |
| Line loss (dB) | 1 | 2 | 4 |
| Antenna gain (dB) | 4 | 6 | 10 |
| Antenna F/B ratio (dB) | 10 | 12 | 14 |
| 1. See Appendix B of the Sixth Report and Order (MM 87-268), adopted April 3, 1997, for a discussion of the dipole factor. | | | |

**Table 13.2 DTV Interference Criteria** (*After* [2].)

| | |
|---|---|
| Co-channel DTV-into-NTSC | 33.8 dB |
| Co-channel NTSC-into-DTV | 2.07 dB |
| Co-channel DTV-into-DTV | 15.91 dB |
| Upper-adjacent DTV-into-NTSC | –16.17 dB |
| Upper-adjacent NTSC-into-DTV | –47.05 dB |
| Upper-adjacent DTV-into-DTV | –42.86 dB |
| Lower-adjacent DTV-into-NTSC | –17.95 dB |
| Lower-adjacent NTSC-into-DTV | –48.09 dB |
| Lower-adjacent DTV-into-DTV | –42.16 dB |

## 13.2 Receiver System Overview

At this writing, second-generation DTV-compliant consumer receivers were entering the marketplace. Because of the rapidly changing nature of the consumer electronics business, this chapter will focus on the overall nature of the DTV receiving subsystem, rather than on any specific implementation.

Figure 13.1 shows a general block diagram of the VSB terrestrial broadcast transmission system. The major circuit elements include:

- Tuner
- Channel filtering and VSB carrier recovery
- Segment sync and symbol clock recovery

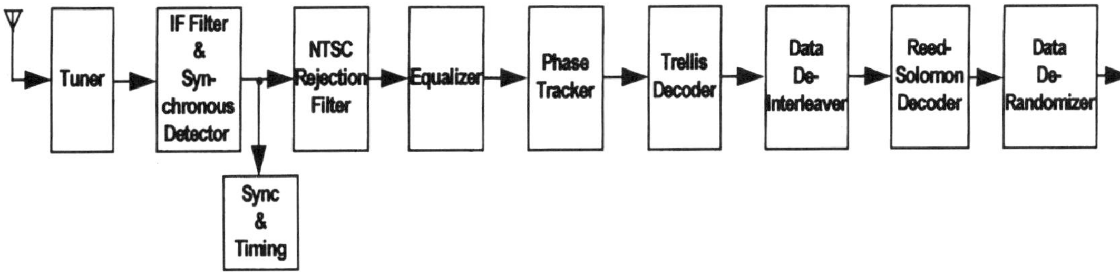


**Figure 13.1** Simplified block diagram of a VSB receiver. (*From* [2]. *Used with permission.*)

- Noncoherent and coherent *automatic gain control* (AGC)
- Data field synchronization
- Interference-rejection filter
- Channel equalizer
- Phase-tracking loop
- Trellis decoder
- Data de-interleaver
- Reed-Solomon decoder
- Data derandomizer
- Receiver loop acquisition sequencing

Descriptions of the major system elements are provided in the following sections.

## 13.2a Tuner

The basic tuner, illustrated in Figure 13.2, receives the 6 MHz signal (UHF or VHF) from an external antenna [2]. The tuner is a high-side injection double-conversion type with a first IF frequency of 920 MHz. This puts the image frequencies above 1 GHz, making them easy to reject by a fixed front-end filter. This selection of first IF frequency is high enough that the input bandpass filter selectivity prevents the local oscillator (978 to 1723 MHz) from leaking out the tuner front end and interfering with other UHF channels; yet, it is low enough for the second harmonics of UHF channels (470 to 806 MHz) to fall above the first IF bandpass. Harmonics of cable channels could possibly occur in the first IF passband but are not a significant problem because of the relatively flat spectrum (within 10 dB) and small signal levels (–28 dBm or less) used in cable systems.

The tuner input has a bandpass filter that limits the frequency range to 50 to 810 MHz, rejecting all other nontelevision signals that may fall within the tuner's image frequency range (beyond 920 MHz). In addition, a broadband tracking filter rejects other television signals, especially

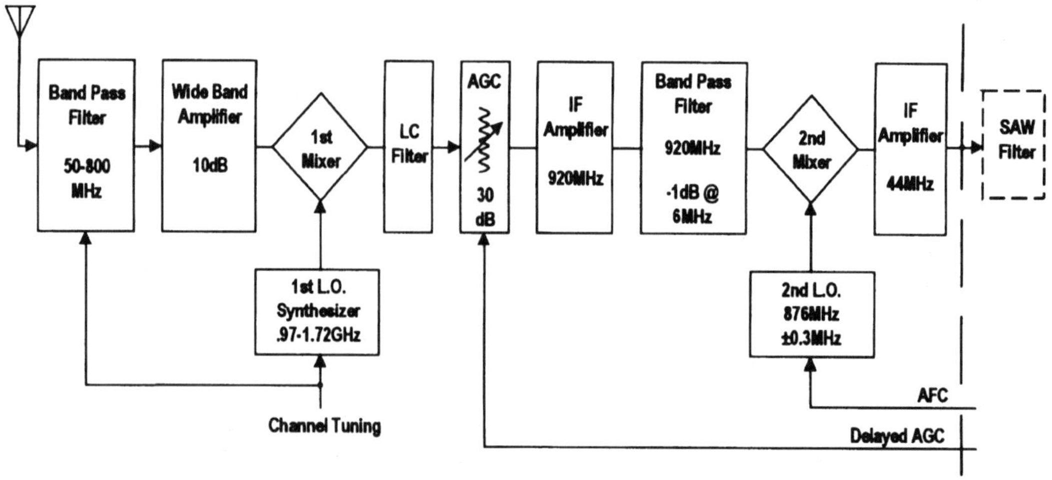


**Figure 13.2** Block diagram of the tuner subsystem. (*From* [2]. *Used with permission.*)

those much larger in signal power than the desired signal. This tracking filter is not narrow, nor is it critically tuned, as is the case of present-day NTSC tuners that must reject image signals only 90 MHz away from the desired channel. Minimal channel tilt, if any, exists because of this tracking filter.

At 10 dB gain, the wideband RF amplifier increases the signal level into the first mixer, and is the dominant determining factor of receiver noise figure (7 to 9 dB over the entire VHF, UHF, and cable bands). The first mixer, a highly linear double-balanced circuit designed to minimize even-harmonic generation, is driven by a synthesized low-phase-noise *local oscillator* (LO) above the first IF frequency (*high-side injection*). Both the channel tuning (first LO) and broadband tracking filters (input bandpass filter) are controlled by a microprocessor. The system is capable of tuning the entire VHF and UHF broadcast bands, as well as all standard, IRC, and HRC cable bands.

The mixer is followed by an LC filter in tandem with a narrow 920 MHz bandpass ceramic resonator filter. The LC filter provides selectivity against the harmonic and subharmonic spurious responses of the ceramic resonator. The 920 MHz ceramic resonator bandpass filter has a –1 dB bandwidth of about 6 MHz. A 920 MHz IF amplifier is placed between the two filters. Delayed AGC of the first IF signal is applied immediately following the first LC filter. The 30 dB range AGC circuit protects the remaining active stages from large signal overload.

The second mixer is driven by the second LO, which is an 876 MHz voltage-controlled *surface acoustic wave* (SAW) oscillator. It is controlled by the *frequency and phase-locked loop* (FPLL) synchronous detector. The second mixer, whose output is the desired 44 MHz second IF frequency, drives a constant-gain 44 MHz amplifier. The output of the tuner feeds the IF SAW filter and synchronous detection circuitry.

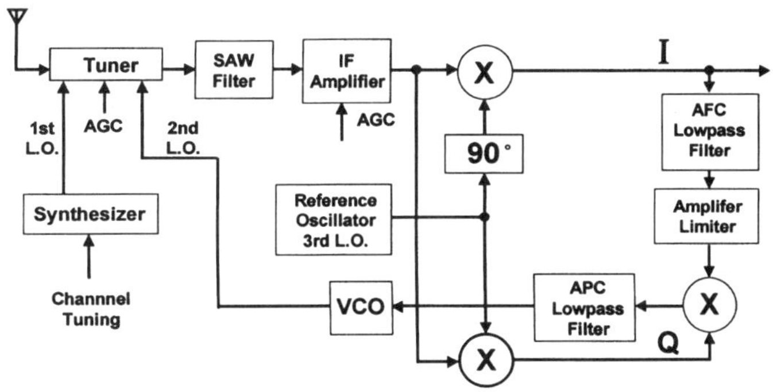


**Figure 13.3** Block diagram of the tuner IF FPLL system. (*From* [2]. *Used with permission.*)

## 13.2b Channel Filtering and VSB Carrier Recovery

Carrier recovery is performed on the small pilot by an FPLL circuit, illustrated in Figure 13.3 [2]. The first LO is synthesized by a PLL and controlled by a microprocessor. The third LO is a fixed-reference oscillator. Any frequency drift or deviation from nominal must be compensated in the second LO. Control for the second LO comes from the FPLL synchronous detector, which integrally contains both a frequency loop and a phase-locked loop in one circuit. The frequency loop provides a wide frequency pull-in range of ±100 kHz, while the phase-locked loop has a narrow bandwidth (less than 2 kHz).

During frequency acquisition, the frequency loop uses both the in-phase *I* and quadrature-phase *Q* pilot signals. All other data-processing circuits in the receiver use only the *I* channel signal. Prior to phase-lock, which is the condition after a channel change, the *automatic frequency control* (AFC) low-pass filter acts on the beat signal created by the frequency difference between the VCO and the incoming pilot. The high-frequency data (as well as noise and interference) is mostly rejected by the AFC filter, leaving only the pilot beat frequency. After limiting this pilot beat signal to a constant amplitude (±1) square wave, and using it to multiply the quadrature signal, a traditional bipolar S-curve AFC characteristic is obtained. The polarity of the S-curve error signal depends upon whether the VCO frequency is above or below the incoming IF signal. Filtered and integrated by the *automatic phase control* (APC) low-pass filter, this dc signal adjusts the tuner's second LO to reduce the frequency difference.

When the frequency difference comes close to zero, the APC loop takes over and phase-locks the incoming IF signal to the third LO. This is a normal phase-locked loop circuit, with the exception that it is biphase-stable. The correct phase-lock polarity is determined by forcing the polarity of the pilot to be equal to the known transmitted positive polarity. Once locked, the detected pilot signal is constant, the limiter output feeding the third multiplier is at a constant +1, and only the phase-locked loop is active (the frequency loop is automatically disabled). The APC low-pass filter is wide enough to reliably allow ±100 kHz frequency pull-in, yet narrow enough to consistently reject strong white noise (including data) and NTSC co-channel interference signals. The PLL has a bandwidth that is sufficiently narrow to reject most of the AM and PM generated by the data, yet is wide enough to track out phase noise on the signal (hence, on the pilot)

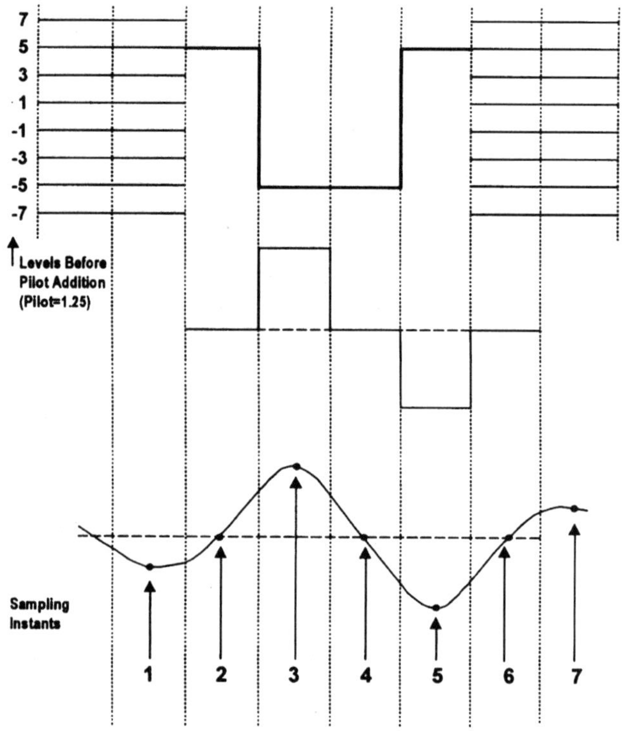


**Figure 13.4** Data segment sync waveforms. (*From* [2]. *Used with permission.*)

to about 2 kHz. Tracking out low-frequency phase noise (as well as low-frequency FM components) allows the phase-tracking loop to be more effective.

## 13.2c Segment Sync and Symbol Clock Recovery

The repetitive data segment sync signals (Figure 13.4) are detected from among the synchronously detected random data by a narrow bandwidth filter [2]. From the data segment sync, a properly phased 10.76 MHz symbol clock is created along with a coherent AGC control signal. A block diagram of this circuit is shown in Figure 13.5.

The 10.76 Msymbols/s (684 ÷ 286 × 4,500,000 Hz) *I* channel composite baseband data signal (sync and data) from the synchronous detector is converted by an A/D device for digital processing. Traditional analog data eyes can be viewed after synchronous detection. After conversion to a digital signal, however, the data eyes cannot be seen because of the sampling process. A PLL is used to derive a clean 10.76 MHz symbol clock for the receiver.

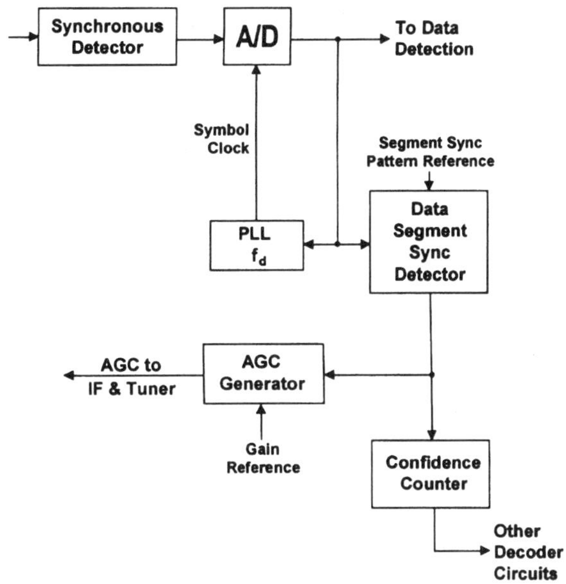


**Figure 13.5** Segment sync and symbol clock recovery with AGC. (*From* [2]. *Used with permission.*)

With the PLL free-running, the data segment sync detector—containing a 4-symbol sync correlator—looks for the two level sync signals occurring at the specified repetition rate. The repetitive segment sync is detected while the random data is not, enabling the PLL to lock on the sampled sync from the A/D converter and achieve data symbol clock synchronization. Upon reaching a predefined level of confidence (using a confidence counter) that the segment sync has been found, subsequent receiver loops are enabled.

## 13.2d Noncoherent and Coherent AGC

Prior to carrier and clock synchronization, noncoherent automatic gain control (AGC) is performed whenever any signal (locked or unlocked signal, or noise/interference) overruns the A/D converter [2]. The IF and RF gains are reduced accordingly, with the appropriate AGC *delay* applied.

When data segment sync signals are detected, coherent AGC occurs using the measured segment sync amplitudes. The amplitude of the bipolar sync signals, relative to the discrete levels of the random data, is determined in the transmitter. After the sync signals are detected in the receiver, they are compared with a reference value, with the difference (error) integrated. The integrator output then controls the IF and *delayed* RF gains, forcing them to whatever values provide the correct sync amplitudes.

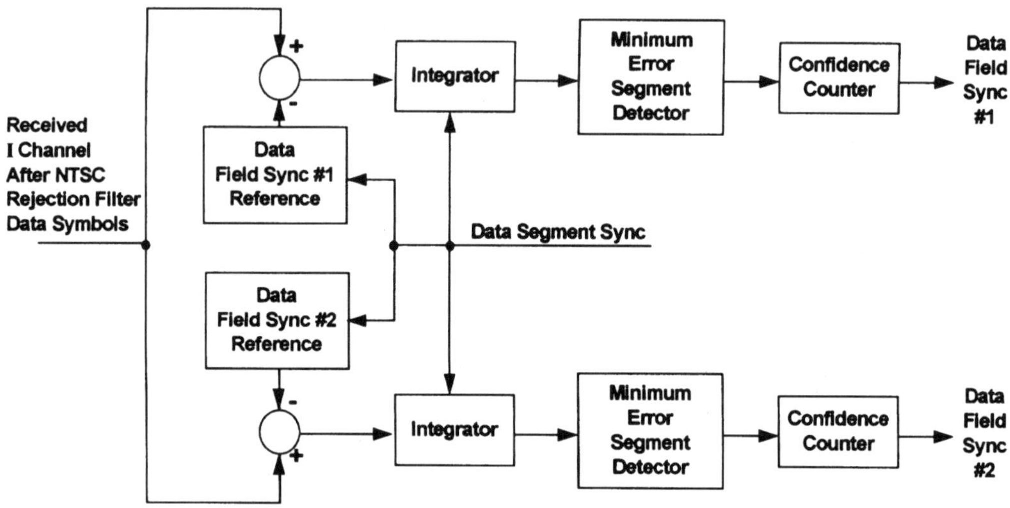


**Figure 13.6** The process of data field sync recovery. (*From* [2]. *Used with permission.*)

## 13.2e Data Field Synchronization

Data field sync detection, shown in Figure 13.6, is achieved by comparing each received data segment from the A/D converter (after interference-rejection filtering to minimize co-channel interference) with ideal field 1 and field 2 reference signals in the receiver [2]. Oversampling of the field sync is not necessary, because a precision data segment and symbol clock already has been reliably created by the clock recovery circuit. Therefore, the field sync recovery circuit knows exactly where a valid field sync correlation should occur within each data segment, and needs only to perform a symbol-by-symbol difference. Upon reaching a predetermined level of confidence (using a confidence counter) that field sync signals have been detected on given data segments, the data field sync signal becomes available for use by subsequent circuits. The polarity of the middle of the three alternating 63-bit *pseudorandom* (PN) sequences determines whether field 1 or field 2 is detected. This procedure makes field sync detection robust, even in heavy noise, interference, or ghost conditions.

## 13.2f Interference-Rejection Filter

The interference-rejection properties of the VSB transmission system are based on the frequency location of the principal components of the NTSC co-channel interfering signal within the 6 MHz television channel and the periodic nulls of a VSB receiver baseband comb filter [2]. Figure 13.7*a* shows the location and approximate magnitude of the three principal NTSC components:

- The visual carrier $V$, located 1.25 MHz from the lower band edge

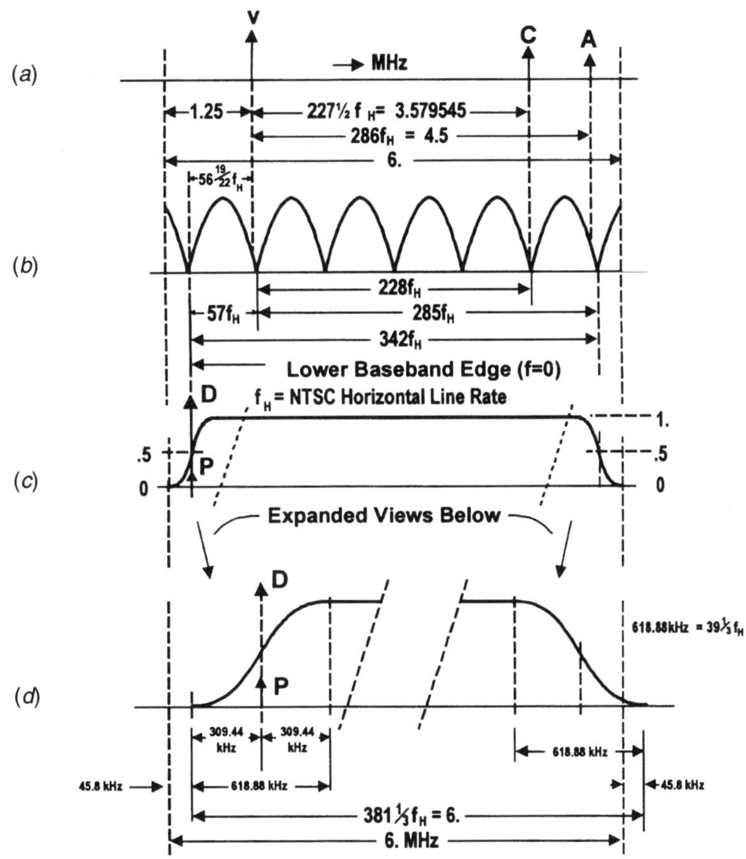


**Figure 13.7** Receiver filter characteristics: (*a*) primary NTSC components, (*b*) comb filter response, (*c*) filter band edge detail, (*d*) expanded view of band edge. (*From* [2]. *Used with permission.*)

- Chrominance subcarrier $C$, located 3.58 MHz higher than the visual carrier frequency
- Aural carrier $A$, located 4.5 MHz higher than the visual carrier frequency

The NTSC interference-rejection filter (a comb filter) is a 1-tap linear feed-forward device, as shown in Figure 13.8. Figure 13.7*b* illustrates the frequency response of the comb filter, which provides periodic spectral nulls spaced $57 \times f_H$ (10.762 MHz/12, or 896.85 kHz) apart. There are seven nulls within the 6 MHz channel. The NTSC visual carrier frequency falls close to the second null from the lower band edge. The sixth null from the lower band edge is correctly placed for the NTSC chrominance subcarrier, and the seventh null from the lower band edge is near the NTSC aural carrier.

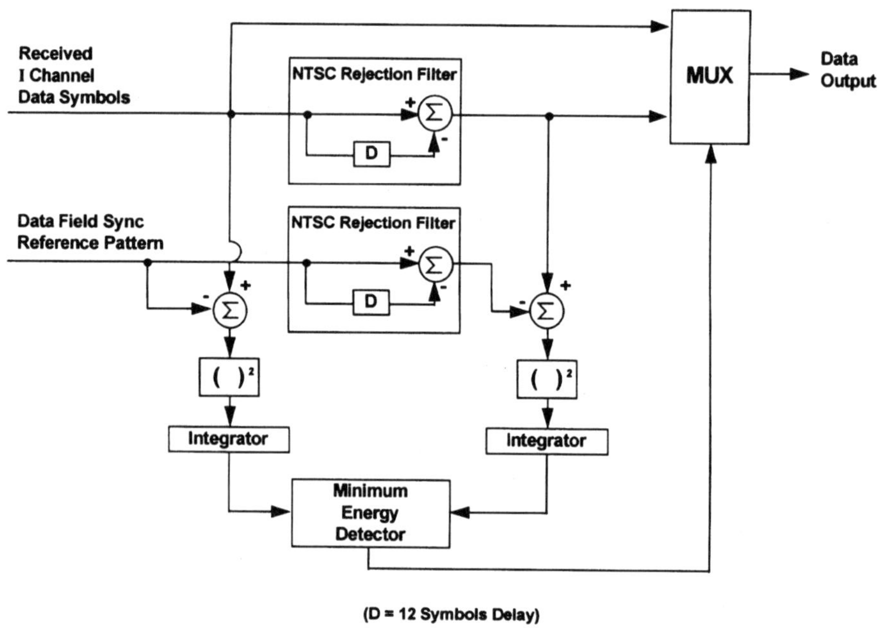


**Figure 13.8** Block diagram of the NTSC interference-rejection filter. (*From* [2]. *Used with permission.*)

Comparing Figure 13.7*a* and Figure 13.7*b* shows that the visual carrier falls 2.1 kHz below the second comb filter null, the chroma subcarrier falls near the sixth null, and the aural carrier falls 13.6 kHz above the seventh null. The NTSC aural carrier is at least 7 dB below its visual carrier.

The comb filter, while providing rejection of steady-state signals located at the null frequencies, has a finite response time of 12 symbols (1.115 μs). Thus, if the NTSC interfering signal has a sudden step in carrier level (low to high or high to low), one cycle of the zero-beat frequency (offset) between the DTV and NTSC carrier frequencies will pass through the comb filter at an amplitude proportional to the NTSC step size as instantaneous interference. Examples of such steps of NTSC carrier are the leading and trailing edge of sync (40 IRE units). If the *desired to undesired* (D/U) signal power ratio is large enough, data slicing errors will occur. Interleaving will spread the interference, however, and make it easier for the Reed-Solomon code to correct them (RS coding can correct up to 10 byte errors per segment).

Although the comb filter reduces the NTSC interference, the data also is modified. The seven data eyes (eight levels) are converted to 14 data eyes (15 levels). This conversion is caused by the *partial response process*, which is a special case of intersymbol interference that does not close the data eye, but creates double the number of eyes of the same magnitude. The modified data signal can be properly decoded by the trellis decoder. Note that, because of time sampling, only the maximum data eye value is seen after A/D conversion.

The detail at the band edges for the overall channel is shown in Figure 13.7*c* and Figure 13.7*d*. Figure 13.7*d* shows that the frequency relationship of 56 × 19/22 × $f_H$ between the NTSC visual carrier and the ATV carrier requires a shift in the ATV spectrum with respect to the nominal channel. The shift equals +45.8 kHz, or about +0.76 percent. This is slightly higher than currently applied channel offsets and reaches into the upper adjacent channel at a level of about –40 dB. If the upper adjacent channel is another DTV channel, its spectrum is also shifted upward and, therefore, no spectral overlapping occurs. If it is an NTSC channel, the shift is below the (RF equivalent of the) Nyquist slope of an NTSC receiver where there is high attenuation, and it is slightly above the customary lower adjacent-channel sound trap. No adverse effects of this shift should be expected. An additional shift of the DTV spectrum is used to track the dominant NTSC interferer, which may be assigned an offset value of –10, 0, or +10 kHz.

NTSC interference can be detected by the circuit shown in Figure 13.8, where the signal-to-interference plus noise ratio of the binary data field sync is measured at the input and output of the comb filter and compared. This is accomplished by creating two error signals. The first is generated by comparing the received signal with a stored reference of the field sync. The second is generated by comparing the rejection filter output with a combed version of the internally stored reference field sync. The errors are squared and integrated. After a predetermined level of confidence is achieved, the path with the largest signal-to-noise ratio (lowest interference energy) is switched in and out of the system automatically.

It is not advisable to leave the rejection comb filter switched in all the time. The comb filter, while providing needed co-channel interference benefits, degrades white noise performance by 3 dB. This occurs because the filter output is the subtraction of two full-gain paths, and as white noise is uncorrelated from symbol-to-symbol, the noise power doubles. There is an additional 0.3 dB degradation resulting from the 12-symbol differential coding. If little or no NTSC interference is present, the comb filter is automatically switched out of the data path. When the NTSC service is phased out, the comb filter can be omitted from digital television receivers.

## 13.2g Channel Equalizer

The equalizer/ghost canceler compensates for linear channel distortions, such as tilt and ghosts [2]. These distortions can originate in the transmission channel or result from imperfect components within the receiver.

The equalizer uses a *least-mean-square* (LMS) algorithm and can adapt on the transmitted binary training sequence, as well as on the random data. The LMS algorithm computes how to adjust the filter taps to reduce the error present at the output of the equalizer. The system does this by generating an estimate of the error present in the output signal, which then is used to compute a cross-correlation with various delayed data signals. These correlations correspond to the adjustment that needs to be made for each tap to reduce the error at the output. The equalizer algorithm can achieve equalization in three ways:

- Adapt on the binary training sequence
- Adapt on data symbols throughout the frame when the eyes are open
- Adapt on data when the eyes are closed (*blind equalization*)

The principal difference among these three methods is how the error estimate is generated.

For adapting on the training sequence, the training signal presents a fixed data pattern in the data stream. Because the data pattern is known, the exact error is generated by subtracting the

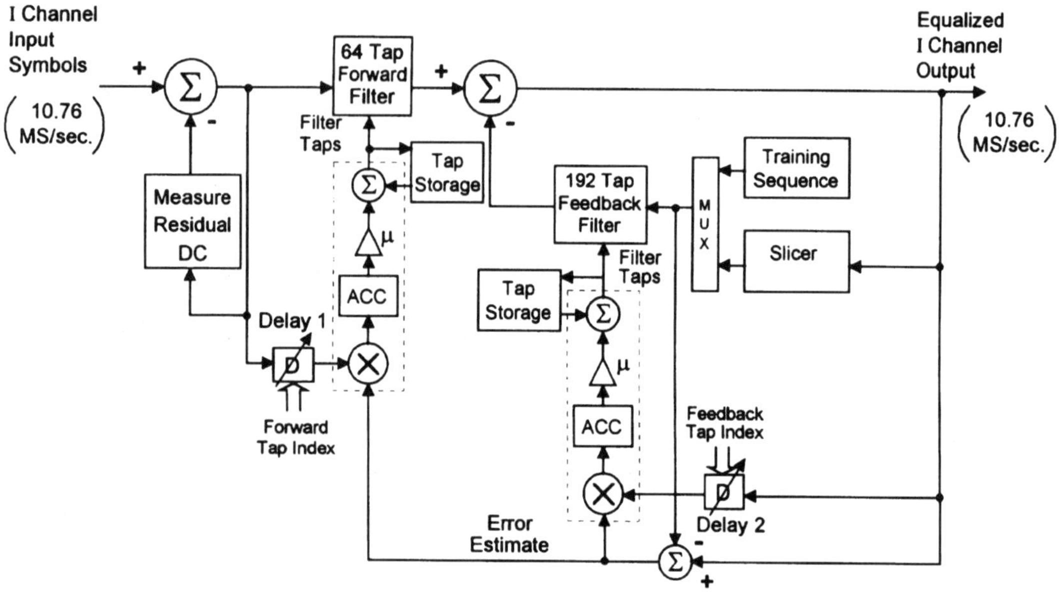


**Figure 13.9** Simplified block diagram of the VSB receiver equalizer. (*From* [2]. *Used with permission.*)

training sequence from the output. The training sequence alone, however, may not be enough to track dynamic ghosts because these conditions require tap adjustments more often than the training sequence is transmitted. Therefore, after equalization is achieved, the equalizer can switch to adapting on data symbols throughout the frame, and it can produce an accurate error estimate by slicing the data with an 8-level slicer and subtracting it from the output signal.

For fast dynamic ghosts such as airplane flutter, it is necessary to use a blind equalization mode to aid in acquisition of the signal. Blind equalization models the multilevel signal as binary data signal plus noise, and the equalizer produces the error estimate by detecting the sign of the output signal and subtracting a (scaled) binary signal from the output to generate the error estimate.

To perform the LMS algorithm, the error estimate (produced using the training sequence, 8-level slicer, or the binary slicer) is multiplied by delayed copies of the signal. The delay depends upon which tap of the filter is being updated. This multiplication produces a cross-correction between the error signal and the data signal. The size of the correlation corresponds to the amplitude of the residual ghost present at the output of the equalizer and indicates how to adjust the tap to reduce the error at the output.

A block diagram of the equalizer is shown in Figure 13.9. The dc bias of the input signal is first removed by subtraction. The dc may be caused by circuit offsets, nonlinearities, or shifts in

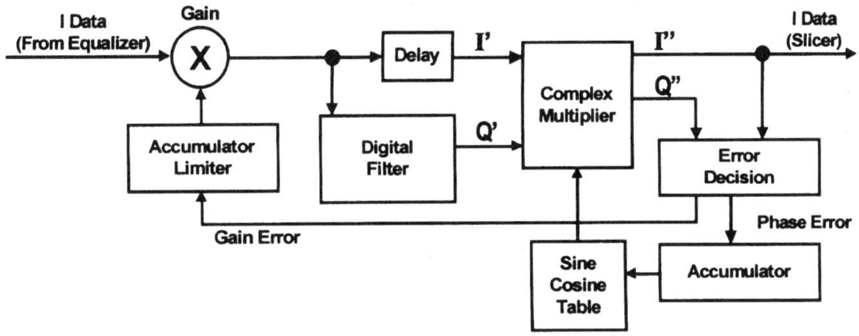


**Figure 13.10** The phase-tracking loop system. (*From* [2]. *Used with permission.*)

the pilot caused by ghosts. The dc offset is tracked by measuring the dc value of the training signal.

The equalizer filter consists of two parts: a 64-tap feed-forward transversal filter followed by a 192-tap decision-feedback filter. The equalizer operates at the 10.762 MHz symbol rate (*T*-sampled equalizer).

The output of the forward filter and feedback filter are summed to produce the output. This output is sliced by either an 8-level slicer (15-level slicer when the comb filter is used) or a binary slicer depending upon whether the data eyes are open. (As pointed out previously, the comb filter does not close the data eyes, but creates twice as many of the same magnitude.) This sliced signal has the training signal and segment sync signals reinserted, as these are fixed patterns of the signal. The resultant signal is fed into the feedback filter and subtracted from the output signal to produce the error estimate. The error estimate is correlated with the input signal (for the forward filter) or with the output signal (for the feedback filter). This correlation is scaled by a step-size parameter and is used to adjust the value of the tap. The delay setting of the adjustable delays is controlled according to the index of the filter tap that is being adjusted.

## 13.2h Phase-Tracking Loop

The phase-tracking loop is an additional decision-feedback loop that further tracks out phase noise that has not been removed by the IF PLL operating on the pilot [2]. Thus, phase noise is tracked out by not just one loop, but two concatenated loops. Because the system is already frequency-locked to the pilot by the IF PLL (independent of the data), the phase-tracking loop bandwidth is maximized for phase tracking by using a first-order loop. Higher-order loops, which are needed for frequency tracking, do not perform phase tracking as well as first-order loops. Therefore, they are not used in the VSB system.

A block diagram of the phase-tracking loop is shown in Figure 13.10. The output of the real equalizer operating on the *I* signal is first gain-controlled by a multiplier, then fed into a filter that recreates an approximation of the *Q* signal. This is possible because of the VSB transmission

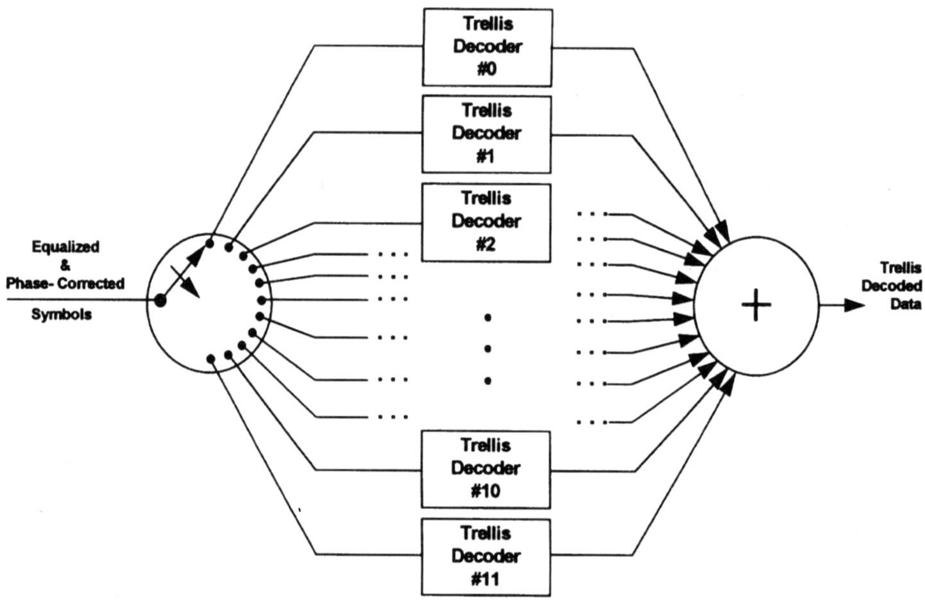


**Figure 13.11** Functional diagram of the trellis code de-interleaver. (*From* [2]. *Used with permission.*)

method, where the *I* and *Q* components are related by a filter function that is almost a Hilbert transform. The complexity of this filter is minor because it is a *finite impulse response* (FIR) filter with fixed antisymmetric coefficients and with every other coefficient equal to zero. In addition, many filter coefficients are related by powers of 2, thus simplifying the hardware design.

These *I* and *Q* signals then are fed into a *de-rotator* (complex multiplier), which is used to remove the phase noise. The amount of de-rotation is controlled by decision feedback of the data taken from the output of the de-rotator. Because the phase tracker is operating on the 10.76 Msymbols/s data, the bandwidth of the phase-tracking loop is fairly large, approximately 60 kHz. The gain multiplier also is controlled with decision feedback.

## 13.2i Trellis Decoder

To help protect the trellis decoder against short burst interference, such as impulse noise or NTSC co-channel interference, 12-symbol code intrasegment interleaving is employed in the transmitter [2]. As shown in Figure 13.11, the receiver uses 12 trellis decoders in parallel, where each trellis decoder sees every 12th symbol. This code interleaving has all the same burst noise benefits of a 12 symbol interleaver, but also minimizes the resulting code expansion (and hardware) when the NTSC rejection comb filter is active.

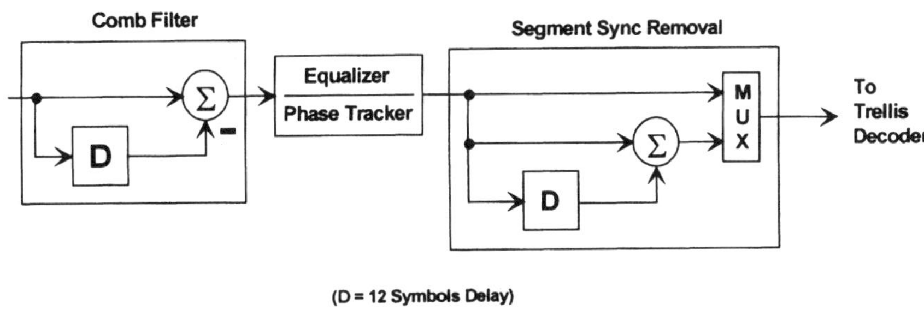


**Figure 13.12** Block diagram of the 8-VSB receiver segment sync suspension system. (*From* [2]. *Used with permission.*)

Before the 8-VSB signal can be processed by the trellis decoder, it is necessary to suspend the segment sync. The segment sync is not trellis-encoded at the transmitter. The segment sync suspension system is illustrated in Figure 13.12.

The trellis decoder performs the task of slicing and convolutional decoding. It has two modes: one when the NTSC-rejection filter is used to minimize NTSC co-channel energy, and the other when it is not used. This is illustrated in Figure 13.13. The insertion of the NTSC-rejection filter is determined automatically (before the equalizer), with this information passed to the trellis decoder. When there is little or no NTSC co-channel interference, the NTSC-rejection filter is not used, and an optimal trellis decoder is used to decode the 4-state trellis-encoded data. Serial bits are recreated in the same order in which they were created in the encoder.

In the presence of significant NTSC co-channel interference, when the NTSC-rejection filter (12-symbol, feed-forward subtractive comb) is employed, a trellis decoder optimized for this partial response channel is used. This optimal code requires eight states. This approach is necessary because the NTSC-rejection filter, which has memory, represents another state machine seen at the input of the trellis decoder. To minimize the expansion of trellis states, two measures are taken: first, special design of the trellis code; and second, 12-to-1 interleaving of the trellis encoding. The interleaving, which corresponds exactly to the 12-symbol delay in the NTSC-rejection filter, ensures that each trellis decoder sees only a 1-symbol-delay NTSC-rejection filter. Minimizing the delay stages seen by each trellis decoder also minimizes the expansion of states. Only a 3.5 dB penalty in white noise performance is paid as the price for having good NTSC co-channel performance. The additional 0.5 dB beyond the 3 dB comb filter noise threshold degradation is the result of the 12-symbol differential coding.

As noted previously, after the transition period, when NTSC is no longer being transmitted, the NTSC-rejection filter and the 8-state trellis decoder can be eliminated from digital television receivers.

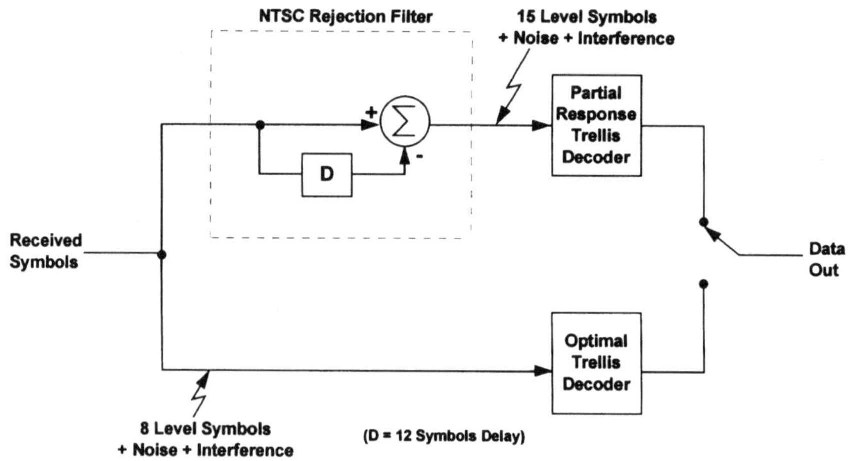


**Figure 13.13** Trellis decoding with and without the NTSC rejection filter. (*From* [2]. *Used with permission.*)

## 13.2j Data De-Interleaver

The convolutional de-interleaver performs the exact inverse function of the transmitter convolutional interleaver [2]. The 1/6-data-field depth and intersegment *dispersion* properties allow noise bursts lasting approximately 193 μs to be handled. Even strong NTSC co-channel signals, passing through the NTSC-rejection filter and creating short bursts due to NTSC vertical edges, are handled reliably because of the interleaving and RS coding process. The de-interleaver uses data field sync for synchronizing to the first data byte of the data field. The convolutional de-interleaver is shown in Figure 13.14.

## 13.2k Other Receiver Functional Blocks

The trellis-decoded byte data is sent to the Reed-Solomon decoder, where it uses the 20 parity bytes to perform the byte error correction on a segment-by-segment basis [2]. Up to 10 byte errors per data segment are corrected by the RS decoder. Burst errors created by impulse noise, NTSC co-channel interference, or trellis-decoding errors are greatly reduced by the combination of the interleaving and RS error correction.

The data is randomized at the transmitter by a pseudorandom sequence (PRS). The derandomizer at the receiver accepts the error-corrected data bytes from the RS decoder and applies the same PRS randomizing code to the data. The PRS code is generated identically as in the transmitter, using the same PRS generator feedback and output taps. Because the PRS is locked to the reliably recovered data field sync (and not some code word embedded within the potentially noisy data), it is exactly synchronized with the data, and it performs reliably.

The receiver incorporates a *universal reset* feature that initiates a number of "confidence counters" and "confidence flags" involved in the lockup process. A universal reset occurs, for

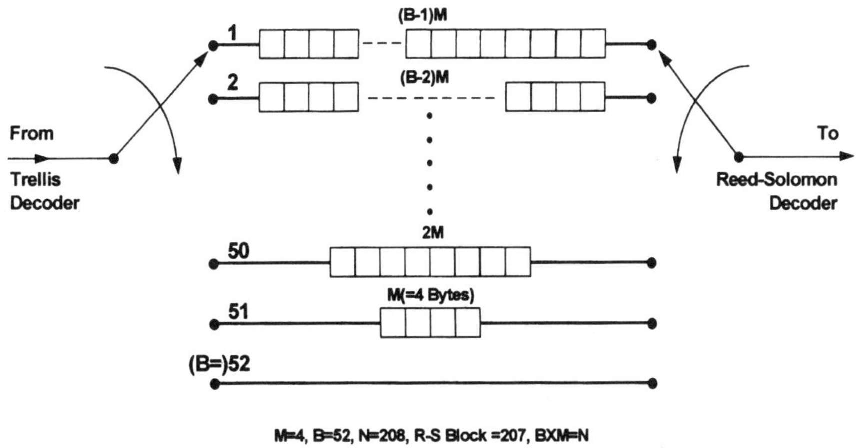


**Figure 13.14** Functional diagram of the convolutional de-interleaver. (*From* [2]. *Used with permission.*)

example, when another station is being tuned or the receiver is being turned on. The various loops within the VSB receiver acquire and lock up sequentially, with "earlier" loops being independent from "later" loops. The order of loop acquisition is as follows:

- Tuner first LO synthesizer acquisition
- Noncoherent AGC reduces unlocked signal to within the A/D range
- Carrier acquisition (FPLL)
- Data segment sync and clock acquisition
- Coherent AGC of signal (IF and RF gains properly set)
- Data field sync acquisition
- NTSC-rejection filter insertion decision made
- Equalizer completes tap adjustment algorithm
- Trellis and RS data decoding begin

Most of the loops mentioned here have confidence counters associated with them to ensure proper operation, but the build-up or letdown of confidence is not designed to be equal. The confidence counters build confidence quickly for quick acquisition times, but lose confidence slowly to maintain operation in noisy environments.

### High-Data-Rate Mode

The VSB digital transmission system provides the basis for a family of DTV receivers suitable for receiving data transmissions from a variety of media [2]. This family shares the same pilot, symbol rate, data frame structure, interleaving, Reed-Solomon coding, and synchronization pulses. The VSB system offers two modes: a simulcast terrestrial broadcast mode and a high-data-rate mode.

Most elements of the high-data-rate mode VSB system are similar or identical to the terrestrial system. A pilot, data segment sync, and data field sync all are used to provide robust operation. The pilot in the high-data-rate mode also adds 0.3 dB to the data power. The symbol, segment, and field signals and rates are all the same, allowing either receiver to lock up on the other's transmitted signal. Also, the data frame definitions are identical. The primary difference is the number of transmitted levels (8 versus 16) and the use of trellis coding and NTSC interference-rejection filtering in the terrestrial system.

The high-data-rate mode receiver is identical to the VSB terrestrial receiver, except that the trellis decoder is replaced by a slicer, which translates the multilevel symbols into data. Instead of an 8-level slicer, a 16-level slicer is used.

## 13.21 Receiver Equalization Issues

The VSB signal contains features that allow the design of receivers that reliably perform the functions of acquiring and locking onto the transmitted signal [2]. The equalization of the signal for channel frequency response and ghosts is facilitated by the inclusion of specific features in the data field sync, as illustrated in Figure 13.15. Utilization of these features is made more reliable by the inclusion of means to first acquire and synchronize to the VSB signal, particularly by the segment sync. The data field sync can then be used both to identify itself and to further perform equalization of linear transmission distortions. The VSB signal also may be equalized by data-based or *blind equalization* methods that do not use the data field sync.

Data field sync is a unique type of data segment in the VSB signal. All payload data in the VSB signal is contained in data segments, which are processed with data interleaving, Reed-Solomon error coding, and trellis coding. The data field sync (and the segment sync portion of every data segment), however, is not processed this way because its purpose is to provide direct measurement and compensation for transmission channel linear distortion. Equalizer training signals consisting of pseudonoise sequences are major parts of the data field sync.

### Equalizer Performance Using Training Signals

Theoretically, in a noise-free signal, ghosts of amplitude up to 0 dB with respect to the largest signal—and within a total window of 63 symbols—can be exactly canceled in one pass using the information in the sequence [2]. Ghosts of any delay can be canceled using the information in the sequence, with a single-pass accuracy of approximately –27 dB, which improves by averaging over multiple passes. The operation of a complete receiver system with 0 dB ghosts may not be achievable because of the failure of carrier acquisition, but operation of the equalizer itself at 0 dB is demonstrable under test by supplying an external carrier.

The number of ghosts to be canceled has, in itself, little effect on the theoretical performance of the system. Theoretical limits to cancellation depend on the amount of noise gain that occurs in compensating the frequency response that results from the particular ghosted signal.

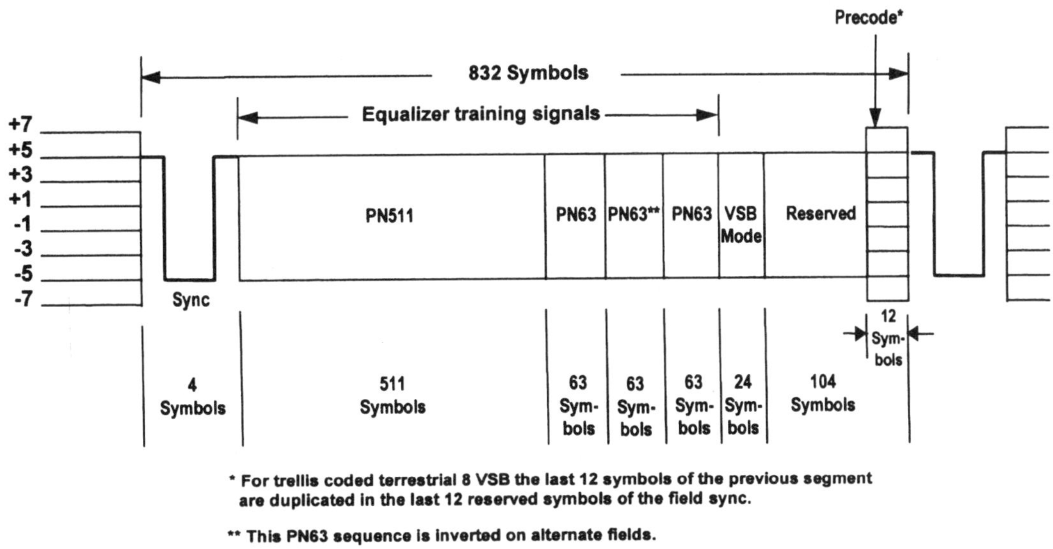


**Figure 13.15** Data field sync waveform. (*From* [2]. *Used with permission.*)

Because the equalizer training signals recur with a period of approximately 24 ms, the receiver cannot perform equalization updates at a faster rate. Therefore, the signal provides information such that the equalization system, theoretically, can have a bandwidth of 20 Hz. Additional constraints are implied by the desire to average out the correlation between PN sequences.

The speed of convergence is not the only important criterion of performance, however. Ultimate accuracy and response to noise also are of importance. Equalization techniques generally proceed by successive approximation to the desired state, and therefore exhibit a convergence time measured as a number of frame periods. If the information in the signal is used in different ways, the speed of operation can be traded off for accuracy and noise immunity. The success of such a tradeoff may depend on nonlinear techniques, such as switching between a quick acquisition mode and slower refinement mode or using varying step sizes in a *steepest-descent* technique.

## Receiver Implementation Using Blind Equalization

*Blind equalization* techniques are not based on a training signal reference, but they may be appropriate for use with the VSB transmission system [2]. Blind equalization techniques are particularly useful when the channel impairments vary more rapidly than the transmission of the training waveform.

As in many modern digital data communication systems, an adaptive equalizer is used in the ATSC system to compensate for changing conditions on the broadcast channel. In communication systems that use an adaptive equalizer, it is necessary to have a method of adapting the equalizer's filter response to adequately compensate for channel distortions. Several algorithms are available for adapting the filter coefficients. The most widely used is the LMS, or least-mean-square method [3].

When the equalizer is started, the tap weights usually are not set to adequately compensate for the channel distortions. To force initial convergence of the equalizer coefficients, a known training signal (that is, a signal known by both the transmitter and receiver) is used as the reference signal. The error signal is formed by subtracting a locally generated copy of the training signal from the output of the adaptive equalizer. When using the training signal, the eye diagram is typically closed. The training signal serves to open the eye. After adaptation with the training signal, the eye has opened, and the equalizer may be switched to a decision-directed mode of operation. The decision-directed mode uses the symbol values at the output of the decision device instead of the training signal.

A problem arises in this scenario when a training signal is not available. In this case, a method of acquiring initial convergence of the equalizer taps and forcing the eye open is necessary. Blind equalization has been studied extensively for QAM systems. Several methods typically are employed: the *constant modulus algorithm* (CMA) and the *reduced constellation algorithm* (RCA) are among the most popular [4, 5]. For VSB systems, however, neither of these methods are directly applicable. CMA relies on the fact that, at the decision instants, the modulus of the detected data symbols should lie on one of several circles of varying diameters. Thus, it inherently relies on the underlying signal to be 2-dimensional. Because VSB is essentially a 1-dimensional signal (at least for the data-carrying portion), CMA is not directly applicable. RCA, on the other hand, relies on forming *super constellations* within the main constellation. The data signal is forced to fit into a super constellation, then the super constellations are subdivided to include the entire constellation. Again, as typically used, RCA implies a 2-dimensional constellation.

Blind equalization can be performed on the VSB constellation, however, using a *modified reduced constellation algorithm* (MRCA). The key part of this modification is to realize the existence of a 1-dimensional version of the RCA algorithm that is appropriate for VSB signals. The MRCA consists of an algorithm to determine appropriate decision regions for a VSB decision device, so as to generate decisions that allow an adaptive equalizer to converge without the use of a training signal.

In VSB systems, the decision regions typically span one data symbol of the full constellation, and the upper and lower boundaries of each decision region are set midway between the constellation points. If these decision regions are used for initial convergence of the equalizer, the equalizer will not converge because, as a result of the presence of intersymbol interference, a significant number of the decisions from the decision device will be incorrect.

To force more correct decisions to be made, an algorithm for determining new upper and lower decision region boundaries was designed. The algorithm clusters the full VSB constellation into several sets, determines upper and lower boundaries for decision regions, and determines appropriate decision device output *symbol* values. These first sets are further divided into smaller sets until each set of symbols contains exactly one symbol, and the decision regions correspond to the standard decision regions for VSB described previously. The function of each stage is to allow for more decisions to be correct, thereby driving the equalizer toward convergence. In this way, each stage in the blind equalization process serves to further open the eye.

In general, the MRCA algorithm consists of clustering the decision regions of the decision device into finer portions of the VSB constellation. The method starts with a binary (2-level) slicer, then switches to a 4-level slicer, then an 8-level slicer, and so on. It should be noted that the MRCA algorithm is applicable to both linear equalization and decision-feedback equalization.

## 13.2m Receiver Performance in the Field

Shortly after the initiation of DTV service, receiver manufacturers began studying the performance of their first-generation designs. One widely-quoted study was an ambitious project undertaken by Zenith [6]. The objectives of the DTV field tests included the following:

- To build a large database of DTV field measurements to refine the propagation models used for DTV field strength prediction, and ultimately to develop a first order prediction of DTV service.
- Gathering performance data on the DTV RF transmission system in a variety of terrain, as well as urban, suburban, and rural environments throughout the U.S.
- Analysis of DTV service availability and system performance in all three FCC-allocated frequency bands (low VHF, high VHF, and UHF).

As of June 1999, twelve field tests had been completed—a total of 2,682 outdoor sites in 9 U.S. cities and 242 indoor sites in 5 cities. All but two of the tests were conducted on UHF channels. Each DTV field test followed a standardized data gathering methodology (described in [6]). Test results indicated that DTV service availability would approach NTSC replication in most instances. Taller transmitter antenna installations were found to have a greater beneficial affect on signal reception than greater transmitter power.

Table 13.3 contains target DTV receiver parameters for reasonable performance in the field (as identified in the field study). These values served as a starting point for receiver designers. As more field test data became available, these target values were expected to be updated.

Extensive laboratory tests were conducted by the Advanced Television Test Center [7] that compared production line receivers under various conditions. Although these tests were not intended to represent either actual reception conditions or a specification for receiver design, they did indicate that the performance of DTV receivers could be improved beyond the Grand Alliance "blue rack" reference receiver. However, the results of these tests also demonstrated the need to further improve the performance of DTV receivers beyond what was current available.

### Canadian DTV Receiver Tests

The Communications Research Center of Canada (CRC) carried out laboratory and field tests to evaluate the performance of ATSC 8-VSB receivers against channel impairments such as noise, multipath, and other distortions. A number of receivers were tested in the laboratory, including consumer models available on the market, prototypes, and professional models. The summary report, presented at the International Broadcasting Convention in September 2000, cautioned that the results were valid only for the particular units used for the tests and that no attempts were made to establish whether the same results would be obtained with another receiver of the same model [8].

All receivers were tested against random noise, and random noise in the presence of static multipath ensembles. As a simplified version of the multipath test, a single echo test was also conducted to verify the delay range of the receiver equalizer, and to observe the sensitivity of the receiver to the phase of an echo.

**Table 13.3 Target DTV Receiver Parameters** (*After* [6].)

| DTV Receiver Parameter | Target Value |
|---|---|
| Dynamic range: minimum level | < –80 dBm |
| Dynamic range: maximum level | > 0 dBm |
| Noise figure | < 10 dB |
| Synchronization white noise (lock) limit | < 3 dB |
| AGC speed (10 dB peak/valley fade) | > 75 Hz |
| White noise threshold of errors | < 15.5 dB |
| Phase noise threshold of errors | > –76 dBc/Hz @ 20 kHz offset |
| Gated white noise burst duration | > 185 μs |
| Co-channel N/D interference @ –45 dBm | < 3 dB, D/U |
| Co-channel D/D interference @ –45 dBm | < 15.5 dB, D/U |
| First adjacent N/D interference @ –60 dBm | < –40 dB, D/U |
| First Adjacent D/D Interference @ –60 dBm | < –28 dB, D/U |
| Inband CW Interference @ –45 dBm | < 10 dB, D/U |
| Out-of-band CW interference @ –45 dBm | < –40 dB, D/U |
| Equalizer length (pre-ghost) | < –3 μs |
| Equalizer length (post-ghost) | > +22 μs |
| Quasi-static multipath (1 μs @ < 0.2 Hz) | < +4 dB, D/U |
| Dynamic multipath: (1 μs @ < 5 Hz) | < +7 dB, D/U |
| Dynamic multipath: (1 μs @ <10 Hz) | < +10 dB, D/U |

The laboratory setup is shown in Figure 13.16. The high-definition video signal (SMPTE 310M) was generated by an MPEG-2 transport stream spooler, and fed into a modulator/exciter and up-converted to channel 54 (710 to 716 MHz).

Specific tests conducted in the study included:

- **Susceptibility to random noise**. Measurements were intended to determine the performance of the receivers under random noise impairment. The effect of random noise on the ATSC 8-VSB receiver was evaluated subjectively at the threshold of visibility (TOV) by observing pictures on a monitor. The average power of the DTV signal was adjusted to three different levels: strong (–28 dBm), moderate (–53 dBm) and weak (–68 dBm), as recommended by FCC/ACATS [9]. For each power level of the DTV signal, random noise was added and increased until TOV was reached. The ratio between the average power of the DTV signal and the random noise (C/N) was measured; the results are summarized in Table.13.4. The value measured by the ATTC for the Grand Alliance [10] prototype (Zenith blue rack) is also included in the table for comparison purposes.
- **Multipath test**. The multipath measurements were intended to determine the robustness of ATSC 8-VSB receivers against random noise in the presence of static multipath (an ensemble of five echoes). TOV was determined for the combinations of multipath used by ATTC/

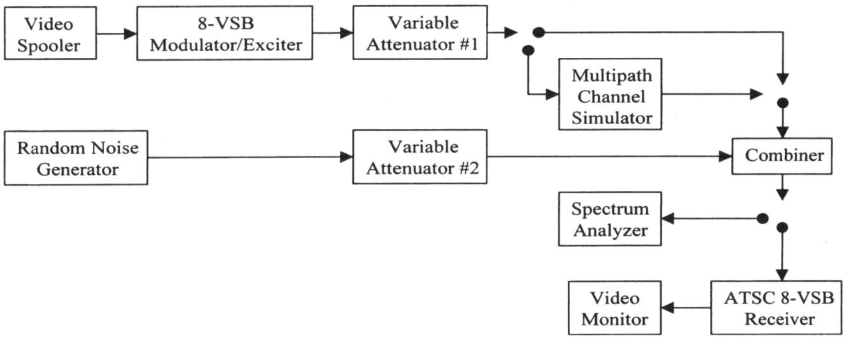


**Figure 13.16** Laboratory setup for receiver evaluation. (*From* [8]. *Used with permission.*)

**Table 13.4 Susceptibility to Random Noise Test Results** (*After* [8])

| Receiver | C | E | F | G | H | Blue Rack |
|---|---|---|---|---|---|---|
| Desired Signal Level | C/N (dB) at TOV | C/N (dB) at TOV | C/N (dB) at TOV | C/N (dB) at TOV | C/N (dB) at TOV | C/N (dB) at TOV |
| Strong: –28 dBm | 16.8 | 15.3 | 15.4 | 15.2 | 16.0 | 15.28 |
| Moderate: –53 dBm | 17.2 | 15.3 | 15.3 | 15.2 | 16.0 | N/A |
| Weak: –68 dBm | 17.8 | 15.3 | 15.8 | 15.0 | 16.7 | N/A |

ACATS [9]. Normally, the main (or desired) signal would be adjusted to the strong signal level (–28 dBm), but because of the poor C/N ratio and insertion loss of the channel simulator, the main or desired signal level was adjusted to –43 dBm.

The reference carrier-to-noise ratio (C/N) with no ghost was measured by increasing the level of the random noise until TOV was reached. The degradation in C/N was then measured the same way for each multipath ensemble.

The phase of each echo was selected arbitrarily for each ensemble and the same values were used for each receiver tested. The C/N measurements relative to the no ghost cases for the receivers tested are presented in Table 13.5. Receivers E, F and G obtained better results than the receiver H and the blue rack. Receiver C could not work without any errors, with or without random noise for any combination of multipath.

- **Single echo without random noise**. A single echo was added to the main signal. The range of operation of all the equalizers under test was covered with an echo delay range of –10 to 24 μs except for receiver G with a range that extended to 44 μs. The step size was 1 μs. For every delay, the relative attenuation of the echo was adjusted until TOV was reached.

  The phase of the echo was varied from 0 degree to 360 degrees at a rotation speed of 0.1 Hz (Doppler). The goal of this test was to quickly determine how the phase of the echo is critical for each receiver, or the worst case. With a phase variation limited to 0.1 Hz, the receivers

**Table 13.5 Susceptibility to Random Noise in the Presence of Static Multipath** (*After* [8])

| Static Multipath Ensemble | Degradation Relative to No Ghost at TOV (dB) for the Following Receiver | | | | | |
|---|---|---|---|---|---|---|
| | C | E | F | G | H | Blue-rack |
| No Ghost | 18.6 | 16.0 | 18.0 | 15.3 | 16.6 | 15.16 |
| A * | Not working | Δ 1.8 | Δ 3.3 | Δ 0.5 | Δ 2.3 | Δ 3.28 |
| B * | Not working | Δ 2.2 | Δ 3.8 | Δ 0.9 | Δ 3.5 | Δ 2.40 |
| C * | Not working | Δ 1.3 | Δ 2.1 | Δ 0.0 | Δ 2.5 | Δ 2.18 |
| D * | Not working | Δ 1.9 | Δ 3.1 | Δ 0.2 | Δ 4.7 | Δ 2.89 |
| E * | Not working | Δ 2.0 | Δ 3.8 | Δ 1.1 | Δ 2.6 | Δ 3.64 |
| F * | Not working | Δ 1.2 | Δ 2.7 | Δ 0.9 | Δ 2.5 | Δ 1.20 |
| G * | Not working | Δ 0.1 | Δ 0.1 | Δ 0.0 | Δ 0.1 | Δ 1.68 |

* Combinations of multipath used by ATTC/ACATS [9]

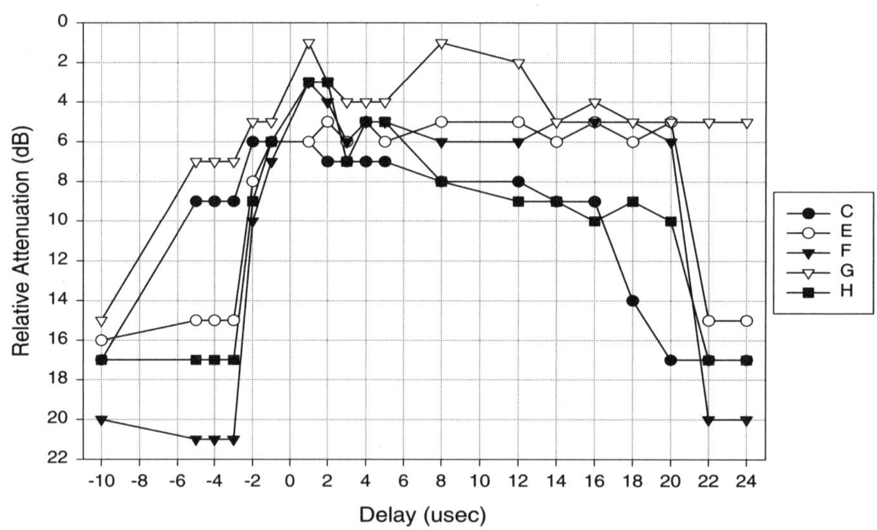


**Figure 13.17**: Single echo without random noise test result. (*From* [8]. *Used with permission.*)

were not sensitive to the Doppler effect and were affected only by the instant phase of the echo. The results are presented in Figure 13.17 and show that all the receivers are sensitive to the phase of the echo. Receivers E, F, and G performed better than receiver C.

- **Single echo with random noise**. This test was designed to verify the susceptibility of the receiver to random noise in the presence of a static single echo. The main difference from the previous test was that both the relative attenuation and phase of the single echo were kept

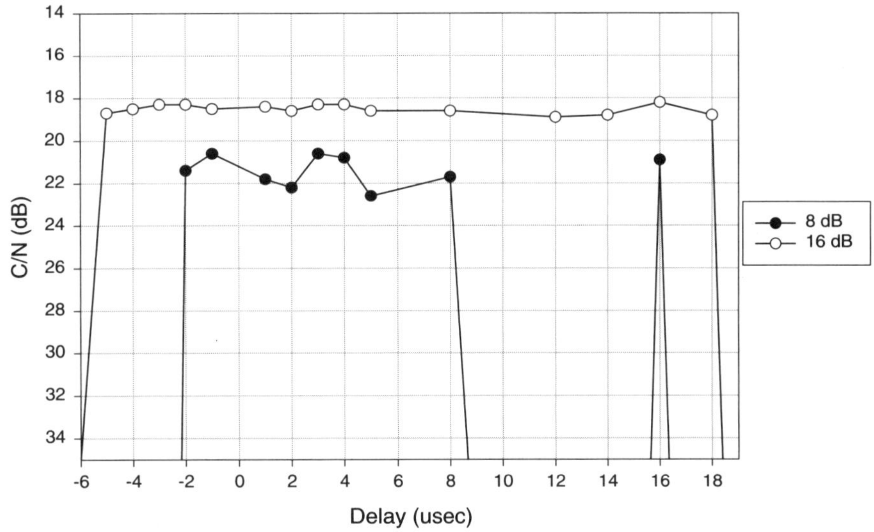


**Figure 13.18** Single echo with random noise test result for receiver C. (*From* [8]. *Used with permission.*)

fixed for the complete range of echo delays, and only the level of the random noise was increased until TOV was reached.

A noticeable performance difference can be observed throughout this test. Figure 13.18 shows the result of one of the receivers with poor performance (receiver C). When the echo amplitude was 16 dB lower than the main signal, the C/N required at TOV was almost stable over the delay range from –5 to 18 μsec. As the amplitude of the echo was increased, the C/N required at TOV increased and started to vary greatly over the delay range. For some delay values, there were errors in the picture even with no noise injected. This is represented by discontinuities in the curves.

The same test was repeated for the receivers E, F, G, and H. The results for the receiver F are presented in Figure 13.19 and show a better performance than receiver C. Receivers F and G worked properly, even for the –4 dB echo, over most of the delay range.

The following broad conclusions were drawn by the authors:

- In the random noise test, the difference between the performance of the worst and the best receivers (C and G respectively) is approximately 2.0 dB at the moderate signal level. For the multipath ensemble, the difference in C/N degradation, between receiver H and receiver G, is approximately 0.1 to 4.5 dB. The older receiver C did not work without any errors for this test.
- In the single echo without random noise test, receiver G did work with an echo from 2 to 7 dB higher than receiver C. This is for the worst case (phase rotation) over the delay range.

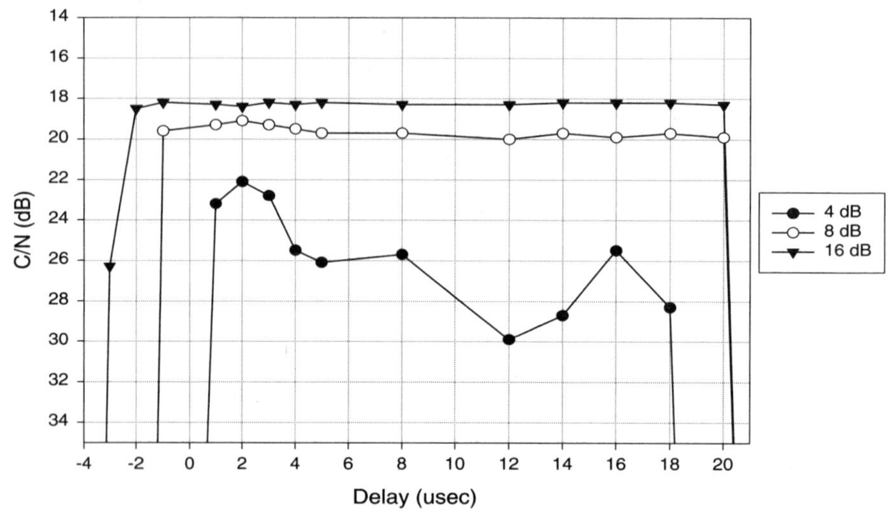


**Figure 13.19** Single echo with random noise test result for receiver F. (*From* [8]. *Used with permission.*)

- In the test with single echo in the presence of random noise, receivers F and G did work properly over most of the delay range for an echo of –4 dB, and this is 4 dB better than receiver E and at least 6 dB better than receiver C.

Further tests were conducted with these receivers for outdoor and indoor reception to determine how much the differences measured in the laboratory could be translated into variations in field performance.

The results of the indoor DTV reception measurements were reported at the IEEE Broadcast Symposium in September 2000. The test program focused on 46 residential homes in the Ottawa metropolitan area. The goal of the measurement program was to gain a better understanding of the following issues [11]:

- The location availability of DTV using simple off-the-shelf indoor antennas
- The types of antennas (active or passive) required for indoor DTV reception
- DTV indoor reception problems caused by low field strength and/or multipath

The tests were conducted using an NTSC signal on channel 65 and a DTV signal on channel 67. Initially, two different DTV receivers were used for the field tests; however, because one receiver consistently outperformed the other, after the first month, the tests were performed with only the better receiver.

The following types of indoor antennas were used at each site:

- An active loop antenna with a gain of 32 dB in the UHF band and another active antenna with a gain of about 30 dB
- A passive log-periodic antenna with a gain of 5 to 7.5 dB across the UHF band

The antennas were placed on a tripod mount 5 ft. above the floor, close to a window. They were then oriented for best DTV reception (that is, highest carrier-to-noise ratio and maximum spectrum flatness on channel 67).

The engineers recorded the DTV and NTSC field strengths at each measurement location and stored a "snapshot" of the DTV signal pass-band using a PC controlled spectrum analyzer. DTV reception at each site was characterized as being either:

- *Reliable* (not sensitive to people moving across the room)
- *Sensitive* to the movement of people across the room
- *Completely unreliable* (frozen and/or broken picture and concomitant loss of audio)

If the TV reception was found to be reliable, then, while the technicians were moving around in the proximity of the antenna (to generate dynamic ghosts), the DTV signal was attenuated until the first visible impairments were observed. This level of attenuation was recorded as the margin relative to the threshold of visibility (TOV). Next, the DTV signal was attenuated further until the *point of unusability* (POU) was determined. NTSC reception was also evaluated using the ITU-R picture impairment scale at each site.

The principal results and conclusions of the field tests were as follows:

- Three sites were rejected because of technical problems.
- DTV signals could reliably be received with both types of antennas (active and passive) in 21 of the 43 sites (49 percent) and with one or the other antenna types in 27 sites (63 percent). For the former sites (21/43), NTSC pictures received on channel 65 were rated between 1.5 and 4.5 with a mean of 3.4 (slightly annoying impairments).
- DTV reception failed using the active antenna at 13/43 sites (30 percent) where the NTSC pictures were rated between 0.5 and 2 with a mean of 0.8 (very annoying impairments). A combination of low field strength and multipath was the most common cause of these failures at 11 of the 13 failed sites (85 percent).
- DTV reception failed with the passive antenna at 17/43 sites (40 percent) where the NTSC pictures were rated between 0.5 and 3 with a mean of 1.3 (very annoying impairments). A combination of low field strength and multipath was again the most common cause of these failures, being present at 13 of these 17 sites (76 percent). Only in very few cases (3/17 or 18 percent) did low field strength alone cause such failures. On the other hand, the DTV failure could be explained by excessive multipath alone in only one case (6 percent) where the corresponding NTSC picture was rated at 3 (slightly annoying impairments).

Based on these figures and a comparison of the median margins to TOV and POU for each type of antenna, the CRC concluded that the benefits of using an active indoor antenna—chiefly boosting the received signal strength—outweighs the potentially negative effects of increased levels of intermodulation distortion contained in the amplifier.

## 13.2n Compatibility Standards

As second generation DTV receivers were making their way to retail outlets, the Consumer Electronics Association (CES) released a set of technical specifications regarding compatibility requirements for DTV receivers and digital cable systems. The document defines the minimum requirements that must be met by digital cable TV systems and digital TV receivers so that the receivers and the cable systems will interoperate to support the following baseline services [12]:

- NTSC analog television signals.
- Digital television programs.
- Signals utilizing a *point-of-deployment* (POD) security module, supplied by the cable TV system operator, to decode scrambled digital television programs that can be authorized by one-way downstream data transmission to the POD module. These services include subscription television programs and pay-per-view programs that are separately ordered by telephone.
- Event information data equivalent to that provided by terrestrial broadcasters to support certain navigation function in the receiver.

There are two fundamental parts to the specification:

- **Part A**—defines the minimum requirements that must be met by the cable TV system
- **Part B**—defines the minimum requirements that must be met by the TV receiver or other cable-compatible consumer device

These two parts go together and must be coordinated. However, each element has been written so that they can be considered as separate specifications.

The CEA requirements are minimum requirements and are not intended to constrain either the cable system operator or the consumer electronics manufacturer from offering and supporting additional services and/or features.

The document contains interface specifications for a number of parameters, including:

- The physical interface, which defines the RF interconnection—including the signal levels, modulation types, and channel plans supported.
- The video formats supported, which includes all of the ATSC Table 3 formats and adds 1440 × 1080, 720 × 480, 544 × 480, 528 × 480, and 352 × 480.
- How to support System Information for digital transport streams and Emergency Alert System messaging.

The specifications also detail the use of POD modules. A POD module is a replaceable security module supplied to the consumer by the cable TV system operator that can be plugged into a cable-compatible digital television receiver or other cable-compatible digital consumer device through a standardized POD interface to provide access to scrambled services. This feature is significant because it allows DTV receivers to be connected directly to a cable system without the use of a set top box.

## 13.2o Digital Receiver Advancements

Experience with first-generation consumer sets demonstrated that improvements were needed to provide reliable service to indoor locations where consumers are currently enjoying NTSC [13]. The DTV reception problems were believed to be the result of the inability of the receiver to decode signals in the presence of long-delay static and dynamic multipath conditions. To combat this problem, a number of second-generation improvements were announced by receiver manufacturers. Examples included the following:

- The Philips TDA8961 demodulator chip with feed-forward adaptive equalization to deal with ghosts from –2.3 to +22.5 μs, with a capability to reach +80 μs using external software. The equalization could use either the 8-VSB training signal, or the data itself (*blind equalization*), to handle complex urban multipath situations. The device also featured low insertion loss NTSC co-channel interference filtering.
- The Samsung KS 1402 decoder chip, which eliminated dynamic multipath impairment in the range of –5.1 to +18.6 μs with a 56-tap feed-forward filter. It was capable of decoding both 8- and 16-VSB, with automatic detection and filtering of an interfering NTSC signal.
- The NxtWave (a spin-off of the David Sarnoff Research Center) NXT2000 receiver subsystem. The ATSC-compliant device performed in either the 8-VSB mode for terrestrial broadcasting or in 64-QAM, 256-QAM, or 16-VSB modes for cable. The NXT2000 was designed to cancel transmission channel impairments such as static and dynamic multipath, phase noise, adjacent or co-channel NTSC interference, and impulse noise. The NXT2000 and the complementary proprietary equalization scheme provided demodulation even when the VSB pilot was destroyed because of severe channel conditions. The equalizer range extended from –4.5 to +44.5 μs. Other features included an integrated FEC and a programmable symbol rate for the different VSB/QAM modes.
- The Motorola MCT2100 8-VSB decoder chip, the result of a partnership between Motorola and Sarnoff. Motorola reported that the MCT2100 completely eliminated multipath issues from the equation in home and pedestrian portable reception. The MCT2100 corrected static multipath with delays of up to 41 μs. It accomplished this by incorporating a full equalizer capable of blind equalization.

From these second-generation efforts, progress in receiver design continued to be made, with most attention given to improved chip sets that form the basis of the major receiver subsystems.

## 13.3 8-VSB Receiver Performance Issues

With the foregoing as a backdrop, the Advanced Television Systems Committee undertook a comprehensive examination of the field performance of the 8-VSB transmission system specified in the DTV standard (document A/53). The work, conducted by the ATSC Task Force on RF System Performance, finished its work in April 2001 and issued a detailed report on its findings. The remainder of this section is adapted from that document [14].

Revision of channel allotments will eventually result in a fully occupied spectrum in the TV broadcasting bands. Until the transition to DTV broadcasting is completed, the spectrum will include NTSC transmissions at relatively high power levels compared to DTV transmissions. Thus far, the only experience with a fully occupied spectrum has been in cablecasting, where the power levels of the channels are uniform and there are no interfering signals except for other TV signals. It is difficult to be certain just how much general improvement may be needed in tuner design so that over-the-air transmissions will be satisfactorily received in the future. Techniques for making such general improvement are, however, already known to receiver designers and are already being implemented in late-generation receivers.

## 13.3a DTV Tuners

A variety of DTV tuners exist in the consumer market [14]. Some tuners utilize a double-conversion architecture to convert VHF or UHF channels to a first IF signal above the UHF reception band and then to a second IF signal having a 44 MHz center frequency. Plural-conversion tuners are less susceptible than single-conversion tuners to interference from signals in "taboo" channels, particularly those containing image frequencies of the channel selected for reception. This is important for achieving satisfactory reception with the revised channel allotments.

Some manufacturers offer single-conversion DTV tuners for converting directly to an IF signal having a 44 MHz center frequency; these are less expensive to manufacture than plural-conversion tuners. Because a single-conversion tuner contains only one oscillator, it typically exhibits less phase noise than a plural conversion tuner.

Designing DTV tuners for off-the-air reception involves a series of compromises to achieve acceptable performance over a wide range of received signal strengths. These compromises are similar to those made in designing tuners for off-the-air NTSC reception. All DTV tuners must be capable of providing linear output signals with $C/(N + I)$ in excess of the 15 dB threshold value that 8-VSB needs at the data slicer in order for a Gaussian channel to be successfully received. In the real world, the reception channel is likely to be Ricean or Rayleigh in nature, so the DTV tuner must deliver linear output signals with substantially higher $C/(N + I)$ in order to accommodate loss of $C/N$ in the filtering used for channel equalization and multipath suppression. $C/(N + I)$ of as much 30 dB may be necessary for satisfactory reception of a Rayleigh channel. Such $C/(N + I)$ is achieved only when the RF signal is received with sufficient strength, but is not so strong as to overload the tuner.

*Adjacent channel interference* (ACI) is caused by a strong adjacent channel overloading an RF amplifier or mixer in the tuner. There are nearly 400 upper or lower first adjacent channels in the DTV allotments, increasing the likelihood that ACI will be a problem in the field. Biasing of the initial stage in the tuner so as to minimize overloading by a strong adjacent channel is not optimal biasing for reducing the noise figure, however. The noise figure in a DTV tuner designed for off-the-air reception is likely to be around 10 dB, not counting the effects of echoes or loss in a downlead from an outside antenna. This is substantially higher than the noise figure of approximately 2 dB for an RF amplifier specifically designed for receiving only weak signals. Designers should consider using a low-noise-figure RF amplifier as an auxiliary preamplifier for current DTV receivers when signal strength is low. Inclusion of a signal strength indicator in future DTV receiver designs would facilitate detection of such circumstances. Another potential improvement is incorporation of means for the receiver to exert automatic control of a remotely located low-noise amplifier.

Tuner input impedance is an area in which significant improvements are possible. It has been anecdotally reported that early DTV receivers had a poor return loss, only slightly greater than 10 dB. In most cases, the reactive component was capacitive and was about 1/2 the magnitude of the real component. In most, but not all cases, the real component of the tuner impedance was closer to 50 Ω than 75 Ω Rarely was the impedance match optimal for the tuned channel.

## 13.3b IF Processing

Receiver IF processing includes all IF filtering and amplification from the first converter output of the tuner to the analog-to-digital converter (ADC) input [14]. Typically, the IF filter that principally determines channel selectivity is a surface acoustic wave (SAW) filter with a flat passband, a constant group delay within the passband, and a 3 dB bandwidth of approximately 6 MHz centered near 44 MHz. Some early DTV receiver designs establish the Nyquist roll-off by IF filtering with somewhat narrower bandwidth SAW filters. As compared to the SAW filter in an NTSC receiver, the SAW filter in a DTV receiver requires more fingers on both transducers to obtain the flatter passband and the steeper skirt slopes for similar adjacent channel suppression. This requires a larger substrate and packaging, so cost is somewhat higher. Current SAW filter designs can suppress *triple transient echoes* (TTE) about 50 dB with approximately 15 dB insertion loss. Acoustic delay is around 1.5 μs, so TTE exhibits about 3 μs delay. TTE is well within equalization range and is too low in amplitude to affect data slicing significantly.

IF processing also includes fixed and variable gain amplifiers to: 1) make up for the losses of the SAW filter, and 2) provide automatic gain control for weak signal levels. In some designs, the IF signal from IF processing is demodulated in the analog regime and the resulting baseband signal is digitized by an analog-to-digital converter (ADC). In most newer designs, the IF signal is digitized by an ADC and is then demodulated in the digital regime. Preferably, the IF processing regulates the IF signal voltage so the full dynamic range of the ADC can be exploited and the full number of bits resolution available from the ADC is secured. For a modern ADC, this voltage is typically between 0.5 and 1.0 Vpp. Thus, the IF dynamic range requirements for DTV receivers differ little from those for NTSC receivers.

### 13.3c Analog-to-Digital Conversion and Digital Demodulation

Most designs for A/53-compliant receivers utilize either 8-bit or 10-bit analog-to-digital converters, followed by purely digital demodulation of the 8-VSB signal to baseband $I$ and $Q$ components [14]. The use of 10-bit-resolution ADCs allows more dynamic range for coping with multipath fading and ACI than 8-bit converters. Additional performance gains can likely be achieved using 12-bit or higher resolution ADCs, presuming these become economical for consumer receivers. The 12-phase trellis coding lends itself to polyphase analog-to-digital conversion, with an oversampling ADC in each phase to gain extra bits of resolution, reducing the quantizing noise.

Recent designs for A/53 receivers employ bandpass subsampling of the 44 MHz IF. Beyond an increase in ADC precision, little performance improvement may be necessary in the bandpass sampling and digital demodulation subsystems. The conversion of 8-VSB signals to double-sideband AM signals prior to demodulation, which suppresses quadrature-phase echoes and helps stabilize carrier synchronization loops, has also come into use. The procedure may simplify some types of channel equalization.

### 13.3d Automatic Gain Control

The AGC circuitry used to regulate gain in the analog portions of a DTV receiver typically encompasses the tuner, the IF processing, and a portion of the demodulator [14]. One purpose of such circuitry is to ensure that the full dynamic range of the analog-to-digital converter is exploited, thus maximizing the actual bit resolution of the digitized DTV signal. In addition to the AGC circuitry used to regulate gain in analog elements, the DTV receiver customarily has AGC circuitry regulating gain in its digital portions, so that the equalized baseband DTV signal is scaled appropriately for data slicing. This is done to minimize the dynamic range required of digital multipliers employed in the equalizer. Analysis of DTV field test data has revealed the need for rapid AGC response times to adapt to broadband ("flat") fading conditions. A closed-loop bandwidth of at least 75 Hz has been suggested for the AGC loop(s) in a home DTV receiver. A mobile receiver will require a higher closed-loop bandwidth, perhaps 200 Hz in its AGC loop(s).

As in NTSC television receivers, it is customary to employ *delayed* AGC in DTV receivers. This technique must be carefully implemented for optimum receiver performance. Delayed AGC maintains the tuner at maximum gain, where it has the best noise figure, under weak signal conditions. As the input RF signal level increases above the minimum, all AGC is performed by the IF processing system. When the IF processing gain approaches its minimum, the tuner gain is reduced to accommodate any further increase in RF input level. Accordingly, the tuner noise figure is degraded only when the signal is moderately strong and there is substantial *C/N* margin; i.e., at an operating point where the tuner noise figure is less important to receiver performance. The *crossover* point at which the tuner takes over the AGC function from the IF processing must be selected not only for noise figure optimization, but also based on receiver overload (AM/AM nonlinearity) and cross-modulation considerations. The preferred crossover point for delayed AGC in a DTV receiver differs from crossover point preferred in an NTSC receiver. In an NTSC receiver, keeping noise in the video more than 40 dB below average luminance is subjectively more important than maintaining the best linearity of video signals, so the RF amplifier stage gain is cut back only when mixer overload is imminent. In a DTV receiver, linearity of the data modulation is critical to accurate data slicing. As long as the signal-to-noise ratio is maintained above threshold, the signal-to-noise requirements are less stringent than for an NTSC receiver. Accordingly, in a DTV receiver it is preferred to begin reduction of RF amplifier gain at a lower-strength input signal than is done in an NTSC receiver. This also reduces the likelihood of ACI in the DTV receiver.

Cross-modulation products can be generated in a mixer when strong adjacent or taboo channels are present and are excessively amplified. As an alternative to delayed AGC for DTV signals, the RF and IF sections may utilize separate AGC systems. With such designs, in order to prevent interfering signals being excessively amplified by the RF amplifier, so as to overload the mixer, the RF AGC is derived from the mixer output signal and is amplified wideband.

## 13.3e Carrier Synchronization

Many A/53-compliant receivers rely to some extent upon the 8-VSB pilot for carrier synchronization [14]. Sole reliance upon the pilot is problematic for robust carrier synchronization in dynamic multipath channels with severe fading, since frequency-selective fades will sweep through the pilot frequency. Because most consumer receivers utilize fully digital processing techniques, pilot tracking loops are typically implemented on-chip with digital phase locked loops (DPLLs), rather than with external analog PLLs. Improving carrier synchronization is the subject of substantial research by 8-VSB receiver designers.

Carrier synchronization in multipath channels is a significant performance issue in DTV receivers. Alternatives to purely pilot-based synchronization exist and are reported to be used in some A/53-compliant receiver designs. Besides pilot tracking loops, decision-directed techniques for suppressed carrier recovery, commonly used in QAM and QPSK systems, are well known. Some receivers incorporate fractional equalizers or complex synchronous equalizers, which provide better tracking of the carrier than can be achieved using the simpler synchronous equalizers of early 8-VSB receiver designs.

Each method of carrier phase and frequency detection is theoretically subject to failure under certain multipath conditions, but a combination of techniques will reduce the likelihood of complete carrier synchronization failure in any particular real-world reception condition. The rapid changes in reception conditions that are encountered by mobile receivers in fast-moving vehicles present particularly difficult challenges to maintaining carrier synchronization.

## 13.3f Symbol Timing Synchronization

Symbol timing synchronization, like carrier synchronization, is another area of active research by receiver manufacturers [1]. Early A/53-compliant receivers relied exclusively upon the *data segment synchronization* (DSS) sequence, which is provided in the A/53 framing structure for this purpose. Newer receivers have either supplemented or replaced the DSS-based methods with *band edge detection*, *spectral bright line detection*, or other techniques designed to improve timing synchronization robustness in the presence of dynamic multipath.

As in the case of carrier synchronization, each method of symbol timing phase and frequency detection is theoretically subject to failure under certain multipath conditions. However, a combination of techniques will reduce the likelihood of complete symbol timing synchronization failure in any particular real-world reception condition. The rapid changes in reception conditions that are encountered by mobile receivers in fast-moving vehicles present particularly difficult challenges to maintaining symbol synchronization.

## 13.3g Adaptive Equalization

Several receiver manufacturers have attempted to address severe Ricean (low *K*-factor) channels and Rayleigh faded channels, in the interest of improving indoor reception robustness in urban environments [14]. Representative examples include adaptive equalizers that can cancel multipath echoes from 5–6 μs pre-echo delay to over 40 μs post-echo delay at amplitudes reaching 0 dB attenuation for single echoes at particular echo delays. Designers have also implemented a variety of equalizer algorithms that track echo dynamics out to 5–20 Hz, but with exponentially decreasing echo amplitude tolerance as echo dynamics increase in frequency. The benefits of these improvements relative to first generation receivers have been clearly demonstrated in laboratory and field tests, but these improvements do not assure robust reception in certain severe Ricean and Rayleigh channels that exist in the real world. Moreover, the positioning of the antenna is critical to maintaining satisfactory reception of DTV signals, presenting an ease-of-use issue in markets where not all DTV signals emanate from the same transmission antenna site.

Besides increasing the ranges of pre-echo advance and post-echo delay that can be accommodated by A/53-compliant receivers, strong echoes—with amplitudes as large as or even slightly larger than the desired signal—must be tolerated to secure reception in many indoor reception environments. Some designers have reported excellent simulation results for severe Ricean channels using discrete Fourier transform (DFT) techniques to measure echoes and to use these measurements for initializing the adaptive equalizer. In order to initialize or re-initialize adaptive equalizer filter coefficients, these methods require measurements of the echo phases at times that known symbol sequences are transmitted. It may be difficult to extract sufficiently accurate echo phase measurements from the existing data field sync (DFS) reference signal for channels with echo delay spreads that are a large percentage of the DFS duration, and it is impossible to extract this information for channels with echo delay spreads that exceed the duration of the DFS.

Manufacturers have recognized that receiver capability to track higher-frequency echo dynamics is important for practical mobile reception. Existing 8-VSB equalizer designs show an exponential decrease in tolerable echo amplitude as a function of Doppler shift. Other techniques for tracking higher frequency dynamics are being developed [15].

### 13.3h Antennas with Electronically Steered Directional Sensitivity

A so-called smart antenna is a device that electronically combines the responses of two similar component antennas to achieve an overall directional antenna response that can be redirected subject to electrical control signals [14]. These control signals can be supplied from a DTV receiver especially equipped to optimize the directionality of the overall antenna response. The receiver will then perform the optimization based on a variety of data including AGC, $C/(N + I)$, and equalizer tap weights. Such smart antennas are expected to contribute to ease of reception of 8-VSB signals, especially in markets where DTV signals are received from different broadcast antenna locations. More rapid changes between channels received from different directions is thus possible. The directionality of a smart antenna could be adjusted automatically by the receiver or manually by remote control. The directionality of an outdoor smart antenna can be changed almost instantly; there is no wait for the slewing of an electromechanical antenna rotor.

The responses of the two similar component antennas can be simultaneously combined in different ways to support concurrent reception of TV channels received from different directions. This concurrent reception can be by a single TV receiver with picture-in-picture capability or can be by different receivers in the household.

### 13.3i Antenna (Spatial) Diversity

Antenna or *spatial diversity* is a technique long used in digital communication systems to combat the effects of multipath interference, but which thus far has not been used to any extent in DTV reception [14]. A spatial diversity receiver employs two or more antennas, and two or more component receivers or significant portions of receivers. The antennas are positioned to be affected differently by multipath signals, so any notches in the frequency spectra of their respective responses can be filled in when they are combined. The diversity receiver may simply select the best signal, or may intelligently combine the signals from two or more antennas prior to decoding. To achieve sufficient de-correlation between the multipath characteristics "seen" by each antenna, the antennas usually are physically separated by at least one-quarter of the RF wavelength. Other ways of achieving spatial diversity are also being explored. Antenna diversity is most likely to be used in mobile applications. Insofar as portable receivers or fixed receivers with indoor or outdoor antennas are concerned, antenna diversity tends to be impractical for low VHF channels, but is quite manageable for UHF channels.

### 13.3j Polarization Diversity

Polarization diversity concerns the simultaneous transmission, reception, or both transmission and reception of a signal in two orthogonal RF polarizations [14]. Such techniques have long been used to improve reception in various communication systems. By using a CP receiver antenna, deep nulls in one polarization component are substantially filled in by the other polarization component of a received CP signal. Improvement in multipath fading is also observed using elliptical polarization (EP). The benefits of CP and EP are achieved without increasing the licensed transmitter output power in either case. The present FCC Rules do not prohibit CP or EP for DTV broadcasting.

### 13.3k Changing Overall Receiver Layout

High VSWR in the downlead connection to the DTV receiver from an outside antenna raises the noise floor of the receiver, as well as reducing the signal power available to it [14]. These effects significantly increase the receiver noise figure. It is difficult to optimize VSWR for different DTV signals with diverse channel frequency allocations, especially when the down-lead connection is driven directly from a wideband outside antenna designed to receive signals over as many as four octaves. Consider the possibilities, however, if the DTV receiver is reconfigured so the front-end tuner and IF amplifier sections are located outdoors with the antenna connected directly to the front-end tuner through a balun, and the IF amplifier sections driving the downlead connected to a further IF amplifier section in the indoor receiver apparatus. The downlead is terminated in characteristic impedance for the IF signal, which prevents echoes in that connection no matter which DTV channel is selected for reception. A demodulator follows the further IF amplifier section in the indoor receiver apparatus. Channel-tuning remote-control signals are transmitted from the indoor receiver apparatus to the outdoor tuner. Frequency multiplexing can be used to transmit these remote-control signals and operating power to the outdoor tuner via the downlead.

In this proposal, plural outdoor tuners facilitate picture-in-picture (PIP) displays, receiving different displays on different monitors, or recording a channel other than the one currently being viewed. The tuners are arranged to drive a shared down-lead connection with different-frequency IF signals. There is a reduction of the number of 6-MHz channels at which the down-lead connection must be terminated with its characteristic impedance in order to prevent echoes, compared to supplying the down-lead connection RF DTV signals that have not been converted in frequency. A lossy transmission line with the same characteristic impedance as the downlead is preferred for terminating the indoor end of the downlead over a broad band of frequencies. A high-input-impedance buffer amplifier with neutralization is preferred for extracting the IF signals from the indoor end of the downlead.

Reconfiguration of the DTV receiver to suit an outdoor antenna would create an additional class of DTV receiver than that used with an indoor antenna and does not help with problems of indoor antenna reception. Additional liability issues of weather-resistance, safety against lightning, proper installation, and so on arise for the manufacturer of the outdoor tuner.

## 13.4 References

1. "Receiver Planning Factors Applicable to All ATV Systems," Final Report of PS/WP3, Advanced Television Systems Committee, Washington, D.C., Dec. 1, 1994.

2. ATSC, "Guide to the Use of the ATSC Digital Television Standard," Advanced Television Systems Committee, Washington, D.C., Doc. A/54, Oct. 4, 1995.

3. Qureshi, Shahid U. H.: "Adaptive Equalization," *Proceedings of the IEEE*, IEEE, New York, vol. 73, no. 9, pp. 1349–1387, September 1985.

4. Ciciora, Walter, et. al.: "A Tutorial on Ghost Canceling in Television Systems," *IEEE Transactions on Consumer Electronics*, IEEE, New York, vol. CE-25, no. 1, pp. 9–44, February 1979.

5. Ungerboeck, Gottfried: "Fractional Tap-Spacing Equalizer and Consequences for Clock Recovery in Data Modems," *IEEE Transactions on Communications*, IEEE, New York, vol. COM-24, no. 8, pp. 856–864, August 1976.

6. Sgrignoli, Gary: "Preliminary DTV Field Test Results and Their Effects on VSB Receiver Design," *ICEE '99*.

7. Einolf, Charles: "DTV Receiver Performance in the Real World," *2000 Broadcast Engineering Conference Proceedings*, National Association of Broadcasters, Washington, D.C., pp. 478–482, 2000.

8. Ledoux, B, P. Bouchard, S. Laflèche, Y. Wu, and B. Caron: "Performance of 8-VSB Digital Television Receivers," *Proceedings of the International Broadcasting Convention*, IBC, Amsterdam, September 2000.

9. FCC/ACATS SSWP2-1306: *Grand Alliance System Test Procedures*, May 18, 1994.

10. Digital HDTV Grand Alliance System: *Record of Test Results*, October 1995.

11. *NAB TV TechCheck*: "Canadians Perform Indoor DTV Reception Tests," National Association of Broadcasters, Washington, D.C., October 9, 2000.

12. *NAB TV TechCheck:* "Digital Cable–DTV Receiver Compatibility Standard Announced," National Association of Broadcasters, Washington, D.C., November 15, 1999.

13. *NAB TV TechCheck*: "New Digital Receiver Technology Announced," National Association of Broadcasters, Washington, D.C., August 30, 1999.

14. ATSC: "Performance Assessment of the ATSC Transmission System, Equipment, and Future Directions," ATSC Task Force on RF System Performance, Advanced Television Systems Committee, Washington, DC, revision: 1.0, April 12, 2001.

15. Megawave Corp.: "Megawave/NAB Joint Technology Development Improved Antennas For NTSC Off-The-Air TV Reception," *NAB Engineering Conference Proceedings*, National Association of Broadcasters, Washington, D.C., 1996.

## 13.5 Bibliography

Hufford, G. A: "A Characterization of the Multipath in the HDTV Channel," *IEEE Trans. on Broadcasting*, vol. 38, no. 4, December 1992.

Miyazawa, H.: "Evaluation and Measurement of Airplane Flutter Interference," *IEEE Trans. on Broadcasting*, vol. 35, no. 4, pp 362–367, December 1989.

Zborowski, R. W.: "Application of On-Channel Boosters to Fill Gaps in DTV Broadcast Coverage," *NAB Engineering Conference Proceedings*, National Association of Broadcasters, Washington, D.C., 2000.

Bendov, O.: "On the Validity of the Longley-Rice (50,90/10) Propagation Model for HDTV Coverage and Interference Analysis," *Proceedings of the Broadcast Engineering Conference*, National Association of Broadcasters, Washington, D.C., 1999.

Plonka, Robert: "Can ATV Coverage be Improved with Circular, Elliptical or Vertical Polarized Antennas?," *NAB Engineering Conference Proceedings*, National Association of Broadcasters, Washington, D.C., 1996.

# Chapter 14

# Consumer Video and Networking Issues

**Skip Pizzi**

**Jerry C. Whitaker, Editor**

## 14.1 Introduction[1]

One of the promises of DTV is a marriage of video and computer technologies in a wide range of consumer devices. To accomplish this goal, several fundamental technologies and standards must be in place, and certain agreements must be reached among the many potential information-providers.

### 14.1a Internet and TV Convergence

Although the Internet and television are viewed on similar video display devices, the difference between them remains profound. The concept of convergence of these mediums is often misunderstood or overplayed. The Internet remains an interactive (2-way) medium carrying largely static content for point-to-point distribution. Meanwhile, television remains a primarily one-way medium with dynamic content distributed in point-to-multipoint fashion.

Interactive television (ITV) is often considered the true convergence point between these systems, but in fact, it can be more accurately viewed as adapting certain features of each service to create a new hybrid medium. ITV takes the rich, dynamic content of television and adds to it the personalization and responsiveness of the Internet. ITV will likely be seen by consumers as neither TV nor the Internet, but a medium unique unto itself.

The first and most obvious method of merging these mediums is simply putting Internet content on a television screen, or conversely, viewing TV content on a computer. Neither of these processes is as simple as it might seem, but the latter is generally more easily accomplished. With the proper hardware, television programs can be fairly successfully viewed on a personal

---

1. Portions of this chapter were adapted from: Pizzi, Skip: "Internet and TV Convergence," in *Interactive TV Survival Guide*, Jerry C. Whitaker ed., McGraw-Hill, New York, N.Y., 2001. Used with permission.

computer screen. On the other hand, putting a Web page from a computer browser output directly onto a TV screen is generally not a satisfying experience. This is because Web content is typically viewed on a computer by a single user positioned at close proximity to the screen, while television is usually watched at far greater distances and often by groups of viewers. This distinction is referred to as the "one-foot vs ten-foot experience" or "lean forward vs. lean back" usage. This implies that when viewing the typical Web page on a TV screen, fonts and graphics are generally too small to be comfortably viewed. Also, the selection of hyperlinks can be difficult with an infrared remote control.

ITV systems have dealt with this issue in different ways. Some systems transcode the Web page content in a specialized server so that it displays more appropriately on a TV screen. Other systems perform this transcoding in the client receiver, or with a combination of client and server-based steps. These transcodings typically involve font substitution to a larger typeface, while retaining as much of the look and feel of the original page as possible. The overall success of the process varies widely. Some systems do a better job than others, and some Web pages lend themselves to such transcoding more readily than others.

In addition, because a mouse is generally not used with ITV receivers, Web-page links are converted to another form of display. Common practice to date has been to highlight each link on the screen one at a time, using the remote control navigation keys to sequence among the links. For example, as a user pushes the "down arrow" key on the ITV remote, the next link down the page will be highlighted. The user steps through the links on the page using the navigation keys until a link that the user wishes to follow is highlighted. Then the user presses a "Go," "OK" or "Enter" key on the remote, and the display switches to the linked page, once again transcoded for TV display.

The ultimate solution to this problem involves the design of Web pages customized for TV display by the content author. While this is rarely done at present, it is likely to become standard practice as the ITV medium takes hold.

## 14.2 IEEE 1394

The interconnection of home information and entertainment devices is a key element in advanced consumer products. IEEE 1394 is an international standard, low-cost digital interface designed to integrate entertainment, communication, and computing electronics into consumer multimedia products [1]. Originated by Apple Computer as a desktop LAN and developed by the IEEE 1394 working group, IEEE 1394 has the following attributes:

- High data rate capabilities—the hardware and software standard can transport data at 100, 200, or 400 Mbits/s
- Physically small—the serial cable can replace larger and more expensive interfaces
- Ease of use—there is no need for terminators, device IDs, or elaborate setup procedures
- *Hot plug* capable—users can add or remove 1394 devices with the bus active
- Scaleable architecture—users can mix 100, 200, and 400 Mbits/s devices on a single bus
- Flexible topology—support is provided for daisy chaining and branching, facilitating true peer-to-peer communication

- Nonproprietary—there are no significant licensing issues for implementation in products or systems

Serial bus management provides overall configuration control of the 1394 bus in the form of optimizing arbitration timing, guarantee of adequate electrical power for all devices on the bus, assignment of which compliant device is the *cycle master*, assignment of the isochronous channel ID, and notification of errors. Bus management is built upon the IEEE 1212 standard register architecture.

There are two types of IEEE 1394 data transfer:

- **Asynchronous**: the traditional computer memory-mapped, load-and-store interface. Data requests are sent to a specific address and an acknowledgment is returned.
- **Isochronous**: data channels provide guaranteed data transport at a predetermined rate. This is especially important for time-critical multimedia data where just-in-time delivery eliminates the need for costly buffering.

Much like LANs and WANs, IEEE 1394 is defined by the high level application interfaces that use it, not a single physical implementation. Therefore, as new silicon technologies allow higher speeds, longer distances, and alternate media, IEEE 1394 can scale to enable new applications.

Perhaps most important for use as a digital interface for consumer electronics is that IEEE 1394 is a peer-to-peer interface. This allows not only dubbing from one camcorder to another without a computer, for example, but also allows multiple computers to share a given camcorder without any special support in the camcorders or computers. All of these features of IEEE 1394 are key reasons why it is a preferred audio/video digital interface.

## Enhancements

A number of extensions to IEEE 1394 were under consideration at this writing by the 1394 Trade Association and the IEEE 1394.1 Study Group. These extensions include the following [2]:

- Gigabit speeds for cables
- 100 Mbits/s for backplane implementations
- Longer-distance cables using copper wire and plastic fiber
- Audio/video command and control protocols
- 1394 to 1394 bus bridges
- IEEE 1394 gateways to communication interfaces, such as ATM

It is believed that ATM and IEEE 1394 will have beneficial effects on each other's markets. It has been predicted that ATM will become the world-wide standard for voice/video/data public switched networks. However, ATM is considered too expensive for devices such as disk drives, cameras, and desktop computers. IEEE 1394, therefore, is being positioned by proponents as a complementary technology, serving as the device interface for ATM systems.

## 14.3 Digital Home Network

It is expected that in the near-term future, data for audio, video, telephony, printing, and control functions are all likely to be transported through the home over a digital network [2]. This network will allow the connection of devices such as computers, digital TVs, digital VCRs, digital telephones, printers, stereo systems, and remotely controlled appliances. To enable this scope of interoperability of home network devices, standards for physical layers, network information, and control protocols need to be generally agreed upon and accepted. While it would be preferable to have a single stack of technology layers, no one selection is likely to satisfy all cost, bandwidth, and mobility requirements for in-home devices.

From a broadcasting perspective, the ability to provide unrestricted entertainment services to consumer devices in the home is a key point of interest. Also, ancillary data services directed to various devices offer significant marketplace promise. Control and protocol standards to enable the delivery of selected programming from cable set-top boxes to DTV sets using IEEE 1394 have been approved by the Consumer Electronics Manufacturers Association (CEMA) and the Society of Cable and Telecommunications Engineers (SCTE). The CEMA standard is EIA-775. The SCTE has approved two complementary standards, DVS 194 and DVS 195 (with copy protection—the *SC* system—and without copy protection).

The standards from these organizations differ in some respects, and the SCTE version has some troublesome aspects for broadcasters. For example, the SCTE standards require the cable set-top box to be the program selection device and only transfer data related to one selected program at a time to the 1394 bus. Accessing data from a broadcast stream that is unrelated to the current program selection on the cable box is also not defined in the SCTE standards.

More generally, the principal physical layer interconnections at the DTV set are expected to be RF (NTSC, VSB, QAM), baseband component (RGB, Y Pr Pb), and digital (1394). Composite video and S-video will be around a long time as well. All of these but 1394 are one-way paths.

Other appliances (or sensors) in the home may use RF on the power lines, dedicated coax, and dedicated twisted pair for control functions. From this plethora of physical layer choices, the signaling defined in the set of 1394 standards appears to be emerging as the choice for high-speed (at the 196 Mbits/s level and higher) local connections and as the backbone for passing information around the home. This physical layer comes in several versions. The most mature technology is the IEEE 1394-1995 standard for communicating over 4.5 m of unshielded twisted pair (UTP). In addition, 60- and 100-m versions have been produced that work over fiber. Commercial products for 1394 over plastic fiber are now available.

On top of the physical layer, network and control layers are needed. This is the part of the communications stack that was defined for the DTV interface (i.e., EIA-775), and was being defined at this writing for long distance 1394 by the Video Electronics Standards Association (VESA) Home Network group (VHN). Their objective was to define a network layer approach to allow seamless operation across different physical layers using Internet protocol (IP). HAVI (Home Audio/Video Interoperability) is a different network layer solution for home networking that has the same unifying objective as VHN. Optimized for the 1394 physical layer, HAVI is focused on audio/video applications and requirements.

On top of the network layer, widely differing approaches exist to the application interfaces, operating system, rendering engines, browsers, and the degree of linkage with the Internet.

### 14.3a PVRs

An important development in the television environment, and one that has strong resonance for ITV applications, is the introduction of the personal video recorder (PVR). Because these devices allow personalization of programming, they are attractive to consumers and offer a fundamental form of interactivity with the content environment for consumers.

These devices allow a substantial amount of television programming to be stored on a hard disk and played back in non-linear fashion. Because a non-linear storage device (a hard drive) is being used, these devices can begin to replay a program while it is still being recorded. This allows a powerful feature of the PVR called *live pause*, by which the device is used as a buffer for live programming. This permits users to pause a program that they are watching as it is currently being broadcast, and return to it a few minutes later without interruption.

PVRs also offer a range program selection processes from simple searching of electronic program guides to intelligent searching for keywords in program titles, subjects, content, talent names, and so on. It is likely that future PVRs will incorporate a browsing ability to seek and download stored content from on-demand libraries, as well as access broadcast program schedules. Many of today's researchers forecast that a significant portion of future television viewing will be from PVRs rather than from real-time broadcast sources.

Some current PVRs, and most future units, also store ITV elements of a program so they can be selectively accessed upon later playback as well. A PVR that can access both broadcast channels (via terrestrial, satellite, or cable) plus the Internet via broadband connectivity will serve as a powerful media collection tool. When interfaced to a home network system, such a storage device could act as the home media server. It is easy to see why the PVR is considered such an important development in the TV environment.

While today's PVRs are standalone devices, those of the future may be part of a PC, a home server, or even a remote storage system in which the actual physical storage is on a server at some service provider's location, with unique personal access by the user. In this respect, the PVR is essentially a user interface that accesses storage locally or on various networks, with the actual storage location of a particular program being transparent to the user. The ultimate vision of this process is the creation of virtual channels, by which the users's PVR build the program stream that the user views from various stored sources.

Numerous business models have been proposed for PVR-based systems. The simplest involves subscription to EPG data (as in current services offered by TiVo, Replay Networks, and WebTV). Future scenarios involve VOD services, selective commercial skipping or replacement, intelligent global searching, automatic highlighting, sharing and transferring of stored videos or clips, rental of off-line storage space, and much more.

## 14.4 Advanced Television Enhancement Forum

The *Advanced Television Enhancement Forum* (ATVEF) was formed in 1997 by a consortium of 14 leading companies in the television and computing industries [3]. This group developed a public, worldwide specification for creating and delivering interactive TV content, based largely on existing content standards from the Internet world. Over 100 other companies have become "adopters" (i.e., licensees) of the format, which allows them to incorporate the standard in products or programs they produce. The ATVEF Specification v 1.1, r26 was finalized and published in 1999, after which the organization sunset itself as an active group. Interested parties are still able to become adopters of the format, however, through the ATVEF licensing office.

A successor group of sorts, the ATV Forum, was formed in 2000, with the primary intention of marketing and expanding the influence of the ATVEF specification. This group includes many of the original founders of ATVEF, but it does not intend to develop further technical specifications. That work has largely been taken up by the SMPTE, under its *Declarative Data Essence* (DDE) specialist group. At press time, SMPTE was preparing a DDE-1 specification, which would extend ATVEF v1.1 into an internationally recognized format.

The ATVEF specification's fundamental premise is the COPE principle: "Create Once, Play Everywhere." The format was therefore written with the intention of allowing ITV authors to write content once and deliver it to a variety of platforms. These platforms include intelligent TV receivers (analog and digital), set-top boxes, and PCs. The format accommodates both one-way and two-way modes of operation, and offers methods of interoperability with all international television formats.

## 14.4a The Specification

The ATVEF Specification v1.1, and its successor SMPTE DDE-1, contain the following primary components:

- Content specifications to establish minimum requirements for intelligent receivers
- Delivery recommendations for the transport of enhanced TV content over various distribution formats
- Bindings for content to the respective delivery formats

A key design point of the specification involved the use of existing content and delivery standards whenever possible. The creation of new methods was done only when necessary to conform to the unique requirements of a broadcast ITV architecture. Therefore, the ATVEF specification references the full existing specifications for the following:

- Hypertext Markup Language (HTML 4.0)
- ECMAScript (ECMA 262)
- Document Object Model (DOM 0)
- Cascading Style Sheets (CSS 1)
- Various other media types

The specification does not set an upper limit on such content, but provides a nominal capability set for content developers.

A second primary design goal of the format was the provision of a single solution that would run on a wide variety of delivery networks. With the proper bindings, it is possible to carry ATVEF-compliant content on either analog or digital television systems, as well as systems that do not support video at all. This includes transmission over terrestrial broadcast, cable and direct-to-home (DTH) satellite systems, as well as over the Internet. Analog TV broadcast systems would typically use the vertical blanking interval for such content, and digital systems could use a variety of delivery methods including data carousels, data transport streams, IP over video, and other applications.

The format also acknowledges the "food chain" of broadcast delivery, so it is designed to allow ATVEF-compliant content to bridge between networks, such as in the case of data on an analog terrestrial broadcast bridging to a digital cable system. This functionality was made possible via the implementation of transport-independent content and the use of IP as the reference binding, by way of the IP bindings already defined for various television systems.

The ATVEF specification does not specify the bindings for each television delivery system, however, preferring to leave that to the standards bodies that controlled each of those formats. It sets as a default the IP-over-TV bindings that each TV broadcast format defines, using the IP Multicast protocol for data transport to the delivery system. The ATVEF specification also includes appendices for the Unidirectional Hypertext Transfer protocol (UHTTP) for optimized IP delivery in a one-way, point-to-multipoint broadcast environment, as well as an example of a transport specific binding set for the NTSC television format.

## 14.4b Content Types

A significant difference between the on-line and television broadcast worlds is the need to accommodate one-way reception of data on at least some interactive TV receivers. This implies that content creators cannot customize their data for different receivers or browsers as they do in the two-way environment of HTTP. One way the ATVEF specification addresses this problem is by declaring a base profile of Multimedia Internet Mail Extension (MIME) file types that must be supported in all ATVEF-compliant receivers. These files types include the following:

- Text/html (HTML 4.0)
- Text/plain
- Text/css (CSS1 only)
- Image/png
- Image/jpg
- Audio/basic

In addition, ATVEF recommends, but does not require, support for the following MIME file types:

- Image/gif
- Audio/wav

The ATVEF specification also accommodates the integration of TV pictures into Web pages. For this, the specifications defines the *tv:URL*, which can be used to identify a specific broadcast channel as a source for a video image to be placed on a Web page. Similarly, the specification's *lid:protocol* supports a locally stored video source. In either of these operations, the video image is scaled and placed in a window the Web page using the HTML <OBJECT> and <IMG> tags. The image can also be placed as a background in a table cell or the entire page, onto which the data of a Web page can be overlaid.

## 14.4c Transport

The ATVEF specification defines two different transport modes, entitled *Transport A* and *Transport B*. Transport A refers to situations where a data-return path (or "back-channel") from the user is available, while Transport B covers one-way broadcast data. Transport A content generally includes only URIs ("links") and triggers, with actual ITV content resources accessed from the Web or other on-line sources. In other words, ATVEF Transport A content directs the ITV receiver to other on-line locations to gather resources (in many cases by caching them in advance), and activate them when the appropriate trigger is received.

ATVEF Transport B, on the other hand, includes all the ITV content resources in the broadcast stream. This implies that significantly greater bandwidth in the broadcast channel may be required for Transport B, but no backchannel is required.

The standard also defines a method for a single ITV program to support both Transport A and B simultaneously. In this case, a program carries enhancement data intended for bidirectional IP-type connections as well as for one-way, broadcast-only receivers, which would allow the content to accommodate a wide range of broadcast delivery systems and receiver types. ATVEF receivers can choose between these formats when both are encountered.

Note that the ATVEF specification only specifies a *content* format. The transport-related elements of the ATVEF specification are presented simply as examples of how that content might be successfully transmitted. Because the ITV industry was in its earliest days when ATVEF was formed, the specification deals with transport issues at some depth, although it takes pains not to *specify* particular transport mechanisms.

## 14.4d Triggers and Announcements

Triggers are real-time events delivered for the enhanced TV program that determine the running time of various data elements, synchronized with reference to the television program they accompany. The ATVEF specification includes details for specifying these "markers" in ITV content.

Various receiver implementations can unilaterally determine how users will enable or disable display of such enhanced TV content. For example, trigger arrival can be used as a signal to the receiver, which can then notify the user of enhanced content availability. Alternatively, the receiver can be set to automatically display such content upon arrival, or to ignore it completely.

In the one-way broadcast mode (Transport type B), *announcement* data is used to initiate or offer enhancements to the TV program. An announcement specifies the location of both the *resource stream* (i.e., the files that provide content) and the *trigger stream* for an enhancement. These files will likely have been cached in the local memory, for which the ATVEF specification requires a minimum of 1 MB be provided in the receiver.

Announcements also provide information about the enhancement's language, its start and stop times, its bandwidth, and its peak storage size needed for incoming resources, along with other data. Among the latter, an option exists for the identification of another broadcast channel for cases in which ATVEF-compliant content is sent separately from the audio/video TV program. The receiver must be able to start receiving data using only the description broadcast in the announcement.

The data format of an announcement follows the ATVEF specification's *Session Description Protocol* (SDP), which is preceded by a *Session Announcement Protocol* (SAP, not to be confused with the *Second Audio Program* of multichannel television audio formats, which shares the same acronym).

An additional time-related element of the ATVEF specification involves how to cancel previously delivered enhancement data and remove it from the local memory after it has expired (for example, after the broadcast program that the interactive data accompanies has concluded).

### 14.4e Bindings

Bindings in the ATVEF specification define how ATVEF-compliant content will run on a given delivery system. The binding may support either or both Transport types A and B. It is necessary that a standard ATVEF binding exist for each transmission format so receivers and broadcast tools can be developed independently but remain interoperable. The binding provides the "glue" between the delivery format's specification and the ATVEF specification, in cases where the network specification does not contain all the necessary information for such data carriage.

Because carriage of IP is already included in the specifications of many media-transmission systems, ATVEF defines its binding to IP as a reference binding, as noted previously. This implies that one way to build an ATVEF binding for a particular network is to simply define how IP runs on that network when associated with a particular video program.

The ATVEF specification also includes a model of how to build a binding specific to a network standard, using NTSC as an example. ATVEF-compliant data bytes are encoded into the vertical blanking interval (VBI) of NTSC fields. For Transport A, Line 21 of the VBI is used, following the Text-2 (T2) service specified in EIA 608 (recommended practices for Line 21 services). The data encoding follows EIA 746A, which describes how to send URLs and related data on Line 21 without interfering with closed captioning and other data that share this VBI line.

For Transport B in the example binding, the aforementioned IP reference binding is applied, using the Internet Engineering Task Force's (IETF) IP-over-VBI recommendation. The NABTS encoding format is used (as is standard practice for NTSC VBI data), with ATVEF data streams constrained to packet addresses 0x4B0 through 0x4BF.

### 14.4f UHTTP

An Appendix to the ATVEF specification defines a one-way, IP-multicast based, resource transfer protocol called the *Unidirectional Hypertext Transfer Protocol* (UHTTP). It is designed to efficiently deliver resource data in a one-way broadcast-only environment. It can be used to carry IP Multicast data on an analog TV signal's vertical blanking interval (IPVBI), or in the MPEG-2 DTV stream, as well as other unidirectional transport systems.

This IP Multicast stream can carry Web pages and their related resources (including scripts) alongside the related TV signal. A session announcement broadcast by the TV service tells the receiver which IP Multicast address and port to monitor for the associated data.

## 14.5 TV Anytime

In 1999, an international industry consortium was formed to set standards for the emerging *personal video recorder* (PVR) industry. The consortium, entitled the TV Anytime Forum (TVAF),

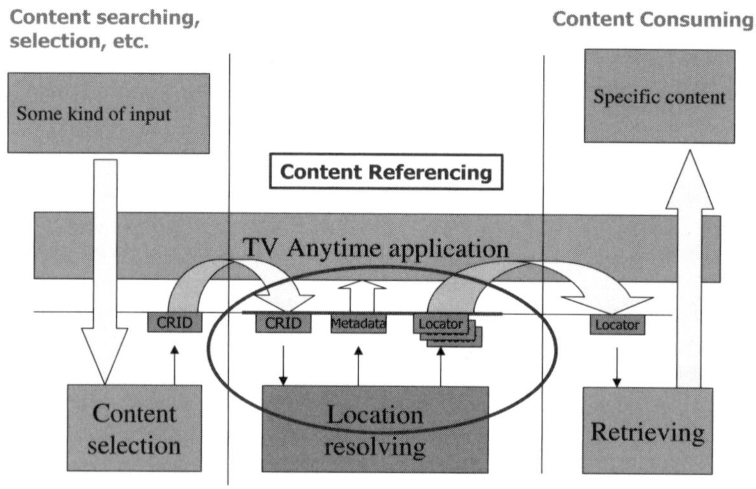


**Figure 14.1** Overview of the TV Anytime Forum realm. (*Courtesy of TV Anytime Forum.*)

set as its goals the establishment of primarily content-related standards, which would allow global interoperability of digital media recording equipment. (See Figure 14.1.)

To accommodate this goal, TVAF is developing (at this writing) three specifications and two recommendations. The specifications cover the areas of content referencing, metadata, and content rights management; while the recommendations deal with business models and system design. TVAF's began issuing its first draft specifications in mid-2000.

## 14.6 DTV Application Software Environment

The DTV Application Software Environment (DASE) is an ambitious four-year effort nearing completion as this book went to press within the ATSC (specialist group T3/S17) to define a platform for advanced receiver functions. As such, it will form the basis for a wide range of new services. The DTV application software environment makes it appear to interactive programming content that it is running on a so-called common receiver. This *common receiver* contains a well-defined architecture, execution model, syntax, and semantics. The benefits of this approach include the following [4]:

- It affords the capability to write content once and run successfully on all compliant receivers
- Consumer electronic equipment manufacturers have the freedom to independently choose the receiver hardware and operating system
- Content and application authors have the assurance that their content will be decoded and displayed on all receiver brands uniformly
- All receivers have certain common functions and features, such as remote control, service selection, MPEG decoding, and so on

The DASE Standard is the next step in the evolution of DTV functionality following the ATSC Data Broadcast Standard (specified in document A/90). The DASE system builds upon A/90, which builds upon the core DTV Standard described in A/53, the Program and System Information Protocol (PSIP) Standard described in A/65, and the Conditional Access Standard described in A/70.

There are certain fundamental building blocks involved in the DASE system, the most basic of which are:

- *Application*—information which expresses behavior (i.e., a program or a document)
- *Application environment*—a system which interprets an application in order to produce a specific behavior (i.e., a program or document processing system)

There are three basic *application approaches* that can be taken to develop an advanced interactive service, specifically:

- **Embedded application approach**, where the application is pre-installed on the receiver. Such a service is generally non-portable and requires re-implementing or porting for new receivers or new technology. As such, it is difficult to change or innovate with new applications. Still, this approach is very stable, with the drawback of only simple features.
- **Thin-client application approach**, where the application is shared between a server and the receiver. The application is executed or interpreted on a remote server. This approach requires a low-latency, high-bandwidth, point-to-point communication channel; it typically does not scale well.
- **Full-client application approach**, where the application is dynamically installed on the receiver through broadcast or a point-to-point channel. The application is executed or interpreted at the receiver. This system requires more resources and greater performance than the thin-client approach, but offers significantly greater flexibility.

A range of challenges exists for implementing any of these approaches, specifically:

- **Installing the application**. If pre-installed (embedded), then it is difficult to innovate. If dynamically installed, then the application must be transmitted (downloaded) to the receiver and prepared for processing in sufficient time for it to be ready for presentation at the intended time. For the DASE standard, the application is downloaded through the broadcast stream.
- **Application form**. There are range of forms that an application can take, essentially dividing into two categories: *procedural* and *declarative*. If procedural, some options include *native compiled code*, *portable byte code* (p-code), and *source code*. If declarative, some options include HTML, XHTML, SMIL, SVG, XML, and MHEG. For DASE, the approach is a standardized form with strict conformance.
- **Environment**. The system designer must identify the "native" resources that an application can reference or utilize. Examples of resources include graphics, video, audio, user input (remote/keyboard), broadcast stream, network, memory, and processor functions. Other issues relate to the method used to reference or utilize these resources. If the mechanism is proprietary, then portability of applications cannot be maintained. For DASE, the approach is a standardized environment with strict compliance.

## 14.7 DTV Product Classification

On August 31, 2000, the Consumer Electronics Association (CEA) announced new definitions and labels for DTV products. The definitions were expected to be incorporated into manufacturers' television marketing materials as DTV receivers continued into the retail channels.

The CEA Video Division Board resolved that analog-only televisions (televisions/monitors with a scanning frequency of 15.75 kHz) should not be marketed or designated to consumers as having any particular DTV capabilities or attributes. In a second related resolution, the Board agreed that the new definitions for monitors and tuners should be used by all manufacturers and retailers to replace general, non-industry terminology like "DTV-ready" or "HDTV-ready." They also defined minimums for HDTV displays as those with active top-to-bottom scan lines of 720 progressive or 1080 interlaced, or higher. Manufacturers were also to disclose the number of active scan lines for a high-definition image within a 16:9 aspect ratio "letter boxed" image area if the unit has a 4:3 HDTV display.

The CEA digital television definitions are as follows [5]:

- **High-definition television** (HDTV): a complete product/system with the following minimum performance attributes: 1) receives ATSC terrestrial digital transmissions and decodes all ATSC Table 3 video formats; 2) has active vertical scanning lines of 720P, 1080I, or higher; 3) capable of displaying a 16:9 image; and 4) receives and reproduces, and/or outputs Dolby Digital audio.

- **High-definition television monitor**: a monitor or display with the following minimum performance attributes: 1) has active vertical scanning lines of 720P, 1080I, or higher; and 2) capable of displaying a 16:9 image. In specifications found on product literature and in owner's manuals, manufacturers were required to disclose the number of vertical scanning lines in the 16:9 viewable area, which must be 540P, 1080I or higher to meet the definition of HDTV.

- **High-definition television tuner:** an RF receiver with the following minimum performance attributes: 1) receives ATSC terrestrial digital transmissions and decodes all ATSC Table 3 video formats; 2) outputs the ATSC Table 3 720P and 1080P/I formats in the form of HD with minimum active vertical scanning lines of 720P, 1080I, or higher, and 3): receives and reproduces, and/or outputs Dolby Digital audio. Additionally, this tuner may output HD formats converted to other formats. The lower resolution ATSC Table 3 formats can be output at lower resolution levels. Alternatively, the output can be a digital bitstream with the full resolution of the broadcast signal.

- **Enhanced definition television** (EDTV): a complete product/system with the following minimum performance attributes: 1) receives ATSC terrestrial digital transmissions and decodes all ATSC Table 3 video formats; 2) has active vertical scanning lines of 480P or higher; and 3) receives and reproduces, and/or outputs Dolby Digital audio. The aspect ratio is not specified.

- **Enhanced definition television monitor**: a monitor or display with the following minimum performance attributes: 1) has active vertical scanning lines of 480P or higher. No aspect ratio is specified.

- **Enhanced definition television tuner**: an RF receiver with the following minimum performance attributes: 1) receives ATSC terrestrial digital transmissions and decodes all ATSC

Table 3 video formats; 2) outputs the ATSC Table 3 720P and 1080I/P and 480P formats with minimum active vertical scanning lines of 480P; and 3) receives and reproduces, and/or outputs Dolby Digital audio. Alternatively, the output can be a digital bitstream output capable of transporting 480P, except the ATSC Table 3 480I format, which can be output at 480I.

- **Standard definition television** (SDTV): a complete product/system with the following performance attributes: 1) receives ATSC terrestrial digital transmissions and decodes all ATSC Table 3 video formats, and produces a useable picture; 2) has active vertical scanning lines less than that of EDTV; and 3) receives and reproduces usable audio. No aspect ratio is specified.
- **Standard definition television tuner**: an RF receiver with the following minimum performance attributes: 1) receives ATSC terrestrial digital transmissions and decodes all ATSC Table 3 video formats; 2) outputs all ATSC table 3 formats in the form of NTSC output; and 3) receives and reproduces, and/or outputs Dolby Digital audio.

These industry standard definitions were intended to eliminate the confusion over product features and capabilities of television sets and monitors intended for DTV applications. The agreement promised to spur the sale of DTV-compliant sets by injecting a certain amount of logic into the marketing efforts of TV set manufacturers. The consumer electronics industry had come under fire early in their DTV product rollouts because of the use of confusing and—in many cases meaningless—marketing terms. For example, terms such as "DTV-ready" means different things to different people.

## 14.7a Cable/DTV Receiver Labeling

On the heels of the CEA DTV product classification agreement, the FCC adopted a Report and Order requiring standardized labeling of DTV receivers that are marketed for connection to cable television systems. The R&O, issued September 14, 2000, specified that such receivers offered for sale after July 1, 2001, must be permanently marked with a label on the outside of the product that reads: "Meets FCC Labeling Standard Digital Cable Ready (DCR) *x*," where *x* = 1, 2, or 3 [6].

The Commission prohibited marketing of receiving devices claimed to be fully compatible with digital cable services unless they have the functionality of one of the three categories defined. The new rules cover any consumer television receiving device with digital signal processing capability that is intended to be used with cable systems. The rules permit marketing devices with less capability, provided full compatibility is not claimed.

DCR 1 refers to a consumer electronics television receiving device capable of receiving analog basic, digital basic, and digital premium cable television programming by direct connection to a cable system providing digital programming. This device does not have an IEEE 1394 connector or other digital interlace. A security card or point of deployment module provided by the cable operator is required to view encrypted programming.

DCR 2, a superset of DCR 1, adds a 1394 digital interface connector. It is clear from the R&O that other digital interfaces also could be present. The FCC noted that connection of a DCR 2 receiver to a digital set-top box (also presumably a level 2 device) may support advanced and interactive digital services and programming delivered by the cable system to the set-top box.

The distinction asserted for DCR 3 is the addition of the capability to receive advanced and interactive digital services. A device with this label is not required to have a 1394 connector. As such services are not defined, the meaning of this distinction is unclear. The FCC did state that additional industry work wa required to design specifications for the DCR 3 category of receivers, and that it would keep the record open in this proceeding, allowing the option to incorporate these specifications into the rules at a later date.

The FCC required the consumer electronics and cable industries to report their progress on developing technical standards in two other areas: direct connection of digital TV receivers to digital cable television systems and the provision of tuning and program schedule information to support on-screen program guides for consumers. The FCC said these two issues had been substantially, but not completely, resolved in an earlier agreement. Reporting requirements were consolidated into a single reporting timetable that began on October 31, 2000, and every six months thereafter until October 2002.

The Commission determined that the copy protection issue raised in this proceeding, in fact, relates to its navigation device rules. For that reason, the copy protection licensing issue was incorporated in the Further Notice of Proposed Rulemaking and Memorandum Opinion and Order/Declaratory Ruling in the navigation devices docket (CS 97-80).

When these rules were announced by the Commission, it was unclear how much the R&O would aid in the delivery of DTV over a cable system to a DTV receiver, other than preventing some grossly misleading marketing practices. The actual connection of the cable systems coaxial cable to the DTV set is not covered. Presumably the products will use the relevant CEA or Society of Cable Television Engineers standards for the coax interface, which were being harmonized as this book went to press. The FCC also stopped short of selecting the standard for transport of signals over the IEEE-1394 physical interface (technically only requiring a connector, not any signaling through this connector). It is unlikely that a manufacturer will put a 1394 connector on a product without some implementation of a protocol using one of the 1394-based protocol standards. Unfortunately, TV receivers with the DCR 2 label are not assured to work with set-top boxes with the same label because of multiple 1394-based protocols.

## 14.8 References


1. Hoffman, Gary A.: "IEEE 1394: The A/V Digital Interface of Choice," 1394 Technology Association Technical Brief, 1394 Technology Association, Santa Clara, Calif., 1999.
2. *NAB TV TechCheck*, National Association of Broadcasters, Washington, D.C., February 1, 1999.
3. "Advanced Television Enhancement Forum Specification," Draft, Version 1.1r26 updated 2/2/99, ATVEF, Portland, Ore., 1999.
4. Wugofski, T. W.: "A Presentation Engine for Broadcast Digital Television," *International Broadcasting Convention Proceedings*, IBC, London, England, pp. 451–456, 1999.
5. *NAB TV TechCheck*: "CEA Establishes Definitions for Digital Television Products," National Association of Broadcasters, Washington, D.C., September 1, 2000.
6. *NAB TV TechCheck:* "FCC Adopts Rules for Labeling of DTV Receivers," National Association of Broadcasters, Washington, D.C., September 25, 2000.

Chapter

# 15

# Cable Television Systems

K. Blair Benson

D. Stevens McVoy, Joseph L. Stern, Charles A. Kase, Wilbur L. Pritchard

## 15.1 Introduction

Cable television systems use coaxial and fiber optic cable to distribute video, audio, and data signals to homes or other establishments that subscribe to the service. Systems with bidirectional capability can also transmit signals from various points within the cable network to the central originating point (*head end*).

Cable distribution systems typically use leased space on utility poles owned by a telephone or power distribution company. In areas with underground utilities, cable systems are normally installed underground, either in conduits or buried directly, depending on local building codes and soil conditions.

While current cable practice involves extensive use of fiber in new construction and upgrades, it is important to understand cable techniques used prior to fiber's introduction. This is partly because a significant portion of cable systems have not yet upgraded to fiber and because these older cable techniques illustrate important technical principles for the service.

### 15.1a Evolution of Cable Television

Cable television began in rural areas in the early 1950s as a means of bringing television service to regions with no broadcast stations. Other early systems brought television reception into mountainous areas where the terrain blocked signals from individual home antennas. These systems typically had a limited capability of five channels and carried only the three major television networks, plus an independent station or two. During the 1960s, cable moved into areas served by broadcast stations but without a full complement of network channels. Twelve to 20 channels were common and were used almost exclusively to carry broadcast television signals. By the mid-1970s, satellite distribution of pay television programming made cable viable in urban areas where a good selection of over-the-air programming already existed. The systems offered a greater variety of viewing choices, including independent stations from distant cities and pay television.

Systems constructed in the 1980s typically had a capacity of 50 to 100 channels and included bidirectional transmission capability. Interactive programming, information retrieval, home monitoring, and point-to-point data transmission were possible, although not widely implemented. Many systems included an institutional network in addition to one or two subscriber networks.

The two main drivers of current progress in the cable industry are the *hybrid fiber/coax* (HFC) architecture and digital video. These technologies have changed cable from a stable, well understood, even stale technology into something with massive possibilities and choices [1].

In the mid 1990's, the vision of what fiber in the cable plant could accomplish inspired the cable industry to dramatically upgrade its physical facilities. Much of that work has already been done. More is underway. The current vision driving progress in the cable industry is digital video compression, which facilitates massive increases in the amount of programming cable can deliver.

## 15.2 Coax-Based Cable Systems

The division between coax-based and fiber-based cable television systems essentially divide conventional from state-of-the-art designs. While the implementation of fiber-based systems, typically built around digital transmission techniques, represent the technologies of choice today, many older systems still exist. It is instructive, therefore, to examine some of the key operational considerations that went into designing these systems.

Cable television systems are generally of two types:

- *Subscriber networks*, which serve primarily residential subscribers. Services carried by subscriber networks are mainly entertainment and information programming.
- *Business networks*, which serve commercial, educational, and governmental concerns. The majority of channels on these systems carry informational programming or data.

### 15.2a Program Sources

Television picture and sound program material provided on cable systems may originate from local or distant broadcast stations or may be relayed by microwave or satellite. A number of advertiser-supported networks available only through cable are distributed by satellite, as are pay television services. In addition, programs can be originated by the cable television operator. Such *local origination* (LO) programming is produced in studio facilities, reproduced using videotape playback equipment, or generated by automated character-generating systems controlled by wire news services and weather instruments, for example. Bidirectional cable also allows "remote" LO, using an upstream channel to transport the signal from the remote site to the head end, where the signal is converted to a downstream channel for distribution to all subscribers.

### 15.2b Channel Capacity

Although the majority of conventional (analog, non-fiber) systems have an upper frequency limit of 300 MHz, many are capable of transmitting signals in the range of 5 to 400 MHz. Systems designed for bidirectional transmission divide the spectrum between the two directions. For subscriber networks, 5 to 35 MHz is used for *upstream transmission* (toward the head end); 50 to

**Table 15.1 Cable Television Channel Frequencies and Designations**

| Frequency, MHz | Channel | Designation |
|---|---|---|
| 5–35 | Reverse transmission | Sub-band HF |
| 54–88 | 2–6 | Low-band VHF |
| 120–170 | 14–22 | Mid-band VHF |
| 174–216 | 7–13 | High-band VHF |
| 216–300 | 23–36 | Super-band VHF |
| 300–464 | 37–64 | Hyper-band VHF |
| 468–644 | 65–94 | Ultra-band VHF |

400 MHz carries *downstream transmissions* (toward the subscriber). For business networks, an equal number of channels in each direction is common. Upstream transmission is usually from 5 to 150 MHz, and downstream channels use frequencies from 200 MHz up.

The upper frequency limit of a given cable system is largely a function of the hybrid devices used in the amplifiers. Amplifiers providing a bandwidth of 500 MHz are common, and higher-frequency devices are readily available. Other components needed for the system, such as passive line splitters, cable, and converters already exist for operation at 500 MHz and higher.

A maximum of 54 television channels can be carried downstream on a conventional 400 MHz system (36 channels for a 300 MHz system). However, FCC regulations prohibit cable systems from using frequencies in the aircraft navigation and communications bands, making several channels unusable. As a result, most 400 MHz systems carry 50 channels (32 channels for 300 MHz systems).

## 15.2c Channel Assignments

Cable television systems use VHF frequencies for all channels provided to subscribers. Bidirectional systems allow transmissions heading upstream on high frequencies (HF). UHF signals, as a rule, are not used on cable. Some confusion arises in that channel numbers normally associated with UHF channels are used in referring to the additional VHF channels. (See Table 15.1.)

Business networks generally allocate a minimum of 20 or 25 channels in the downstream direction and 15 or 20 channels to upstream service

## 15.2d Elements of a Cable System

Because cable television evolved from a specialized system for transmitting numerous television channels in a sealed spectrum, rather than a general-purpose communications medium, the topology or layout of the network was customized for maximum efficiency. The topology that has evolved over the years is called *tree-and-branch* architecture. Many small and intermediate sized systems fit this model. When analyzed, most large systems can be seen as having evolved from this basic prototype. Typical cable television systems are comprised of four main elements (Figure 17.4.1):

- A head end, the central originating point of all signals carried

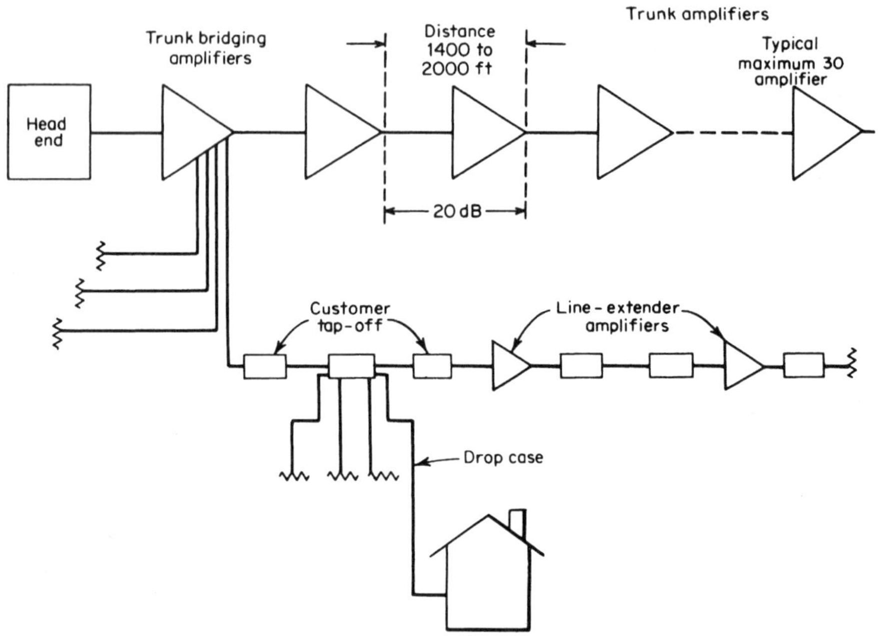


**Figure 15.1** Typical cable distribution network. (*From* [2]. *Used with permission.*)

- *Trunk system*, the main artery carrying signals through a community
- *Distribution system*, which is bridged from the trunk system and carries signals into individual neighborhoods for distribution to subscribers
- *Subscriber drops*, the individual lines to subscribers' television receivers, fed from taps on the distribution system

In a subscriber's home, the drop may terminate directly into the television receiver on 12-channel systems, or into a converter where more than 12 channels are provided. Most modern receivers and videocassette recorders are "cable-ready" and include the necessary converters to access the additional system channels. Systems providing pay services may require a *descrambler* (or other form of converter) in the subscriber's home to allow the viewer to receive these special services. Newer cable systems use addressable converters or deseramblers, giving the cable operator control over the channels received by subscribers. Such control enables impulse viewing or pay-per-view pay television service without a technician visiting the home to "install" the special service.

## Interactive Service

While the main purpose of cable television is to deliver a greater variety of high-quality television signals to subscribers, a growing interest is found in interactive communications, which allow subscribers to interact with the program source and to request various types of information. The growth in the capabilities of the Internet has led to renewed interest in this feature of cable TV. An interactive system also can provide monitoring capability for special services such as home security. Additional equipment is required in the subscriber's home for such services. Monitoring requires a home terminal, for example, whereas information retrieval requires a modem (cable or conventional land-line) for data transmission.

## Head End

The head end of a cable system is the origination point for all signals carried on the downstream system. Signal sources include off-the-air stations, satellite services, and terrestrial microwave relays. Programming may also originate at the head end. All signals are processed and then combined for transmission via the cable system. In bidirectional systems, the head end also serves as the collection (and turnaround) point for signals originated within the subscriber and business networks.

The major elements of the head end are the antenna system, signal processing equipment, pilot-carrier generators, combining networks, and equipment for any bidirectional and interactive services. Figure 15.2 illustrates a typical analog system head end.

A cable television antenna system includes a tower and antennas for reception of local and distant stations. The ideal antenna site is in an area of low ambient electrical noise, where it is possible to receive the desired television channels with a minimum of interference and at a level sufficient to obtain high-quality signals. For distant signals, tall towers and high-gain, directional receiving antennas may be necessary to achieve sufficient gain of desired signals and to discriminate against unwanted adjacent channel, co-channel, or reflected signals. For weak signals, low-noise preamplifiers can be installed near the pickup antennas. Strong adjacent channels are attenuated with bandpass and bandstop filters.

Satellite earth receiving stations, or *television receive only* (TVRO) systems, are used by most cable companies. Earth station sites are selected for minimum interference from terrestrial microwave transmissions, which share the 4 GHz spectrum, but an unobstructed line-of-sight path to the desired satellites also is necessary. Most earth stations use parabolic receiving antennas, 6 to 23 ft in diameter, a low-noise preamplifier at the focal point of the antenna, and a waveguide to transmit the signal to receivers.

## Signal Processing

Several types of signal processing are performed at the head end, including:

- Regulation of the signal-to-noise ratio at the highest practical value
- Automatic control of the output level of the signal to a close tolerance
- Reduction of the aural carrier level of television signals to avoid interference with adjacent cable channels
- Suppression of undesired out-of-band signals

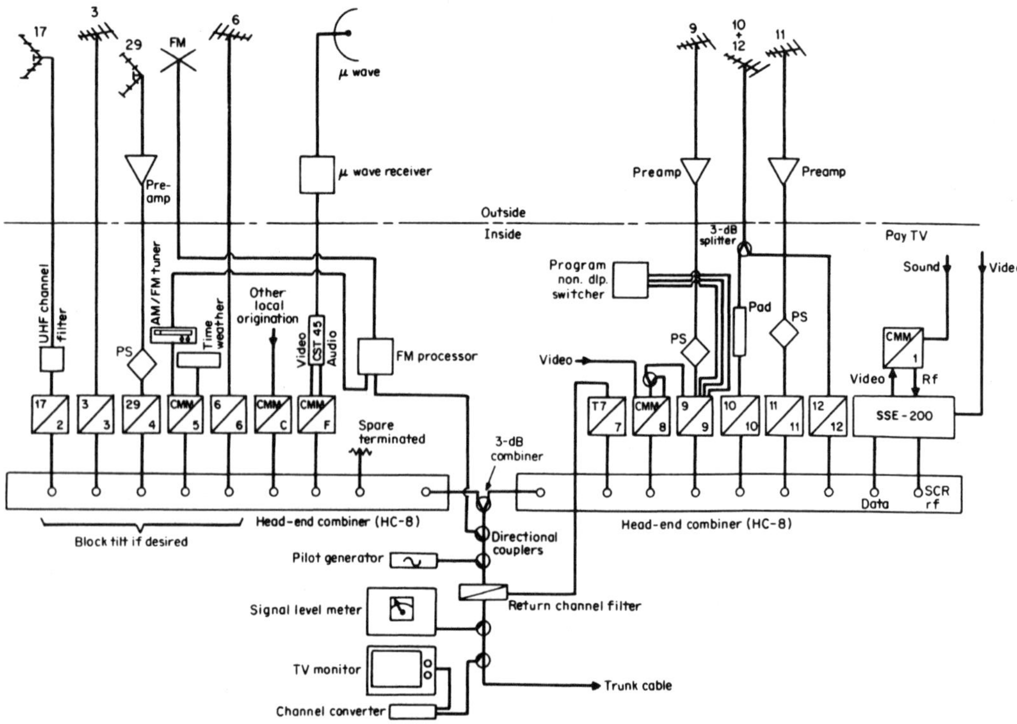


**Figure 15.2** A typical cable system head end. (*From* [2]. *Used with permission.*)

Received signals are converted as necessary to different channel frequencies for cable use. Such processing of off-air signals is accomplished with heterodyne processors or demodulator/modulator pairs. Signals from satellites are processed within a receiver and placed on a vacant channel by a modulator. Similarly, locally originated signals are converted to cable channels with modulators.

Pilot-carrier generators provide precise reference levels for automatic level control in trunk-system amplifiers. Generally, two reference pilots are used, one near each end of the cable spectrum. Combining networks group the signals from individual processors and modulators into a single output for connection to the cable network.

In a bidirectional system, a supervisory computer system is located at the head end. The configuration of the computer varies with the type of service to be offered and can range from a small microprocessor and display terminal to multiprocessor minicomputers with many peripherals. Such computers control the flow of data to and from terminals located within the cable television network.

Interactive services require one or more data receivers and transmitters at the head end. Polling of home terminals and data collection are controlled by the computer system. Where cable networks are used for point-to-point data transmission, modems are required at each end location, with RF-turnaround converters used to redirect incoming upstream signals back downstream.

Most cable television systems use computerized switching systems to program one or more video channels from multiple-program sources. In addition, computer-controlled alphanumeric character generators are used for development of automated news, weather, stock market, and other information channels.

Business networks on cable systems require switching, processing, and turnaround equipment at the head end. Video, data, and audio signals that originate within the network must be routed back out over either the business or the subscriber network. One method of accomplishing this is to demodulate the signals to the base band, route them through a switching network, and then remodulate them onto the desired network. Another method is to convert the signals to a common intermediate frequency, route them through an RF switching network, and then up-convert them to the desired frequency with heterodyne processors.

In larger systems, different areas of the network may be tied together with a central head end using supertrunks or multichannel microwave. Supertrunks or microwave systems are also used where the pickup point for distant over-the-air stations is located away from the central head end of the system. In this method, called the *hub system*, supertrunks are high-quality, low-loss trunks, which often use FM transmission of video signals or feed-forward amplifier techniques to reduce distortion. Multichannel microwave transmission systems may use either amplitude or frequency modulation.

For new construction, fiber optic lines are used almost exclusively for trunking. Fiber also typically replaces cable during system retrofit upgrading, required to support new user features such as additional channels, cable modems, and Internet telephony.

## Heterodyne Processors

The heterodyne processor, shown in Figure 15.3, is the most common type of analog head-end processor. The device converts the incoming signal to an intermediate frequency, where the aural and visual carriers are separated. The signals are independently amplified and filtered with automatic level control. The signals are then recombined and heterodyned to the desired output channel.

Prior to cable delivery systems, U.S. television licensing practices avoided placing two stations on adjacent channels. As a result, many television receivers were not designed to discriminate well between adjacent channel signals. When used with cable television, such receivers can experience interference between the aural carrier of one channel and the visual carrier of the adjacent channel. Channel processors alleviate this problem by changing the ratio between the aural and visual carriers of each channel on the cable. The visual-to-aural carrier ratio is typically 15 to 16 dB on cable television systems. At one time, the aural carrier of broadcast stations was only 10 dB lower than the visual carrier, but many stations now operate with the aural signal 20 dB below visual.

Channel conversion allows a signal received on one channel to be changed to a different channel for optimized transmission on the cable. Processors are generally modular, and with appropriate input and output modules, any input channel can be translated to any other channel of the cable spectrum. Conversion is usually necessary for local broadcast stations carried on cable networks. If the local station were carried on its original carrier frequency, direct pickup of the station off the air within the subscriber's receiver would create interference with the signal from the cable system.

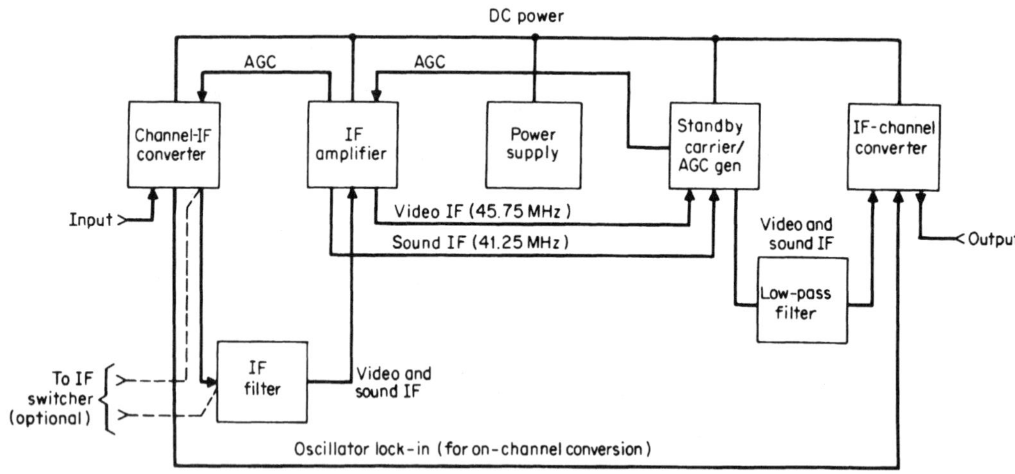


**Figure 15.3** Block diagram of a typical heterodyne-type head-end processor. (*Courtesy of Jerold.*)

The visual-signal intermediate-frequency (IF) passband of a typical heterodyne processor is shown in Figure 15.4. The heterodyne processor is designed to reproduce the received signal with a minimum of differential phase and amplitude distortion. This is best accomplished with a flat passband. Note that the visual carrier is positioned within the flatter portion of the passband response curve. Television demodulators place the visual carrier at a point 6 dB down on the response curve. For this reason, the heterodyne processor has better differential phase characteristics than a modulator/demodulator pair.

An integral circuit of heterodyne processors is a *standby carrier generator.* If the incoming signal fails for any reason, the standby signal is switched in to maintain a constant visual carrier level, particularly on cable systems that use a television channel as a pilot carrier for trunk automatic level control. Such standby carriers can include provisions for modulation from a local source for emergency messages.

## Demodulator/Modulator Pair

Demodulator/modulator pairs are commonly used to convert satellite or microwave signals to cable channels. The demodulator is essentially a high-quality television receiver with base-band video and audio outputs. The modulator is a low-power television transmitter. This approach to channel processing provides increased selectivity, better level control, and more flexibility in the switching of input and output signal paths, compared to other types of processors.

In the demodulator (Figure 15.5), an input amplifier, local oscillator, and mixer down-convert the input signal to an intermediate frequency. Crystal control or phase-locked loops maintain stable frequencies for the conversion. In the IF amplifier section, the aural and visual signals are separated for detection. In comparing the IF response curve of a demodulator (Figure 15.6) with that of the heterodyne processor (Figure 15.4), we find that the visual carrier is located on the

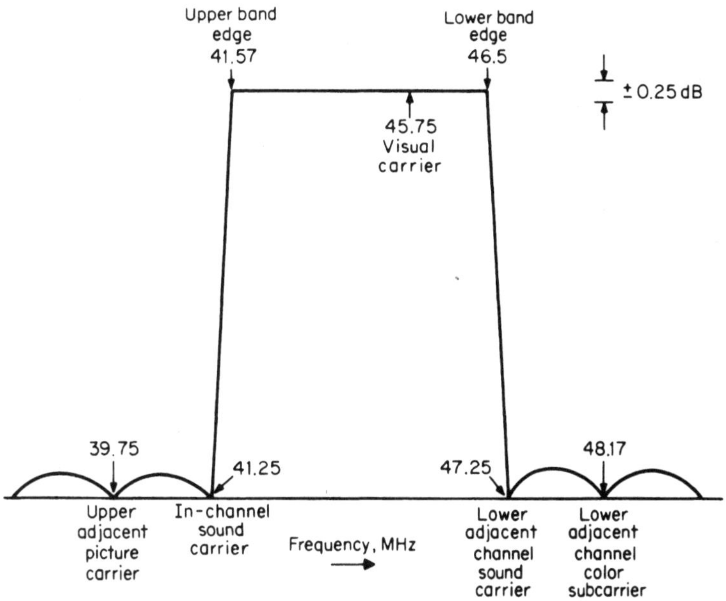


**Figure 15.4** Typical idealized video-IF response curve of a heterodyne processor. Note that the visual carrier is located on the flat-loop portion of the curve in order to provide improved phase response compared to television receivers. (*From* [2]. *Used with permission.*)

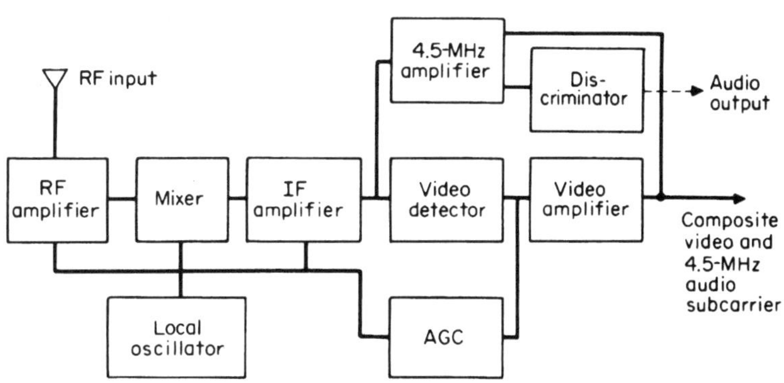


**Figure 15.5** Block diagram of demodulator portion of a demodulator/modulator pair. (*From* [2]. *Used with permission.*)

edge of the passband. The 4.5-MHz intercarrier sound is taken off before video detection, amplified, and limited to remove video components. The sound is often maintained on a 4.5-MHz

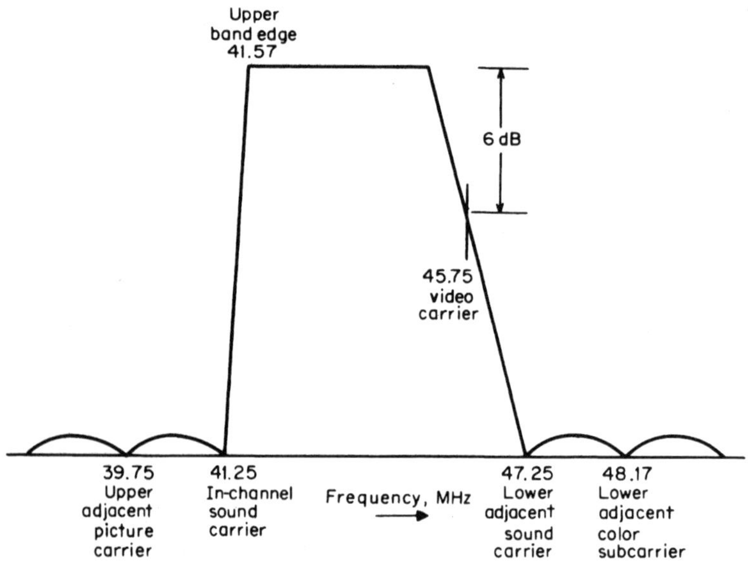


**Figure 15.6** The idealized video-IF response curve of a demodulator. The video carrier is located 6 dB below the maximum response. (*From* [2]. *Used with permission.*)

audio subcarrier for remodulation.

Demodulators are designed to minimize phase and amplitude distortion with linear detector circuits. However, quadrature distortion is inherent in systems that use envelope detectors with vestigial sideband signals. Such distortion can be corrected with video processing circuitry.

In the modulator (Figure 15.7), a composite input signal is applied to a separation section, where video is separated from the sound subcarrier. The video is amplified, processed for dc restoration, and mixed with a carrier oscillator to an IF frequency. An amplifier boosts the signal to the required power level, and a vestigial sideband filter is used to trim most of the lower sideband for adjacent-channel operation.

Audio, if derived from the composite input, is separated as a 4.5 MHz signal from the visual information, filtered, amplified, and limited before being mixed with the visual carrier. If audio is provided separately to the modulator, an FM modulation circuit generates the 4.5 MHz aural subcarrier to be combined with the visual carrier. After the two are combined, they are mixed upward to the desired output channel frequency and applied to a bandpass filter to remove spurious products of the upconversion before being applied to the cable system.

## Single-Channel Amplifiers

The single-channel or *strip amplifier* is designed to amplify one channel and include bandpass and bandstop filters to reject signals from other channels. In simplest form, the unit consists of an amplifier, filter, and power supply. More elaborate designs include automatic gain control (AGC) and independent control of the aural and visual carrier levels. (See Figure 15.8.)

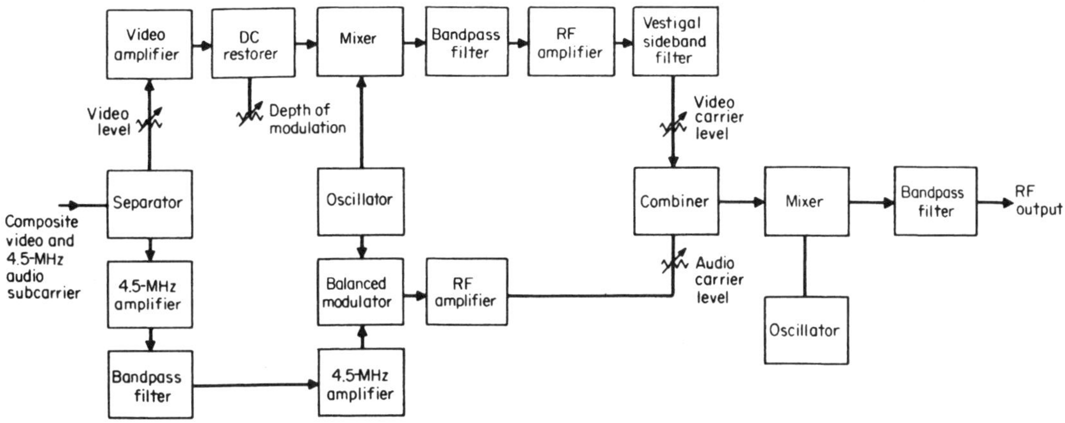


**Figure 15.7** Block diagram of a conventional analog modulator. (*From* [2]. *Used with permission.*)

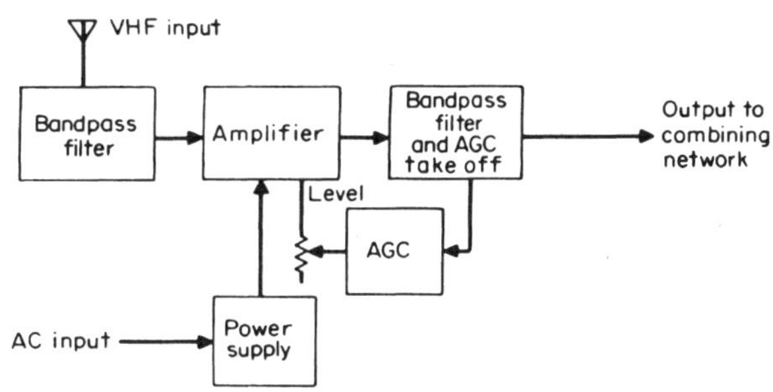


**Figure 15.8** Block diagram of a single-channel amplifier with AGC. (*From* [2]. *Used with permission.*)

Single-channel amplifiers are useful where the desired signal levels are relatively high and undesired signals are low or absent. Selectivity is low compared with other processor types, and the units generally lack the independent control of carrier levels, the ratio between carriers, independent AGC of the carriers, or limiting for the aural carrier. They cannot be used for channel conversion, and they usually present difficulties in adjacent-channel applications because of their limited selectivity.

## Supplementary Service Equipment

Most cable television systems have the ability to distribute other signals besides television. They may offer additional services to subscribers, such as FM broadcast radio, data transmission, interactive signaling, and specialized audio and video transmissions.

FM radio service on cable systems is commonly provided only in the standard FM broadcast band from 88 to 108 MHz. Heterodyne processors are usually used to place an FM radio signal

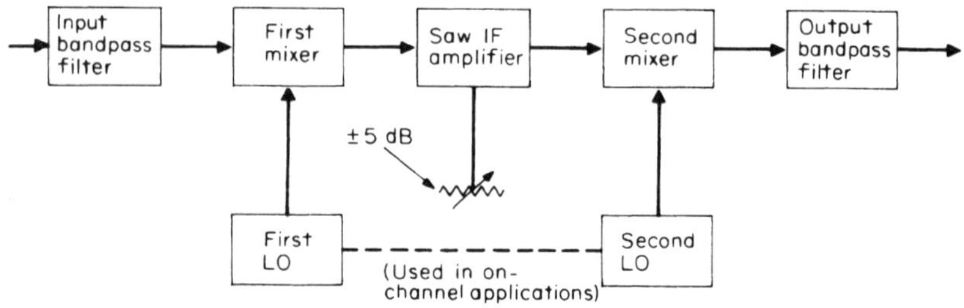


**Figure 15.9** FM radio heterodyne processor block diagram.

into the cable system spectrum at a frequency where it will not interfere with direct pickup of over-the-air signals. This process, in general terms, is illustrated in Figure 15.9.

Data transmission on a cable television network requires modems designed for the particular bandwidth, frequency, and data transmission speed required. Many cable systems dedicate a portion of the spectrum on a subscriber network or install a separate institutional network for data transmission. Generally, these systems are bidirectional to allow interactive data transmission and signaling. Modems for cable system applications are available for data speeds to 1.544 Mbits/s and higher.

## Head-End Signal Distribution

The trunk system (Figure 15.10) is designed for bulk transmission of multiple channels. The trunk may connect numerous distribution points and may interconnect widely spaced sections of a cable system or more than one cable system. A trunk system spanning long distances without intermediate distribution is often referred to as a *super trunk*. Fiber has become the medium of choice for trunk systems, displacing coaxial cable in new installations.

The coaxial cable used in trunk systems is made with a solid aluminum shield, a gas-injected polyethylene, low-loss dielectric insulator, and a copper or copper-clad aluminum-center conductor. Underground cable has an additional outer polyethylene jacket. Under the jacket is a flooding compound to seal small punctures that might occur during or after installation. Trunk cables are usually 3/4 to 1-in. in diameter and typically have a loss of about 1 dB/100 ft at 400 MHz.

For long-distance trunking with minimum distortion, frequency-modulation (FM) techniques are preferred. While FM requires a larger portion of the available bandwidth than amplitude modulation, the distortion in the delivered signal is considerably lower. Thus, FM techniques permit longer trunk lengths.

## Amplification

Amplifiers in the cable system are used to overcome losses in the coaxial cable and other devices of the system. From the output of a given amplifier, through the span of coaxial cable (and an equalizer to linearize the cable losses), and to the output of the next amplifier, unity gain is

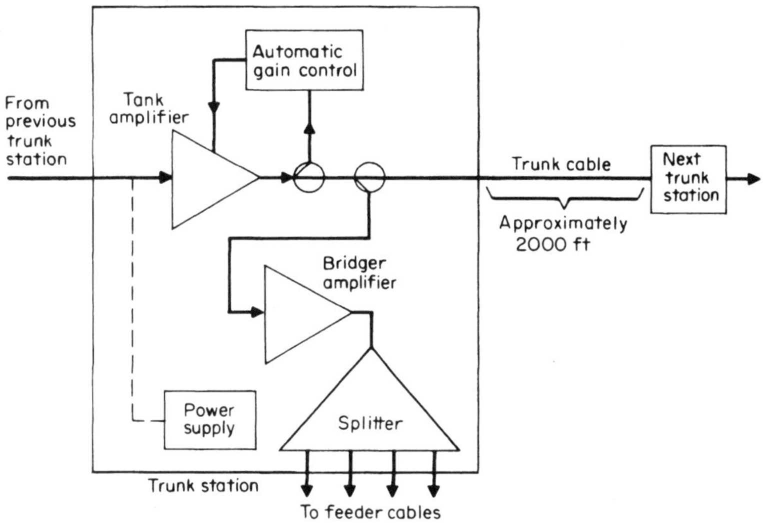


**Figure 15.10** A basic trunk system. (*From* [3]. *Used with permission.*)

required so that the same signal level is maintained on all channels at the output of each amplifier unit.

Standard cable system design places repeater amplifiers from 1400 to 2000 ft apart, depending on the diameter of the cable used. This represents an electrical loss of about 20 dB at the highest frequency carried. Systems with amplifier cascades to 64 units are possible, depending on the number of channels carried, the performance specifications chosen, the modulation scheme used, and the distortion-correction techniques used. Table 15.2 tabulates representative figures of distortion versus the number of amplifiers in cascade.

## Bridging Amplifier

Signals from a trunk system are fed to the distribution system and eventually to subscriber drops. A wide-band directional coupler extracts a portion of the signal from the trunk for use in the bridge amplifier. The bridge acts as a buffer, isolating the distribution system from the trunk, while providing the signal level required to drive distribution lines.

## Subscriber-Area Distribution

As shown in Figure 15.11, distribution lines are fed from a bridging station. Distribution lines are routed through the subscriber area with amplifiers and tap-off devices to meet the subscriber density of the area. The cable for distribution lines is identical to that used for trunks, except that its diameter is commonly 1/2 or 5/8-in. with a resulting increase in losses, typically to 1.5 dB/100 ft at 400 MHz.

**Table 15.2 Distortion-vs.-Number of Amplifiers Cascaded in a Typical Cable System**

| Number of Amplifiers in Cascade | Cross Modulation, dB | Second Order, dB | S/N, dB |
|---|---|---|---|
| 1 | –96 | –86 | 60 |
| 2 | –90 | –81.5 | 57 |
| 4 | –84 | –77 | 54 |
| 8 | –78 | –725 | 51 |
| 16 | –72 | –68 | 48 |
| 32 | –66 | –63.5 | 45 |
| 64 | –60 | –59 | 42* |
| 128 | –54* | –54.5* | 39 |
| * Lower limit of acceptable system performance. | | | |

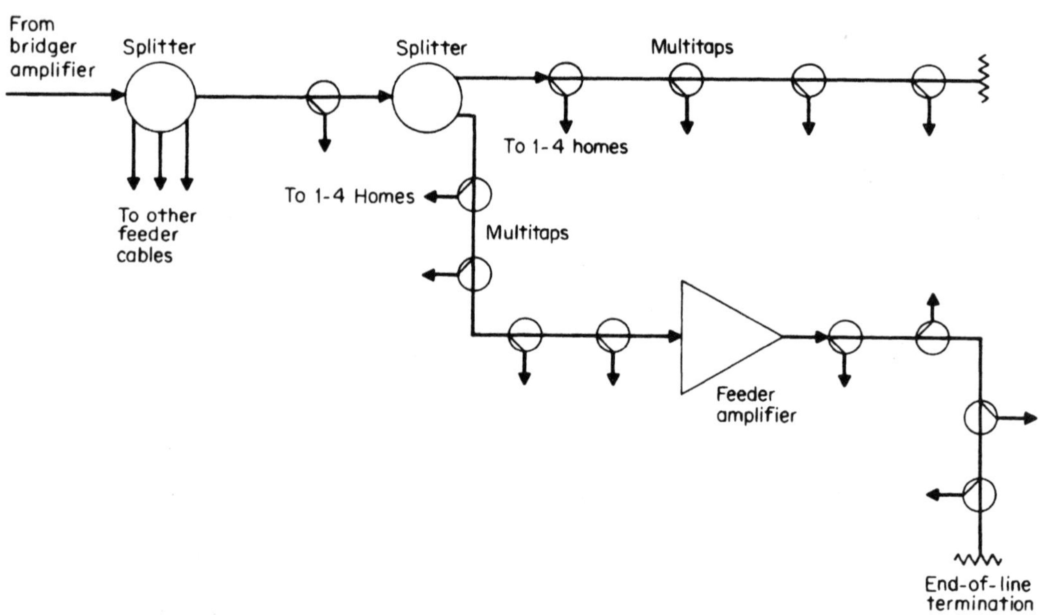


**Figure 15.11** Distribution and feeder system. (*From* [3]. *Used with permission.*)

As the signal proceeds on a distribution line, attenuation of the cable and insertion losses of tap devices reduce its level to a point where line-extender amplifiers may be required. The gain of line extenders is relatively high (25 to 30 dB), so generally no more than two are cascaded, because the high level of operation creates a significant amount of cross-modulation and inter-

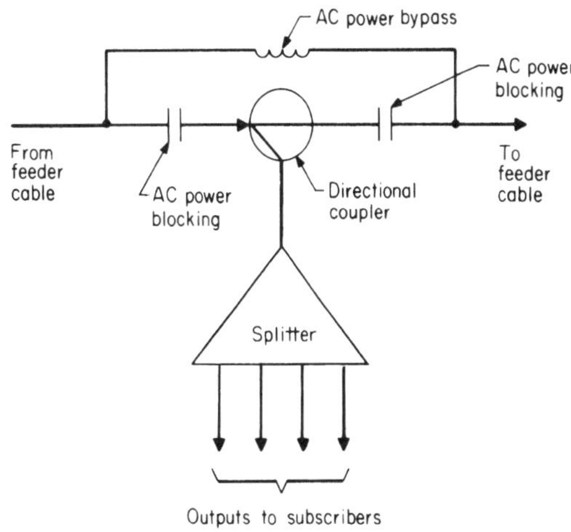


**Figure 15.12** Multitap functional block diagram.

modulation distortion. These amplifiers typically do not use automatic level-control circuitry to compensate for variations in cable attenuation caused by temperature changes. By limiting the system to no more than two line extenders per distribution leg, the variations are not usually sufficient to affect picture quality.

Two types of tap-off devices are commonly used:

- The *directional-coupler multitap*, which provides for connections from two to eight subscribers per tap. Most manufacturers offer a series of taps, each given a value based on the amount of signal taken from the distribution line. (See Figure 15.12.)
- Individual *pressure taps*, which fasten directly to the distribution cable. These devices find limited use today. Pressure taps tend to create signal reflections, because their construction and installation onto the cable simulate a distortion in the center conductor-to-shield distance.

A tap removes an appropriate amount of energy from the distribution cable through a directional coupler, then splits the energy into several paths. Each path proceeds through a drop into a subscriber's home. The value of each tap is selected to produce an adequate signal level for a good S/N at the subscriber's receiver, but not so high as to create intermodulation in the receiver. Each tap introduces some attenuation into the distribution system, related to the amount of signal tapped off for subscribers.

By extracting the signal through a directional coupler, a high attenuation from the tap outlets back into the cable reduces the interference that originates in a subscriber drop cable from reaching subscribers served by other multitaps. The directional characteristics of the taps also prevent reflected signals from traveling along the distribution system and causing ghosting on subscribers' receivers.

The splitter portion of a multitap is a hybrid design that introduces a substantial amount of isolation from reflections or interference coming from a subscriber drop, both to the distribution system and to other subscribers connected to the same tap.

The tap is the final service point immediately prior to a subscriber's location, so it becomes a convenient point where control of channels authorized to individuals can be applied. An *addres-*

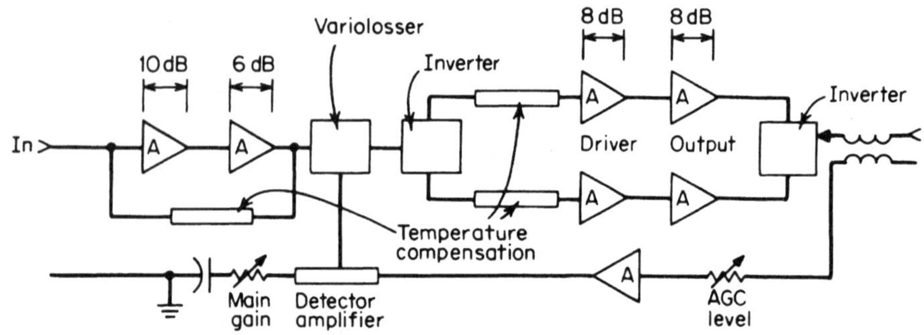


**Figure 15.13** Functional block diagram of a trunk-AGC amplifier and its gain distribution. (*From* [2]. *Used with permission.*)

*sable tap*, controlled from the system head end, can be used instead of the standard multitap to enable access to basic cable services or to individual channels. When activated, the switch and related components can enable or disable a channel, a group of channels, or the entire spectrum.

## Trunk and Distribution Amplifiers

Because the trunk is the main artery of a cable television system, trunk amplifiers must provide a minimum of degradation to the system. A typical amplifier station is of modular design with plug-in units for trunk amplification, automatic level control, bridging, and signal splitting to feed distribution cables. The modules receive power from the coaxial cable. A power supply in each amplifier housing taps power from the cable to supply dc voltage to operate the other modules. The power is inserted onto the coaxial cable at points where a connection to the local power utility is convenient. A step-down supply is used to insert 30 V ac or 60 V ac onto the coaxial cable through a passive inductive network.

Figure 15.13 shows a block diagram of the typical trunk amplifier with provisions for automatic level control and the distribution of gain. The first stage provides low-noise operation, while remaining stages produce gain with low cross-modulation. Trunk amplifiers use hybrid circuits, combining integrated circuits and discrete components in a single package. Such hybrid packages are available with a variety of gain and performance characteristics.

## Level Control

Because the attenuation characteristics of coaxial cable changes with ambient temperature, trunk amplifiers include automatic level-control (ALC) circuits to compensate for the changes and to maintain the output level at a predetermined standard. Automatic control depends on the use of pilot carriers. One is at the lower end of the spectrum, while the other is at the upper limit. Special-purpose pilot carriers also can be placed on the system at the head end. Some system designs use video channels as pilots. ALC circuits sample the trunk amplifier output at the pilot frequencies and feed a control voltage back to diode attenuators at an intermediate point in the amplifier circuit. The upper pilot adjusts the gain of the amplifier, while the lower pilot controls

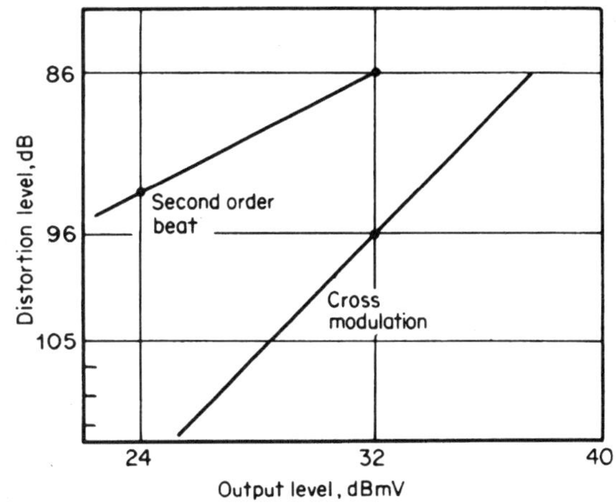


**Figure 15.14** Cross-modulation and second-order beat distortion-vs.-output level of a trunk amplifier. (*From* [2]. *Used with permission.*)

the slope of an equalizer. This method is effective in maintaining the entire spectrum at a constant output level.

A second design of trunk amplifier uses feed-forward distortion correction and permits carriage of a larger number of channels (for equivalent gain) than the standard trunk amplifier.

## Cross-Modulation

The maximum output level permissible in cable television amplifiers depends primarily on *cross-modulation* of the picture signals. Cross-modulation is most likely in the output stage of an amplifier, where levels are high. The cross-modulation distortion products at the output of a typical amplifier (such as Figure 15.13) are 96 dB below the desired signal at an output level of +32 dB mV. In the system shown in Figure 15.13, gains of the third and fourth stages are each 8 dB. With an amplifier output level of +32 dB mV, the output of the third stage would be +24 dB mV, and that of the second stage, 16 dB mV.

Figure 15.14 expresses the relationship between cross-modulation distortion and amplifier output level. A 1 dB decrease in output level corresponds to a 2 dB decrease in cross-modulation distortion.

Cross-modulation increases by 6 dB each time the number of amplifiers in cascade doubles. The distortion also increases by 6 dB each time the number of channels on the system is doubled. Experience has shown that cable subscribers find a cross-modulation level of –60 dB objectionable. Therefore, based on a cross-modulation component of –96 dB relative to normal visual carrier levels in typical trunk amplifiers, 64 such amplifiers can be cascaded before cross-modulation becomes objectionable.

## Frequency Response

The most important design consideration of a cable amplifier is frequency response. A response that is flat within ±0.25 dB from 40 MHz to 450 MHz is required of an amplifier carrying 50 or more 6 MHz television channels to permit a cascade of 20 or more amplifiers. Circuit designs

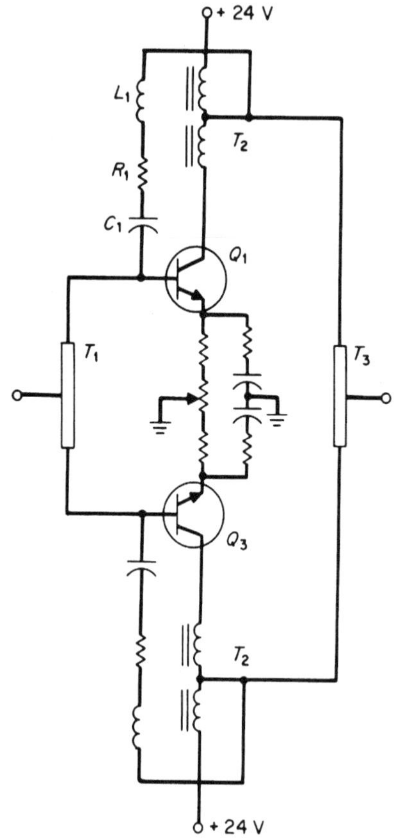


**Figure 15.15** High-performance, push-pull, wide-band amplifier stage for a trunk amplifier. (*From* [2]. *Used with permission.*)

must pay special attention to high-frequency parameters of the transistors and associated components, as well as circuit layout and packaging techniques.

The circuit given in Figure 15.15 represents an amplifier designed for flat response. The feedback network ($C_1$, $R_1$, $L_1$) introduces sufficient negative feedback to maintain a nearly constant output over a wide frequency range. Collector transformer $T_2$ is bifilar-wound on a ferrite core and presents a constant 75 Ω impedance over the entire frequency range. Splitting transformers $T_1$ and $T_3$, of similar construction, have an additional function of introducing 180° phase shift for the push-pull transistor pair $Q_1$ and $Q_2$, while maintaining 75 Ω input and output impedances.

## Second-Order Distortion

When cable television systems carried only 12 channels, those channels were spaced to avoid the second harmonic of any channel from falling within any other channel. The expansion of systems to carry channels covering more than one octave of spectrum causes second-order beat distortion characteristics of amplifiers to become important. Push-pull amplifier circuits, as shown in Figure 15.15, provide a common method of reducing amplifier second-order distortion. Combining several of these circuit building blocks into an amplifier effectively cancels second-order distortion if the 180° phase relationship is maintained over the full amplifier bandwidth. The gain of both halves of the amplifier must be equal and the individual transistors must be optimized for maximum second-order linearity.

Second-order distortion increases by approximately 1.5 dB for each 1 dB in amplifier output level, as given in Figure 15.14. Referring to Table 15.2, each amplifier introduces a distortion level of approximately –86 dB relative to the visual carriers. Because second-order beats greater than –55 dB become objectionable, approximately 64 amplifiers can be cascaded before that level is attained.

## Noise Figure

The typical trunk amplifier hybrid module has a noise figure of approximately 7 dB. Equalizers, band-splitting filters, and other components preceding the module add to the overall station noise figure, producing a total of approximately 10 dB.

A single amplifier produces an S/N of approximately 60 dB for a typical signal input level. Subjective tests show that an S/N of approximately 43 dB is acceptable to most viewers. Again, based on Table 15.2, noise increases 3 dB as the cascade of amplifiers is doubled. Therefore, a cascade of 32 amplifiers, resulting in a signal-to-noise ratio of 45 dB, is a commonly used cable television system design specification.

## Hum Modulation

Because power for the amplifiers is carried on coaxial cable, some hum modulation of the RF signals occurs. This modulation results from power-supply filtering and saturation of the inductors used to bypass ac current around active and passive devices within the system. A hum modulation specification of –40 dB relative to visual carriers has been found to be acceptable in most applications.

## Subscriber-Premises Equipment

The output of the tap device feeds a 75 Ω coaxial drop cable into the home. The cable, generally about 1/4-in. in diameter, is constructed with an outer polyethylene jacket, a shield of aluminum foil surrounded by braid, a polyethylene dielectric insulator, and a copper-center conductor. A loss at 400 MHz of 6 dB/100 ft is typical.

For reasons of safety and signal purity, isolation between the subscriber's television receiver and the distribution cable is required. In the multitap, blocking capacitors prevent the ac current used to power amplifiers from reaching the subscriber drop. As the drop enters the subscriber's home, a grounding connection, required by local regulations, protects the television receiver from power surges, such as those caused by lightning discharges, and protects the subscriber from possible shock by voltages that might otherwise be present on the shield of the drop cable. Finally, at the subscriber's receiver, the signal from the unbalanced 75 Ω coaxial cable is converted to a 300 Ω balanced output before connection to the tuner section of the receiver. This matching transformer provides a further level of isolation to energy along the sheath of the cable.

## Converters and Descramblers

Prior to the manufacture of cable-ready television receivers and VCRs, systems with more than 12 channels required the use of a converter at the subscriber's location. A nontunable block converter, which translates a group of channels to a different portion of the spectrum, is probably the least expensive type of converter. Such devices are used primarily when only a small number of channels are to be added to the system.

Most converters used today are tunable units, with microprocessor control of the frequency-synthesizing circuit referenced to a crystal. The viewer selects a channel (usually with a hand-held remote control unit), which is converted to a standard VHF frequency (usually channel 3 or 4). In such cases, the television receiver remains tuned to "channel 3," and all tuning is done with the converter.

Cable-ready tuners avoid the need for a converter by including additional tuning capabilities into the receiving equipment. As the viewer selects the desired channel, digital switching sets the input control lines of a phase-locked-loop frequency synthesizer. When selected, any of the additional channel signals are converted to the normal television IF amplifier frequency band.

Prior to cable-ready receivers, the use of nonstandard channels and converters provided an easy method of limiting access to special channels. Now, some method of scrambling or other control must be used to manage access. Early approaches included suppression of the horizontal and vertical synchronizing pulses of the controlled program signal during transmission at the head end. At the same time, a pilot signal that contained information about the missing sync was also transmitted on the cable. In a descrambler, located in the subscriber's home, the pilot information provided the key to correctly re-create the synchronization.

A greater degree of security for scrambled signals can be attained through digital methods and base-band scrambling. Numerous concepts have been implemented, including the inversion of lines of video on a random basis, shuffling of lines of video, and conversion of the video into a digital bit stream. These methods, requiring that signals be demodulated into audio and video, are more expensive in terms of equipment costs, but they present a number of possible advantages as well, because remote control of the audio level is possible, as is the incorporation of videotext or other types of data decoding circuits.

By including an addressable switch into a set-top descrambler (Figure 15.16), the cable operator can use computer control from the head end to restrict access to channels, depending on the tiers of service a subscriber has chosen. In a bidirectional cable system, pay-per-view impulse viewing is relatively easy to implement as well. A subscriber need only key in an appropriate code to request reception of a special program. The head-end computer, constantly polling all converter/descrambler units, detects the request, enables the special channel for the duration of the program request, and subsequently places an entry into the automated billing system for that subscriber.

## Off-Premises Systems

The off-premises approach is compatible with recent industry trends to become more consumer electronics friendly and to remove security-sensitive electronics from the customer's home [1]. This method controls the signals at the pole rather than at a decoder in the home. This increases consumer electronics compatibility because authorized signals are present in a descrambled format on the customer's drop. Customers with cable-compatible equipment can connect directly to

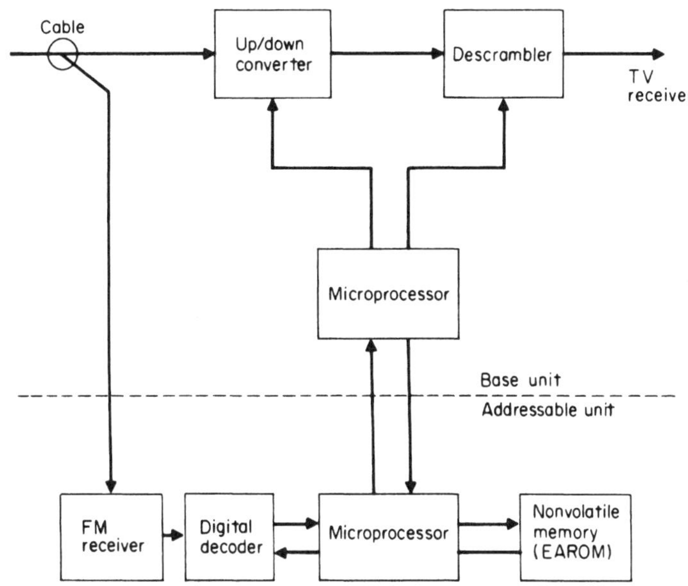


**Figure 15.16** Addressable-converter block diagram. (*Courtesy of Scientific Atlanta.*)

the cable drop without the need for converter/descramblers. This allows the use of all VCR and TV features.

Signal security and control in the off-premises devices take different forms. Nearly all off-premises devices are addressable. Specific channels or all channels are controlled remotely.

*Interdiction technology* is one approach to plant program security. In this format, the pay television channels to be secured are transported through the cable plant in the clear (not scrambled). The security is generated on the pole at the subscriber module by adding interference carrier(s) to the unauthorized channels. An electronic switch is incorporated, allowing signals to be turned off. In addition to being consumer electronics friendly, this method of security does not degrade the picture quality on an authorized channel.

## Bidirectional Cable Television Systems

Although a bidirectional cable system could be constructed using separate cables for the upstream and downstream signal paths, the costs are prohibitive. More practical is a single-cable system in which the direction of the signal is related to frequency, as outlined previously in this chapter. Using this approach, called *subsplitting*, appropriate circuits must be integrated into all amplifiers in the system. Each amplifier station now has two amplifier sections, one for operation in each direction, with band-splitting and combining filters used to direct signals to the appropriate amplifier modules.

Bidirectional operation introduces a number of new problems into the cable system. Probably the greatest difficulties encountered stem from transmitters external to the system operating on frequencies between 5 and 35 MHz. That includes citizen's band, amateur radio, and shortwave broadcast, as well as some public-service applications. Ingress of signals from these transmitters is difficult to prevent and interferes with reliable upstream transmission. Some of these problems can be reduced by allowing the use of remotely controlled switchers at bridger amplifier locations. In effect, only certain portions of the distribution system can send signals upstream at any given time, which places definite limitations on the effectiveness of bidirectional operation.

## 15.3 Fiber-Based Cable Systems

The last few years have brought exciting trends employing new technologies for cable television [1]. Fiber is now installed to upgrade older systems and as part of rebuilds and new builds. The old trunk system of long cascades of amplifiers is now considered obsolete. Work on new amplifier technologies will allow a realization of cable's inherent bandwidth, which exceeds 1 GHz.

A bandwidth of 1 GHz contains 160 slots of 6 MHz spectrum. These can be allocated to NTSC, DTV, and to new services. The most exciting potential lies with utilizing video compression technology to squeeze six to twelve NTSC-like quality signals in a 6 MHz slot. This opens the door for hundreds of channels. *Near video on demand* (NVOD) becomes practical; the most popular movies could be repeated every few minutes to minimize the wait time before a movie starts. The average wait time could be made shorter than the trip to the video store, and the subscriber does not have to make a second trip to return the movie. A microprocessor can keep track of which channel to return to should the subscriber wish to take a break. It is possible to design systems that behave as if they are switched even though they remain more like a traditional cable tree-and-branch structure.

Currently, analog video fiber technology is preferred for cable applications, although occasionally cable systems use digital video. The focus of current research is to optimize fiber optic transmitters, amplifiers, and receivers.

Digital fiber links are used when video signals are partially transported in a digital common carrier network, The video interfaces used operate at a DS-3 rate of 45 Mbits/s, which routes efficiently through common carrier *points of presence* and switching networks.

### 15.3a Hybrid Fiber/Coax

The lasers that drive fiber optic links are expensive, and the receivers that convert the optical energy into video signals also are relatively expensive [1]. To be practical, these components must serve hundreds of subscribers each so that the costs can be shared. This is accomplished with a layout in which fibers feed small coaxial systems, which in turn serve a few hundred to a few thousand subscribers.

The hybrid fiber/cable (HFC) architecture has made it possible to cost effectively increase bandwidth, signal quality, and reliability while reducing maintenance costs and retaining a *craft-friendly* (easily serviced) plant. HFC makes two-way service practical. The bandwidth of coaxial cable has no sharp cut off. In a coax-based system, it is the cascade of amplifiers that limits the practical bandwidth. Twenty to forty amplifiers in cascade not only reduces bandwidth, but also constitutes a considerable reliability hazard. Overlaying low-loss fiber over the trunk portion of the plant eliminates the trunk amplifiers. This, in turn, leaves only the distribution portion of the plant with its relatively short distances and only two or three amplifiers. Wider bandwidth is thus facilitated.

Two way operation in a fiber system is quite practical for two reasons. First, the fiber itself is not subject to ingress of interfering signals. Second, the cable system is broken up into a large number of small cable systems, each isolated from the others by its own fiber link to the head-end. If ingress should cause interference in one of these small cable systems, that interference will not impair the performance of the other portions of the cable plant.

### 15.3b HFC Architectures

A traditional HFC architecture is shown in Figure 15.17 [4]. Broadcast analog video is transmitted by an externally modulated 1550 nm transmitter from the head end to the hub. A redundant transmitter and path are included for signal protection. Narrowcasting services (data, voice, and possibly digital video) are typically transmitted via a *synchronous digital hierarchy* (SDH) link, although other options include ATM, IP, or proprietary baseband digital transport. At the hub, the digital narrowcast signals are processed for transmission over the HFC network by the appropriate interface unit. A *cable modem termination system* (CMTS) converts the downstream data signal to QAM, and QPSK demodulates the upstream signal, as well as supplying media access control. The CMTS and cable modems may be based on ether the DOCSIS or DAVIC standards, or use proprietary protocols. The *host digital terminal* (HDT) performs the same tasks for cable telephony. Video servers are shown at the head end, but could also be located in the hub to provide digital video and VOD service. After processing and QAM modulation, the narrowcast signals are combined with the broadcast analog signal and fed to 1310 nm transmitters, which in turn feed multiple or individual optical nodes. Each optical node typically serves 500 to 1000 subscribers, but nodes serving as few as 100 homes are becoming common. The fewer homes served per node, the more dedicated bandwidth is available per subscriber. Because each hub serves on the order of 50,000 subs, the hubs must be relatively large buildings in order to house all of the equipment and fiber connections.

Several drawbacks exist for operators deploying traditional double-hop architectures for high-bandwidth subscribers. Aside from the difficulty and expense of locating and building a large hub site, there are the functional problems of redundancy and the sheer numbers of fibers necessary to feed 500 or so nodes from each hub. For services such as telephony, most operators require both equipment and fiber path redundancy down to at least the 500 to 1000 home level. This is difficult to achieve in the double-hop architecture unless there are redundant receivers in each node fed by redundant fiber rings and transmitters. This arrangement provides redundancy down to the 100 home level, which might be considered overkill, and is definitely expensive. A star architecture from the hub to the nodes would mitigate the redundancy problem, but would be prohibitively expensive in terms of fiber cable, since there would be a unique fiber cable to each node.

An additional disadvantage of the double-hop architecture is the high cost of SONET or proprietary digital transport as bandwidth demand increases beyond 1 Mbits/s per home. For a hub serving 50,000 subscribers, providing just 2 Mbits/s to each requires 100 Gbits/s from the head-end to the hub. This corresponds to 10 OC-192 terminals in both the head end and the hub, a great expense in both equipment and space.

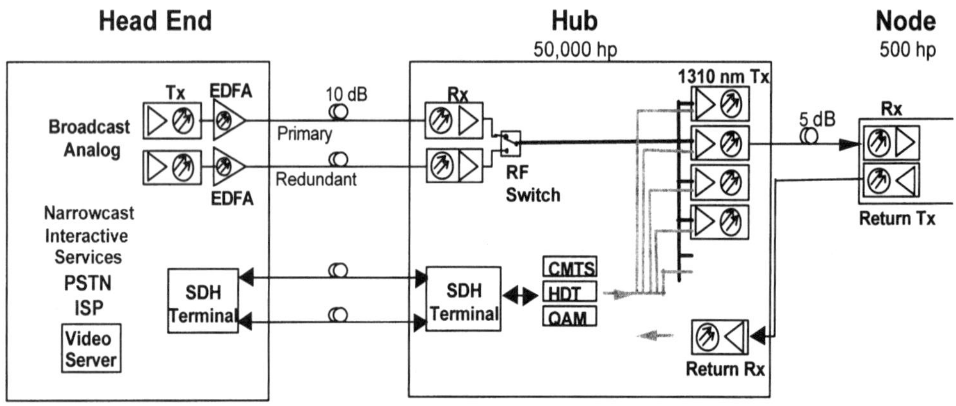


**Figure 15.17** A traditional double hop HFC architecture. (*After* [4].)

## DWDM Architectures

A "hubless" architecture is shown in Figure 15.18 [4]. In this case, most of the equipment formerly located in the hub (CMTS, HDT, and video servers) has been pulled back to the head end. Because the forward path signals are passively transmitted through the hub, it is easy to replace the large hub site with a small, inexpensive cabinet or vault. In Figure 15.18, four cabinets serve the 50,000 subs formerly served by a single large hub. Placing all of the equipment at the head end has the additional advantage of lowering operational costs and permitting less total equipment to be deployed in the initial stages. In the traditional architecture, it is necessary to locate at least one CMTS, HDT, and possibly a video server at each hub, regardless of how limited the demand. With the CMTS and HDT pulled back to the head end, this equipment can be shared among multiple cabinets/hubs when demand is low.

The traditional architecture cost increases rapidly with increasing bandwidth demand, but the *dense wavelength division multiplexing* (DWDM) architecture only requires additional, and relatively inexpensive, DWDM transmitters. Similar to SDH, the DWDM architecture also provides path and equipment protection via redundant optical amplifiers and optical switches. However, it does not provide all of SDH's performance monitoring capabilities. Therefore, for certain high-priority services, such as telephony, a hybrid approach is possible. The low bandwidth, high-priority services like telephony can be transmitted to the hub/cabinet via SDH, while the high bandwidth, non-lifeline services such as VOD can be transmitted via DWDM.

## Deep Fiber-Node Segmentation

The number of subscribers served by an optical node for a typical HFC system has steadily decreased over the last decade [4]. Formerly, 1000 to 2000 home nodes were the norm, but today new builds and upgrades are more likely to be in the 100 to 500 homes per node range. Pushing fiber deeper into the network results in improved bandwidth, performance, and reliability, particularly due to the reduction RF active elements in the coax plant. Completely eliminating RF amplifiers from the system is an excellent goal, since the system would have a 33 to 50 percent

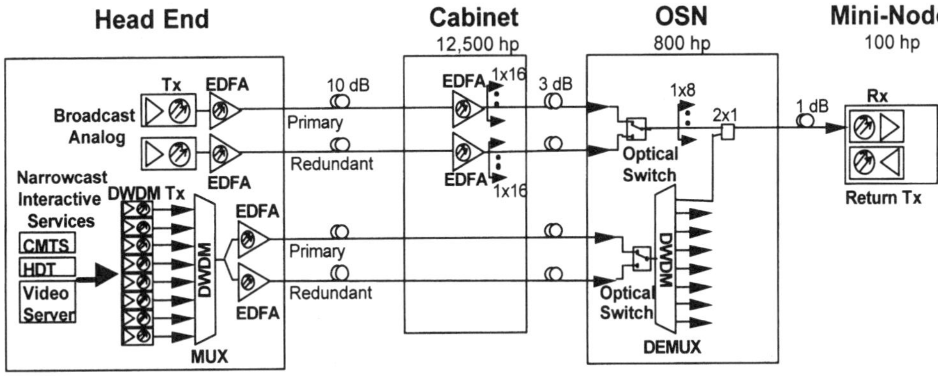


**Figure 15.18** A hubless, DWDM, Deep Fiber Architecture. (*After* [4].)

less active components resulting in greater reliability and reduced power consumption. Completely passive coax networks are possible for cases of very high subscriber density, such as multiple dwelling units (MDU's) with +200 subscribers per mile. However, for more typical densities of 100 subs/mile, it is necessary to segment down to approximately 35 homes per optical node in order to create a completely passive coax network when transmitting a full complement of 60 to 80 analog channels. This assumes a four-output node with 51 to 53 dBmV outputs. Unfortunately, for most cable plant, it is difficult or impossible to effectively use all four outputs from the node, and there is not a tremendous cost difference between nodes utilizing 2, 3, or 4 ports. Although it is generally better to push fiber as deep as possible, 35 homes per node is probably not a cost-effective alternative at this time. A practical trade-off is to employ a four-output node passing 100 subscribers and add one line extender to each node output. This provides plenty of RF level at the home, and up to 50 Mbits/s of dedicated downstream bandwidth. With only a single line extender in each path, high performance based primarily on the optical link(s) is easily obtained, and higher order modulation formats like 256-QAM and even 1024-QAM can be supported.

## Application Considerations

Although DWDM has recently been applied in cable systems, most implementations demultiplex the signal at the hub and feed the individual wavelengths to large nodes [4]. This configuration saves fiber between the head end and hub, but only provides one dedicated narrowcast wavelength per 1000 homes. In the DWDM deep fiber architecture shown in Figure 15.18, DWDM is extended one level further into the network to an *optically scalable node* (OSN) that may be either strand- or cabinet-mounted. The signal at the OSN is further split to feed 100 home mini-nodes. Starting at the headend, the narrowcast signals are transmitted via directly modulated DWDM transmitters, whose wavelengths correspond to those of the ITU grid. The wavelengths are multiplexed together, optically amplified by an *erbium doped fiber amplifier* (EDFA), pass through the hub site, and are demultiplexed at the OSN. In this ultimate configuration, a dedicated wavelength serves each mini-node, providing approximately 300 MHz of narrowcasting

256-QAM channels (~2 Gbits/s or 20 Mbits/s per subscriber). Eventually, as the broadcast analog spectrum is gradually reduced and the narrowcast spectrum increased, the entire 50 to 860 MHz band can be utilized to deliver 50 Mbits/s per subscriber. Initially, however, the system can be scaled such that each wavelength is shared among all the mini-nodes fed by a particular OSN. A simple optical splitter is deployed in the OSN instead of a DWDM demultiplexer. As bandwidth demand increases, additional DWDM transmitters are added at the head end, and DWDM demultiplexers replace the optical splitters in the OSN.

The broadcast analog video is redundantly transmitted via an externally modulated 1550 nm transmitter from the head end to a hub, where it is amplified and split to feed the OSN's. The broadcast and demultiplexed narrowcast signals are combined and then transmitted over the same fiber to a single receiver in the mini-node. The signals are combined via a simple 2:1 optical coupler, or by a 2:1 DWDM multiplexer, depending on the available loss budget. If a DWDM mux is necessary because of link budget considerations, then another option is to leave the broadcast and narrowcast signals separate, and transmit them over individual fibers to separate receivers in the mini-node. The cost of a 2:1 DWDM mux and 10 km of fiber is greater than the cost of an additional receiver. In addition, avoiding the 2:1 combiner more easily enables the equipment associated with a 1:16 split to be packed into the OSN, as opposed to only serving 8 mini-nodes from each OSN. An additional advantage of this approach is that it eliminates potential problems associated with the CSO and CTB from the narrowcast signals interfering with the broadcast signal when using a single receiver.

The plant costs of the deep fiber, DWDM architecture are greater than those of traditional HFC architectures, due to the expanded number of nodes (one per 75 to 100 homes), but this cost increase is mostly offset by the decrease in RF amplifiers and power supplies. The elimination or minimization of SDH or ATM transport systems from the head end to hub actually makes the hubless deep fiber system more cost-effective than the traditional network. The primary benefit of these next generation architectures is the massively scalable bandwidth, along with the operational savings in power and RF plant and hub maintenance, and the increased network reliability due to deeper redundant fiber and fewer total network elements. In addition, the hubless architecture typically can be deployed much more quickly, an important advantage in today's competitive service provider environment.

### 15.3c Return Path

Traditionally, multiple return paths are simply combined such that the upstream signal from thousands of subscribers is fed to a single CMTS or HDT [4]. This results in high noise levels, and severely limits the available upstream bandwidth. As service penetration increases, operators must be able to segment the upstream to serve smaller numbers of subscribers. Several methods exist for segmenting the return path, thus providing dedicated upstream bandwidth to customers. For the traditional double hop system, upstream DWDM provides basically the mirror image of the DWDM downstream. ITU return path transmitters in the mini-node transmit back to the OSN, where the signal is DWDM muxed with the signals from the other mini-nodes served by the OSN. Because the signals are analog, it is often necessary to amplify with an EDFA before transmitting back to the head end in order to maintain acceptable performance. At the head end, the signals are demultiplexed and fed to individual return path receivers. The DWDM upstream option provides excellent segmentation. But it is not scalable, since the required ITU lasers in every mini-node and the cabinet EDFA's combine to make the initial system deployment relatively expensive. The system does provide excellent return path bandwidth of up to approximately 100 Mbits/s, assuming 16-QAM modulation. This corresponds to 1 Mbits/s peak rate per subscriber, which may be more than necessary in the early stages of deployment.

## 15.4 Regulatory Issues

Cable television (also called CATV or community antenna television) was developed in the late 1940's for communities unable to receive TV signals because of terrain or distance from TV stations [5]. Cable television system operators located antennas in areas with good reception, picked up broadcast station signals, and then distributed them by coaxial cable to subscribers for a fee.

In 1950, cable systems operated in only 70 communities in the United States. These systems served about 14,000 homes. By October 1998 there were more than 10,700 systems serving more than 65 million subscribers in more than 32,000 communities. Cable systems are operating in every state of the U.S. and in many other countries, including Austria, Canada, Belgium, Germany, Great Britain, Italy, Japan, Mexico, Spain, Sweden, and Switzerland.

Most cable systems are technically capable of offering between 36 and 60 channels. Channel capacity in the industry has increased dramatically in recent years; some systems now offer in excess of 100 channels. Most cable subscribers receive service from a system offering more than 54 channels.

The channel capacity of a cable system makes it possible for a cable television operator to provide many services. In addition to over-the-air television broadcast signals, most systems also offer diverse program services, including—for example—news, weather, business information, movies, sports, special entertainment features, and programming designed for specific audiences such as children, and specific ethnic and racial groups. Within the past few years, some cable systems have begun offering a full-range of telecommunications services, including high-speed Internet access and local telephone service.

### 15.4a Initial Jurisdiction and Rules

The Federal Communications Commission first established rules in 1965 for cable systems that received signals by microwave antenna [5]. In March 1966, the Commission established rules for all cable systems (whether or not served by microwave). The Supreme Court affirmed the Commission's jurisdiction over cable in *United States vs. Southwestern Cable Co.*, 392 U.S. 157 (1968). The Court ruled that "the Commission has reasonably concluded that regulatory authority over CATV is imperative if it is to perform with appropriate effectiveness certain of its responsibilities." The Court found the Commission needed authority over cable systems to assure the preservation of local broadcast service and to effect an equitable distribution of broadcast services among the various regions of the country.

In March 1972, new rules regarding cable television became effective. These rules required cable television operators to obtain a certificate of compliance from the Commission prior to operating a cable television system or adding a television broadcast signal. The rules applicable to cable operators fell into several broad subject areas:

- Franchise standards

- Signal carriage
- Network program nonduplication and syndicated program exclusivity
- Nonbroadcast or cablecasting services
- Cross-ownership
- Equal employment opportunity
- Technical standards

Cable television operators who originated programming were subject to equal time, Fairness Doctrine, sponsorship identification, and other provisions similar to rules applicable to broadcasters. Cable operators were also required to maintain certain records and to file annual reports with the Commission concerning general statistics, employment, and finances.

In succeeding years, the Commission modified or eliminated many of the rules. Among the more significant actions, the Commission deleted most of the franchise standards in 1977, substituted a registration process for the certificate of compliance application process in 1978, and eliminated the distant signal carriage restrictions and syndicated program exclusivity rules in 1980. In 1983, the Commission deleted its requirement that cable operators file financial information. In addition, court actions led to the deletion of the pay cable programming rules in 1977.

### 15.4b 1984 Congressional Policy and Rules

In October 1984, the U.S. Congress amended the Communications Act of 1934 by adopting the Cable Communications Policy Act of 1984 [5]. The 1984 Cable Act established policies in the areas of ownership, channel usage, franchise provisions and renewals, subscriber rates and privacy, obscenity and lockboxes, unauthorized reception of services, equal employment opportunity, and pole attachments. The new law also defined jurisdictional boundaries among federal, state and local authorities for regulating cable television systems.

### 15.4c 1992 Congressional Policy and Rules

Following the 1984 Cable Act, the number of households subscribing to cable television systems increased, as did the channel capacity of many cable systems [5]. However, competition among distributors of cable services did not increase, and, in many communities, the rates for cable services far outpaced inflation. Responding to these concerns, Congress enacted the Cable Television Consumer Protection and Competition Act of 1992. The 1992 Cable Act mandated a number of changes in the manner in which cable television is regulated.

In adopting the 1992 Cable Act, Congress stated that it wanted to:

- Promote the availability of diverse views and information
- Rely on the marketplace to the maximum extent possible to achieve such availability
- Ensure cable operators continue to expand their capacity and program offerings
- Ensure cable operators do not have undue market power
- Ensure consumer interests are protected in the receipt of cable service

The Commission has adopted various regulations to implement these goals.

## 15.4d 1996 Congressional Policy and Rules

In adopting the Telecommunications Act of 1996, Congress noted that it wanted to provide a pro-competitive, de-regulatory national policy framework designed to accelerate rapidly private sector deployment of advanced telecommunications and information technologies and services to all Americans by opening all telecommunications markets to competition [5]. The Commission has adopted regulations to implement the requirements of the 1996 Act and the intent of Congress.

## 15.4e Cable Television Defined

Cable television is a video delivery service provided by a cable operator to subscribers via a coaxial cable or fiber optic link [5]. Programming delivered without a wire via satellite or other facilities is not "cable television" under the Commission's definitions.

A *cable television system operator* is any person or group of persons who provides cable service over a cable system and directly or through one or more affiliates owns a significant interest in such cable system, or who otherwise controls or is responsible for, through any arrangement, the management and operation of such a cable system.

*Cable service* is the transmission to subscribers of video programming, or other programming service. This definition includes any subscriber selection required in choosing video programming or other programming service.

A *cable system* is a facility—consisting of a set of closed transmission paths and associated signal generation, reception, and control equipment—that is designed to provide cable service which includes video programming and which is provided to multiple subscribers within a community. This term does not include a facility that serves only to retransmit the television signals of one or more television broadcast stations; a facility that serves subscribers without using any public right-of-way; a facility of a common carrier which is subject in whole or in part, to the provisions of Title II of the Communications Act, except that such facility shall be considered a cable system to the extent such facility is used in the transmission of video programming directly to subscribers; unless the extent of such use is solely to provide interactive on demand services; an open video system; or any facilities of any electric utility used solely for operating its electric utility system.

Cable services are often provided in *tiers*. A tier is a category of cable service or services provided by a cable operator for which a separate rate is charged by the cable operator. There are three types of cable service:

- Basic service
- Cable programming service
- Per-channel or per-program (sometimes called *pay-per-view*) service

Basic service is the lowest level of cable service a subscriber can buy. It includes, at a minimum, all over-the-air television broadcast signals carried pursuant to the *must-carry* requirements of the Communications Act, and any public, educational, or government access channels required by the system's franchise agreement. It may include additional signals chosen by the operator. Basic service is generally regulated by the local franchising authority (the local or state entity empowered by federal, state, or local law to grant a franchise to a cable company to operate in a given area).

Cable programming service includes all program channels on the cable system that are not included in basic service, but are not separately offered as per-channel or per-program services. Pursuant to a 1996 federal law, the rates charged for cable programming services tiers provided after March 31, 1999 are not regulated. There may be one or more tiers of cable programming service.

Per-channel or per-program service includes those cable services that are provided as single-channel tiers by the cable operator, and individual programs for which the cable operator charges a separate rate. Neither of these services is regulated by the local franchising authorities or the Commission.

## 15.4f Signal Carriage Requirements

The 1992 Cable Act established new standards for television broadcast station signal carriage on cable systems [5]. Under these rules, each local commercial television broadcast station was given the option of selecting mandatory carriage (must-carry) or retransmission consent (may carry) for each cable system serving the same market as the commercial television station. The market of a television station is established by its Area of Dominant Influence (ADI), as defined by Arbitron and/or modified by the Commission. Every county in the country is assigned to an ADI, and those cable systems and television stations in the same ADI are considered to be in the same market. Upon the request of a television station or a cable system, the Commission has the authority to change the ADI to which a station is assigned. As a result of Arbitron abandoning the television research business, the Commission has determined that, effective January 1, 2000, the market of a television station shall be its Designated Market Area (DMA) as determined by Nielsen Media Research.

Every three years, each local commercial television station has the right to elect either must-carry or retransmission consent. The initial election was made on June 17, 1993, and was effective on October 6, 1993. All subsequent elections occurred (and will continue to occur) every three years.

Generally, if a local commercial television station elects must-carry status, it is entitled to insist on cable carriage in its local market. Each cable system with more than 12 channels must set aside up to one-third of its channel capacity for must-carry stations. For example, if a cable system has 60 channels, it must set aside 20 of those channels for must-carry stations. If there are 25 stations in the market which elected must-carry, the cable operator may choose 20 to carry. On the other hand, if only 15 stations elected must-carry in the market, the cable system would have to carry all 15 of these stations. A must-carry station has a statutory right to a channel position, usually its over-the-air channel number, or another channel number on which it has historically been carried.

A cable system is not permitted to carry a commercial station without the station's consent. Therefore, if the local commercial television station elects retransmission consent, the cable system must obtain that station's consent prior to carrying or transmitting its signal. Except for "superstations," a cable system may not carry the signal of any television broadcast station that is not located in the same market as the cable system without that broadcaster's consent. Superstations are transmitted via satellite, usually nationwide, and the cable system may carry such stations outside their local market without their consent. The negotiations between a television station and a cable system are private agreements which may, but need not, include some form of compensation to the television station such as money, advertising time, or additional channel access.

## Noncommercial Educational Television Stations

Every cable system across the country must carry at least one local noncommercial educational (NCE) station. A noncommercial station which places a Grade B signal over a cable system's principal headend, or whose city of license is within fifty miles of the cable system's principal headend, is considered "local" for this purpose. Cable systems with more than 36 channels may be required to carry all local noncommercial educational television stations that request carriage. Any cable system operating in a market where no local NCE station is available is required to import one NCE station.

## Low Power Television Stations

The 1992 Cable Act provides mandatory carriage for "qualified" low power television stations (LPTV) in certain situations. A LPTV station has to meet certain qualifications specified in the Act and incorporated into the Commission's Rules, before it is qualified for the right to must-carry. If a LPTV is qualified, it may assert must-carry rights, and, provided the cable operator has not met its mandatory carriage obligations, the LPTV station must be carried. Otherwise, a LPTV station must negotiate for carriage under the retransmission consent provisions or under the leased commercial access provisions.

## Radio Programming

While the 1992 Cable Act's must-carry provisions only apply to local commercial and noncommercial educational television stations, the Act's retransmission consent provisions apply to all commercial broadcast stations. Many cable systems carry radio stations as an "all-band" offering, meaning that as with any standard radio receiver, all stations which deliver a signal to the antenna are carried on the system. The Commission only requires consent from those radio stations within 57 miles of the cable system's receiving antenna. Thus, even though a cable operator's antenna may pick up a station's signal, operators are not required to obtain the consent of stations outside of the 57 mile zone unless the station affirmatively seeks retransmission consent.

## Manner of Carriage

Subject to the Commission's network nonduplication, syndicated exclusivity, and sports broadcasting rules, cable systems must carry the entirety of the program schedule of every local television station carried pursuant to the mandatory carriage provisions or the retransmission consent provisions of the 1992 Cable Act. A broadcaster and a cable operator may negotiate for partial carriage of the signal where the station is not eligible for must carry rights, either because of the station's failure to meet the requisite definitions or because the cable system is outside the station's market. In those situations where the carriage in the entirety rule applies, the primary video and accompanying audio of all television broadcast stations must be carried in full, without alteration or deletion of their content. Ancillary services such as closed captioning and program-related material in the vertical blanking interval must be carried. However, other information contained in the vertical blanking interval need not be carried.

## 15.4g Technical Standards

To ensure the delivery of satisfactory television signals to cable subscribers, the Commission has adopted various technical performance standards that local franchising authorities are generally authorized to enforce through their franchising process. Cable operators are required to establish a complaint resolution process and to advise their subscribers about it annually.

In addition to establishing new standards for the delivery of color signals and of closed captioned data, the Commission generally preempted conflicting local standards. Cable systems with fewer than 1,000 subscribers, as well as those serving rural areas, may negotiate with their respective franchising authorities for certain lower standards.

## 15.5 References


1. Ciciora, Walter S.: "Cable Television," in *NAB Engineering Handbook*, 9th ed., Jerry C. Whitaker (ed.), National Association of Broadcasters, Washington, D.C., pp. 1339–1363, 1999.

2. Fink, D. G., and D. Christiansen (eds.): *Electronic Engineer's Handbook*, 2nd ed., McGraw-Hill, New York, 1982.

3. Baldwin, T. F., and D. S. McVoy: *Cable Communications*, Prentice-Hall, Englewood Cliffs, N.J., 1983.

4. Bonang, C., and C. Auvray-Kander: "Next Generation Broadband Networks for Interactive Services," in Proceedings of IBC 2000, International Broadcasting Convention, Amsterdam, 2000.

5. "Cable Television Information Bulletin," Federal Communications Commission, Washington, D.C., June 2000.

Chapter

# 16

# Satellite Delivery Systems

Carl Bentz, K. Blair Benson

## 16.1 Introduction

The first commercial satellite transmission occurred on July 10, 1962, when television pictures were beamed across the Atlantic Ocean through Telstar 1. The launch vehicle, however, lacked sufficient power to place the spacecraft into a stationary position. Three years later, after considerable progress in the development of rocket motors, INTELSAT saw its initial craft, Early Bird 1, launched into a geostationary orbit, and a rapidly growing communications industry was born. In the same year, the USSR inaugurated the Molnya series of satellites, which traveled in more elliptical orbits, to better meet the needs of that nation and its more northerly position. The Molnya satellites were placed in orbits inclined about 64° relative to the equator, with an orbital period half that of the earth.

From these humble beginnings, satellite-based audio, video, and data communications have emerged to become a powerful force in communications across all types of industries, from military logistics and support to consumer television.

## 16.2 The Satellite Communications Link

A satellite relay system involves three basic elements. On the ground is an up-link transmitter station, which beams signals to the satellite. The satellite, as the space segment of the system, receives, amplifies, and retransmits the signals back to earth. The down-link receiving station completes the system and with the up-link forms the earth segment. (See Figure 16.1.)

Two frequency bands are assigned for standard video satellite communications. C-band operates on 6 GHz frequencies for transmission of information to the satellite. The electronics package aboard the craft converts the signals to 4 GHz frequencies for the return to earth. At much higher frequencies, the Ku-band up-link transmits in the region of 14 GHz, with conversion at the satellite to down-link channels at 12 GHz. Satellites may provide service for either C- or Ku-band, or both bands on the same spacecraft. For military and maritime applications, other frequencies are typically used.

Because of the high frequencies used in the system, satellite communications are considered a microwave radio service. This introduces some concerns with regard to satellite-link operations, including the following:

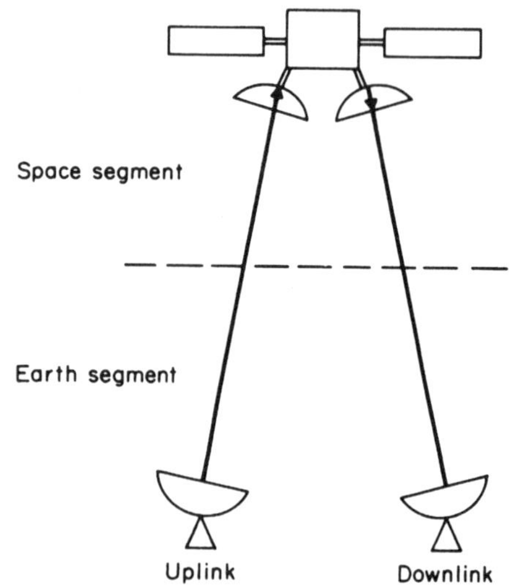


**Figure 16.1** The satellite communications system consists of the up-link and down-link, which form the earth segment, and the satellite, which forms the space segment.

- A line-of-sight path between the transmitter and the receiver is required.
- Attenuation of signals resulting from meteorological conditions, such as rain and fog, is problematic for Ku-band operation, but less troublesome for C-band operation.

Arrangements for shielding the antennas from terrestrial interference include siting the antenna in a depression in the terrain, either constructed or natural, or building a wall of material that is opaque to microwave energy around the antenna. Any foliage at the antenna site threatens possible attenuation effects if it extends into the beam path of the antenna or through scattering of the signal energy by individual leaf surfaces.

Adequate received signal strength, based on the inverse-square law, requires highly directional transmitting and receiving antennas, which, in turn, need significant accuracy of aiming. Most antennas are parabolic in form and can be moved for best signal reception. The development of the phased array introduced a steerable radiation pattern from a relatively fixed, flat-surface antenna primarily for Ku-band use.

The effects of galactic and thermal noise sources on low-level signals require electronics for satellite service with exceptionally low noise characteristics.

## 16.2a The Satellite

It is convenient to discuss the satellite communications link by starting with its middle point, the satellite itself. By understanding the relay station requisites for receiving and transmitting, it is easier to realize the requirements placed on the up-link and down-link sections. Like all relay stations, the communications spacecraft contains antennas for receiving and retransmission. A *low-noise amplifier* (LNA) boosts the signal from the antenna before frequency conversion for transmission. A *high-power amplifier* (HPA) feeds the signal to a directional antenna with a radiation pattern designed to cover a prescribed area of the earth. The electronics receives power generated with solar cells and buffered with storage batteries, which are recharged by the array of solar devices. The batteries are the only source of power when the satellite passes into the earth's shadow at certain times of the year. Finally, a guidance system stabilizes the attitude of the craft, because as it rotates around the earth, the craft would otherwise begin to tumble and its antennas would no longer point toward the earth. As a part of the guidance system, the craft also contains rocket engines for *station keeping*—maintaining its position in the geostationary arc. The overall mass and volume of all components of the craft must be limited, however, largely because of launch considerations.

### Antennas

The antenna structure of a communications satellite consists of several different antenna sections. One receives signals from earth; another transmits those signals back to earth. The transmitting antenna may be constructed in several sections to carry more than one signal beam. Finally, a receive-transmit beacon antenna provides communication with a ground station to control the operation of the satellite. The control functions include turning parts of the electronics on and off, adjusting the radiation pattern, and maintaining the satellite in its proper position.

The design of the complex antenna system for a satellite relies heavily on the horizontal and vertical polarizations of signals to keep incoming and outgoing information separated. Multiple layers in carefully controlled thicknesses of reflecting materials, which are sensitive to signal polarizations, can be used for such purposes. Also, multiple-feed horns can develop more beams to earth. Antennas for different requirements may combine several antenna designs, but nearly all are based on the parabolic reflector, because of two fundamental properties of the parabolic curve:

- Rays received by the structure that are parallel to the feed axis, are reflected and converged at the focus
- Rays emitted from the focal point are reflected and emerge parallel to the feed axis

Special cases may involve spherical and elliptical reflectors, but the parabolic is most common.

## 16.2b Transponders

From the antenna, a signal is directed to a chain of electronic equipment known as a *transponder*. This unit contains the circuits for receiving, frequency conversion, and transmission. Some of the first satellites launched for INTELSAT and other systems contained only one or two transponder units. However, initial demand for satellite link services forced satellite builders to develop more economical systems with multiple-transponder designs. Such requirements meant refinements in circuitry that would allow each receiver-transmitter chain to operate more efficiently (less power draw). Multiple-transponder electronics also meant a condensation of the physical volume needed for each transponder with as small as possible an increase in the overall size and mass of the satellite to accommodate the added equipment.

General purpose satellites placed in orbit now have 12 or 24 transponders for C-band units, each with 36-MHz bandwidths. However, wide variations in the number of transponders and in their operating bandwidths, exist. Some Ku-band systems use fewer transponders with wider bandwidths, but others—particularly those designed for use in Europe—have 40 or more transponders of 27 MHz bandwidth. Identification of the transponders is described by an agreed-upon *frequency plan* for the band.

Users of satellite communication links are assigned to transponders generally on a lease basis, although it is possible to purchase a transponder. The cost, however, is prohibitive for most professional video customers. Assignments usually leave one or more spare transponders aboard each craft, allowing for standby capabilities in the event of a transponder failure.

By assigning transponders to users, the satellite operator simplifies the design of up-link and down-link facilities. An earth station controller can be programmed for the transponders of interest. For example, a television station may need access to several transponders from one satellite. The operator enters only the transponder number (or carrier frequency) of interest. The receiver handles retuning of the receiver and automatic switching of signals from the dual-polarity feed horn on the antenna. Controllers can also be used to move an antenna from one satellite to another automatically.

Each transponder has a fixed center frequency and a specific signal polarization according to the frequency plan. For example, all odd-numbered transponders might use horizontal polarization, while even-numbered ones might be vertically polarized. Excessive deviation from the center carrier frequency by one signal does not produce objectionable interference between transponders and signals because of the isolation provided by cross-polarization. This concept extends to satellites in adjacent parking spaces in the geosynchronous orbit. Center frequencies for transponders on adjacent satellites are offset in frequency from those on the first craft. An angular offset of polarization affords additional isolation. The even and odd transponder assignments are still offset by 90° from one another. As spacing is decreased between satellites, the polarization offset must be increased to reduce the potential for interference.

While this discussion has centered on the down-link frequency plan for a C-band satellite, up-link facilities follow much the same plan, except that up-link frequencies are centered around 6 GHz. A similar plan for Ku-band equipment uses up-linking centered around 14 GHz and down-link operation around 12 GHz.

## 16.2c Power Sources

Providing power to the electronics aboard the spacecraft is a major engineering challenge. Nuclear power has not been utilized because many questions regarding integration of a small reactor into a satellite system remain unsolved. Fortunately, constant evolution of the storage battery has provided the space program with various types of renewable power devices.

During the launch phase, control circuitry obtains power from rechargeable batteries on the craft. After the final orbital position is achieved, the control station activates the deployment of arrays of solar cells. When the solar panels are functional, all operational power is derived from the panels, as well as current to recharge the batteries for backup during times of eclipse, caused when the earth blocks illumination of the panels from direct sunlight.

The solar cells most commonly used are silicon with boron or arsenic doping to create *pn* junctions. The light-to-electricity conversion efficiency of this material is approximately 10 percent. Aluminum or gallium arsenide cells with approximately 18 percent efficiency are also used. The array is created by connecting cells in series to achieve a larger voltage and a number of series groups in parallel for greater current capability. An array of 2000 cells covering 1 $m^2$ produces approximately 100 W.

With any type of photovoltaic material, the output of a cell is determined by a number of factors. Primary among them are:

- The surface area of the cell exposed to light

- The average solar flux per unit area (light intensity)
- The temperature of the cell material
- The variation of the light source from a position perpendicular to the cell surface

The initial capabilities, as well as the end-of-life capacity, of a solar power source must take these factors into account and allow for them by designing in accordance with worst-case conditions. One such consideration is the proper positioning of the solar array when operation of the satellite is initiated. Positioning remains important at all times: The variation of solar energy with the seasons varies approximately 8 percent between the solar equinoxes and solstices, because of the varying angle of incidence of the sunlight. This seasonal change produces a variation in temperature as well, and over time, the bombardment by proton radiation from solar flares reduces the efficiency of electric current production.

Two types of solar panels are in general use. One is a flat configuration arranged as a pair of deployable panel wings that extend to either side of the satellite body. In the other type, the solar cells are attached to the barrellike body of the satellite.

Each approach has advantages and disadvantages. On the winglike panels, approximately 18,000 cells are used to maintain a 1200 W output capability at end-of-life. The other configuration uses 80,000 cells mounted on a 2.16 m diameter cylinder that is 5.5 m in height. The wing approach is the more efficient of the two, with an end-of-life (7 years) rating between 15 and 25 W/kg. The spinning cylindrical array is rated at 6 to 10 W/kg.

The wing keeps all cells illuminated at all times, producing more electric current. However, all cells are subject to a higher rate of bombardment by space debris and by solar proton radiation, and to a higher average temperature (approximately 60°C) because of constant illumination.

The rotating-drum design places a small portion of all cells perpendicular to the sun's rays at any one time. Cells that are not perpendicular to the light source produce power as long as they are illuminated, but the amount is reduced and depends on the angle of light incidence. The individual cells are not all exposed in one direction simultaneously and are less likely to be struck by space debris. Also, because cells on the drum are not exposed to sunlight at all times, their average temperature is significantly less, approximately 25°C, which extends their operating lifetime.

The output current, voltage, and power from the solar panels undergo variation, depending on the operating conditions. Without a means of controlling or regulating the power, the operation of the electronics package would also vary. An unregulated power bus is simpler and takes less of the allowable mass, or *mass budget*, of the craft. System electronics circuitry must include on-board regulators to maintain consistent RF levels. The alternative is to use bus regulation with more complex circuitry, a somewhat lower output power, and reduced reliability because of the additional components in the system.

Some variation can be accommodated by the storage batteries in either case. The most serious need for batteries following insertion into orbit of the geosynchronous satellite occurs approximately 84 days per year. The earth passes between the sun and the satellite, causing an eclipse. The blackout period can last as long as 70 min within one day. During these periods, the need for a rechargeable supply is essential. Nickel-cadmium batteries have played a major role in powering such spacecraft.

Power to the electronics on the craft requires protective regulation to maintain consistent signal levels. Most of the equipment operates at low voltages, but the final stage of each transponder chain ends in a high power amplifier. The HPA of C-band satellite channels may include a *traveling-wave tube* (TWT) or a *solid-state power amplifier* (SSPA). Ku-band systems rely primarily on TWT devices. Traveling-wave tubes, similar to their earth-bound counterparts, klystrons, require multiple voltage levels. The filaments operate at *5* V, but beam-focus and electron-collection electrodes in the device require voltages in the hundreds and thousands of volts. To develop such a range of voltages, the power supply must include voltage converters.

From these voltages, the TWT units produce output powers in the range of 8.5 to 20 W. Most systems are operated at the lower end of the range for increased reliability and greater longevity of the TWTs. In general, lifetime is assumed to be 7 years.

At the receiving end of the transponder, signals coming from the antenna are split into separate bands through a channelizing network, which directs each input signal to its own receiver, processing amplifier, and HPA. At the output, a combiner brings all the channels together again into one signal to be fed to the transmitting antenna.

## 16.2d Control of Satellites

The electronics and attitude control section of the satellite performs a number of key tasks, including:

- Turning transponders on and off, as directed from the ground station
- Adjustments of antennas to keep the beams focused to form the footprint of signal levels designated for each satellite system
- Controlling the small package of rocket engines in the craft to maintain the correct parking position in the geosynchronous orbit

Much of the attitude control is microprocessor-controlled, but the ground station can override on-board equipment to perform necessary corrections.

Communications with the satellite from the ground often use other frequencies in addition to those relayed through the transponders. Common arrangements include a 148–149.9 MHz up-link and 136–138 MHz down-link in VHF or a 2025–2.120 GHz up-link and 2.2–2.3 GHz down-link in S-band. Control information and telemetry data reporting on the status of the satellite system to the ground station are transmitted as encoded digital signals. Because of the proximity of satellites, strict protocols are needed to avoid one satellite from responding when the neighboring device is being interrogated. Protection against erroneous control signal reception is improved through narrow-band reception, or wide-band reception combined with spread-spectrum transmission and digitally encoded data.

### Positioning and Attitude Control

Maintaining the position of the spacecraft in its orbit is called *station keeping*. With the altitude of the craft at about 23,000 mi, some motion of the craft is possible without requiring major redirecting of earth station antennas. In most cases, a window of sides approximately 0.1° in length as seen from earth (equivalent to 75 km) is considered acceptable. If the craft moves outside the window, reception S/N is degraded.

Movement of a craft from its assigned location is caused by forces of the surrounding environment. One significant factor is gravitational effects. It is the normal pull of gravity upon the fast-moving craft (centrifugal force) that keeps the satellite in its orbit. The force of gravity has been found to vary at sonic locations around the equator. As a result, satellites positioned near

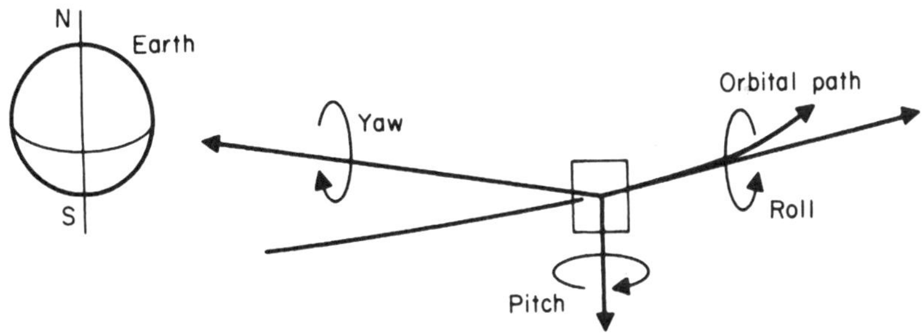


**Figure 16.2** Attitude of the spacecraft is determined by pitch, roll, and yaw rotations around three reference axes.

the longitudes of gravitational anomalies become advanced or retarded from their correct location, and corrections are periodically required.

Other factors to be accounted for are the gravitational attractions of the moon and sun. The pull of the moon is approximately three times stronger than that of the sun. Most of the error to the orbit caused by these bodies is perpendicular to the equatorial plane, that is, in the inclination of the orbit with respect to the equator. Slight errors in the longitude of the craft result from solar and lunar attractions, but these are compensated with a change in the dimensions of the orbit.

Lesser factors that disturb the position of the spacecraft and may require occasional corrections include:

- The pressure on the solar power panels by the solar radiation flux
- The force caused by radio frequency power radiated from the satellite
- The magnetic field of the earth
- The self-generated torques related to antenna displacements, the solar arrays, and changing masses of on-board fuel supplies

The results of meteorite impacts on spacecraft are considered, but have not been experienced significantly by any satellite to date.

The *attitude* of the spacecraft is defined as the combination of pitch, roll, and yaw. These are motions that can occur without causing a change in the orbital position of the satellite. As illustrated in Figure 16.2:

- *Pitch* is the rotation of the craft about an axis perpendicular to the plane of the orbit
- *Roll* is rotation of the craft about an axis in the plane of the orbit
- *Yaw* is rotation about an axis that points directly toward the center of the earth

Satellite attitude is determined by the angular variation between satellite body axes and these three reference axes.

Attitude control involves two functions. First, it is necessary to rotate that part of the satellite that must point toward the earth around the pitch axis. Rotation must be precisely one revolution per day or 0.25° per minute. Second, the satellite must be stabilized against torque from any other source. To perform these functions, numerous detectors measure the sense of the satellite with respect to earth horizons, the sun, specified stars, gyroscopes, and radio frequency signals. Attitude correction systems may involve a control loop with the ground control station or may be designed as a closed loop by storing all required attitude reference information within the satellite itself.

Relative to the first function listed, satellites can be spin-stabilized gyroscopically by rotating the body of the satellite in the range of 30 to 120 times per minute. This effectively stabilizes the craft with respect to the axis of the spin and is used with cylindrical-type satellites. Although a relatively simple approach, it has a major drawback—the antenna would not remain pointed toward earth if it were mounted on the rotating portion of the craft. The solution is to de-spin the section holding the antenna at one revolution per day about the pitch axis to maintain a correct antenna heading.

## 16.2e Satellite Transmission Modes for Television

Analog and digital formats are used for the transmission of television signals, but the use of analog frequency modulation is by far the most common [1]. The advantages of FM for satellite transmission are:

- It minimizes the effects of nonlinearities in the transmission channel
- It is immune to AM noise
- Power-limited systems can take advantage of the wider bandwidth to increase the C/N performance
- Various processing techniques can be employed to optimize video transmission, e.g., multiplexing, preemphasis, and threshold extension

In addition to the deviation of the carrier by the modulating signal, it is usually necessary to subject the main carrier to a low-frequency deviation. This spreads the high concentration of carrier and sideband energy over a larger range of the spectrum and permits higher satellite ERP without exceeding the FCC's limit on watts/meter$^2$/kHz down-link power density.

Preemphasis and deemphasis is employed in FM systems for the transmission of video to compensate for the increase in thermal noise with increasing frequency. CCIR Recommendation 567 specifies a standard 75 μs preemphasis, for example.

A characteristic of FM is that the detected S/N for the video signal is higher than its C/N. This difference is the *FM improvement factor*; satellite transmission takes advantage of this improvement, provided the received C/N is greater than the receiver operating threshold.

*Threshold extension demodulation* (TED) is a common technique used in FM receivers to reduce video impulse noise when the C/N drops below the receiver's operating threshold. Above threshold, the receiver acts like a standard discriminator; when C/N drops below threshold, TED circuitry automatically switches to a narrow bandwidth.

## 16.2f The Up-Link

The ground-based transmitting equipment of the satellite system consists of three sections: baseband, intermediate frequency (IF), and radio frequency (RF). The baseband section interfaces various forms of incoming signals with the design of the satellite being used. Signals provided to the baseband section may already be in a modulated form with modulation characteristics decided or determined by the terrestrial media bringing signals to the up-link site. Depending on the nature of the incoming signal—voice, data, or video—some degree and type of processing will be applied. In many cases, multiple incoming signals can be combined into a single up-link signal through multiplexing.

When the incoming signals are in the correct format for passage through the satellite, they are applied to an FM modulator, which converts the entire package upward to an intermediate frequency of 70 MHz. The use of an IF section has several advantages. First, a direct conversion between the baseband and the output frequency presents difficulties in maintaining frequency stability of the output signal. Second, any mixing or modulation step has the potential of introducing unwanted by-products. Filtering in the IF may be used to remove spurious signals resulting from the mixing process. Third, many terrestrial microwave systems include a 70 MHz IF section. If a signal is brought into the up-link site by terrestrial microwave, it becomes a simple matter to connect the signal directly into the IF section of the up-link system.

From the 70 MHz IF section, the signal is converted upward again, this time to the output frequency of 6 GHz (or 14 GHz) before application to an HPA. Several amplifying devices are used in HPA designs, depending on the power output and frequency requirements. For the highest power level in the C- or Ku-band, klystrons are employed. Different devices are available with pulsed outputs ranging from 500 W to 5 kW and a bandwidth capability of 40 MHz or greater. This means that a separate klystron is required for each 40 MHz wide signal to be beamed upward to a transponder.

Another type of vacuum power device for HPA designs is the traveling-wave tube While similar in some aspects of operation to a klystron, the TWT is capable of amplifying a band of signals at least 10 times wider than the klystron. Thus, one TWT system can be used to amplify the signals sent to several transponders on the satellite. With output powers from 100 W to 2.5 kW, the bandwidth capability of the TWT offsets its higher price than the klystron.

Solid-state amplifiers, based on field-effect transistor (FET) and other technologies, can be used for both C- and Ku-band up-link HPA systems. The cost of the devices is rather high, but the wide-band capabilities are good and reliability is excellent. These devices are useful in amplifying signals directed at more than one transponder.

### 16.2g The Down-Link

Satellite receiving stations, like up-link equipment, interface ground-based equipment to the satellite. To serve this function, the earth station—as a TVRO (television receive only) or a TR (transmit-receive) facility—consists of the receiving antenna with a low-noise amplifier. A 4 GHz (or 12 GHz) tuner, a 70 MHz IF section, and a baseband output stage complete the necessary equipment package.

Antennas for ground-based receiving stations fall into several categories. First, the antenna used for transmitting to the satellite can also be used as a receiving antenna. Unlike terrestrial two-way radio systems, where one station transmits and then the other, the satellite system participates in both functions simultaneously. This feature is possible because of the offset between up-link and down-link frequencies. To make operation possible, the circulator separates and directs the transmitted and received signals to the correct locations.

Receiving antennas for commercial applications, such as radio/TV networks, CATV networks, and special services or teleconferencing centers, range from 7 to 10 m in diameter for C-band operation, while the Ku-band units can be smaller. Antennas for consumer and business use are even smaller, depending on the type of signal being received and the quality of the signal to be provided by the down-link installation. If the signals being received are strictly digital in nature, smaller sizes are sufficient. If TV signals are of interest and "broadcast quality" is required, a larger-size reflector is typically used.

Antenna size for any application is determined primarily by the type of transmission, the band of operation, the location of the receiving station, typical weather in the receiving station locale, and the quality of the output signal required from the down-link. As indicated previously, digital transmissions allow a smaller main reflector to be used, because digital systems are designed with error correction. The data stream periodically includes information to check the accuracy of the data and, if errors are found, to make corrections. If errors are too gross for error correction, the digital circuitry provides error concealment techniques to hide the errors. Absolute correction is less critical for entertainment-type programming functions, such as television and audio programming. The nature of the application also helps to determine if the antenna must be strictly parabolic, or if one of the spherical types, generally designed for consumer use, will be sufficient.

Regarding the frequency band of operation, the lower the signal frequency, the larger the antenna reflector must be for the same antenna gain. However, there are compensating factors such as the required output signal quality, station location, and local environment. Generally, the gain figure and directivity of larger reflectors are greater than those of smaller reflectors.

One of the most critical sections of the receiver is the low-noise amplifier or low-noise converter, the first component to process the signal following the antenna. The amplifier must not add noise to the signal, because once added, the noise cannot be removed. LNAs are rated by their *noise temperature*, usually a number around 211K. This is equivalent to –62°C. The cost of LNAs increases significantly as the temperature figure (and noise) goes down.

Following the LNA, the receiver tuner converts signals to a 70 MHz IF frequency. As with the up-link equipment, an output at 70 MHz is useful to connect to terrestrial microwave equipment if desired. A second conversion takes video, audio, or data to baseband signals in a form most convenient for those using the communications link.

## 16.3 Performance Considerations

The performance of any communications link, irrespective of its bandwidth, is determined by the carrier-to-noise ratio (C/N) [2]. In the case of a broadcast satellite system, that ratio is determined by the following elements:

- The *effective isotropic radiated power* (EIRP) of the satellite
- Frequency of operation
- Home terminal figure of merit (G/T)
- Degradations resulting from rain

For a fixed G/T, improving the system performance requires increasing the EIRP, as illustrated in Figure 16.3. The curves show that the C/N depends on the strength of the transmitter

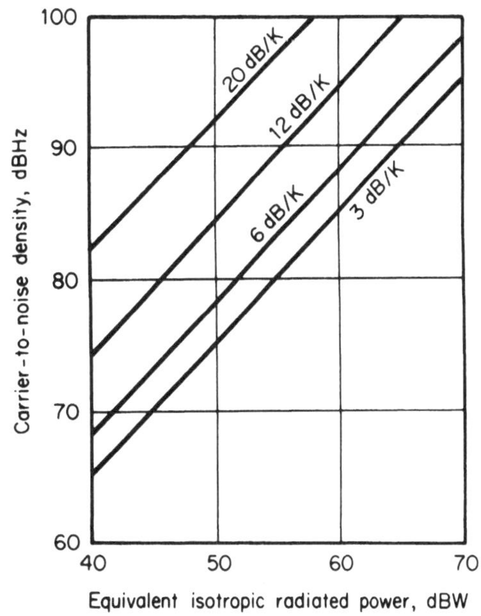


**Figure 16.3** Carrier-to-noise ratio is determined by the spacecraft's effective isotropic radiated power (EIRP) and the home receiver figure of merit (G/T). For a fixed receiver quality, C/N can be improved by increasing the EIRP. This may be accomplished by either increasing the satellite transmitter power or reducing coverage area (increasing antenna gain). These curves assume an operating frequency of 12 GHz. For a typical receiver noise figure of 3.0 dB, the antenna diameters would range from 0.36 m for the 3 dB/K case to 2.5 m for the 20 dB/K case. (*After* [2].)

power and on the size of the reception area. Increasing the transmitter power can be costly, whereas limiting the coverage reduces the service and the income derived from it. An optimum consumer television picture for an analog system would require a C/N of 90 dBHz.

Because the picture quality is related to C/N, a common method of improving the picture is to use higher figures of merit for the home terminal. This can be achieved with a larger antenna that has a higher gain or with an LNA with lower noise-temperature characteristics.

The relationship between C/N and home-terminal antenna size is independent of carrier frequency. It is independent because the antenna gain of the home terminal depends on the square of the frequency in exactly the same way as the free-space loss. Therefore, in the equation for transmission power, the two variables effectively cancel each other, all other things being equal. On the other hand, for a given power level in a satellite and a given coverage area, the move to higher carrier frequencies requires larger antennas to achieve the same picture quality because of atmospheric and rain losses. Thus, to maintain picture quality with atmospheric losses, high-power signals have to be used in the satellite to accommodate the smaller home-receiver antennas.

FM carrier transmission is necessary in order to conserve satellite-transmitter power and maintain an acceptable S/N in the received television picture. A basic tradeoff between carrier-to-noise ratio and the available radio frequency bandwidth per channel is crucial to determining the modulation scheme. If a weighted video S/N is used, the desired picture quality can be plotted against the available RF bandwidth, as given in Figure 16.4. For an excellent consumer analog video image, the range for S/N is 45 to 50 dB. For an acceptable picture, it can be as low as 40 dB. If an FM C/N of 10 dB can be met, tremendous improvements in picture quality are possible with only small increases in power by going to broader bandwidths.

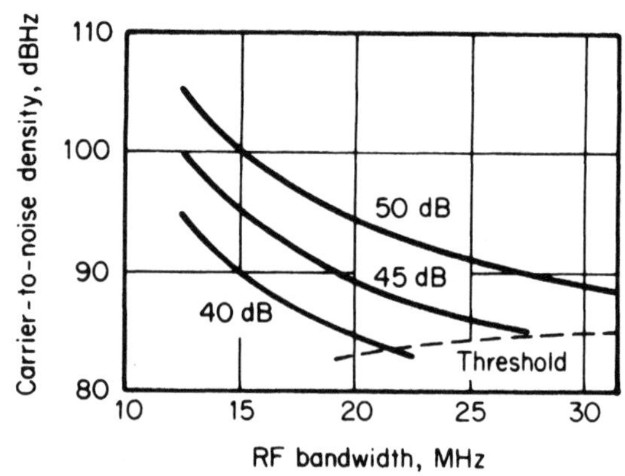


**Figure 16.4** When the video S/N is used as a parameter, the requirement for C/N decreases as bandwidth increases. In FM systems, the bandwidth can be increased and that ratio decreased up to the point where the FM threshold is reached. In a practical system, where rain attenuation must be accommodated, a good design would allow a margin of several decibels above the threshold point for clear-weather performance. (*After* [2].)

## 16.3a Home Terminal Variables

To deliver a quality picture to an analog or digital home terminal, two random variables must be addressed:

- The subjective perception of picture quality
- Attenuation caused by rain

Both of these elements ultimately determine how successfully the system performs.

Studies of subjective evaluations of picture quality have been made worldwide, with the results hinging on video S/N for analog systems and the video compression technology used for digital systems, when operating above the "cliff effect". (Subjective evaluation of digital coding schemes is addressed in Section 18.) For an analog system, an average ratio of peak-to-peak signal-to-weighted noise level of 40 dB is acceptable to a majority of viewers.

Cable systems are typically designed for reception from satellites with picture S/N of 50 dB, although much lower values are usually satisfactory for a home terminal. The smallest terminals and the worst conditions of high rain and low angles of elevation decrease the minimum ratio to the 30 dB level, a very poor-quality picture.

### Rain Attenuation

Attenuation caused by rain can be a difficult problem to deal with. Both the rate and volume of rain vary extensively around the U.S. and elsewhere. A rain attenuation model, drawn up by the National Aeronautics and Space Administration for the 12 GHz broadcast band, summarizes the different climates in the United States. The hardest to deal with is that of the southeast, particularly in Louisiana and Florida.

To illustrate how the climate curves reflect the video S/N, these parameters are plotted in Figure 16.5 against a percentage of time for each climate type. The picture quality in the worst climate would be usable until the C/N dropped below the FM threshold. This would happen about 0.2 percent of the time in the southeastern U.S. This amounts to almost 18 h a year—enough to prove frustrating to viewers, who would inevitably have to acquire larger and more expensive antennas.

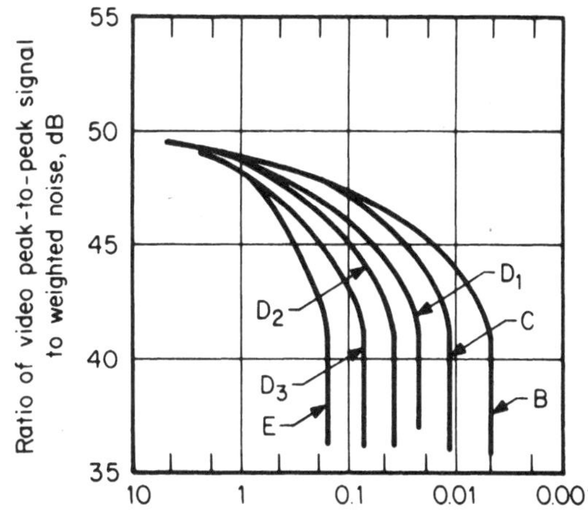


**Figure 16.5** In the 12-GHlz band, video performance will degrade during rain. This example shows performance for a system designed to have a S/N of 50 dB during clear weather. Climate types range from cold and dry areas such as Maine (B) to tropical areas such Florida and Louisiana (E). (*After* [2].)

As mentioned previously, a critical characteristic for receiving high quality pictures from a direct-broadcast satellite is the figure of merit (G/T) of the home terminal. At a particular frequency, this figure simply depends on the size of the antenna and on the system noise temperature of the LNA/receiver system. The system noise temperature, for its part, is affected by rain. Rain not only weakens the signal but also increases the antenna temperature. This temperature must be added to the LNA temperature to set the system noise temperature. Variation of the antenna temperature is particularly distressing because the system temperature margin that can account for a high G/T is obliterated just when the margin is needed the most. System temperature in clear weather and antenna size are, thus, not perfectly interchangeable.

For the same ratio, a larger antenna is better than a lower clear-weather system temperature, because the increased gain of a larger antenna is not affected by rain.

A plot of the figure of merit versus antenna diameter for different low-noise amplifier temperatures in both clear weather and heavy rain yields the results shown in Figure 16.6. A 1.8 m antenna with a 480K temperature and a 1 m antenna with a 120K temperature both give the same clear-weather figure of merit: 17 dB/K. However, in heavy rain the larger antenna G/T deteriorates to only 16 dB, whereas the smaller antenna system drops all the way to 13 dB.

## 16.4 Digital Transmission

Many satellite systems are used for the transmission of digital information [1]. A typical digital transmission up-link consists of a modem, upconverter, and power amplifier. The modem converts digital information to and from a modulated carrier. The center-frequency of the modulated carrier positions the signal within a satellite transponder. The upconverter translates the modulated carrier to a satellite frequency and, thus, selects the transponder of the satellite.

The modem is the earth station component used to convert digital information to a format suitable for transmission by satellite. The modem accepts a digital data input signal and outputs

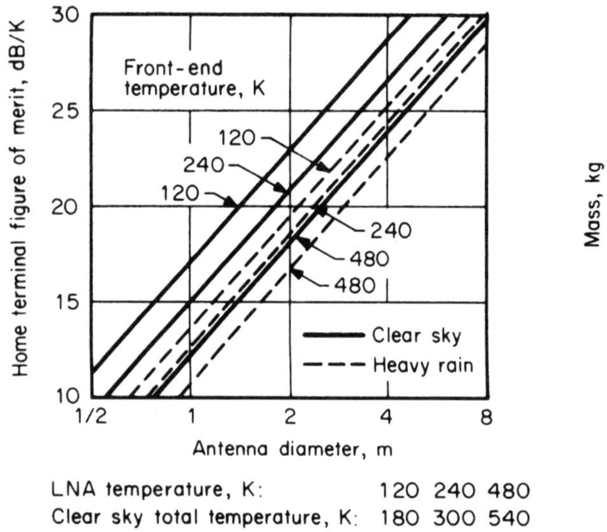


**Figure 16.6** Receiving station G/T is a function of antenna diameter, receiver temperature, and antenna temperature. In heavy rain, the antenna temperature increases from a clear-weather value of perhaps 50K to a maximum value of 290K (ambient). (*After* [2].)

an intermediate frequency, typically a range centered on either 70 or 140 MHz, containing the modulated digital information.

Transmitted digital data is first applied to the encoder section of the modem for *forward error correction* (FEC) encoding. This process appends additional bits to the original information to provide error detection and correction capabilities. The resulting signal is scrambled using a standard algorithm to ensure random data.

The aggregate data (i.e., original data plus error correction bits) is applied to the modulator for frequency modulation onto an IF carrier. The IF carrier is selectable, typically in the range of 50 to 90 MHz for 70 MHz operation, or 100 to 180 MHz for 140 MHz operation. The center frequency of the modem modulator is tuned to position the signal within the satellite transponder.

A satellite digital transmission system is characterized by the following parameters:

- **Data rate**. The data rate is the number of bits per second transmitted by the modem, typically ranging from 32 kbits/s to 3 Mbits/s, selectable in increments as small as 1 bit/s.
- **Data interface**. Modems normally support a number of data interfaces. The data interface refers to the connector and signal levels, such as DS1, CEPT, EIA-422, V.35, and MIL-188/114.
- **Code rate**. The code rate refers to the FEC encoding scheme used. In some modems, the code rate is selectable. The code rate configuration is referred to as "*m*/*n*." In this context, *m* refers to the number of original bits per block of transmitted bits; *n* refers to the number of original bits plus error correction bits per block of transmitted bits. Thus, a code rate of 3/4 means that for every three data bits, four data bits are transmitted. Thus a 1024 kbits/s modem operating with a code rate of 3/4 would transmit 1365 kbits/s over the satellite channel.
- **Modulation system**. The modulation scheme refers to the method of indicating data bits. Two common modulation schemes are employed in satellite transmission systems: *bi-phase*

*shift keying* (BPSK) and *quadrature phase shift keying* (QPSK). These modulation schemes generate a periodic set of phase shifts referred to as *symbols*.

The symbol rate (i.e., the number of symbols per second) and data rate determines the amount of bandwidth required in the channel.

In BPSK, two phase-shifts are used to represent two states. For this case, the symbol rate is equal to the transmission rate. The QPSK scheme uses four phase shifts, thus transmitting two bits per symbol. Therefore, QPSK uses a symbol rate that is half the transmission rate. QPSK requires less bandwidth than BPSK but demands increased performance from the channel.

## 16.5 References

1. Cook, James H., Jr., Gary Springer, Jorge B. Vespoli: "Satellite Earth Stations," in *NAB Engineering Handbook*, Jerry C. Whitaker (ed.), National Association of Broadcasters, Washington, D.C., pp. 1285–1322, 1999.

2. Kase, C. A., and W. L. Pritchard: "Getting Set for Direct-Broadcast Satellites," *IEEE Spectrum*, vol. 18, no. 8, pp. 22–28, 1981. © 1981 IEEE.

Chapter

# 17

# Receiver Antenna Systems

K. Blair Benson

L. H. Hoke, Jr., L. E. Donovan, J. D. Knox, D. E. Manners, W. G. Miller, R. J. Peffer, J. G. Zahnen, U. L. Rohde

## 17.1 Introduction

The antenna system is one of the most important circuit elements in the reception of television signals. The ultimate quality of the picture and sound depends primarily on how well the antenna system is able to capture the signal radiated by the transmitting antenna of the broadcast station and to feed it, with minimum loss or distortion, through a transmission line to the tuner of the receiver.

In urban and residential areas near the television station antenna, where strong signals are present, a compact set-top telescoping rod or "rabbit ears" for VHF and a single loop for UHF usually are quite adequate. The somewhat reduced signal strength in suburban and rural areas generally requires a multielement roof-mounted antenna with either a 300 Ω twin-lead or—to reduce pickup of interference from nearby sources such as automobile ignition—a shielded 75Ω coaxial transmission line to feed the signal to the receiver.

Fringe areas, where the signal level is substantially lower, usually require a more complicated, highly directional antenna, frequently on a tower, to produce an even marginal signal level. The longer transmission line in such installations may dictate the use of an all-channel low-noise preamplifier at the antenna to compensate for the loss of signal level in the line.

### 17.1a Basic Characteristics and Definitions of Terms

Antennas have a number of key characteristics that define their ability to receive energy from a radiated field. These are as follows:

- *Wavelength*—the distance traveled by one cycle of a radiated electric signal. The frequency of the signal is the number of cycles per second. It follows that the frequency $f$ is inversely proportional to the wavelength $\lambda$. Both wavelength and frequency are related to the speed of light. Conversion between the two parameters can be made by dividing either into the quantity of the speed of light $c$ (i.e., $\lambda \times f = c$). The velocity of radio waves in air is essentially the same as that of light in free space, which is $2.9983 \times 10^{10}$ cm/s, or for most calculations, $3 \times 10^{10}$ cm/s (186,000 mi/s).

- *Radiation*—the emission of coherent modulated electromagnetic waves in free space from a single or a group of radiating antenna elements. Although the radiated energy, by definition, is composed of mutually dependent electric and magnetic field vectors having specific magnitude and direction, conventional practice in television engineering is to measure and specify radiation characteristics in terms of the electric field.
- *Polarization*—the angle of the radiated electrical field vector in the direction of maximum radiation. If the plane of the field is parallel to the ground, the signal is *horizontally polarized*; if at right angles to the ground, it is *vertically polarized*. When the receiving antenna is located in the same plane as the transmitting antenna, the received signal strength will be maximum. If the radiated signal is rotated at the operating frequency by electrical means in feeding the transmitting antenna, the radiated signal is *circularly polarized*. Circularly polarized signals produce equal received signal levels with either horizontal or vertical polarized receiving antennas. (See Figure 17.1.)
- *Beamwidth*. In the plane of the antenna, beamwidth is the angular width of the directivity pattern where the power level of the received signal is down by 50 percent (3 dB) from the maximum signal in the desired direction of reception. Using Ohm's law ($P = E^2/R$) to convert power to voltage, assuming the same impedance $R$ for the measurements, this is equal to a drop in voltage of 30 percent (Figure 17.2).
- *Gain*—the signal level produced (or radiated) relative to that of a standard reference dipole. Gain is used frequently as a figure of merit. Gain is closely related to directivity, which in turn is dependent upon the radiation pattern. High values of gain usually are obtained with a reduction in beamwidth. Gain can be calculated only for simple antenna configurations. Consequently, it usually is determined by measuring the performance of an antenna relative to a reference dipole.
- *Input impedance*—the terminating resistance into which a receiving antenna will deliver maximum power. Similar to gain, input impedance can be calculated only for very simple formats, and instead is usually determined by actual measurement.
- *Radiation resistance*. Defined in terms of transmission, using Ohm's law, as the radiated power $P$ from an antenna divided by the square of the driving current $I$ at the antenna terminals.
- *Bandwidth*—a general classification of the frequency band over which an antenna is effective. This requires a specification of tolerances on the uniformity of response over not only the effective frequency spectrum but, in addition, over that of individual television channels. The tolerances on each channel are important because they can have a significant effect on picture quality and resolution, on color reproduction, and on sound quality. Thus, no single broadband response characteristic is an adequate definition because an antenna's performance depends to a large degree upon the individual channel requirements and gain tolerances, and the deviation from flat response over any single channel. This further complicates the antenna-design task because channel width relative to the channel frequency is greatly different between low and high channels.

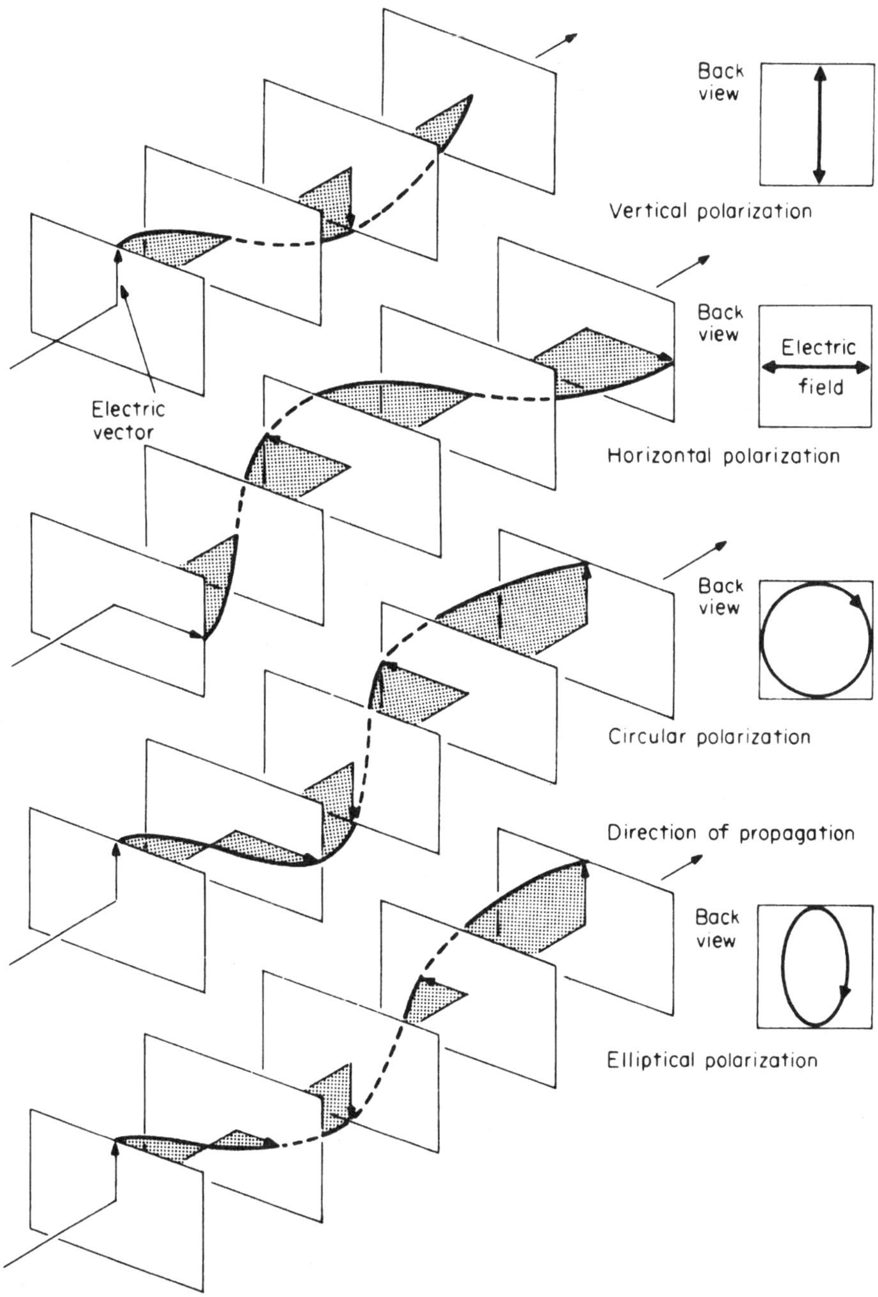


**Figure 17.1** Polarization of the electric field of the transmitted wave

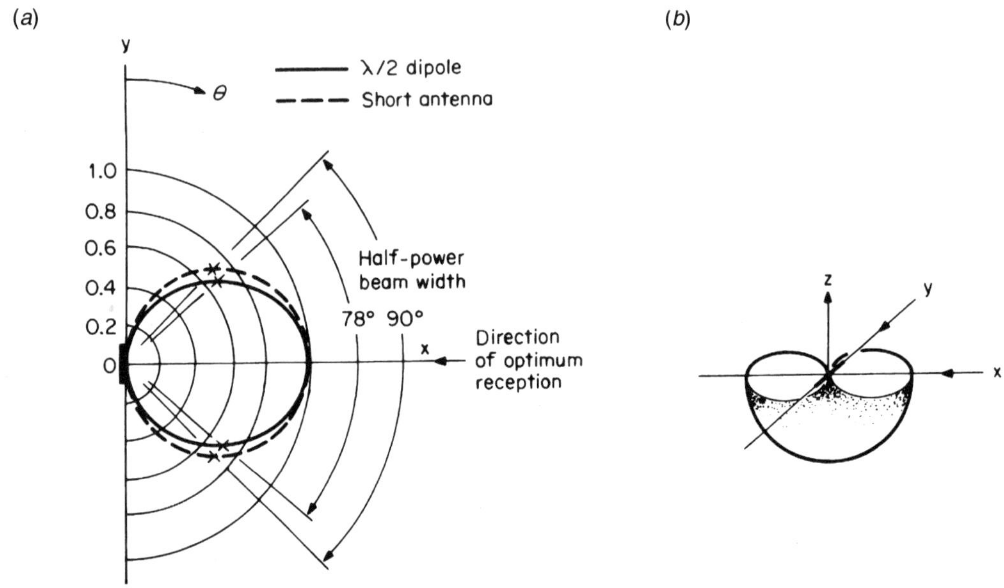


**Figure 17.2** Normalized *E*-field pattern of λ/2 and short dipole antennas: (*a*) plot representation, (*b*) three-dimensional view of the radiation pattern.

The foregoing problem is readily apparent by a comparison of low VHF and high UHF channel frequencies with the 6 MHz single-channel spectrum width specified in the FCC Rules. VHF Channel 2 occupies 11 percent of the spectrum up to that channel, while UHF Channel 69 occupies only 0.75 percent of the spectrum up to Channel 69. In other words, a VHF antenna must have a considerably greater broadband characteristic than that of a UHF antenna.

## 17.2 VHF/UHF Receiving Antennas

In strong-signal urban areas, a set-top antenna consisting of rabbit ears for VHF and loops for UHF with a foot or so of 300 Ω twin-lead connecting to the receiver antenna terminals usually is adequate if ghosts resulting from multipath reception are not present. In suburban and rural areas, a roof-mounted antenna with extra elements to increase the signal level is required. In weak-signal fringe areas, beyond the FCC-specified "Grade B" contour area of 2.5 mV/m covered by the transmitter, a high outdoor antenna and low-noise signal preamplifier may be necessary at the antenna in order to boost the signal above that of any stray pickup from interference signals such as auto ignition and CBs, and to a level required by the receiver for stable scanning and color synchronization.

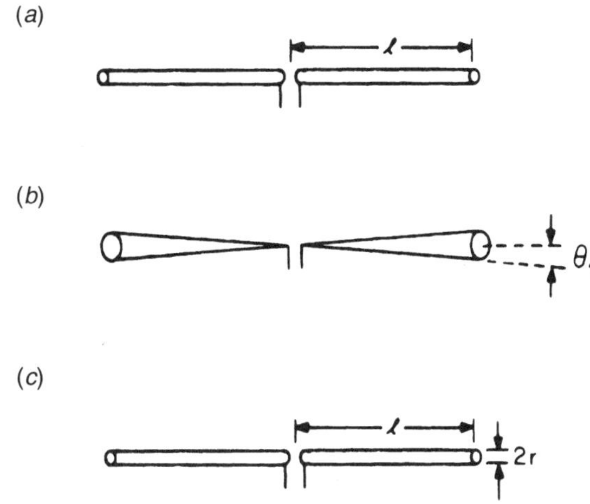


**Figure 17.3** Basic dipole antenna designs: (*a*) half-wave dipole, (*b*) conical dipole, (*c*) cylindrical dipole.

## 17.2a Dipole Antenna

The dipole receiving antenna is not only the simplest type suitable for feeding a transmission line but also is the basis for other more complex designs [1, 2]. It consists of two in-line rods or wires with a total length equal to a half-wave at the primary band it is intended to cover. In Figure 17.3*a*, the two quarter-wave elements *l* are connected at their center to a two-wire transmission line. Normally a 300 Ω balanced transmission line is connected at the center. However, because the radiation resistance of a half-wave dipole is 73 Ω, use of the usual 300 Ω balanced transmission line will result in a mismatch and some loss in transfer of received signal power. In addition, if the receiving end of the line is not a matching impedance of approximately 73 Ω, there will be reflections up and down the line. These will show up in the received picture as horizontal smear for short lines or ghosts displaced to the right in the case of longer lines. This picture distortion is avoided in receiver design, where the tuner input usually is unbalanced 75 Ω coaxial cable, by the use of a balanced-to-unbalanced (*balun*) coupler. A balun consists of a compact transformer with closely coupled input and output coils wound on a powdered-iron magnetic core. Early designs used a compact coil of small-gauge parallel wires spaced for a 300 Ω impedance and wound on a small rod. Figure 17.4 shows some common balun implementations [3].

Alternatively, a coaxial transmission line can be used to feed the tuner 75 Ω input directly. In this case, a balun is mounted at the antenna to transform the balanced dipole signal to the unbalanced coaxial transmission line and the antenna's balanced impedance of 73 Ω is closely matched to the 75 Ω transmission line. Dissimilar circuits can be matched by a resistive network, but with a resultant loss in signal. A typical case would be between a 300 Ω lead-in from an antenna to a 75 Ω coaxial section of a building distribution system.

The bandwidth of a dipole antenna can be increased by increasing the diameter of the elements through the use of cones or cylinders rather than wires or rods, as illustrated in Figures 17.3*b* and *c*. This also increases the impedance of the antenna so that it more nearly matches the balanced 300 Ω transmission line. The 3 dB, or half-power, beam width, as shown in Figure 17.2, is 78°.

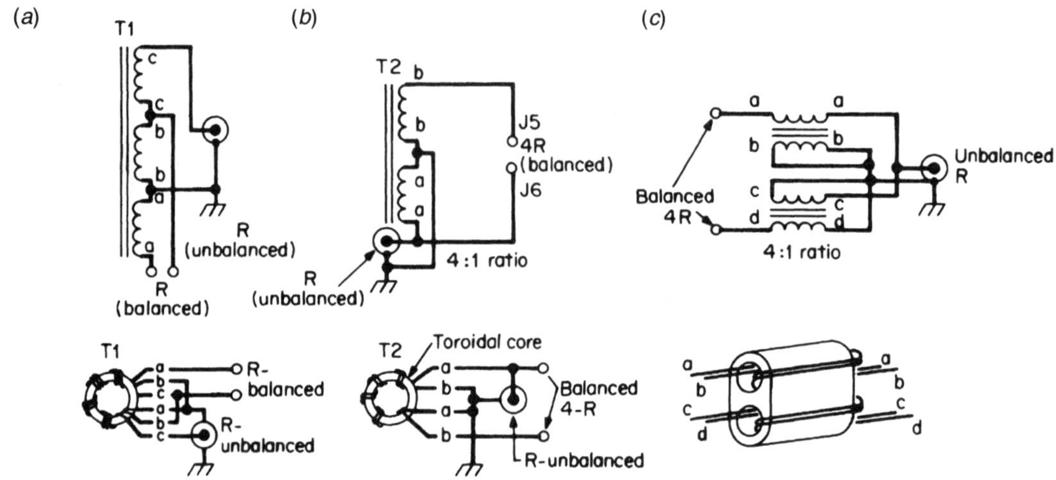


**Figure 17.4** Balun transformers: (*a*) balanced to unbalanced, 1/1, (*b*) balanced to unbalanced, 4/1, (*c*) balanced to unbalanced on symmetrical balun core, 4/1. (*Illustrations* a *and* b *after* [4].)

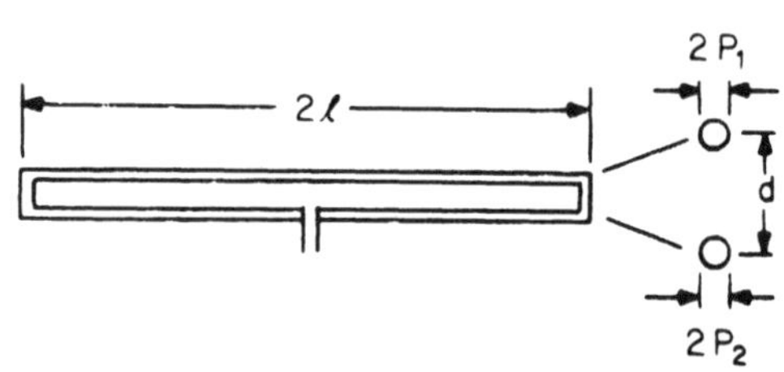


**Figure 17.5** The folded dipole antenna.

If the total length of the dipole is much less than a half wave, the characteristics become those of a "short dipole" with a substantially lower resistance and high capacity.

## Folded Dipole

An increase in bandwidth and an increase in impedance can be accomplished through the use of the two-element folded-dipole configuration shown in Figure 17.5. The impedance can be increased by a factor of as much as 10 by using rods of different diameter and varying the spacing. The typical folded dipole with elements of equal diameter has a radiation impedance of 290 Ω, four times that of a standard dipole and closely matching the impedance of 300 Ω twin lead.

The quarter-wave dipole elements connected to the closely coupled half-wave single element act as a matching stub between the transmission line and the single-piece half-wave element. This broadbands the folded-dipole antenna to span a frequency range of nearly 2 to 1. For example, the low VHF channels, 2 through 6 (54–88 MHz), can be covered with a single antenna.

### V Dipole

The typical set-top rabbit-ears antenna on VHF Channel 2 is a "short dipole" with an impedance of 3 or 4 Ω and a capacitive reactance of several hundred ohms. Bending the ears into a V in effect tunes the antenna and will affect its response characteristic, as well as the polarization. A change in polarization may be used in city areas to select or reject cross-polarized reflected signals to improve picture quality.

The *fan dipole*, based upon the V antenna, consists of two or more dipoles connected in parallel at their driving point and spread at the ends, thus giving it a broadband characteristic. It can be optimized for VHF reception by tilting the dipoles forward by 33° to reduce the beam splitting (i.e., improve the gain on the high VHF channels where the length is longer than a half-wave).

A ground plane or flat reflecting sheet placed at a distance of 1/16 to 1/4 wave behind a dipole antenna, as shown in Figure 17.6, can increase the gain in the forward direction by a factor of 2. This design is often used for UHF antennas, e.g., a "bow-tie" or cylindrical dipole with reflector. For outdoor applications, a screen or parallel rods can be used to reduce wind resistance. At 1/4-wave to 1/2-wave spacing, the antenna impedance is 80 to 125 Ω respectively.

## 17.2b Quarter-Wave Monopole

The *quarter-wave monopole* above a ground plane, as shown in Figure 17.7, is another antenna derived from the elementary dipole. It is supplied with many personal portable television receivers in which the set itself acts as a ground plane. Although the monopole is intended to receive vertically polarized signals, it usually is moved in its swivel joint to either a horizontal position or to some other angle to a best compromise between direct and reflected signals. The theoretical resistance characteristic of a monopole with an infinite ground plane is 37 Ω

## 17.2c Loop Antenna

The *loop antenna* set-top configuration—in effect, a form of the folded dipole—is used for UHF reception. Analyzing it as a single-turn magnetic-field pickup loop, the radiation resistance can be calculated and found to vary over a more than 10-to-1 range. The values for a typical 7-in diameter wire loop for several UHF channels are given in Table 17.1.

## 17.2d Multielement Array

Multielement arrays can be used to achieve higher gain and directivity [2]. One popular design for single-channel reception is the *Yagi-Uda array*, illustrated in Figure 17.8. The receiving element is 1/2 wavelength for the center of the band covered. The single reflector element is slightly longer, and the three director elements slightly shorter, all spaced approximately 1/4 wavelength from each other. Typically, the bandwidth is only one or two channels. However, this can be increased by trading-off of a slightly longer reflector and slightly shorter directors from theoretical for a slight loss in gain. On the other hand, broadbanding a single-channel five-element Yagi to cover all the low-band VHF channels, for example, will reduce the gain figure from 9 dB to between 4 and 6 dB. (See Figure 17.9.)

Another technique for broadening the bandwidth is to add shorter elements in groups between the directors. For example, in a low-band VHF Yagi, elements can be added for high-band VHF

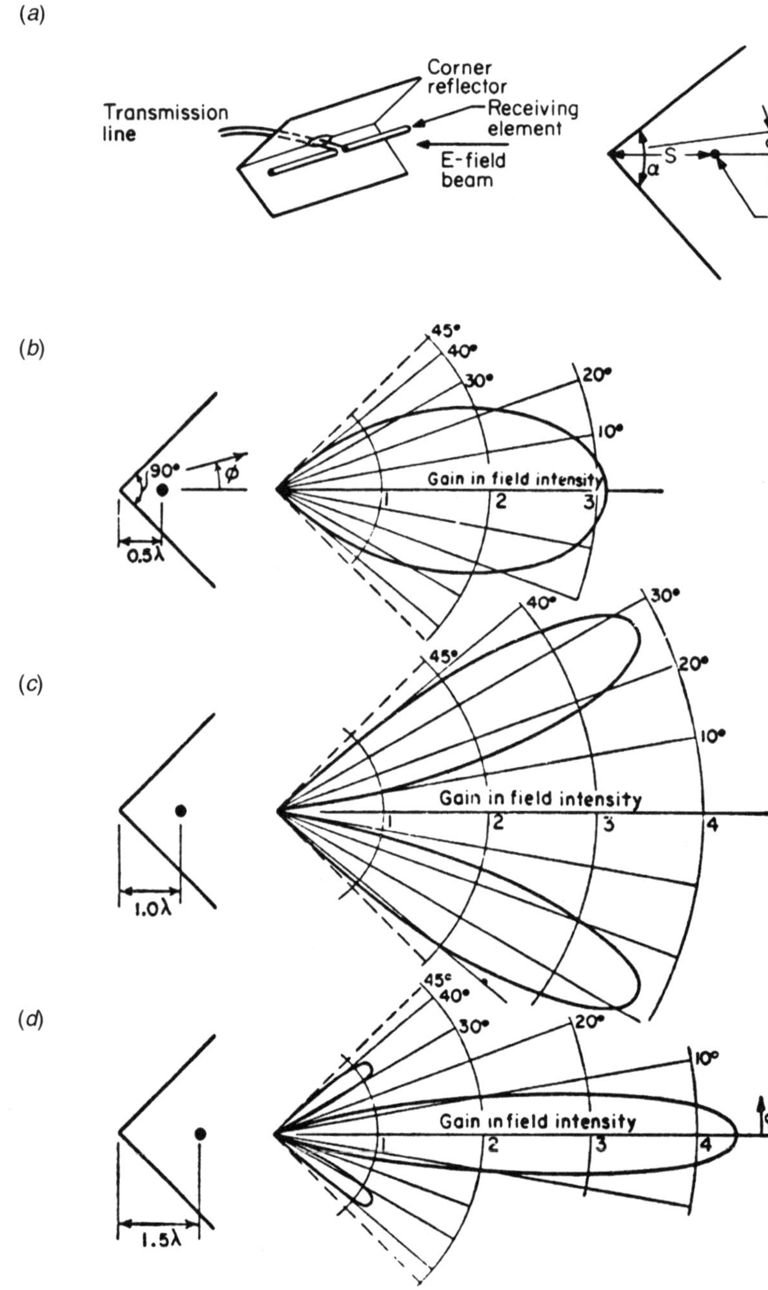


**Figure 17.6** The corner-reflector antenna: (*a*) mechanical configuration, (*b*) calculated pattern of a square corner-reflector antenna with antenna-to-corner spacing of 0.5 wavelength, (*c*) pattern at 1.0 wavelength, (*d*) pattern at 1.5 wavelength. (*Illustrations* b–d *from* [5]).

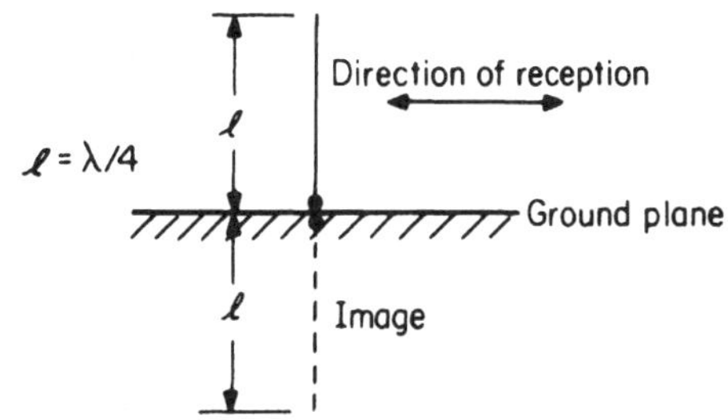


**Figure 17.7** Vertical monopole above a ground plane.

**Table 17.1 Radiation Resistance for a Single-Conductor Loop Having a Diameter of 17.5 cm (6.9-in) and a Wire Thickness of 0.2 mm (0.8-in)**

| Channel | Frequency, MHz | Radiation Resistance, Ω |
|---|---|---|
| 14 | 473 | 108 |
| 42 | 641 | 367 |
| 69 | 803 | 954 |
| 83 | 887 | 1342 |

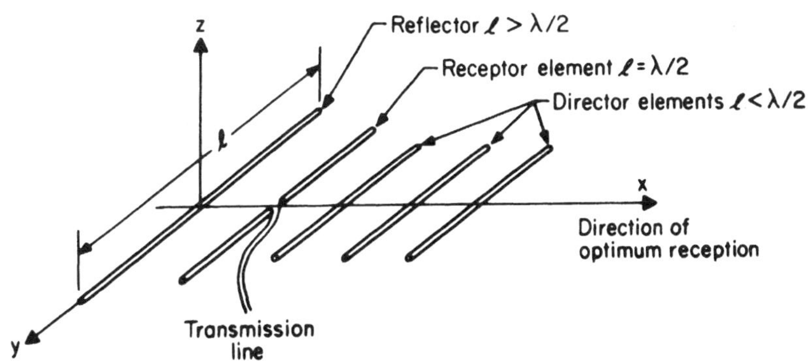


**Figure 17.8** Five element Yagi-Uda array.

channels. The gain of the Yagi over a standard dipole for up to 15 elements and beamwidth directivity are listed in Table 17.2.

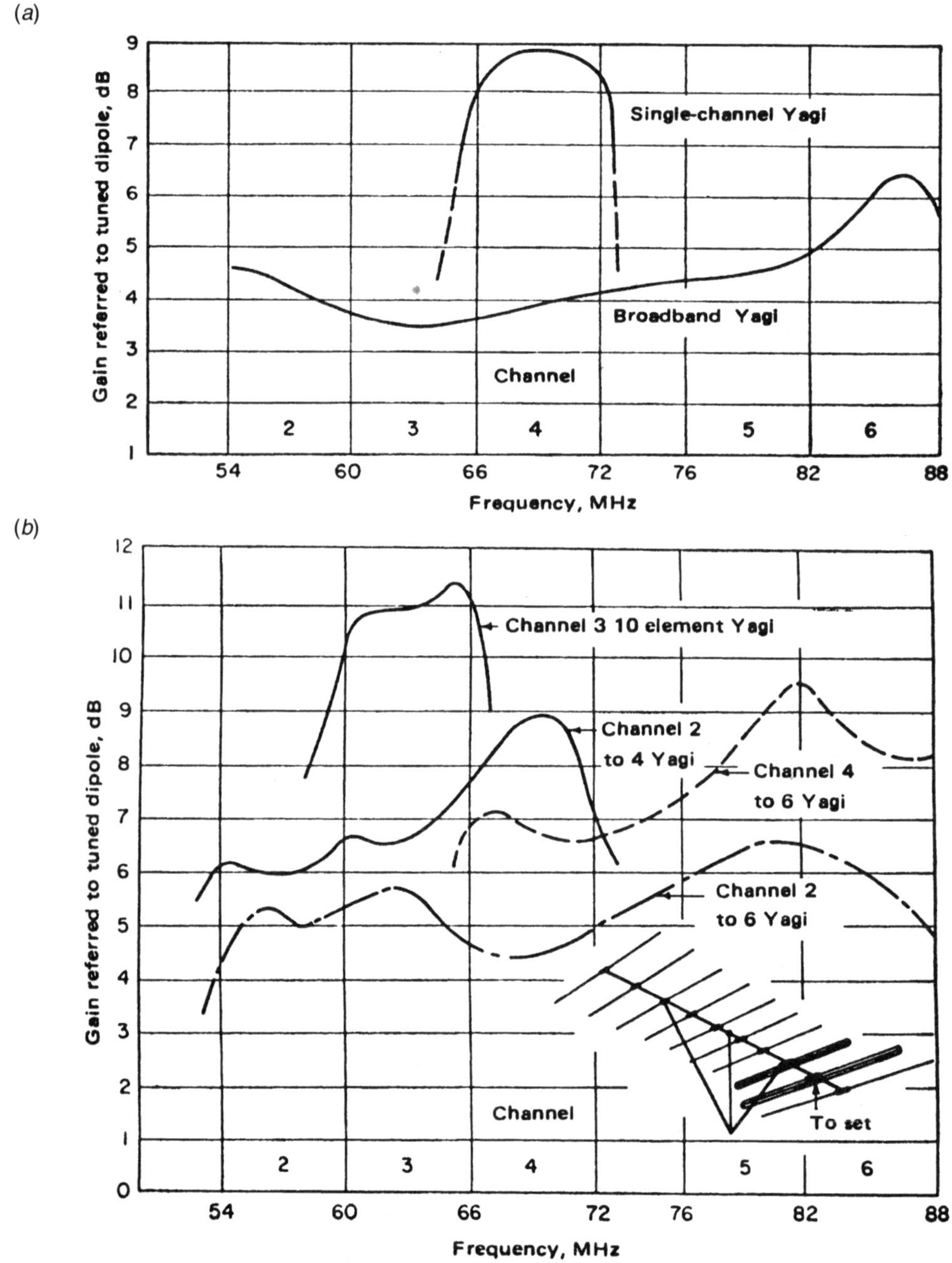


**Figure 17.9** Gain-vs.-bandwidth for a Yagi antenna: (*a*) measured gain of a five-element single-channel Yagi and a broadband Yagi, (*b*) measured gain of three twin-driven 10-element yagi antennas and a single-channel 10-element Yagi. (*From* [6].)

**Table 17.2 Typical Characteristics of Single-Channel Yagi-Uda Arrays**

| Number of Elements | Gain, dB | Beam Width, degrees |
|---|---|---|
| 2 | 3–4 | 65 |
| 3 | 6–8 | 55 |
| 4 | 7–10 | 50 |
| 5 | 9–11 | 45 |
| 9 | 12–14 | 37 |
| 15 | 14–16 | 30 |

## 17.3 DBS Reception

The ground antenna for analog television satellite reception usually consists of a parabolic reflector of 3–5 m in diameter for the C band and 1–1.5 m for the DBS Ku-band. A *low-noise amplifier* (LNA) with a gain of 40 dB or more and a noise figure of less than 1.5 dB at 120 K is placed at the antenna. Double-conversion receivers are used to obtain the necessary gain and selectivity.

### 17.3a Antenna Designs

Consumer satellite receiving antennas are based on several basic parabolic reflector designs. The parabolic reflector exhibits high gain (referenced to an ideal isotropic antenna). As illustrated in Figure 17.10, the basic configurations are:

- *Prime focus, single parabolic reflector.* This antenna places the pickup element and LNA in front of the reflector precisely at the focal point of the parabola. Because the struts that support the LNA assembly are located directly within the received beam, every effort is made to design these components with as little bulk as possible, yet physically strong enough to withstand adverse weather conditions.
- *Double reflector.* The primary reflector is parabolic in shape while the subreflector surface, mounted in front of the focal point of the parabola, is hyperbolic in shape. One focus of the hyperbolic reflector is located at the parabolic focal point, while the second focal point of the subreflector defines the position for the pickup/LNA.
- *Offset reflector.* This design removes the pickup/LNA and its support from the radiated beam. Although the reflector maintains the shape of a section of a parabola, the closed end of the curve is not included. The LNA assembly, while still located at the focal point of the curve, points at an angle from the vertex of the parabola shape.

Antenna type and size for any application is determined by the mode of transmission, band of operation, location of the receiving station, typical weather in the receiving station locale, and the required quality of the output signal. Digital transmissions allow a smaller main reflector to be used because the decoding equipment is usually designed to provide error correction. The data stream periodically includes information to check the accuracy of the data, and if errors are found, to make corrections. Sophisticated error concealment techniques make it possible to hide

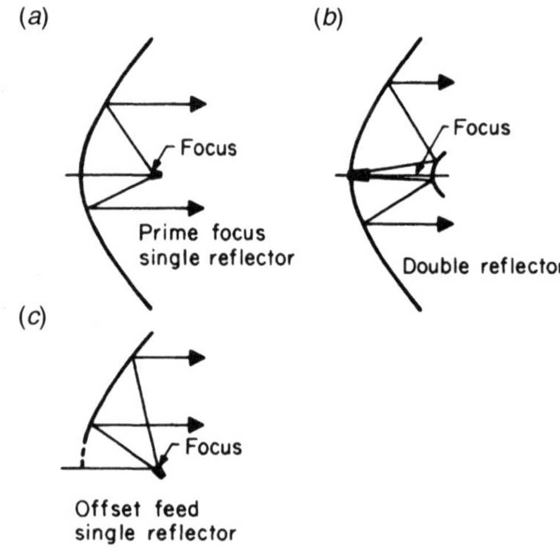


**Figure 17.10** Basic types of satellite receive antennas: (*a*) prime focus, (*b*) double reflector, (*c*) offset feed single reflector.

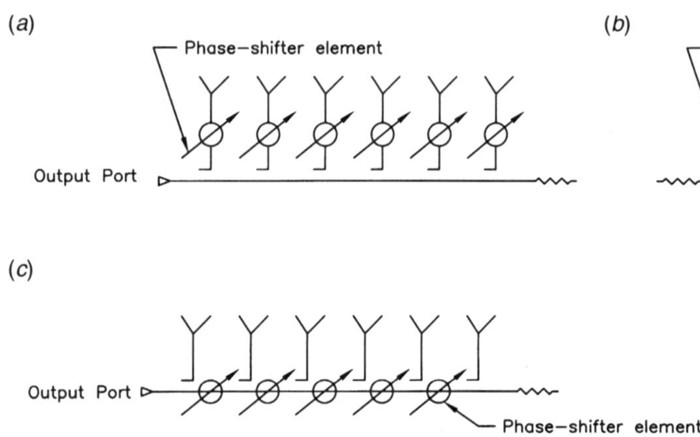


**Figure 17.11** Phased array antenna topologies: (*a*) end collector, (*b*) center collector, (*c*) series phase shifter.

errors that are too large for correction. For entertainment programming, absolute correction is less critical and gives way primarily to concealment techniques.

## Phased Arrays

For some reception applications, the parabolic antenna is being superseded by a flat multielement antenna known as the *planar array.* Ku-band antenna assemblies measuring no more than a foot square have been found to provide acceptable DBS reception under a variety of weather conditions.

The design principle is based upon that of large phased arrays used for many years in radar applications. Rather than collecting the radiated signal by focusing a parabolic reflector on a signal antenna, the electric signal from each antenna is amplified and timed by a filter network so that all signals are in phase at the summing point. The basic approach is shown in Figure 17.11. A more sophisticated design employs a second smaller antenna to provide an error signal to operate automatic circuits to align the antenna for maximum signal.

Printed circuit board manufacturing techniques are used to etch the antenna elements on a flat panel, along with the coupling amplifier and phasing network for each antenna.

## 17.4 Diversity Reception[1]

The term *diversity reception* refers to a receiving process that uses more than one transmission of the same information to obtain a better result than can be achieved in a single transmission. The first use of diversity radio reception probably occurred the first time that an operator received a garbled message and asked for a repeat. This form of diversity, which relies on transmissions repeated after a delay, is known as *time diversity,* since the second channel may use the same transmission medium during a later time interval. Simple *requests for retransmission* (RQs) sufficed until commercial long-distance transmission began to use HF radio for machine telegraphy or AM radiotelephony.

Multipath fading of HF channels resulted in the early use of *space diversity,* in which multiple antennas sufficiently separated could provide the diverse channels (References [7] and [8]). Later it was determined that diversity for narrow-band telegraph channels could be provided by sending the same information over two separate channels with sufficient frequency separation. This is known as *frequency diversity.* Another form of diversity useful at HF when there is not sufficient space available (space diversity requires separation of antennas by a distance of many wavelengths to be effective), is *polarization diversity.* This technique uses two antennas that respond to vertical and horizontal polarization, respectively, of an incoming wave, and is possible because the two components of the wave tend not to fade simultaneously. Circularly polarized antennas can also provide clockwise and counterclockwise polarizations for polarization diversity. At higher frequencies, where circular polarization of antennas is more common, fades between different polarizations tend to be highly correlated, so that polarization diversity is ineffective.

A higher-power transmitter or longer-time transmission is required for time and frequency diversity than for space or polarization diversity to provide comparable performance. This is because in the former cases, the energy available from the transmitter must be divided among the multiple channels, whereas in the latter cases the diverse channels can be derived from a single transmission. In all cases, the effectiveness of diversity reception in improving transmission performance is dependent upon the independence of fading among the diverse signals and the method by which the receivers use these signals. Complete independence is assumed in most analyses. However, in practice there is often some degree of correlation. Analyses have showed

1. This section was adapted from: Rohde, Ulrich L., and Jerry C. Whitaker: *Communications Receiverss—Principles and Design*, 3rd. ed., McGraw-Hill, New York, N.Y., 2000. Used with permission.

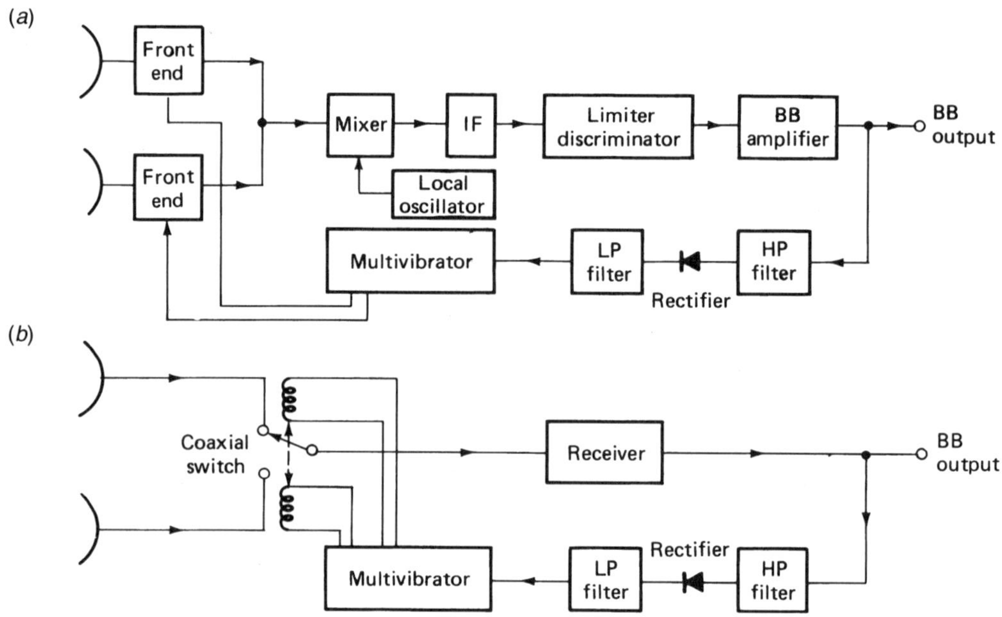


**Figure 17.12** Block diagram of an antenna-switching diversity system: (*a*) front-end switching, (*b*) antenna switching. (*After* [9].)

that the correlation can be substantial (greater than 0.5) without causing major reduction in the diversity advantage.

Diversity techniques can be divided into switching and combining approaches. The former attempts to select for output the channel that has the higher overall level. The S/N would be a better measure, but it is much more difficult to determine. Because the system design should provide for a signal that on average has substantially higher amplitude than the noise, selection of the channel with larger amplitude provides a substantial improvement much of the time. The simplest system of this sort uses antenna switching, as shown in Figure 17.12. Whenever the signal drops below a predetermined level, the system switches to another antenna and if necessary and possible, to still others until a channel above the threshold is encountered.

The simple switching system, while providing adequate signal most of the time, occasionally may lose signal for a short time while finding a suitable antenna. For more than two antennas, the signal selected may not necessarily be the best available from the several antennas. The system can be improved by having a second receiver that is continually scanning the several channels and recording the levels, so that switching may be accomplished to the strongest channel on a regular basis without waiting for the current channel to fade below a satisfactory level. This control system and the second receiver introduce more complexity than the simple system, so that one of the other techniques may prove equal or better.

When multiple receivers are available, they may all be controlled to provide equal gain. The strongest-level output is then selected. A technique similar to the one given in [7] may be used for AM signals. This is illustrated in Figure 17.13 for two receivers, but it can be further

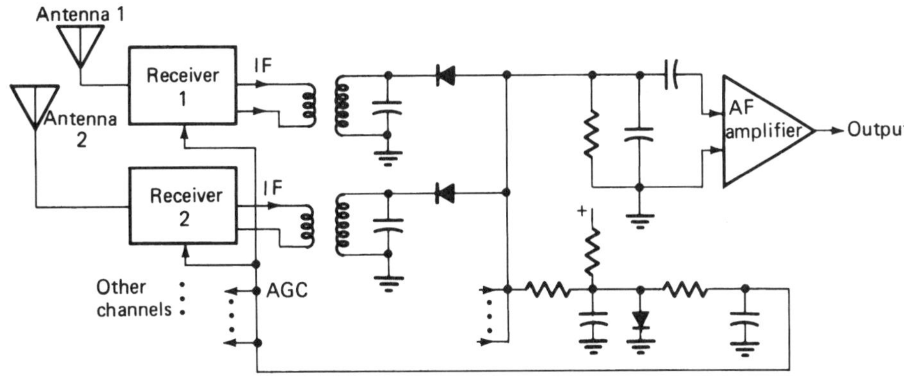


**Figure 17.13** Switching diversity circuit for AM audio frequency signals.

extended. Switching is accomplished by using diode demodulators feeding a common load circuit. The strongest signal develops a demodulated voltage level across the load that prevents conduction in the diodes driven by weaker-amplitude signals. Under the condition that two or more of the receivers produce amplitudes that are nearly the same, all of these stronger signals contribute to the generation of the demodulated voltage, which prevents contribution by the weaker signals. For AM signals, the carrier level provides a dc component across the load, which can be used to develop a common AGC voltage for all the receivers. This maintains their gains equal, even though some fading occurs in the strongest signal, requiring a gain change to provide a constant output. For other modulation types, this simple combination technique is not generally possible, since the diode envelope detector is not an appropriate demodulator for such signals. In such cases, the amplitude-sensing, switching, demodulation, and AGC channels can be separated as shown in Figure 17.14.

Diversity system analysis is generally based on the assumption that multipath fading has an essentially Rayleigh distribution. The justification of this assumption is doubtful in many applications, since it depends on the law of large numbers, and often only two or three multipath components are present. A good summary discussion of the theory of different diversity schemes, along with a careful consideration of assumptions, is given in [9]. The results presented in the following are based on that reference.

What we have called switching diversity is called *selection diversity* in [9]. Figure 17.15 shows the probability that the output S/N will be below various levels when different numbers of diverse channels with identical noise and Rayleigh fading are switched so that the output with best S/N is selected. This curve is based on the fact that for the output to drop below a level S/N, all of the channels must do so. The Rayleigh distribution for the single channel is

$$P_1(x < S_1/N_1) = 1 - \exp(-S_1/N_1) \tag{17.1}$$

For all $n$ channels to be less than $S/N$, when the $S/N$ distributions are equal and independent

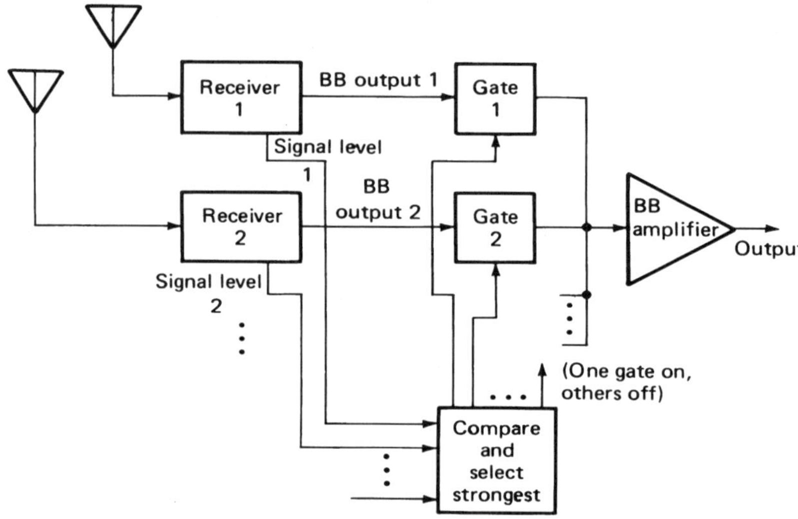


**Figure 17.14** Switching diversity circuit for FM channels.

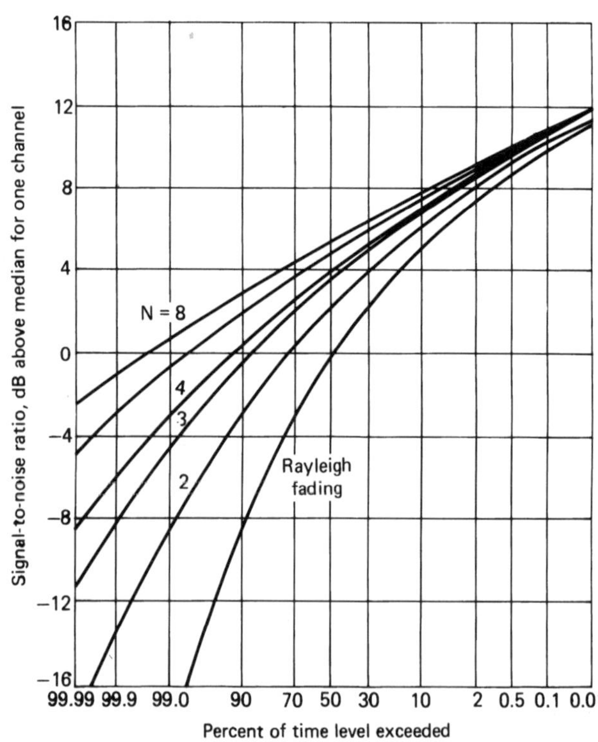


**Figure 17.15** Selection diversity output distributions for *n* independent diversity channels with equal S/N.

$$P_n(x < S/N) = [1 - \exp(-S_1/N_1)]^n \tag{17.2}$$

In combining techniques, the signals from all of the channels are multiplied by a weighting function and then added. The principle is based on the assumption that the signal amplitudes are all in phase so that they add arithmetically, while the noise amplitudes are all independent and add in an rms fashion. The two combining techniques in use are simple addition and *maximal-ratio combining*. In the first case, all weighting functions have the same constant value. In the latter case, they follow a particular law, based on channel measurements.

In switching diversity, the switches can be located either prior to demodulation or after it without affecting the output statistics, as long as the sensing circuits can function adequately. In practice, switching prior to demodulation, where the phases of the RF signals may vary, can cause undesirable switching transients. In the case of combining diversity, however, it is obvious that in combining before demodulation, the random phase differences among the channels will defeat the purpose of adding the weighted signal amplitudes, whereas combining after modulation does not suffer from this problem. The two techniques of combining are generally referred to as *predetection* and *postdetection combining*, respectively.

If the modulation is nonlinear, the advantage of the overall improvement in S/N from combining will be reduced by postdetection combining since the relative relationships of signal and noise amplitudes can be modified in the demodulation process. With FM, for example, when the S/N drops below threshold, the demodulated S/N rises rapidly. Predetection combining can reduce the probability that the composite signal fails to exceed the threshold, whereas in postdetection combining the individual signals may drop below threshold frequently, thus degrading the composite output signal. Predetection combining requires that the phases of the several signals be made the same before combining, hence increasing the complexity of the combiner. Figure 17.16 is a block diagram of an equal-gain (equal weighting in all channels) predetection combiner. In the case of the equal-gain combiner, if the noise voltages in each channel are assumed to have equal independent gaussian distributions, and the noise amplitudes independent Rayleigh distribution variables, the distribution for the combined S/N is given in Brennan [10], Equation (41). The resultant distributions for several orders of diversity are given in Figure 17.17.

A block diagram of maximal-ratio combining, introduced to radio communications by Kahn [12], is shown in Figure 17.18. This differs from the equal-gain combiner in that the weighting functions, instead of being the same and constant in all channels, are now weighted by the factor $x_i/\sigma^2$, where $x_i$ are the values of the $i$th signal amplitude and $\sigma$ is the rms value of the channel gaussian noise, again assumed equal in all the channels. In this case, the resultant combined distribution function is easy to integrate in terms of tabulated functions (incomplete gamma function) and can be expressed as a sum of simple functions

$$P(x < S/N) = 1 - \left\{ \sum_{k=0}^{n-1} \left[ \frac{(S/N)^k}{k!} \right] \right\} e^{-S/N} \tag{17.3}$$

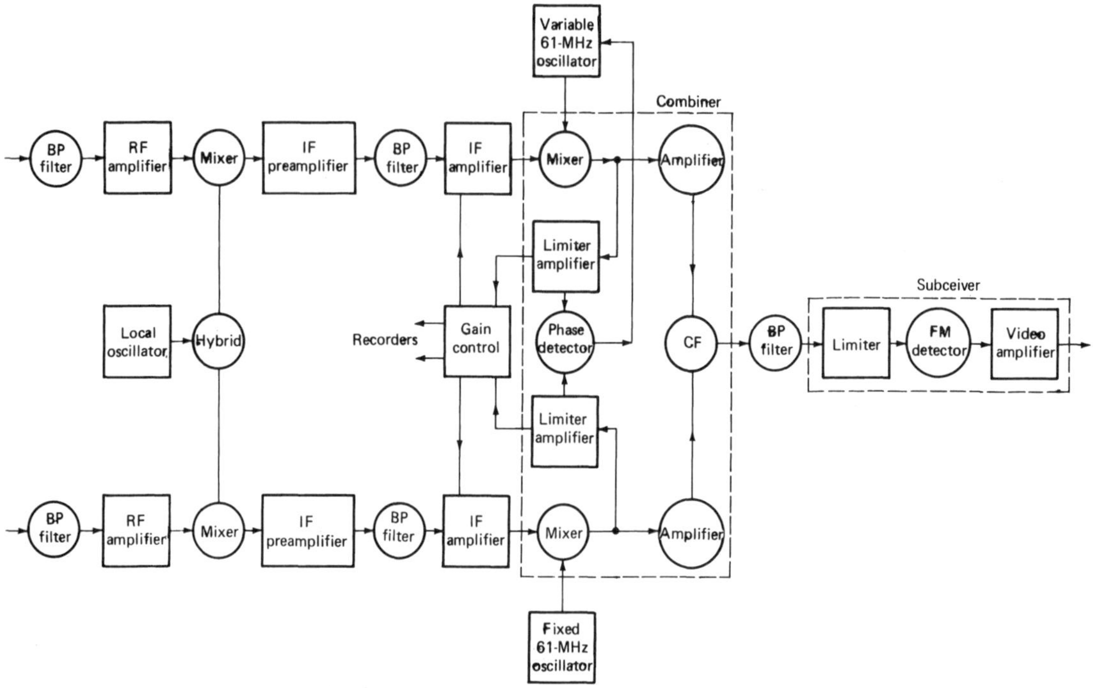


**Figure 17.16** Block diagram of an equal-gain predetection combiner. (*From* [10]. *Used with permission.*)

$$P(x < S/N) = \left\{ \sum_{k=n}^{\infty} \left[ \frac{(S/N)^k}{k!} \right] \right\} e^{-S/N} \tag{17.4}$$

Figure 17.19 shows the resulting distribution functions for various numbers of diversity channels.

In Figures. 17.15, 17.17, and 17.19, we note that the largest increment of diversity improvement occurs between no diversity and dual diversity, and the improvement increment gradually decreases as $n$ grows larger. This is most obvious for switched diversity, where at 0.01 availability the improvement from $n = 1$ to 2 is about 8 dB, from $n = 2$ to 3, it is 3.7 dB, and from $n = 3$ to 4, it is 1.7 dB. This same trend is indicated in the improvement of average S/N level with diversity order, as shown in Figure 17.20. All of the foregoing theoretical estimates have assumed independent Rayleigh fading in the channels. The effects of other possible fading distributions have not been explored extensively, since experimental information on distributions of fading is limited and in many cases the data fits a Rayleigh distribution reasonably well over a limited range. Some calculations have been made on the effect of correlation between Rayleigh fading channels (Figure 17.21).

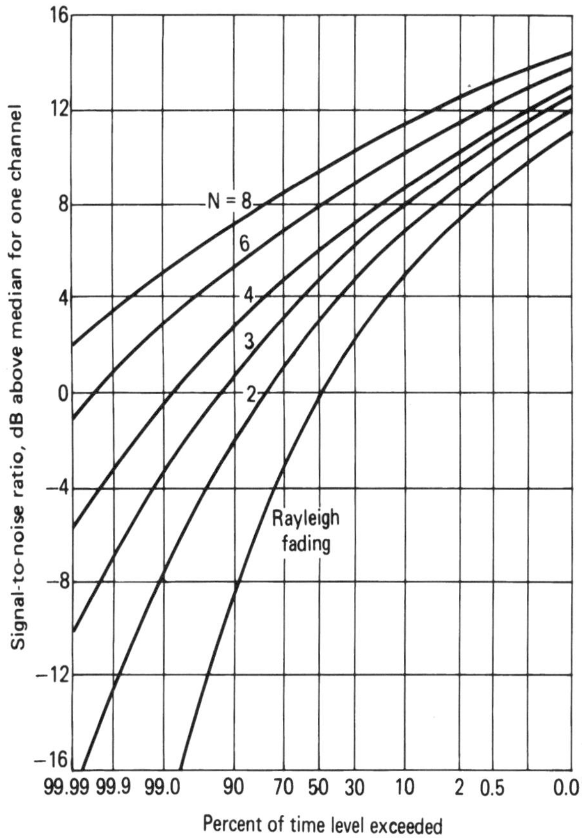


**Figure 17.17** Equal-gain combiner diversity output distributions for *n* diversity channels having equal S/N.

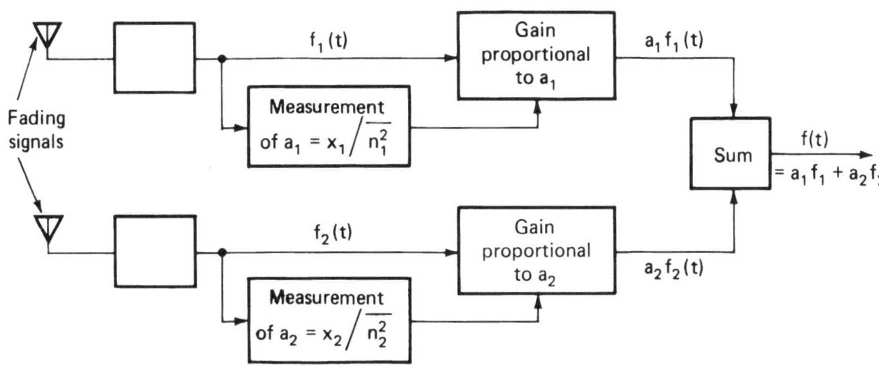


**Figure 17.18** Block diagram of a maximal-ratio diversity combiner. (*From* [11]. *Used with permission.*)

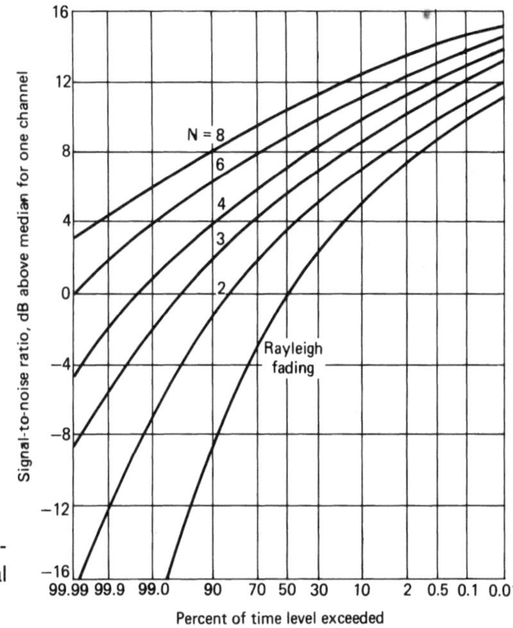


**Figure 17.19** Maximal-ratio combiner distributions for *n* diversity channels having equal S/N.

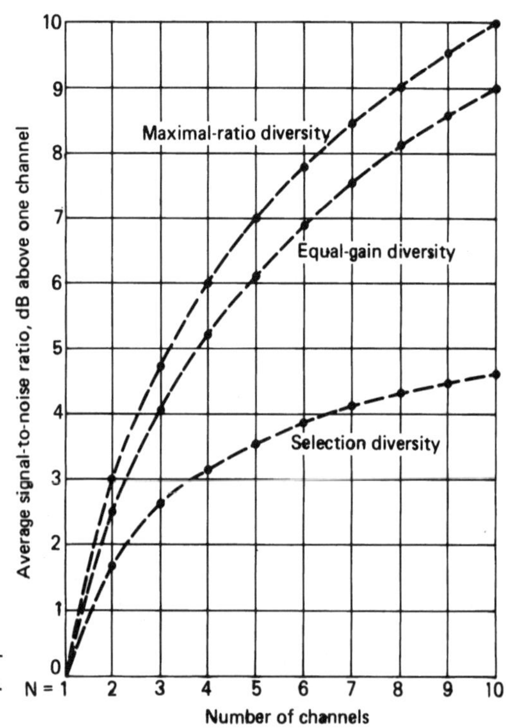


**Figure 17.20** Diversity improvement in average S/N for different diversity techniques. (*From* [10]. *Used with permission.*)

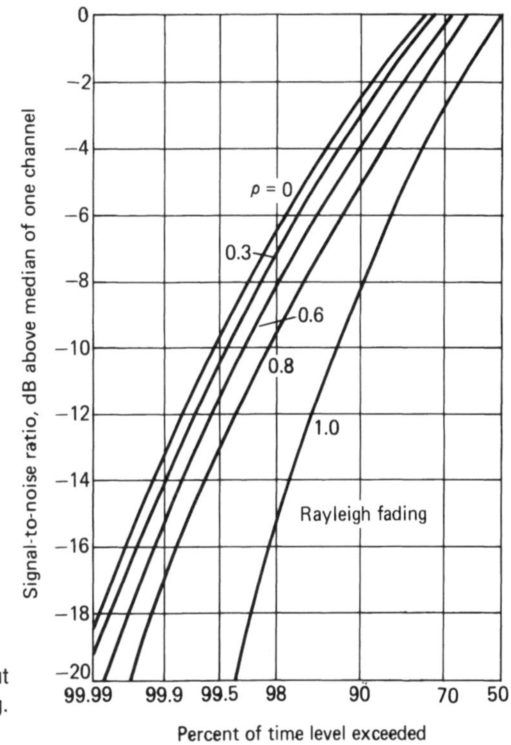


**Figure 17.21** Dual switching diversity output distribution for correlated Rayleigh fading. (*From* [10]. *Used with permission.*)

All of the techniques described here have been found effective in various multipath situations, although in some cases it is difficult to distinguish whether the implementation produces switching or combining diversity. Diversity is only of value for the relatively rapid fading caused by multipath. This kind of fading is encountered extensively at HF and for tropospheric and ionospheric scatter propagation at higher frequencies. Similarly, mobile vehicles passing through a static multipath field at VHF and UHF encounter such fading as a result of their motion. However, the long-term fading, which occurs as a result of diurnal, seasonal, or sunspot cycle variations, affects all direct radio channels between two points essentially equally, and hence cannot be improved by diversity reception.

## 17.5 Transmission Lines

The basic transmission line equations are derived for distributed parameters *R* (series resistance), G (shunt conductance), *L* (series inductance), and *C* (shunt capacitance). All are defined per unit length of the line. Useful relationships are given in Table 17.3. Relationships between the voltage standing-wave ratio (VSWR) and wave transmission and reflection are given in Table 17.4. The nomograph of Figure 17.22 provides a numerical solution for these equations. The Smith chart

**Table 17.3 Summary of Transmission Line Equations**

| Quantity | General line expression† | Ideal line expression† |
|---|---|---|
| Propagation constant | $\gamma = \alpha + j\beta = \sqrt{(\mathrm{R} + j\omega L)(G + j\omega C)}$ | $\gamma = j\omega\sqrt{LC}$ |
| Phase constant $\beta$ | $\mathrm{Im}\ \gamma$ | $\beta = \omega\sqrt{LC} = 2\pi/\lambda$ |
| Attenuation constant $\alpha$ | $\mathrm{Re}\ \gamma$ | 0 |
| Impedance characteristic | $Z_0 = \sqrt{\dfrac{R + j\omega L}{G + j\omega C}}$ | $Z_0 = \sqrt{\dfrac{L}{C}}$ |
| Input | $Z_{-l} = Z_0 \dfrac{Z_r + Z_0 \tanh \gamma l}{Z_0 + Z_r \tanh \gamma l}$ | $Z_{-1} = Z_0 \dfrac{Z_r + jZ_0 \tan \beta l}{Z_0 + jZ_r \tan \beta l}$ |
| Of short-circuited line, $Z_r = 0$ | $Z_\alpha = Z_0 \tanh \gamma l$ | $Z_\alpha = jZ_0 \tan \beta l$ |
| Of open-circuited line, $Z_r = \infty$ | $Z_\alpha = Z_0 \coth \gamma l$ | $Z_\alpha = -jZ_0 \cot \beta l$ |
| Of line an odd number of quarter wavelengths long | $Z = Z_0 \dfrac{Z_r + Z_0 \coth \alpha l}{Z_0 + Z_r \coth \alpha l}$ | $Z = \dfrac{Z_0^2}{Z_r}$ |
| Of line an integral number of half wavelengths long | $Z = Z_0 \dfrac{Z_r + Z_0 \tanh \alpha l}{Z_0 + Z_r \tanh \alpha l}$ | $Z = Z_r$ |
| Voltage along line | $V_{-l} = V_i(1 + \Gamma_0 e^{-2\lambda l})$ | $V_{-l} = V_i(1 + \Gamma_0 e^{-2+\beta l})$ |
| Current along line | $I_{-l} = I_i(1 - \Gamma_0 e^{-2\gamma l})$ | $I_{-l} = I_i(1 - \Gamma_0 e^{-2j\beta l})$ |
| Voltage reflection coefficient | $\Gamma = \dfrac{Z_r - Z_0}{Z_r + Z_0}$ | $\Gamma = \dfrac{Z_r - Z_0}{Z_r + Z_0}$ |

†$l$ = length of transmission line.

*Source:* From J. Feinstein, "Passive Microwave Components," in D. Fink and D. Christiansen (eds.), *Electronic Engineers Handbook,* McGraw-Hill, New York, 1982, chap. 9.

(Figure 17.23) can be used to graphically determine the input impedance of a transmission line of a known length with a known resistive or reactive load.

Coaxial shielded cable and several types of parallel line or balanced twin lead are commonly used between the antenna and receiver. Figure 17.24 shows the insertion loss for five types of line. Typically, the VSWR for the shielded coaxial lines is in the range of 1.3 to 1.5 in the VHF band and 1.1 to 1.2 in the UHF band. The unshielded twin lead also has low VSWR (1.2 to 1.5) in the VHF band, but increases to 2.0 in the UHF band.

Shielded twin-lead VSWR has values up to 2.5 in the VHF band and up to 3 at UHF frequencies.

Although the twin-lead transmission line has lowest insertion loss under ideal conditions, the presence of metal and moisture, as well as outdoor aging, will cause the electrical properties to deteriorate, and—in fact—insertion loss may even exceed that of coaxial lines. Coaxial cable is preferred in many installations because it can be routed along the antenna mast adjacent to

**Table 17.4 Important Relations in Low-Loss Transmission Lines**

| Equation | Explanation |
|---|---|
| $r = \dfrac{1 + \lvert\Gamma\rvert}{1 - \lvert\Gamma\rvert}$ | $r$ = VSWR |
| $\lvert\Gamma\rvert = \dfrac{r - 1}{r + 1}$ | $\lvert\Gamma\rvert$ = magnitude of reflection coefficient |
| $\Gamma = \dfrac{R - Z_0}{R + Z_0}$ | $\Gamma$ = reflection coefficient (real) at a point in a line where impedance is real ($R$) |
| $r = \dfrac{R}{Z_0}$ | $R > Z_0$ (at voltage maximum) |
| $r = \dfrac{Z_0}{R}$ | $R < Z_0$ (at voltage minimum) |
| $\dfrac{P_t}{P_i} = \lvert\Gamma\rvert^2 = \left(\dfrac{r - 1}{r + 1}\right)^2$ | $P_r$ = reflected power<br>$P_i$ = incident power |
| $\dfrac{P_t}{P_i} = 1 - \lvert\Gamma\rvert^2 = \dfrac{4r}{(r + 1)^2}$ | $P_t$ = transmitted power |
| $\dfrac{P_b}{P_m} = \dfrac{1}{r}$ | $P_b$ = net power transmitted to load at onset of breakdown in a line where VSWR = $r$ exists<br>$P_m$ = same when line is matched, $r = 1$ |
| $\dfrac{\alpha_t}{\alpha_m} = \dfrac{1 + \Gamma^2}{1 - \Gamma^2} = \dfrac{r^2 + 1}{2r}$ | $\alpha_m$ = attenuation constant when $r = 1$, matched line<br>$\alpha_t$ = attenuation constant allowing for increased ohmic loss caused by standing waves |
| $r_{max} = r_1 r_2$ | $r_{max}$ = maximum VSWR when $r_1$ and $r_2$ combine in worst phase |
| $r_{min} = \dfrac{r_2}{r_1}\ r_2 > r_1$ | $r_{min}$ = minimum VSWR when $r_1$ and $r_2$ are in best phase |
| $\lvert\Gamma\rvert = \dfrac{\lvert X\rvert}{\sqrt{X^2 + 4}}$ | Relations for a normalized reactance $X$ in series with resistance $Z_0$ |
| $\lvert X\rvert = \dfrac{r - 1}{\sqrt{r}}$ | |
| $\lvert\Gamma\rvert = \dfrac{\lvert B\rvert}{\sqrt{B^2 + 4}}$ | Relations for a normalized susceptance $B$ in shunt with admittance $Y_0$ |
| $\lvert B\rvert = \dfrac{r - 1}{\sqrt{r}}$ | |

*Source:* From J. Feinstein, "Passive Microwave Components," in D. Fink and D. Christiansen (eds.), *Electronic Engineers' Handbook*, McGraw-Hill, New York, 1982, chap. 9.

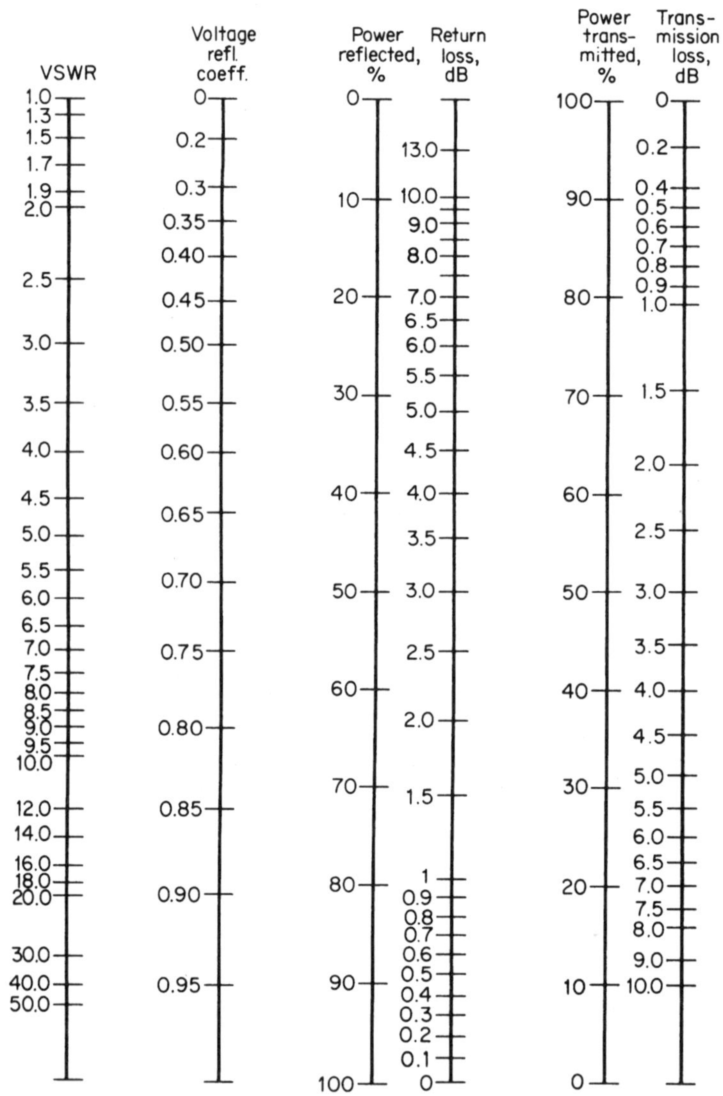


**Figure 17.22** Nomograph for transmission and deflection of power at high-voltage standing-wave ratios (VSWR). (*From: Feinstein, J.: "Passive Microwave Components,"* Electronic Engineers' Handbook, *D. Fink and D. Christiansen (eds.), Handbook, McGraw-Hill, New York, N.Y., 1982.*)

metallic structural members and directly through a hole in the house wall with no need for spacers, which are required with twin lead.

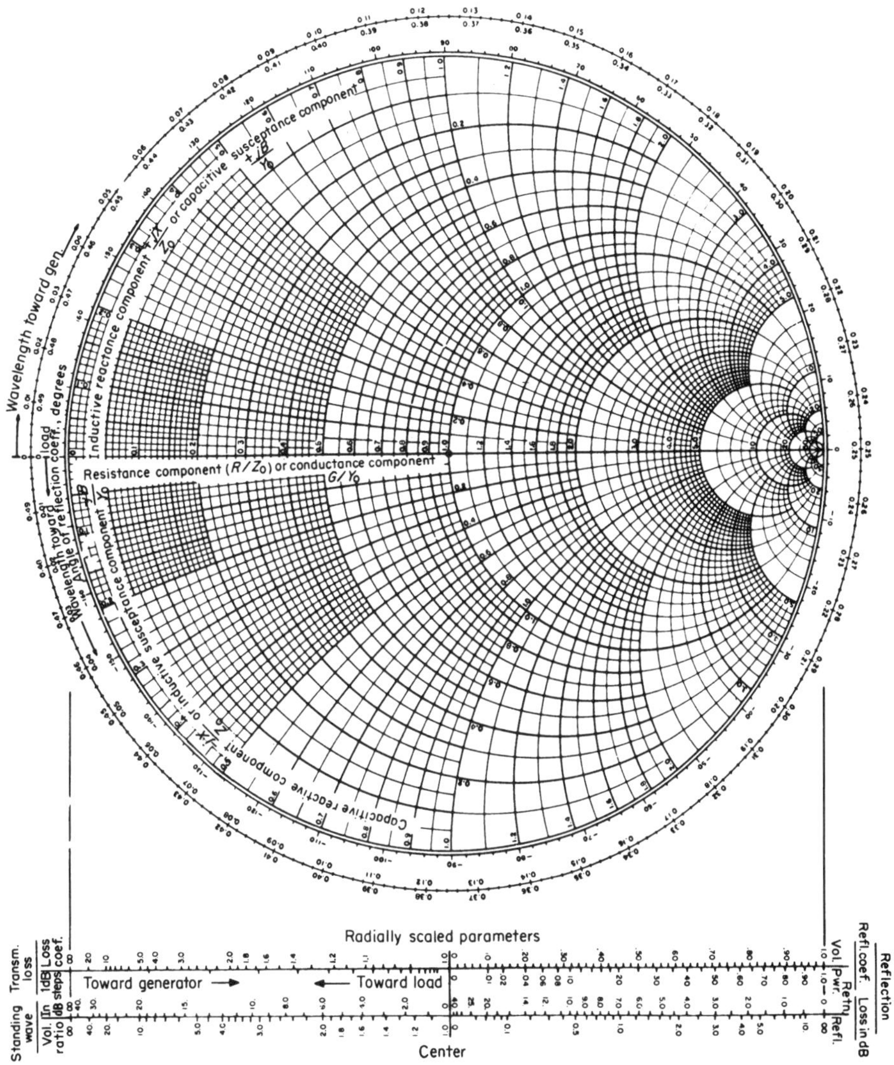


**Figure 17.23** The Smith Chart.

## 17.6 Active Antenna Systems

An active antenna system consists of a preamplifier integrated with an antenna structure. When a full-sized antenna is used, this combination will provide a signal level that is increased by the gain factor of the amplifier over a similar passive antenna. The use of a preamplifier with a much smaller antenna structure ($L < \lambda/2$) can also provide satisfactory reception to sites at which the space for a full-sized VHF antenna is not available.

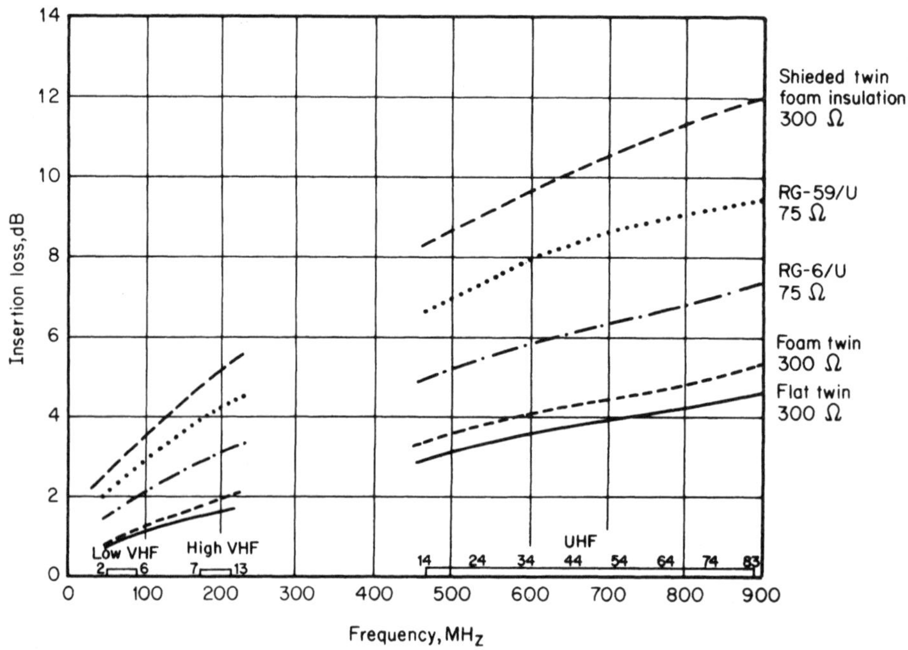


**Figure 17.24** Insertion loss for common transmission lines.

Five major parameters determine the system performance.

- Input and output impedance (300 or 75 Ω, or a combination of both) that must be selected to match the antenna and transmission line characteristics.
- Gain, defined as power output to the transmission line divided by power available from the antenna source.
- Noise figure.
- VSWR caused by mismatch between the source (antenna) and preamplifier input.
- Distortion level or maximum undesired signal level capability, frequently stated as *third-order intercept point* of the amplifier.

## 17.7 References


1. Elliot, R. S.: *Antenna Theory and Design*, Prentice-Hall, Englewood Cliffs, N.J., pg. 64, 1981.
2. Lo, Y. T.: "TV Receiving Antennas," in *Antenna Engineering Handbook*, H. Jasik (ed.), McGraw-Hill, New York, N.Y., pp. 24–25, 1961.

3. Grossner, N.: *Transformers for Electronic Circuits*, 2nd ed., McGraw-Hill, New York, pp. 344–358, 1983.

4. *Radio Amateur's Handbook*, American Radio Relay League, Newington, Conn., 1983.

5. Kraus, J. D.: *Antennas*, McGraw-Hill, New York, N. Y., Chapter 12, 1950.

6. Jasik, H.: *Antenna Engineering Handbook*, McGraw-Hill, New York, Chapter 24, 1961.

7. Beverage, H. H., and H. O. Peterson: "Diversity Receiving System of RCA Communications, Inc., for Radiotelegraphy," *Proc. IRE*, vol. 19, pg. 531, April 1931.

8. Peterson, O. H., H. H. Beverage, and J. B. Moore: "Diversity Telephone Receiving System of RCA Communications, Inc.," *Proc. IRE*, vol. 19, pg. 562, April 1931.

9. Mack, C. L.: "Diversity Reception in UHF Long-Range Communications," *Proc. IRE*, vol. 43, October 1955.

10. Brennan, D. G.: "Linear Diversity Combining Techniques," *Proc. IRE*, vol. 47, pg. 1075, June 1959.

11. Altman, F. J., and W. Sichak: "A Simplified Diversity Communication System," *IRE Trans.*, vol. CS-4, March 1956.

12. Kahn, L. R.: "Ratio Squarer," *Proc. IRE*, vol. 42, pg. 1704, November 1954.

Chapter

# 18

# Content Distribution

Chandy Nilakantan, Kishore Manghnani, Scott Johnson

Jerry C. Whitaker, Editor

## 18.1 Introduction

The Internet is not just ubiquitous, its use for conducting commerce is becoming pervasive. Great strides have been made in the areas of providing widespread connectivity, increased capacity for moving richer content and adequate security for maintaining privacy. One of the largest remaining challenges is scalability—particularly for distributing and delivering mixed media content to large numbers of locations and users in a predictable and scalable manner. The Internet has proven to be a great transaction network but falls woefully short when it comes to delivering services to even moderately sized communities or subscriber bases. On the other hand, present-day broadcast infrastructures, satellite, cable, and terrestrial networks possess built-in scalability and have the means to deliver broadband services to communities of any size, spread over any geography, very effectively.

## 18.2 Trends in Internet Content

The growth of data traffic over the Internet is driven by a number of factors, including:

- The increase in the number of consumers accessing Internet on a more frequent basis
- Increase in the number of businesses using the Internet not only to reach customers, but also to conduct business transactions with partners, vendors, and suppliers
- Proliferation of new wired and wireless Internet enabled devices
- Increase in the availability and use of broadband Internet access services
- Proliferation of increasingly complex multimedia-rich content
- Emergence of new Internet-based applications designed to simultaneously reach large audiences

Content on the Internet is evolving rapidly from simple, static Web pages to complex graphics and audio/video streams. Delivery of this new class of content requires significantly more bandwidth. In addition, while the Internet has historically been used for one-to-one exchanges between a user and a host server, new applications are emerging, such as online learning and live webcasts, that require rich content to reach millions of users simultaneously. These applications require *one-to-many exchanges,* and are placing great demands on the Internet infrastructure in terms of bandwidth, scalability, and predictability of content delivery.

## 18.2a Limitations of Internet Infrastructure

The current Internet infrastructure is not designed to efficiently handle the delivery of rich, multimedia content to a large number of simultaneous users. There are two fundamental reasons for this:

- The Internet is a *point-to-point* network. The Internet is built to handle movement of data from one point to another. When the same data needs to be sent to several locations, additional copies of the same data are sent, separately, on each path that connects the sender to the individual receivers. As the number of simultaneous receiver locations increases to millions, the Internet infrastructure hits a scalability wall.
- The Internet is a *best-efforts* network. The path between each user's personal computer and the computer that holds the requested content is a series of interconnected computer networks. Data moves through these networks in a hop-by-hop manner. Not all networks have the same capacity to carry data and not all devices that interconnect these networks have the same capacity to process the data at similar speeds. If any of the intermediate hops experiences congestion, as is the case when a large number of users try to access the same content, some of the data packets can be dropped arbitrarily or delayed. These errors introduce a degree of unpredictability and unreliability in content delivery. When the content includes audio and/or video streams, this loss or delay of data packets can severely impact the listening and viewing quality.

In recent years, large investments have been made to improve the physical infrastructure of the Internet and to increase the amount of available bandwidth. Advances in networking technologies such as gigabit and terabit-speed routing, optical networking, broadband access, Web caching, and content replication have delivered solutions that greatly increase the Internet's capacity to carry high bandwidth content from point-to-point. For example:

- *Last mile access*: Users enjoy rich content by using access technologies like DSL and cable modems, which have greatly advanced the carrying capacity of the "last mile."
- *High speed routers*: Advances in ASIC technology and high-performance software enable routers to handle complex network topologies and to process IP packets at gigabit and terabit speeds. ISPs can deploy these routers and provision faster links to the core.
- *Optical networking*: Wave division multiplexing (WDM) and *dense wave division multiplexing* (DWDM) have increased the available bandwidth in the core of the Internet by several orders of magnitude.

In spite of increased last mile bandwidth—faster routers moving packets more efficiently and quickly through the network and a high capacity core—there continues to be a lack of predictability of response time for transactions on the Web. When the same content is being requested by a large number of users, the congestion point is no longer the network—it is the data center that serves the content. New strategies are now employed to speed up the overall response time of the Internet, by moving the content closer to the edge of the network and therefore requiring fewer hops from the user.

- *Web caching*. Web caching can be used to pre-fetch commonly requested content and have it stored in locations that are closer to the user, thus eliminating the need for every transaction to be serviced by the data centers. Web caching can be used to effectively decongest data centers, but it only works when the content being served is *static*. In other words, if the content changes frequently, then the caches have to be replenished frequently as well, and this can result in stale caches and unpredictable delays.
- *Content replication*. Content replication is an effective method for improving response time to transactions, particularly when one can predict which content will be in demand and at what locations. Content that changes frequently, such as real-time stock updates, news highlights, and live Web events can be replicated to many servers at locations close to users. Content serving occurs locally, and therefore delivery is more predictable.

## 18.2b Emergence of Content Distribution Networks

*Content distribution networks* (CDNs) aim to help Web site owners deliver their content to consumers and businesses with greater performance and higher reliability. The underlying method used by content distribution networks is replication; by making copies of the content available closer to where the users are, the overall Web access response time is reduced greatly. There are two predominant modes of operation for CDNs:

- Data center based operation, where the content distribution network provider manages the service out of private facilities or co-located facilities. Typically, in this mode, content providers or website owners are the direct beneficiaries of the service.
- ISP *point-of-presence* based operation, where the content distribution network provider's service is oriented towards improving the performance of the Web cache located at the ISP. In this mode, the ISP is the direct beneficiary of the service.

The network infrastructures used by these service providers to move the content to the edge locations vary greatly, from shared to dedicated, and from point-to-point to point-to-multipoint.

All of the content distribution networks in operation today are designed for specific types of content, and specific types of services that are delivered based on that content. For example, some of these networks are intended for streaming media services only and therefore transport streaming audio and video. Others are for speeding up Web response and transporting Web objects. Still others are for delivering broadcast television and radio services and transporting digital audio and video programs. The type of services associated with the content they carry dictates the underlying network architectures employed by these networks.

But, clearly, content types are converging rapidly. Television broadcast programs are showing up on Web sites and Web-based content is being integrated into TV programs. But the networks that transport this converged content are ill prepared for the quality of service required to meet the demands of the applications that employ this content.

### 18.2c Broadcast Internet

Unlike the Internet, broadcast networks have been optimized for the transmission of rich content to large numbers of users in a predictable, reliable, and scalable manner. The advantages they bring to the infrastructure are many:

- Broadcast networks are designed to carry rich, multimedia content. Traditional broadcast networks—including television, cable, and satellite networks—have been explicitly designed to deliver high-quality, synchronized audio and video content to a large population of listeners and viewers. In addition, over the last five years many of these networks have migrated from analog to digital transmission systems, thus greatly enhancing their ability to carry new types of digital content, including Internet content.
- Broadcast networks are inherently scalable. By virtue of their point-to-multipoint transmission capability, broadcast networks are inherently scalable. It takes no more resources, bandwidth or other provisions, to send content to a million locations as it does to one, as long all of the receiving locations are within the transmission footprint of the broadcast network. In contrast, with the traditional Internet, each location that is targeted to receive the content will add to the overall resources required to complete the transmission.
- Broadcast networks offer predictable performance. Again, by virtue of the point-to-multipoint nature of transmission on broadcast networks, there are no variances in the propagation delay of data throughout the network, regardless of where a receiver is located. This inherent capability assures a uniform experience to all users within the broadcast network.

By enabling broadcast networks to connect with the Internet, the *Broadcast Internet* offers a cost-effective, reliable, and seamless path for delivering multimedia rich content to large numbers of users and service providers simultaneously. Pan-continent broadcast networks, such as digital satellite systems, can be used to distribute content to a very large, and potentially highly dispersed, set of locations. Other types of broadcast networks with smaller footprints, such as digital TV networks, can be used for local content distribution. The content is received and cached locally at these locations, which will—in turn—serve the content to users that connect either directly to that location or through a service provider that uses the location for content serving. The content is served to the users using the traditional mechanisms employed on the Internet. For example, instead of the ISPs replenishing their local caches on an on-demand basis, they can adopt the strategy of pre-filling their caches based on analysis of the type of content most frequently requested by the communities they serve. Content that is dynamic in nature, such as audio/video streaming and stock tickers, will be updated on the local caches at the appropriate frequency, keeping the cached content current and providing better overall quality of service to the users. Figure 18.1 illustrates the application of this principle to a cable-based delivery system. Figure 18.2 expands the concept to a variety of distribution media.

The major design considerations for building Broadcast Internet-ready broadcast facilities are:

- Multi-vendor interoperability. Any equipment that is added to the head-end to incorporate Internet data must work seamlessly with all existing equipment and must fit within the operational framework used by the facility. This requires any new equipment to support standard interfaces and protocols, and follow the guidelines set by standards organizations that cover both the digital video and IP domains.

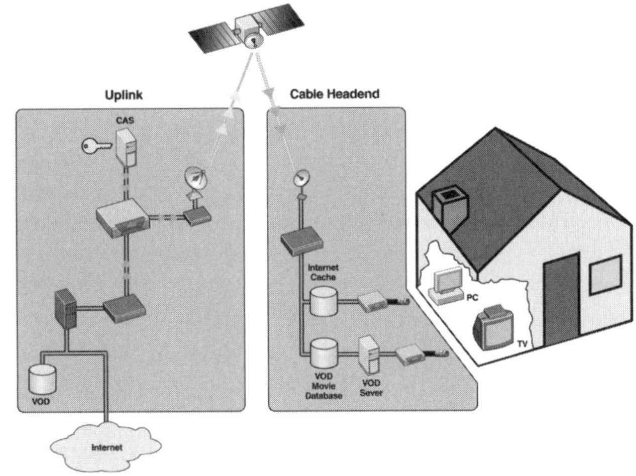


**Figure 18.1** A high-bandwidth data distribution system incorporating cable TV distribution to end-users.

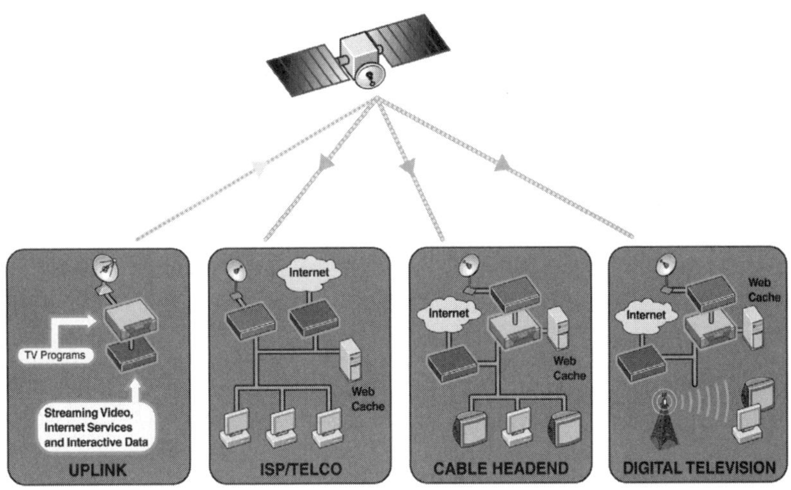


**Figure 18.2** High-speed data delivery using a variety of last-mile systems.

- Bandwidth optimization. Managing the available bandwidth in optimal fashion is a very important design consideration. There is considerable variability in the use of bandwidth by compression and multiplexing equipment, caused either by the inherent inefficiencies of the equipment, the nature of the content, or the service mix. The head-end architecture must provide the ability to capitalize on this variability to maximize the use of the bandwidth in an opportunistic manner. There are two strategies that are commonly applied for allocating bandwidth to data services in an environment that includes legacy video services as well: 1) *pre-allocated bandwidth*, where the data services are treated just like another "channel" or "bundle of channels" and the bandwidth is reserved for the exclusive use of these services; 2) *opportunistic bandwidth*, where the data services are treated as "trickle services", where there are no specific time of delivery requirements for the service. In this case, bandwidth can be optimized by not pre-allocating the bandwidth for the data services and taking advantage of the variability in the bandwidth use by the legacy video programs to "fill the blanks" with data.

- Flexibility. The nature and type of Internet services that will be required to be supported in any head-end is likely to change rapidly, as is evidenced by the rapid rate of technological and application advancement in the Internet in general. The head-end architecture must easily adapt to new applications requirements without having to undergo major design changes.

Currently, applications that incorporate data can be classified as following:

- Pure data services. These services use Internet or private data as the sole content. There are no legacy, television-oriented digital audio/video elements in the content. This type of content is typically targeted towards personal computers and integrated set-top boxes that have an operating environment similar to personal computers and are enabled for interactive services on the television. The content preparation, packaging, scheduling, and delivery systems used for these services can be independent of those used for the legacy video services.
- Loosely co-related data services. These services use Internet or private data to enhance or extend a legacy, television-oriented program, or to provide a parallel co-related content stream that is independent of the video stream, but scheduled for play-out simultaneously. The parallel content stream can be viewed in real time or stored for later viewing. The content preparation, packaging, scheduling, and delivery systems used for these services must have the ability to interface and interact with those used for the legacy video services.
- Enhanced TV. These services use Internet or private data to enhance a legacy, television-oriented program in a highly synchronized manner. The synchronization can be either presentation time-based or frame content-based. The content preparation, packaging, scheduling, and delivery systems for these services must be fully integrated into those used for legacy video services.

## 18.2d DTV Content Distribution Models

There are three different models that can be employed for content distribution over DTV:

- **End-to-end DTV data broadcasting**. In this model, the service provider creates services by aggregating content from both the TV domain and the Internet domain, distributes these services over a satellite or land-based network to television stations in the targeted regions. The television stations—either in real time or in a *store-and-forward* manner—then deliver the services. The customers for such services would be consumers.

- **Data distribution to Internet service providers.** This model is suitable for local distribution of content, particularly multicast services and streaming media, to Internet service providers that then deliver that content to consumers over traditional land-based last-mile connections. This model can be very effective for simulcast services—programs that are viewable on the TV as well as on a PC, with enhancements specific to the PC.
- **DTV as last mile**. In this model, the DTV spectrum is used as a last mile in a content distribution network. Content providers lease or acquire on demand last mile bandwidth for delivering Internet based services.

## 18.3 PC-Based DTV Receivers

The personal computer (PC) can initially play a major role in stimulating the growth of digital television. A huge installed base (over 350 million units), open standards, and low cost makes the PC an ideal platform for deploying DTV in consumer homes during the initial ramp- up phase. The PC monitor, with its million-pixel resolution, is already capable of displaying high-definition TV (HDTV) images. Additionally, the deployment cost is significantly lower on a PC platform. An HDTV PC add-in card can cost considerably less than an HDTV set, making it attractive to consumers.

Many consumers are currently using analog TV cards to watch TV on a PC screen. This represents a sizable market segment that has already combined the TV and PC experiences. The market for PC analog TV cards has grown steadily over the last few years with shipments expected to reach 4 million units annually as this book went to press. With its rich viewing experience and enhanced interactive data services, a DTVPC card is a more compelling solution than current analog TV tuner cards for PCs. The DTV experience is significantly enhanced compared with analog TV, as "cinematic" entertainment is combined with the real-time interactivity of the PC. A potential analog TV card buyer would be inclined to spend a little more for a combination card that supports both analog and DTV broadcasts in order to avoid the risk of quick obsolescence. Also, the consumer can sample the HDTV experience on a PC monitor before committing to a sizeable investment in an HDTV set for the family room. Broadcasters will be encouraged to create more DTV content if the installed base of DTVPC cards grows, which will help the DTV industry as a whole.

### 18.3a Features Required in a DTVPC Solution

There are specific requirements a tuner card must have in order to be suitable for DTV/PC convergence, including:

- Dual-input tuners capable of supporting multiple DTV formats must be used.
- Because the bulk of TV programming will be analog for the next few years, a DTV card must be able to support existing analog TV broadcasts.
- The DTV card must support multi-sync monitors and common PC display resolutions such as $1024 \times 768$, $800 \times 600$, and $640 \times 480$. In the U.S., it must be able to decode and output all 18 ATSC formats, so it can display any DTV broadcast, regardless of the transmission format of any particular station (Table 18.1).

**Table 18.1 The Basic 18 ATSC Display Formats**

| Vertical lines | Pixels | Aspect Ratio | Picture Rate |
|---|---|---|---|
| 1080 | 1920 | 16:9 | 60I, 30P, 24P |
| 720 | 1280 | 16:9 | 60I, 30P, 24P |
| 480 | 704 | 16:9 and 4:3 | 60P, 60I, 30P, 24P |
| 480 | 640 | 16:9 | 60P,60I, 30P, 24P |

### Support of PC Screen Formats

Unlike the existing NTSC analog standard that only supports one mode of operation, the ATSC digital broadcast standard defines multiple modes. These features allow the broadcaster to specify which mode they wish to use based on their own criteria. A broadcaster may wish to send a standard definition picture to the viewer to limit the bandwidth required to send the picture or they may wish to send a full high-definition picture for optimum quality. This means the viewer must have a television capable of displaying these multiple formats.

One of the more convenient aspects of the DTVPC solution is that the PC monitor is ideally suited for display of HDTV. The PC monitor is capable of responding to a wide range of screen formats and resolution, including the various screen formats and resolutions required to support HDTV. This means that most consumers will not have to by a new monitor to display HDTV.

For example, consider the HDTV screen display formats. A 720P DTV stream has a screen format of 1280 × 720 @ 60Hz. Most PC monitors easily work within these parameters for the screen display. The same is true for the other screen formats specified by the ATSC DTV standard. Most PC monitors support formats such as 720P, 480P and 480I, and many are capable of 1080I format.

## 18.3b Basic Feature Set

All of the features that we are currently seeing in set top box implementations are possible in DTV-PC solution. For example, the ability to pause a live broadcast while the viewer is otherwise occupied is implemented with a *time shift* function. In time shift recording, the system needs a fast access large storage medium to save the program for later display. Current set top boxes may include a hard drive in the system for this purpose. This is easily implemented in the DTVPC system because of the existing presence of the system hard drive.

Some of the primary functions and their benefits are:

- **Multicast.** Some broadcasters are already planning to multicast two or more choices of programming. This is particularly useful for sporting events where multiple games are carried by the same broadcaster. Instead of the broadcaster selecting the game that will be shown in a particular region of the country, the viewer will have the opportunity to select the game they wish to see, or to see all of them at the same time.
- **Record/playback**. These functions are identical to the like-named functions of VCRs or tape recorders where the viewer has the opportunity to save a broadcast transmission for viewing at a later time. Instead of using a tape to store the audio and video content, the DTVPC will use the hard drive of the PC to store the recorded information.

- **Pause/time shift**. This function is similar to the pause function of a VCR or recorder, except in this case the viewer has the opportunity to pause a live broadcast and then resume the playback while the live broadcast is still underway. The DTVPC solution is capable of this type of functionality because of its high-speed peripherals, in this case the internal hard drive. Most hard drives can handle data fast enough to be able to record and playback simultaneously.
- **Data broadcasting**. Technically a DTV broadcast is a simple continuous stream of data. Imagine a very fast network connection sending pictures, sounds, multimedia games, and illustrated articles—all related to the television program the viewer is watching. The consumer can still passively watch TV, but can also customize the experience and make it their own.
- **Interactive TV**. Digital television is heading toward a convergence with computers—and students, sports fans, news junkies, and anyone with an interest in anything will get more out of television. The market potential for interactive data services on TV is huge, with applications ranging from on-line shopping and video-on-demand to viewer participation in real-time sporting events. Real-time interactivity allows the viewer to purchase music CDs during an MTV broadcast or play along with the contestants in a game show.

DTVPC cards represent an ideal element in the interactive TV market. This is due to the fact that most PCs already have the most important function required to implement interactive TV—a mechanism for the viewer to provide feedback. The Internet connection found on most of today's PCs is the medium by which viewers will communicate back to the service provider with information ranging from response to game show questions to political polling data.

## 18.3c Practical Implementation Considerations

There are three basic ways to bring the DTV experience to the PC platform:

- Software DTV decoding
- Hardware assisted software DTV decoding
- Hardware DTV decoding

Each of these approaches will be examined in the following sections.

### Software DTV Decoding

The software decoding approach forces all of the processing work on the system CPU. There are three components in the software-based solution: a receiver card, a high-powered CPU, and the graphics chip. The receiver card, consisting of a DTV tuner and a vestigial sideband (VSB) demodulator, receives the DTV broadcast signal and passes it to the CPU for further processing. The CPU demultiplexes the transport signal and decodes the DTV data using software algorithms.

The software approach strains the CPU tremendously and takes up most of the bandwidth for 1080I resolution. A Pentium III 750 MHz CPU combined with a high-powered graphics chip can barely decode a 480P DTV stream. The current crop of high-end processors is necessary to be capable of decoding true HDTV formats such as 720P and 1080I. Software decoding of true HDTV formats demands CPUs with speeds as high as 1 GHz for software-only decoding. Table 18.2 maps the decoding workload.

**Table 18.2 Software-Only DTV Decoding Workload**

| | |
|---|---|
| **DTV Hardware** | Over-the-air reception<br>VSB demodulation |
| **System CPU** | ATSC demultiplex<br>Full MPEG-2 decoding<br>AC3 decoding<br>Applications overhead |
| **Graphics Hardware** | Screen display |

This solution has the advantage of working with any graphics controller, however the software approach will not work with the huge installed base of lower-speed PCs.

## Hardware Assisted Software DTV Decoding

This approach relies on a combination of CPU power and motion-compensation hardware found in graphics chips. There are three components in the hardware assisted software-based solution: a receiver card, a high-powered CPU, and the graphics chip. The receiver card, consisting of a DTV tuner and 8-VSB demodulator, receives the DTV broadcast signal and passes it to the CPU for further processing. The CPU demultiplexes the transport signal and decodes the DTV data using software algorithms. The graphics chip assists the CPU with its on-chip motion compensation and inverse discrete cosine transformation (IDCT) logic.

The approach also strains the CPU and takes up a significant amount of the CPU bandwidth. A Pentium III 550 MHz CPU combined with a high-powered graphics chip can decode a 1080I DTV stream. The current crop of high-end processors is necessary to be capable of decoding true HDTV formats such as 720P and 1080I.

Like the software-only approach, this approach will not work with the huge installed base of lower-speed PCs. The installed base of PCs with speeds less than 400 MHz is well over 200 million. These PCs do not have the bandwidth to decode even the 480P transport stream. Dedicated hardware DTV decoding is the only solution for these PCs. (See Table 18.3.)

## Hardware DTV Decoding

In this implementation, a dedicated hardware decoder is used for enabling DTV on a PC. A typical system is illustrated in Figure 18.3. The DTV decoder does not rely on the CPU or the graphics chip for decoding the DTV stream. There is very little drain on CPU bandwidth as the add-in card decodes the complete MP@HL MPEG stream. All 18 ATSC DTV formats can be supported at full frame rates. A hardware-based implementation will work with most PCs—even a 166 MHz Pentium PC. This makes hardware DTV decoding an ideal solution for the large installed base of PCs. The workload division for this approach is given in Table 18.4.

A DTVPC solution must meet the following criteria for it to be widely accepted among consumers.

- Single-board solution for true HDTV and legacy analog TV reception
- Simple and easy installation
- Consumer price points for mass-market acceptance

**Table 18.3 Hardware Assisted DTV Decoding Workload**

| | |
|---|---|
| **DTV Hardware** | Over-the-air reception<br>VSB demodulation |
| **System CPU** | ATSC demultiplex<br>Partial MPEG-2 decoding<br>AC3 decoding<br>Applications overhead |
| **Graphics Hardware** | Screen display<br>IDCT<br>Motion compensation |

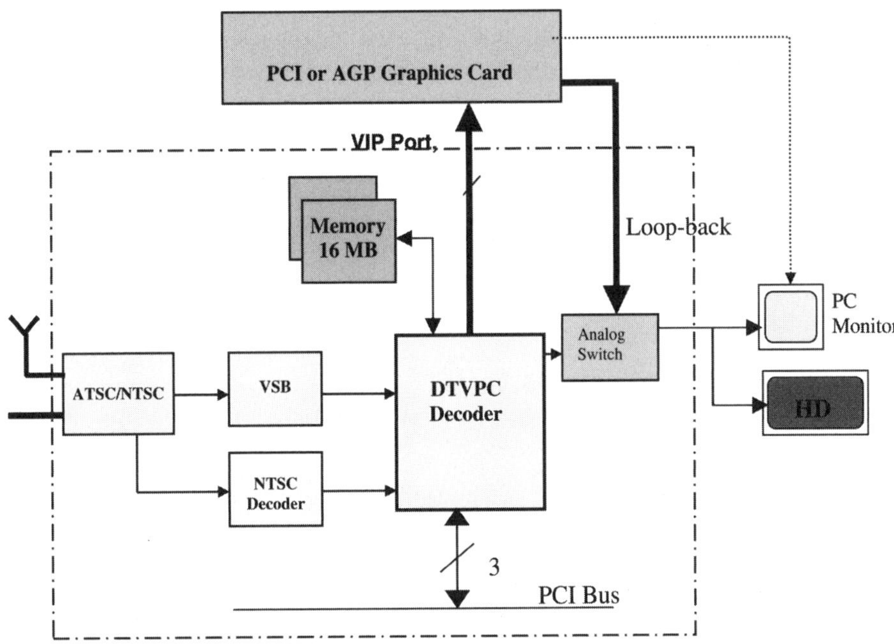


**Figure 18.3** Single-board hardware-based DTV decoder.

- Support of PC screen formats

The DTV card must also support multi-sync monitors and common PC display resolutions such as 1024 × 768, 800 × 600, and 640 × 480. If the board can support DVD playback in addition to ATSC and NTSC, the value proposition for the consumer is significantly enhanced.

The PCI DTVPC tuner card shown in Figure 18.3 is implemented in a single-chip decoder solution. The card has 16 MB of SDRAM memory to support true HDTV display formats (up to 720P and 1080I). An analog loop-back cable connects the DTV card to the graphics subsystem.

**Table 18.4 Hardware DTV Decoding Workload**

| | |
|---|---|
| **DTV Hardware** | Over-the-air reception<br>VSB demodulation<br>ATSC demultiplex<br>Full MPEG-2 decoding<br>Screen display |
| **System CPU** | AC3 decoding<br>Applications overhead |

The analog cable brings the RGB video data and horizontal and vertical sync from the graphics card to the tuner card. The analog multiplexer selects between two different sources: the DTVPC decoder and the graphics card. The DTVPC decoder can be programmed to control the analog switch. Regardless of the incoming video format, this implementation allows full-screen HDTV broadcast to be displayed on the PC monitor when the decoder video output is selected.

This implementation also allows a video window to be displayed on the graphics desktop. The decoder can simultaneously output CCIR 601 compatible digital video with the high-definition analog video. The CCIR 601 video data may be pumped into the video capture port of the graphics card via the *video interface port* (VIP) of the PCI bus. The graphics card can overlay the incoming video data in a video window on top of the graphics. The viewer can select either full-screen HDTV output or can view the DTV programming in a video window while working on a PC application.

## 18.4 Interactive Videoconferencing

With desktop computers nearly as ubiquitous in business these days as telephones, the time has arrived for the next big push in telecommunications—interactive desktop videoconferencing. Interaction via video has been used successfully for many years to permit groups of persons to communicate from widely-distant locations. Such efforts have usually required some degree of advance planning and specialized equipment ranging from custom-built fiber or coax services to satellite links. These types of applications will certainly continue to grow, as the need to communicate on matters of business expands. The real explosion in videoconferencing, however, will come when three criteria are met:

- Little—if any—advance planning is needed
- No special communications links need be installed to participate in a conference
- Participants can do it from their offices

The real promise of videoconferencing is to make it as convenient and accessible as a telephone call. While that day is not here just yet, it will come.

### 18.4a Infrastructure Issues

Clearly, interactive desktop videoconferencing is the next big thing for business. The challenge of videoconferencing is not the video itself. Required equipment includes one or more cameras, microphones, displays, and the equipment to control them. All of this is important in any type of video conference, and strides are being made in this area all the time. Automatic control of camera movements and intelligent audio switching are just two of the innovative technologies that have been developed to a high level of sophistication. The primary remaining challenge is getting the signal to and from the video equipment on each end. To accomplish this, we need to look toward computer networks and common carriers.

## 18.4b Full Motion Video

The ideal for any videoconferencing system is the highest possible video quality, in terms of both frame rate and image quality. Some market analysts estimate that as much as 70% of the videoconferencing market falls into this category, partly as a result of the nature of the application, but also because of the fact that most people have been inadvertently trained to expect television quality when they use videoconferencing equipment. This is due—of course—to the prevalence of TV in our society.

Despite the obvious focus on video, audio quality is usually the most important element of a videoconferencing system because it is the audio that conveys most of the information. The whole idea of a videoconference is to facilitate communication among participants. Research has shown that if the audio is clear and audible, the next area on which users focus is the video quality.

A popular myth in the videoconferencing industry is that all systems that conform to established videoconferencing standards (H.320, H.323, and H.324) offer pretty much the same video quality. There may be, however, a significant difference in the video quality of these services as offered by different vendors. Sometimes this difference is a function of the cost of the system, and sometimes it is price-independent—meaning that two vendors could be charging the same price for their product but have very different video quality.

Image quality for a videoconferencing system is largely determined by the quality of the codec being used and the bandwidth allocated to the session. Moreover, much of the codec quality is determined not simply by how much hardware is being used and how fast it is, but by the sophistication of the algorithms that run on the hardware.

## 18.4c Applicable Standards

Standards for hardware and software are the keys to desktop videoconferencing. The whole idea of making videoconferencing convenient depends on systems from different vendors working together. It would—of course—make no sense to buy a telephone on which you could call only certain phones. Although this is certainly an over-simplification of the interface difficulties involved, the point is that the success of videoconferencing will be driven not so much by the attributes of one device over another, but more by which systems will work together.

It is fair to point out, however, that just because there is an industry standard, it does not mean that there will be complete compatibility; problems do arise in the field. Still, standardization must be the driving force behind new product development.

The videoconferencing standards developed by the ITU include the following:

- **H.320**: This was the initial videoconferencing standard issued in 1992. It deals only with ISDN-based systems.

- **H.323**: A LAN-based videoconferencing specification. Because of the great progress made recently in moving video over LANs, H.323 is seen as an important component in video-based communications within corporations. Proponents have predicted the replacement of text-based e-mail with video-based e-mail in the not too distant future.
- **H.324**: A standard designed for use with conventional twisted-pair analog lines using 28.8 kbits/s (or faster) modems and sophisticated compression techniques. H.324 was released in 1996 with the hope that it would ensure that all videoconferencing systems, whether standalone or computer-based, would be able to talk to each other.

H.324 merits some additional discussion. The standard specifies a common method for video, voice, and data to be shared simultaneously over a single analog line. Audio data is compressed to 6 kbits/s with the remaining 22 kbits/s (for a 28.8 kbits/s modem) allocated to the video signal. In recognition of the importance of the audio link in a videoconference, if bandwidth problems arise during a session, the audio is protected and the video is allowed to degrade. Because most link problems are transient in nature, under this scenario, the picture would begin to break up or otherwise deteriorate, but normal conversations would continue. After the full bandwidth of the channel was restored, the video would return to normal.

Within each of the videoconferencing standards are a number of related ITU standards that specify key operating parameters. For example, under the umbrella of H.324 are the following:

- H.263, which relates to video coding
- H.245, a protocol package
- H.223, which covers the multiplexing video, audio, and user data
- T.120, which addresses user data, whiteboard, file transfer, and other functions

This so-called standards-within-a-standard approach facilitates rapid product development and ease of interconnection among different systems and applications.

### 18.4d Desktop Systems—Hardware and Software

Implementation of a videoconferencing system on the desktop is a party to which both hardware and software must be invited. There are obvious hardware peripherals that must be present for the system to work, such as a camera, microphone, speaker, and network connection (or modem). The key element of the system is the codec, which can be implemented either in hardware or software. Not surprisingly, the hardware approach is typically faster but more expensive. Software-only codecs, however, have come a long way, and the continuing speed improvements in personal computer CPUs capitalize upon these improvements.

The attractiveness of a software codec is that it can run on any machine, and there is no add-on hardware to purchase. Software codecs also can be updated easily by changing the software driver. Naturally, the performance of the software codec is a function of the type and speed of the processor.

The software aspect of videoconferencing also includes considerations for general office application programs. Interface capabilities with such common desktop apps as Microsoft PowerPoint, Excel, and NetMeeting offer a host of user benefits. For example, a PowerPoint slideshow can be viewed during the videoconference and changes made in real-time. NetMeeting and similar programs that comply with the T.120 standard for data collaboration allow interoperability among distant locations and different vendors.

Control and coordination of a videoconference has been made much easier thanks to new, smart peripheral and supervisory systems. Items such as touch-screens, auto-tracking cameras, and participant responses systems permit users to focus on communicating ideas rather than on producing the videoconference. For example, students may electronically "raise their hand" by pressing a button on a microphone. This notifies the instructor that someone has a question and directs the camera to zoom-in on the student. Through such face-to-face interaction, students and instructors can build relationships that make distance learning more responsive and effective.

### Small Group-vs.-Large Group Systems

The nature of a desktop videoconferencing system is that it is usually restricted to one or perhaps two individuals at each terminal. For larger groups, roll-around portable systems and theater-type installations are options. Until recently, the theater-type or dedicated videoconferencing center defined what a video conference was all about: bringing together large groups of people in distant locations. While these applications still exist in large numbers, the primary growth of videoconferencing involves smaller groups.

Portable, roll-around systems offer the benefit of easy setup for small conference rooms and group workspaces. Such packaged systems fill the growing need for group collaboration among distant points.

## 18.4e System Implementations

Although any given videoconferencing session must be designed to meet the particular needs of the subject matter, there are three basic topologies that can be applied to the application:

- *Point-to-point*, the simplest arrangement where two individuals are interconnected in real-time (Figure 18.3*a*).
- *Broadcast*, with a single origination point and multiple receiving points (Figure 18.3*b*)
- *Multicast*, where all (or at least some) of the participating individuals can communicate with each other and/or with the group as a whole (Figure 18.3*c*).

## 18.4f Making Connections

Advances in codec technology have largely eliminated the need for dedicated leased telephone lines. POTS (*plain old telephone service*) twisted-pair lines can deliver video frame rates of 6 to 15 frames per second (f/s), which is sufficient for applications where there is little movement by participants. This situation, of course, essentially describes desktop videoconferencing. With an ISDN line and appropriate hardware, frame rates of 20 to 30 f/s can be realized.

The wide variety of interconnection systems used in business today has led videoconference system designers to offer a range of connection options. Some of the more common options include LAN and WAN (wide-area network) based on one or more of the following:

- POTS
- ISDN

(*a*)

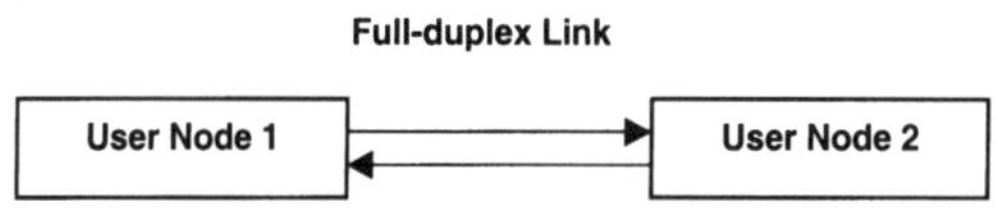


(*b*)

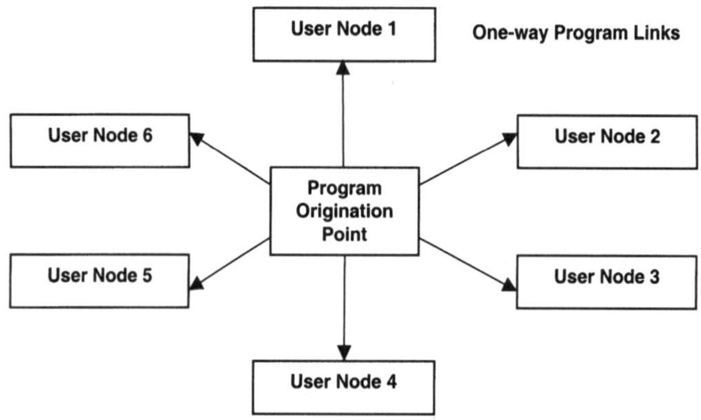


(*c*)

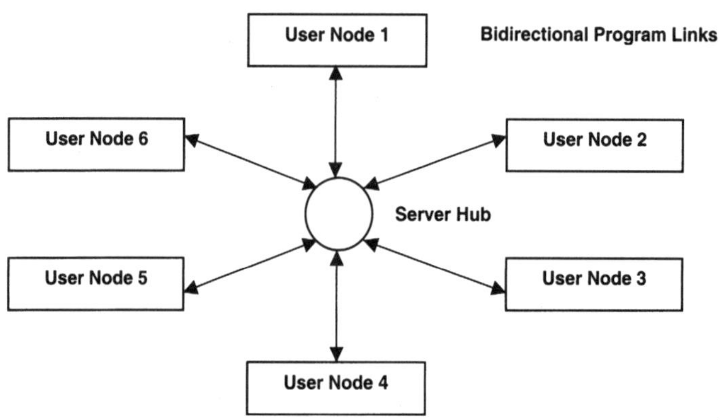


**Figure 18.3** Primary videoconferencing modes: (*a*) point-to-point, (*b*) broadcast, (*c*) multicast.

- Digital subscriber link (DSL)
- Ethernet
- Token Ring
- ATM (asynchronous transfer mode)
- Internet

Not surprisingly, interconnecting between divergent systems can be problematic. Conversion devices and systems are available, however, to facilitate the transfer of data from one system to another. The widespread use of Internet Protocol (IP) has made this technology the prime contender for desktop videoconferencing, and some implementations of conventional, theater-style, conferencing, as well.

Of the modem-function connection options available today, ISDN offers perhaps the best package when measured by the yardsticks of price, performance, availability, and flexibility. ISDN utilizes existing twisted-pair phone lines. ISDN is purely digital; it is not translated into audio-frequency tones as used by analog modems. This makes ISDN more reliable than POTS, allows much higher bandwidth, and greatly reduces the susceptibility to noise. Another digital telco service, DSL, has emerged as a powerful tool for videoconferencing. DSL and ISDN share a number of features, although DSL offers a considerable improvement in bandwidth.

End-user equipment includes a *network terminator* (NT) and any number of *terminal adapters*. The NT acts as the bridge between the phone company network and the end user. The terminal adapter is the final piece of equipment used for the application, such as an ISDN phone, fax machine, or ISDN modem. A terminal adapter refers to any device that is used to generate traffic on a line.

An ISDN connection to the Internet, or any other network for that matter, is a dial-up process. In contrast, a DSL system can be configured in an "always connected" mode. Optimized for use on the Internet, DSL also provides for dynamic IP address assignment.

Chapter

# 19

# Interactive Television

**Jerry C. Whitaker, Editor**

## 19.1 Introduction

The television industry is entering a new era in service to the consumer. Built around two-way interactive technologies, the rollout of the digital television (DTV) infrastructure opens up a new frontier in communication. Two worlds that were barely connected—television and the Internet—are now on the verge of combining into an entirely new service: namely, interactive television. Thanks to the ongoing transition of television from analog to digital, it is now possible to combine video, audio, and data within the same signal.

This combination leads to powerful new applications, with limitless possibilities of great commercial potential [1]. For example, computers can be turned into traditional TV receivers and digital set-top boxes can host applications such as interactive TV, e-commerce, and customized programming.

## 19.2 Interactive TV Implementation

Two terms are commonly used to describe the emerging advanced television environment: *interactive television* and *enhanced television*. For the purposes of this book, we will use the term interactive television (ITV) and define it as anything that lets a consumer engage in action with the system using a remote control or keyboard to access new and advanced services. General categories of use include the following:

- The ability to select movies for viewing at home
- E-mail and on-line chat
- News story selection and archive
- Stock market data, including personal investment portfolio performance in real-time
- Enhanced sports scores and statistics on a selective basis
- On-line real-time purchase everything from groceries to software without leaving home

There is no shortage of reasons why ITV is quickly gaining momentum—and will continue to do so as new technologies take hold. The backdrop for ITV growth comes from both the market strength of the Internet and the technology foundation that supports it. With the rapid adoption of digital video technology in the cable, satellite, and terrestrial broadcasting industries, the stage is now set for the creation of an ITV segment that meets the tests of sound economic principles and introduces to a mass consumer market a whole new range of possibilities.

For example, services will soon be available that offer interactive features for game shows, sports and other programs, interactive advertising, e-mail, and Internet access as a package deal. Rather than concentrating just on Web services, the goal is to deliver a better television experience. An important component of such services is the hard-drive-based video recorder. It is practical, with video compression, to store as much as 30 hours of television programs locally at the consumer's receiver. As the price/performance ratio of hard drives continues to move forward, even more storage will be practical. This capability opens the doors to numerous innovations and features never before possible.

Other services of the DTV era offer customized channel guides and make it easy to search for shows that subscribers want to watch, or to select future TV shows to record. Such services also make it possible for consumers to create their own customized channels filled with their favorite TV shows or custom channels that contain movies by favorite actors or directors.

Soon the entire video distribution system will handle digital media only [2]. These digital media offer content providers and consumers much higher picture quality. Unlike analog, this quality is ensured no matter how many copies are made. The digital content can be distributed in an MPEG-2 compressed format over high-bandwidth digital satellite, cable, terrestrial broadcast, Internet, or by fixed media. In the home it can be played via set-top boxes, other consumer electronic devices, or PCs, all connected by an IEEE 1394 high speed home network bus.

The potential benefits of the digital environment go beyond the picture quality. The data packets associated with digital content give the content provider access to more functions and business opportunities. For example, the video can be tagged with index pointers that allow the viewer to jump to a particular scene or to links that access supplementary information on the local disk or anywhere on the Internet. *Conditional access* methods allow the content provider to control viewing according to their business paradigms; pay-per-view can be enhanced by controlling who views and how much viewing they are entitled to, what price they pay depending on loyalty and other personal criteria, when and/or where they are, and so on.

Digital media of this type requires an investment in new infrastructures at the content provider, distributor, and consumer. This infrastructure will help content providers and distributors take advantage of the ongoing business opportunities created by the new digital media.

## 19.2a Copy Protection

With digital technology, every copy of a program is a perfect copy [2]. The digital era introduces new challenges to content producers and content providers. Their intellectual property needs to be protected so that their investments are not compromised. Also, in order for consumer electronics devices to be successful in the marketplace, they must be relatively low cost and easy to use. And yet, most consumers are not pirates and any scheme protecting the content must also take into account the perceived rights of the consumers and the ease with which they obtain, store, and continue to use the media they purchase.

## 19.2b Overview of Internet Protocols

The Internet architecture uses three core transport protocols [1]:

- *Internet Protocol* (IP)—a network protocol that defines, among other things, the structure of an IP *datagram*. A datagram has a payload and a header. The header contains a source and a destination address that has global significance. These addresses allow datagrams to be individually routed throughout the entire network on a "best-effort" basis. There is actually no guarantee at the IP layer that a datagram is correctly transmitted or that a sequence of datagrams will arrive at their destination in the same order that they were sent. Above the IP protocol, the User Datagram Protocol and Transport Control Protocol serve to provide end-to-end functionality between a pair of host systems.
- *User Datagram Protocol* (UDP)—a service above the IP layer that provides port multiplexing and a checksum that may or may not be used. The UDP does not require the addressee to acknowledge whether or not data transmitted actually arrived properly. As such, it is not a reliable protocol, but it is still to be preferred in the context of real time streaming applications, such as video and audio, when the ability to time packet transmission against an external clock source is of much greater importance than avoiding possible data losses.
- *Transport Control Protocol* (TCP)—a service that adds more functionnalities than UDP at the price of a longer header. Along with IP, TCP guarantees a reliable, serialized, and full duplex channel. A stream of bytes generated by the sender will be passed across the Internet by TCP so that it is presented to the addressee as the same sequence of bytes and in the same order as the one generated by the sender. Unlike UDP, TCP requires a return path to send acknowledgment messages.

TCP and IP are commonly paired for networking purposes, referred to as TCP-IP.

## 19.2c Data Broadcasting

Data broadcasting (*datacasting*) offers the potential for entirely new revenue sources [3]. With modern encoders, even a high-definition (HD) broadcast requires only 16 to 18 Mbits/s of digital bandwidth, out of the 19.38 Mbits/s available in a 6 MHz broadcast band under the ATSC DTV standard. This leaves a significant amount of bandwidth that can be used for arbitrary data. Figure 19.1 illustrates one possible scenario.

There are basically two ways in which broadcasters can use this available bandwidth to generate additional revenue:

- Use the excess bandwidth to broadcast data that enhances the appeal of their TV programming and/or TV advertising in an attempt to attract more advertising dollars.
- Lease the excess bandwidth to other enterprises who want to distribute data to large numbers of users in the broadcaster's viewing area.

In practice, both of these approaches will likely be used to varying degrees by different broadcasters.

For datacasting applications targeted to consumers, a key requirement for success is that large numbers of consumers have a DTV receiver that can receive and use the broadcast data. Utilizing broadcast data requires not only standards for encoding the data in the broadcast stream, but also standards for applications in the receiver to operate on the data. This could be in the form of

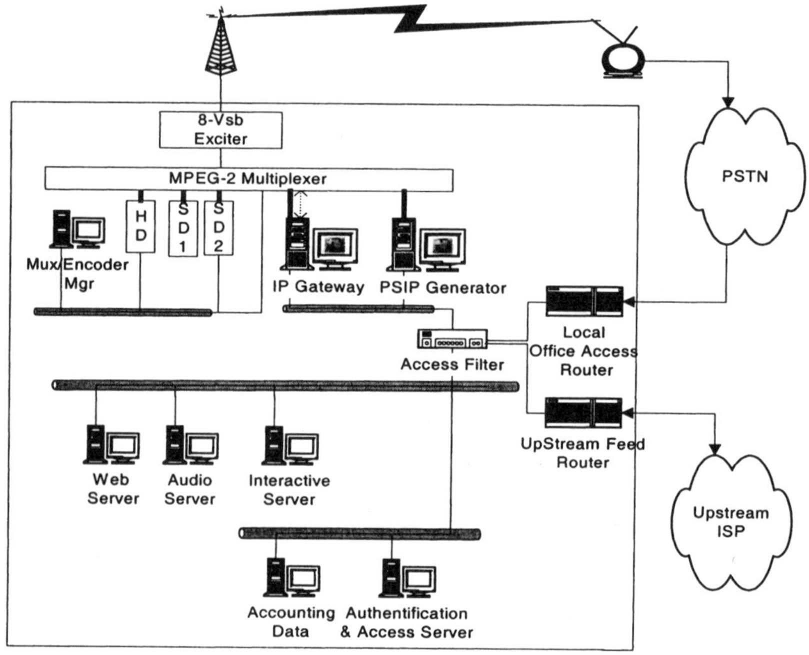


**Figure 19.1** A broadcasting and datacasting terrestrial station. (*After* [1].)

specifications for one or more standard applications in the receiver, such as a standard HTML engine. Alternatively, it could be in the form of specifications for a standard execution environment in the receiver, so that a variety of applications could be downloaded from the broadcast stream and executed.

The standardization requirement can be relaxed somewhat for applications targeted to consumers with PCs and add-in DTV cards because custom software to support such applications can be downloaded from a Web site. However, such applications are still only feasible when large numbers of consumers have DTV cards.

Datacast applications targeted to consumers can be further classified by the degree of *coupling* to the normal TV programming as follows:

- *Tightly coupled data* are intended to enhance the TV programming in real time. The viewer tunes to the TV program and receives the data enhancement along with it. In many cases the timing of the display of the broadcast data is closely synchronized with the video frames being shown.

- *Loosely coupled data* are related to the program, but are not closely synchronized with it in time. For example, an educational program might send along in the broadcast some supplementary reading materials or self-test quizzes. These might not even be viewed at the same time as the TV program. They may be saved in the DTV receiver and perused later.

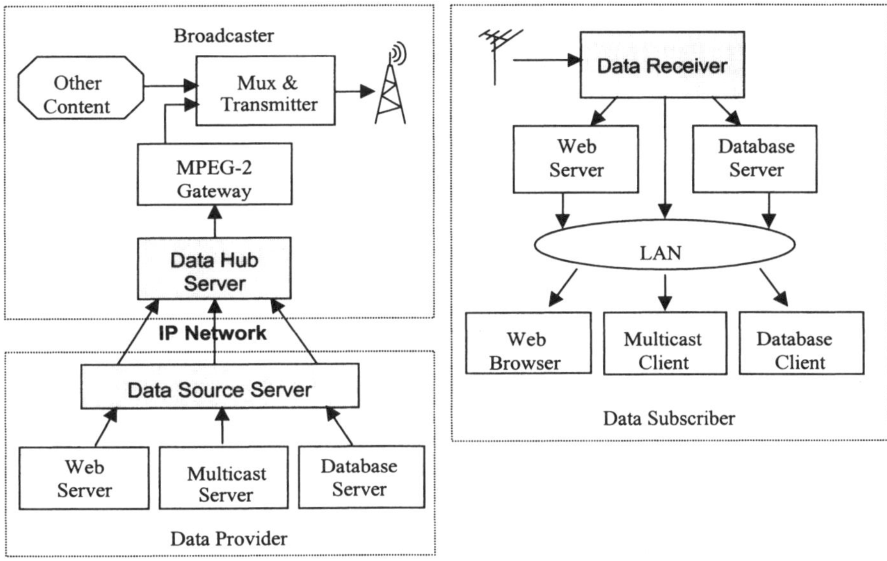


**Figure 19.2** Three-component architecture for datacasting. (*After* [3].)

- *Non-coupled data* are typically contained in separate "data-only" virtual channels. They may be data intended for real-time viewing, such as a 24-hour news headline or stock ticker service, or they may be data intended for use completely outside the DTV context.

## Data Broadcast Architecture

A datacasting system must meet a number of requirements in order to properly support enterprise-to-enterprise data broadcasting [3]. One key factor in the ability of the system to meet these diverse requirements is the underlying system architecture. In particular, it should explicitly recognize and support the roles of data provider, broadcaster, and data subscriber.

The example system shown in Figure 19.2 consists of three primary components:

- *Data source server*, which allows the data provider to specify the detailed scheduling for retrieval and broadcast of individual data items. This has advantages for both the data provider and the broadcaster. The data provider can change the scheduling at any time without involving the broadcaster in the process. The broadcaster does not have to devote any clerical resources to this task. The specifications can be transferred automatically from the data source server to the data hub server for use as needed.
- *Data hub server*, which receives data items from the data source servers and turns them over to the MPEG-2 gateway according to the schedules obtained from the data source servers. The MPEG-2 gateway simply encodes the data into MPEG-2 transport packets and sends them to a multiplexor for actual insertion into the broadcast stream.
- *Data receiver*, which extracts the data from the broadcast stream and applies decryption, decompression, and forward error recovery as needed. It provides the data subscriber with a

menu of the authorized items in the broadcast stream (generated from specifications provided by the data source server and transmitted in the broadcast stream), and allows the data subscriber to select which ones to actually extract and which ones to ignore.

## 19.3 Advanced Television Enhancement Forum

The Advanced Television Enhancement Forum (ATVEF) is a cross-industry group formed to specify a single public standard for delivering interactive television experiences that can be authored once—using a variety of tools—and deployed to a variety of television, set-top, and PC-based receivers [4]. Version 1.1 is a foundation specification, defining the fundamentals necessary to enable creation of HTML-enhanced television content so that it can be reliably broadcast across any network to any compliant receiver.

The ATVEF specification for enhanced television programming uses existing Internet technologies. It delivers enhanced TV programming over both analog and digital video systems using terrestrial, cable, satellite, and Internet networks. The specification can be used with both one-way broadcast and two-way video systems, and is designed to be compatible with all international standards for both analog and digital video systems.

The ATVEF specification consists of three principal parts:

- Content specifications to establish minimum requirements for receivers
- Delivery specifications for transport of enhanced TV content
- A set of specific bindings

A central design point of the ATVEF document was to use existing standards wherever possible and to minimize the creation of new specifications. The content creators in the group determined that existing Web standards, with only minimal extensions for television integration, provide a rich set of capabilities for building enhanced TV content in today's marketplace. The ATVEF specification references full existing specifications for HTML, ECMAScript, DOM, CSS, and media types as the basis of the content specification. The guidelines are not a limit on what content can be sent, but rather are intended to provide a common set of capabilities so that content developers can author content once and reproduce it on a wide variety of players.

Another key design goal was to provide a single solution that would work on a wide variety of networks. ATVEF is capable of running on both analog and digital video systems as well as networks with no video capabilities at all. The specification further supports transmission across terrestrial broadcast, cable, and satellite systems, and the Internet. In addition, it will bridge between networks; for example, in a compliant system, data on an analog terrestrial broadcast will bridge to a digital cable system. This design goal was achieved through the definition of a transport-independent content format and the use of IP as the reference binding. Because IP bindings already exist for each of these video systems, ATVEF can take advantage of a wealth of previous work.

The specification defines two transports—one for broadcast data and one for data pulled through a return path. While the ATVEF specification has the capability to run on any video network, a complete specification requires a specific binding to each video network standard in order to ensure true interoperability.

Reference and example bindings also are specified in the document, although it is assumed that appropriate standards bodies will define the bindings for each video standard—PAL, SECAM, DVB, ATSC, and others.

There are many roles in the production and delivery of television enhancements. The ATFEF document identifies three key roles:

- **Content creator**. The content creator originates the content components of the enhancement including graphics, layout, interaction, and triggers.
- **Transport operator**. The transport operator runs a video delivery infrastructure (terrestrial, cable, or satellite) that includes a transport mechanism for ATVEF data.
- **Receiver**. The receiver is a hardware and software implementation (television, set-top box, or personal computer) that decodes and plays ATVEF content.

A particular group or company may participate as one, two or all three of these roles. Additional information on ATVEF can be found in Chapter 14.

## 19.4 DTV Application Software Environment

As discussed briefly in Chapter 14, the ATSC T3/S17 specialist group was charged with defining a DTV *application software environment* (DASE) for broadcast interactive applications [5]. This environment makes it appear to interactive programming content that it is running on a *common receiver*. This common receiver contains a well-defined architecture, execution model, syntax, and semantics. The benefits of this approach include the following:

- It affords the capability to write content once and run successfully on all compliant receivers
- Consumer electronic equipment manufacturers have the freedom to independently choose the receiver hardware and operating system
- Content and application authors have the assurance that their content will be decoded and displayed on all receiver brands uniformly
- All receivers have certain common functions and features, such as remote control, tuning, MPEG decoding, and so on

A functional block diagram of the DASE receiver platform is shown in Figure 19.3. The *back channel* shown in the diagram is to be supported in a future version of DASE, commonly described as DASE-2. The DASE application environment is illustrated in Figure 19.4; the division between the declarative environment and the procedural environment can be seen.

The API is a platform-independent abstraction of receiver software libraries and built-in functions such as remove control input, network communications, graphics, and other basic features. Specifically, the DASE API categories are:

- **Network communication**—navigation, event information, program selection, and data broadcast services
- **Content management**—audio/video decoder, media decoder and playback control, audio control, video presentation, presentation and synchronization, and decoder/player synchronization services

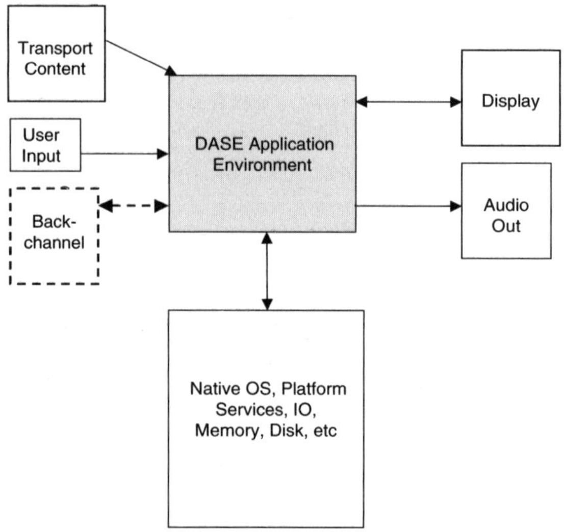


**Figure 19.3** The basic DASE receiver platform.

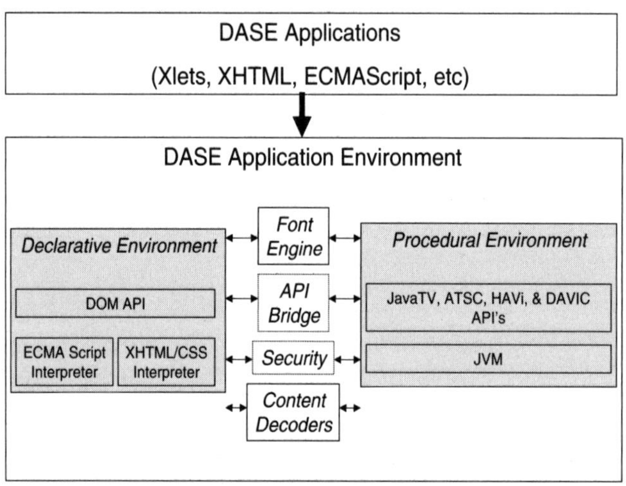


**Figure 19.4** The DASE application environment.

- **Presentation and user interface**—graphics presentation, font management, color management, and user input services
- **Application and resource management**—application lifecycle, application registration, application verification, and application state (diagnostic) services

- **Security management**—authentication, security (policy), and signature checking (certificate exchange) services

The architecture of the DASE declarative environment is built around six basic elements:

- *Markup language*, which defines the structure of a presentation
- *Style language*, which defines the presentational aspects such as position, color, and so on
- *Event model*, which defines a means for interaction with a presentation
- *Document object model*, which defines a means for programmatically manipulating a presentation
- *Media types*, which defines what type of content may be used in a presentation
- *Locations*, which define how to find resources

## 19.4a Evolution of the DASE System

Because of the complexity of DASE, the rollout of this technology has been partitioned into multiple *levels*. The basic specification is identified as DASE Level 1; subsequent releases are designed to build upon this foundation. Specifically:

- **DASE Level 1**—provides for local interaction of enhanced television. This level is the basic foundation of the system. It is broadcast-only in scope; there is no return channel. Examples of DASE-1 applications include play along games (such as *Jeopardy* or *The Price is Right*), "for more information" services (such as sports statistics, product information, local weather, and traffic updates), and a rudimentary "mini" program guide.
- **DASE Level 2**—is expected to provide remote interaction of interactive television. This version will build upon DASE Level 1. It will provide for a return channel and enhanced security framework with digital signature capability and return channel encryption. Interactive television commerce (T-commerce) applications become practical. Examples of DASE-2 applications might include community gaming (where users play against a community of players, gambling (where legal), instant produce purchase, coupon printing, and a full featured program guide.
- **DASE Level 3**—is expected to provide Internet-enabled services, or *Internet television*. This version builds upon Level 2 and facilitates general Internet content. As such, DASE Level 3 must handle invalid, non well-formed content to be interoperable with the existing Web practice. Examples of DASE-3 applications include Internet browsing, general Web access, Internet commerce, banking, investment management, and so on.

There are a number of deployment challenges for the DASE suite of standards. End-to-end issues include metadata management, format conversion, and synchronization requirements. From the standpoint of interoperability, issues include conformance requirements and compliance testing. From the distribution standpoint, issues involve the preferred authoring standard (i.e., will authors create native DASE content format or other content to be transcoded into DASE format?) and distribution (i.e., will non-terrestrial media—cable and satellite—distribute DASE content?).

Harmonization of the DASE system with other technologies has been an on-going priority for the ATSC T3/S17 specialist group. Substantial efforts have been made to make DASE compatible—to the extent possible—with the Multimedia Home Platform (MHP) of DVB and the Open Cable platform (OCAP) of Cable Labs. Although many of the operational details differ, the general approach and key technology choices are identical.

## 19.5 References


1. Clement, Pierre, and Eric Gourmelen: "Internet and Television Convergence: IP and MPEG-2 Implementation Issues," *Proceedings of the 33rd SMPTE Advanced Motion Imaging Conference*, SMPTE, White Plains, N.Y., February 1999.
2. Erez, Beth: "Protecting content in the Digital Home," *Proceedings of the 33rd SMPTE Advanced Motion Imaging Conference*, SMPTE, White Plains, N.Y., pp. 231–238, February 1999.
3. Thomas, Gomer: "ATSC Datacasting—Opportunities and Challenges," *Proceedings of the 33rd SMPTE Advanced Motion Imaging Conference*, SMPTE, White Plains, N.Y., pp. 307–314, February 1999.
4. "Advanced Television Enhancement Forum Specification," Draft, Version 1.1r26 updated 2/2/99, ATVEF, Portland, Ore., 1999.
5. *NAB TV TechCheck*: National Association of Broadcasters, Washington, D.C., February 1, 1999.

Chapter

# 20

# CRT Display Devices


Donald L. Say, R. A. Hedler, L. L. Maninger, R. A. Momberger, J. D. Robbins

Jerry C. Whitaker, Editor


## 20.1 Introduction

Many types of color CRTs have been developed for video, data, and special display applications, including:

- The shadow-mask tube
- The parallel-stripe tube
- The voltage-penetration tube

Most color tubes used for consumer and professional applications fall into three size categories:

- 19-in diagonal
- 21-in diagonal
- 25-in diagonal

Within each category are four primary grades of resolution, based on the center-to-center spacing between phosphor dots of the same color (*pitch*):

- Low resolution: dot pitch 0.44 to 0.47 mm
- Medium resolution: dot pitch 0.32 to 0.43 mm
- High resolution: dot pitch 0.28 to 0.31 mm
- Ultrahigh resolution: dot pitch 0.21 mm (or less) to 0.27 mm

### 20.1a Shadow-Mask CRT

The shadow-mask CRT is the most common type of color display device. As illustrated in Figure 20.1, it utilizes a cluster of three electron guns in a wide neck, one gun for each of the colors—red, green, and blue. All the guns are aimed at the same point at the center of the shadow-mask,

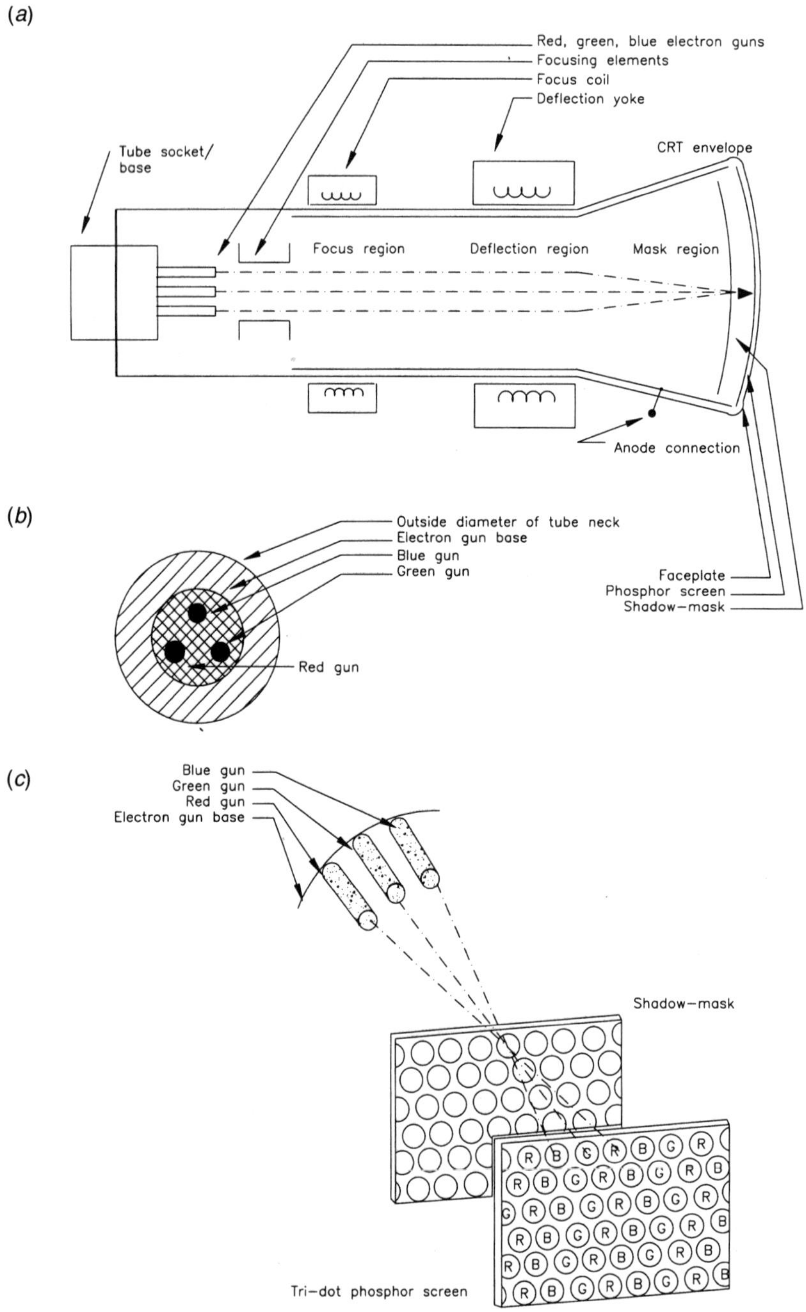


**Figure 20.1** Basic concept of a shadow-mask color CRT: (*a*) overall mechanical configuration, (*b*) delta-gun arrangement on tube base, (*c*) shadow-mask geometry.

which is an iron-alloy grid with an array of perforations in triangular arrangement, generally spaced 0.025-in between centers for entertainment television. For high-resolution studio monitor or computer graphic monitor applications, color CRTs with shadow-mask aperture spacing of 0.012-in center to center or less, are readily available. This triangular arrangement of electron guns and shadow-mask apertures is known as the *delta-gun* configuration. Phosphor dots on the faceplate just beyond the shadow-mask are arranged so that after passing through the perforations, the electron beam from each gun can strike only the dots emitting one color.

All three beams are deflected simultaneously by a single large-diameter deflection yoke, which is usually permanently bonded to the CRT envelope by the tube manufacturer. The three basic phosphors together are designated P-22, individual phosphors of each color being denoted by the numbers P-22R, P-22G, and P-22B. Most modern color CRTs are constructed with rare-earth-element-activated phosphors, which offer superior color and brightness compared with phosphors previously used.

Because of the close proximity of the phosphor dots to each other and the strict dependence on angle of penetration of the electrons through the apertures, tight control over electron optics must be maintained. Close attention also is paid to shielding the CRT from extraneous ambient magnetic fields and to degaussing of the shield and shadow-mask (usually carried out automatically when the equipment is switched on).

Even if perfect alignment of the mask and phosphor triads is assumed, the shadow-mask CRT still is subject to certain limitations, mainly with regard to resolution and brightness. Resolution restriction results from the need to align the mask apertures and the phosphor dot triads; the mask aperture size controls the resolution that can be attained by the device.

Electron-beam efficiency in a shadow-mask tube is low, relative to a monochrome CRT. Typical beam efficiency is 10 percent; considering the three beams of the color tube, total efficiency is approximately 30 percent. By comparison, a monochrome tube may easily achieve 80 percent electron-beam efficiency. This restriction leads to a significant reduction in brightness for a given input power to the shadow-mask CRT.

## 20.1b Parallel-Stripe Color CRT

The parallel-stripe class of CRT, such as the popular *Trinitron* (Sony), incorporates fine stripes of red-, green-, and blue-emitting phosphors deposited in continuous lines repetitively across the faceplate, generally in a vertical orientation. (See Figure 20.2.) This device, unlike a shadow-mask CRT, uses a single electron gun that emits three electron beams across a diameter perpendicular to the orientation of the phosphor stripes. This type of gun is called the *in-line gun*. Each beam is directed to the proper color stripe by the internal beam-aiming structure and a slitted aperture grille.

The Trinitron phosphor screen is built in parallel stripes of alternating red, green, and blue elements. A grid structure placed in front of the phosphors, relative to the CRT gun, is used to focus and deflect the beams to the appropriate color stripes. Because the grid spacing and stripe width can be made smaller than the shadow-mask apertures and phosphor dot triplets, higher resolutions may be attained with the Trinitron system.

Elimination of mask transmission loss, which reduces the electron-beam-to-luminance efficiency of the shadow-mask tube, permits the Trinitron to operate with significantly greater luminance output for a given beam input power.

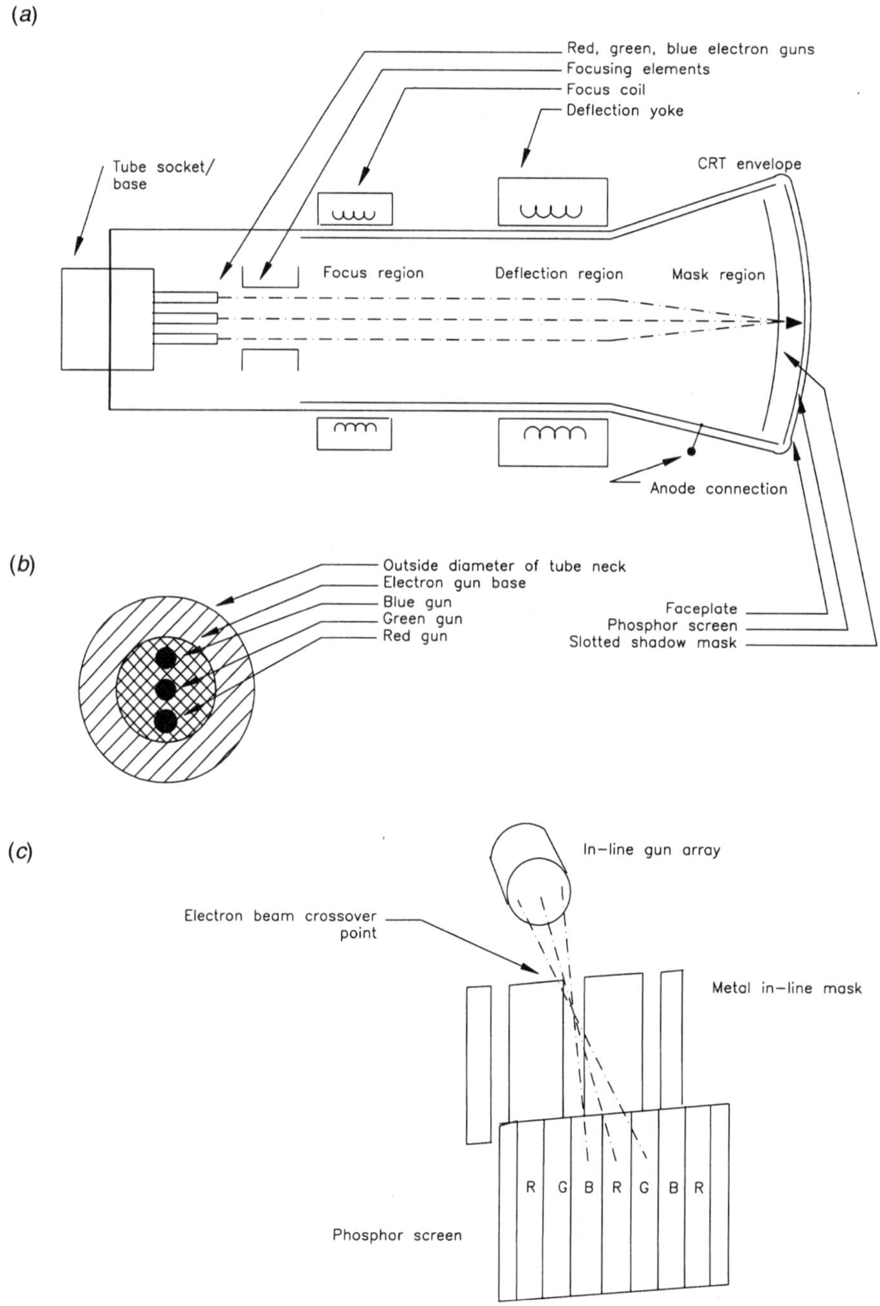


**Figure 20.2** Basic concept of the Trinitron color CRT: (*a*) overall mechanical configuration, (*b*) in-line gun arrangement on tube base, (*c*) mask geometry.

The in-line gun is directed through a single lens of large diameter. The tube geometry minimizes beam focus and deflection aberrations, greatly simplifying convergence of the red, green, and blue beams on the phosphor screen.

## 20.2 Tube Geometry

Figure 20.3 illustrates the shadow-mask geometry for a tube at face center using in-line guns and a shadow-mask of round holes. As an alternative, the shadow-mask may consist of vertical slots, as shown in Figure 20.4. The three guns and their undeflected beams lie in the horizontal plane, and the beams are shown converged at the mask surface. The beams may overlap more than one hole, and the holes are encountered only as they happen to fall in the scan line. By convention, a beam in the figure is represented by a single straight line projected backward at the incident angle from an aperture to an apparent *center of deflection* located in the *deflection plane*. In Figure 20.3, the points B′ , G′ , and R′ , lying in the deflection plane, represent such apparent *centers of deflection* for blue, green, and red beams striking an aperture under study. (These deflection centers move with varying deflection angles.) Extending the rays forward to the faceplate denotes the printing location for the respective colored dots (or stripes) of a tricolor group. Thus, centers of deflection become color centers with a spacing $S$ in the deflection plane. The distance $S$ projects in the ratio $Q/P$ as the dot spacing within the trio. Figure 20.3 also shows how the mask-hole horizontal pitch $b$ projects as screen horizontal pitch in the ratio $L/P$. The same ratio applies for projection of mask vertical pitch $a$. The $Q$-space (mask-to-panel spacing) is optimized to obtain the largest dots that are theoretically possible without overlap. At panel center, the ideal screen geometry is then a mosaic of equally spaced dots (or stripes).

The stripe screen shown in Figure 20.4 is used extensively in color CRTs. One variation of this stripe (or line) screen uses a cylindrical faceplate with a vertically tensioned grill shadow-mask without tie bars. Prior to the stripe screen, the standard construction was a tri-dot screen with a delta gun cluster, as shown in Figure 20.5.

### 20.2a Guard Band

The use of *guard bands* is a common feature for aiding purity in a CRT. The guard band, where the lighted area is smaller than the theoretical tangency condition, may be either *positive* or *negative*. In Figure 20.3, the leftmost red phosphor exemplifies a positive guard band; the lighted area is smaller than the actual phosphor segment, accomplished by mask-hole-diameter design. Figure 20.4, on the other hand, shows negative guard band (NGB) or *window-limited* construction for stripe screens. Vertical black stripes about 0.1 mm (0.004-in) wide separate the phosphor stripes, forming windows to be lighted by a beam wider than the window opening by about 0.1 mm. Figure 20.5 shows NGB construction of a tri-dot screen.

### 20.2b Shadow-Mask Design

The shadow-mask for a color CRT typically is constructed of 0.13 mm low-carbon sheet steel that is chemically etched to the desired pattern of apertures using photoresist techniques. The photographic masters are made by a precision laser plotter. The completed flat mask is then

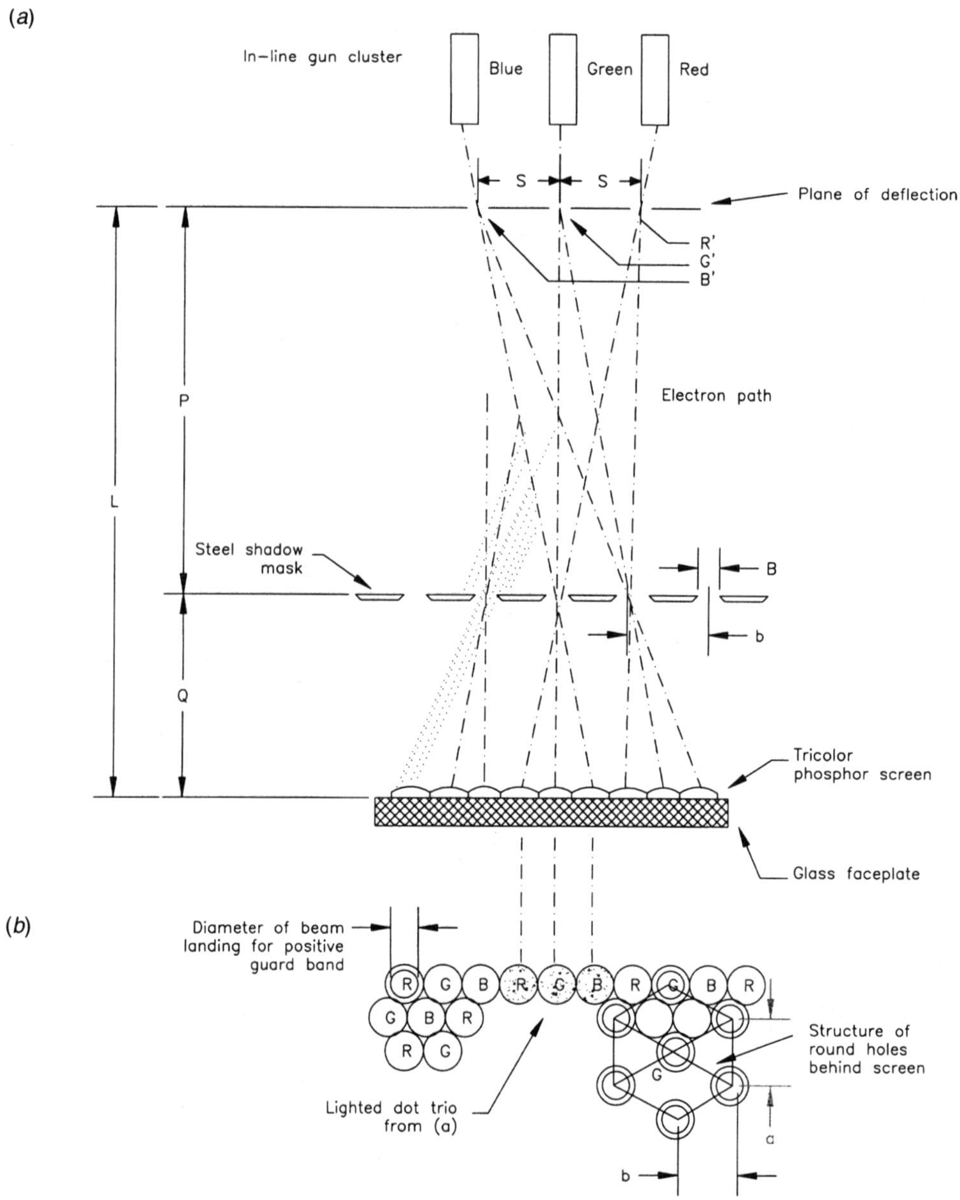


**Figure 20.3** Shadow-mask CRT using in-line guns and round mask holes: (*a*) overall tube geometry, (*b*) detail of phosphor dot layout.

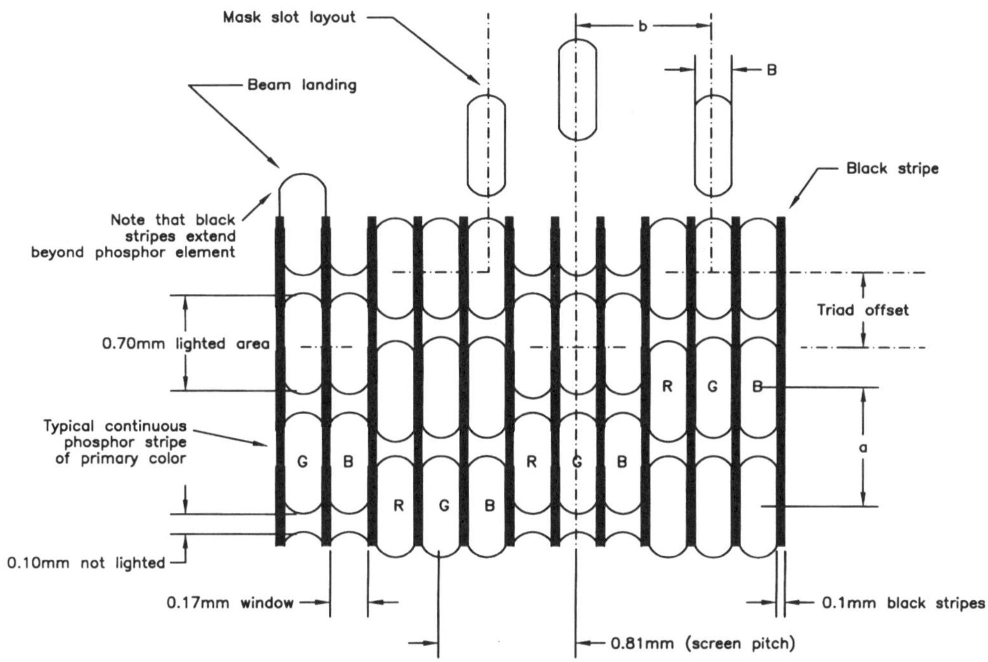


**Figure 20.4** Shadow-mask CRT using in-line guns and vertical stripe mask holes.

press-formed to a contour approximately concentric to the faceplate. Mask-to-panel distance (*Q-spacing*) is locally modified to achieve optimum nesting of screen triplets.

The array layout for a typical round-hole shadow-mask is shown in Figures 20.3 and 20.5. The round holes, numbering approximately 440,000 in a conventional consumer television display, are placed at the vertices and centers of iterated regular hexagons (long-axis vertical).

As illustrated in Figure 20.5, each aperture is tapered to present a more sharply defined limiting-aperture plane to an angled incident beam. This construction increases transmission efficiency and prevents color desaturation resulting from electrons being scattered by sidewalls of the apertures. Apertures are graded radially to smaller diameters at screen edge, because that is where the trio configuration and beam quality are less ideal, and registry is more critical. At the tube center, shadow-mask transmission is typically 16 percent for nonmatrix construction and 22 percent for *black matrix* shadow-mask construction.

The black matrix screen is designed to overcome the loss in luminance and resultant brightness caused by use of a neutral density filter as the faceplate. Such a filter may be used to increase the contrast ratio of the device. This condition arises as a result of the large amount of area in the viewing surface that is not covered by any phosphor. The need to align the phosphor dots with the mask holes to improve convergence can lead to 50 percent of the surface merely

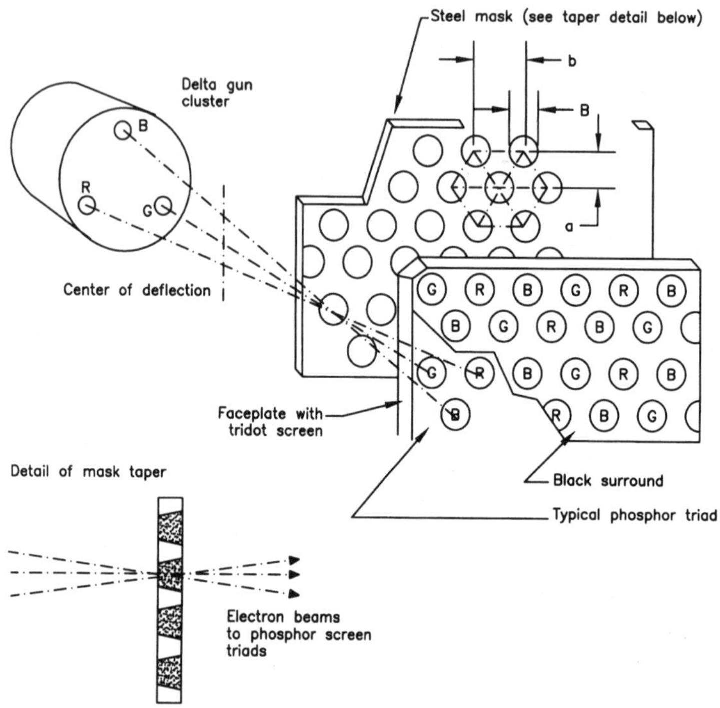


**Figure 20.5** Delta gun, round-hole mask negative guard band tri-dot screen. The taper on the mask holes is shown in the detail drawing only.

reflecting ambient light (in a medium-quality consumer television display). The black matrix screen covers this area with black, thus reducing backscattered and reflected light by the same amount as the faceplate. This scheme reduces the loss in the faceplate without affecting contrast. A brighter displayed picture results.

The beam triad pattern may become distorted near the screen edges, resulting in poorer *packing factors* (*geometric nesting*) for the related phosphor dots. For delta-gun systems, the beam triad triangles compress radially at all screen edges because of foreshortening. For in-line gun dot screens, on the other hand, the beam triad represents three circles in a horizontal line. Using compass points to designate axes and areas, there is no foreshortening at N and S; at E and W, the mask-panel spacing can be sufficiently increased to restore nesting quality. Near the screen corners, however, there will be some rotation of the line trio, thereby demanding smaller holes, dots, and beam landings; the result is less efficient nesting.

For a slot mask, the relationship between $a$ and $b$, horizontal and vertical pitch, can be chosen by the designer. Only the horizontal pitch will affect $Q$-space. The factors affecting the choice of horizontal pitch include:

- Display resolution
- Practical mask-panel spacing
- Attainable slot widths
- Ability to manufacture masks and screens with reasonable yields

The main considerations relating to the choice of vertical pitch and tie bars (or *bridges*) include:

- Avoidance of moiré pattern
- Strength of the tie bars
- Transmission efficiency

Vertical screen stripes may be regarded as aligned oblong dots that have been merged vertically. As a result, there is no color-purity or registration requirement in the vertical direction, which simplifies the design and manufacture of display devices.

## Resolution and Moiré

Resolution is a measure of the definition or sharpness of detail in a displayed image. It can be measured vertically and horizontally. The layout and center-line spacings of the mask apertures are designed to provide sufficient horizontal and vertical lines in the pattern to ensure that they are not the limiting factor in resolution, compared with the number of raster lines. In addition, the number of lines running horizontally in the layout pattern is chosen to avoid moiré, fringes resulting from a "beat" with the scan lines. (This latter criterion is not applicable to the cylindrical grill structure of continuous slits.)

For round-hole masks, the effective number of pattern lines running horizontally, allowing for the staggered pattern, is typically about 2.25 times the number of picture horizontal scan lines. In a conventional television display, there are approximately 440,000 dot trios on the screen, compared with about 200,000 pixels in the raster. These criteria apply to tube sizes from 19- to 25-in (48- to 63-cm) screen diagonal. For smaller tubes, the pattern is made relatively coarser to avoid excessively small apertures and to avoid a moiré pattern in the display. The hexagonal pattern used to lay out the centers of round holes is favorable for increasing the number of pattern lines, reducing the likelihood of moiré, and for nesting of screen dots. The zigzag, or staggering, of alternate columns effectively increases the number of columns and rows, as long as the pattern is significantly finer than the raster scan.

For slot apertures, spacings between columns and the resulting number of columns are chosen for resolution and cosmetic appearance. The tie bars or bridges, which are only about 0.13 mm (0.005-in) in width, are placed according to moiré and strength considerations and are not thought to be significant for resolution in most applications.

## Mask-Panel Temperature Compensation

As mentioned previously, color CRTs for consumer, computer, and industrial applications are shifting toward larger, flatter physical dimensions and higher resolution. Manufacturing such devices for good color purity and stability is a major engineering challenge.

Conventional printing of screen patterns requires that the mask assembly be removed and reinserted without loss of registry. In the most common mask-suspension system, the interior skirt of the glass faceplate has three or four protruding metal studs that engage spring clips welded to the mask frame. Because the shadow-mask may intercept perhaps 75 percent of the beam current, it is subject to thermal expansion during operation. Misregistration would occur if thermal correction were not provided. This temperature compensation is accomplished by automatically shifting the mask slightly closer to the screen surface by thermal expansion of a bimetal structure incorporated into each spring support, or by a designed lever action resulting from the transverse expansion. The exact mechanism used varies from one tube manufacturer to another.

Because the shadow-mask is one of the most critical components in determining CRT performance, the material used to form the structure is an important design parameter. *Aluminum killed* (AK) steel has been extensively employed as mask material because of its etching characteristics, low cost, and relatively good magnetic properties. To build a wide-aspect-ratio (HDTV) CRT at a size of 30 in or larger requires mask thickness of approximately 0.2 mm [1]. The pitch of the device must be 70 to 100 percent higher than that of a conventional CRT. Because of the practical limits of etching technology, mask transmission is significantly lower than for a conventional tube of similar size. Consequently, the percentage of electron-beam bombardment onto the mask will be significantly greater, requiring higher beam current to achieve sufficient brightness. This requirement contributes to color-purity problems when the beam is mislanded onto the screen because of mask thermal expansion.

To overcome the effects of thermal expansion, an iron-nickel alloy was introduced as a substitute to AK steel. *Invar*, a *Fe-36Ni* alloy, exhibits a thermal expansion coefficient 10 times lower than that of AK steel. The effects of misregistration resulting from thermal expansion, therefore, are significantly reduced, as illustrated in Figure 20.6. Because of the relatively low modulus of elasticity of Invar (40 percent less than that of AK steel), however, manufacturing difficulties included mask-forming and mask-mounting to minimize springback effects and microphonics, respectively.

## Tension Mask

Mask stability also may be improved by utilizing the *tension mask* method, in which the mask is tensioned during manufacture to a predetermined stress [2]. When the proper level of tension is applied to the mask, thermal impact on the structure reduces the tension to a certain extent, but will not cause displacement of the aperture features. Therefore, the effects on color purity resulting from thermal expansion are minimized, as shown in Figure 20.7.

The *taut shadow-mask* is a variation on the basic tension mask concept. This design departs from the conventional domed shadow-mask approach and, instead, stretches the shadow-mask into a perfectly flat contour, as illustrated in Figure 20.8 [3]. This design provides greater power-handling capability and is less susceptible to *Z*-axis vibration. The taut shadow-mask typically achieves a fourfold increase in maximum beam current for a specified color purity.

## Magnetic Shielding

Because of the earth's magnetic effect, electron-beam deflection will change, to some extent, depending on the location and orientation of the CRT. This effect influences color purity. An external or internal magnetic metal shield is included, therefore, to compensate for the effect of the earth's magnetic field and stray magnetic fields. The low-carbon steel shadow-mask itself also contributes to the shielding. The shield does not extend into the yoke field because this would result in loading. Also, external shields must be clear of the anode button area. With the

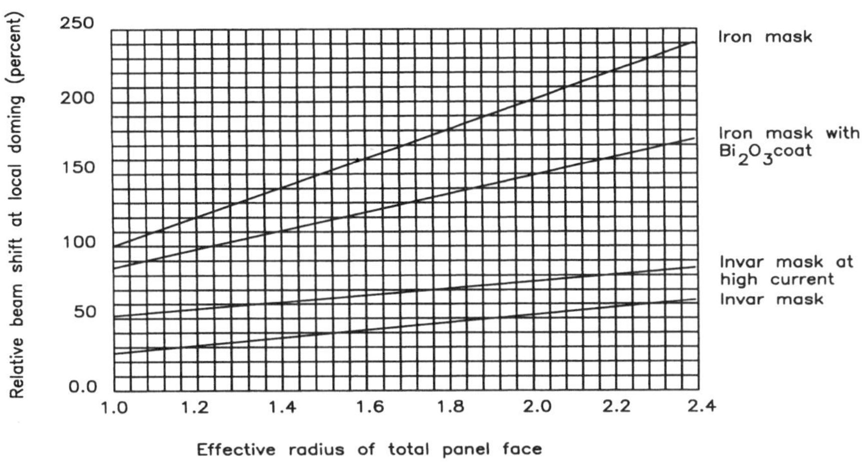


**Figure 20.6** The relationship between beam shift at local doming and the effective radius of the faceplate with various mask materials. (*After* [1]).

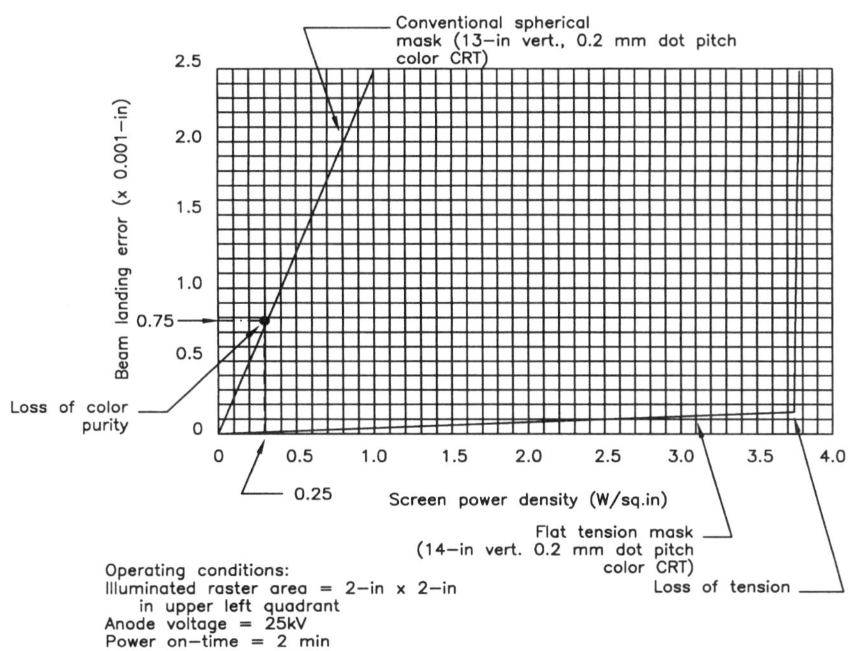


**Figure 20.7** Comparison of small-area-mask doming for a conventional CRT and a flat-tension-mask CRT. (*After* [2].)

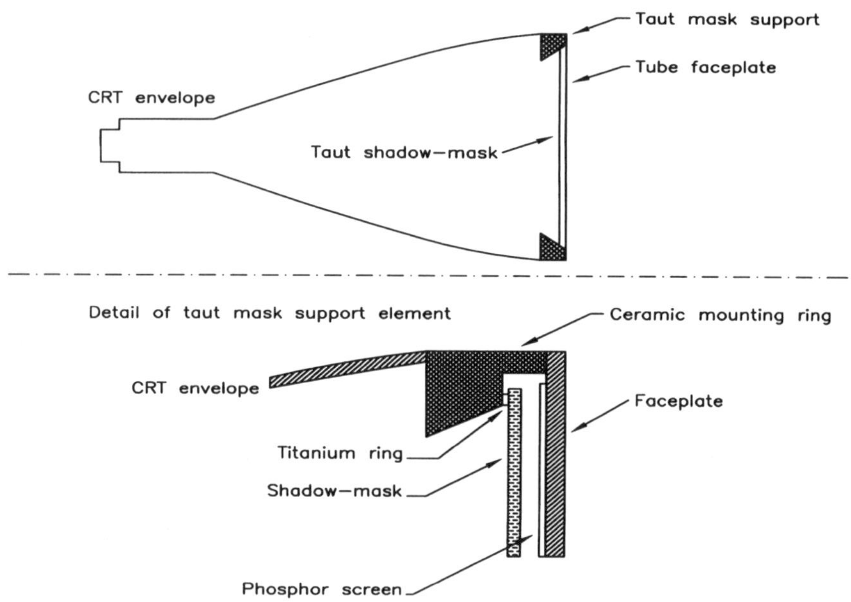


**Figure 20.8** Mechanical structure of a taut shadow-mask CRT. (*After* [3].)

display orientation in the chosen compass direction, the shield and tube must be thoroughly degaussed to gain full purity. This treatment not only removes any residual magnetization of the shield and shadow-mask, but also induces a residual static magnetic field that bucks the ambient magnetic field.

### X-Radiation

The shadow-mask color tube operates with extremely high anode voltages (typically 20 to 30 kV). The possibility of x-rays must be considered for the safety of technicians and end-users. The color-tube envelope is made from glass that has been formulated for x-ray-absorbing characteristics. By closely controlling glass thickness, the glassmaker controls the extent of x-ray absorption. Also, the power levels at which the tube is operated must be controlled. Tube-type data sheets provide the relevant absolute maximum ratings (for voltages and currents) that will keep x-radiation within accepted safety levels.

## 20.2c Screen Size

Consumer demand for ever-larger picture sizes brings to the forefront serious limitations concerning not only the weight and depth of the CRT, but also the higher power and voltage requirements. These limitations are reflected in the sharply increasing costs of receiver cabinets to

**Table 20.1 Color CRT Diagonal Dimension vs. Weight**

| Diagonal Visible (in) | Weight (kg) | Weight (lb.) |
|---|---|---|
| 19 | 12 | 26 |
| 25 | 23 | 51 |
| 30 | 40 | 88 |

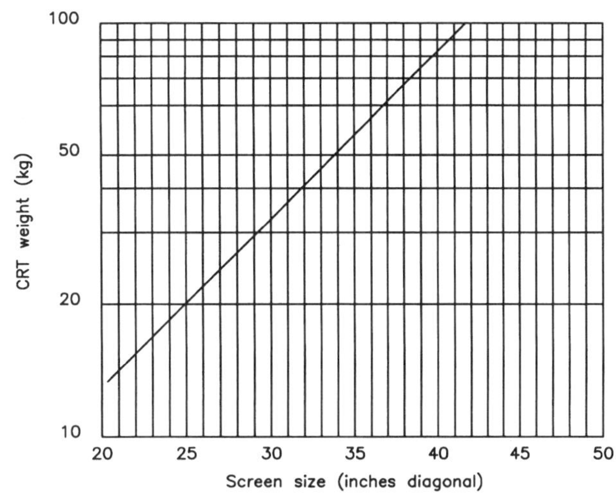


**Figure 20.9** Relationship between screen size and CRT weight. (*After* [4].)

accommodate the larger tubes (a major share of the manufacturing costs) and in more complex circuitry for high-voltage operation.

To withstand the atmospheric pressures on the evacuated glass envelope, the CRT weight increases exponentially with the viewable diagonal. Typical figures for television receiver tubes designed for a 4:3 aspect ratio are shown in Table 20.1. Figure 20.9 charts the relationship. Nevertheless, manufacturers have continued to meet the demand for larger screen sizes with larger direct-view tubes. Examples include an in-line gun, 110° deflection tube with a 35-in-diagonal screen (Mitsubishi and Matsushita). In the Trinitron configuration, 37-, 38-, and 43-in-diagonal tubes have been produced (Sony).

Because of the weight of 35-in and larger tubes and the depth of the required receiver cabinets, tubes of that size are of questionable practicality for home use. Consequently, a 27-in tube is probably the largest size suitable for most home viewing situations.

CRTs designed for HDTV applications suffer an additional weight disadvantage because of the wide aspect ratio of the format (16:9). Furthermore, large-diameter electron guns and small deflection angles (90°) often are employed to improve resolution, making the tube necks fat and giving the devices greater depth and weight.

**Table 20.2 Comparative Resolution of Shadow-Mask Designs** (*After* [5].)

| 19V (48 cm) NGB tube type | Mask Material (mm)[1] | Vertical Pitch (mm)[1] | Center Hole Diameter (mm)[1] | Screen Vertical Pitch (mm) | $N_t$, Trios in Screen | $N_r$ = sq. rt. ($NT$/1.33) |
|---|---|---|---|---|---|---|
| Conventional | 0.15 | 0.56 | 0.27 | 0.60 | 400,000 | 500 lines |
| Monitor | 0.15 | 0.40 | 0.19 | 0.43 | 800,000 | 775 lines |
| High resolution | 0.13 | 0.30 | 0.15 | 0.32 | 1,400,000 | 1025 lines |
| 1 Flat shadow mask | | | | | | |

## 20.2d Resolution Improvement Techniques

In a shadow-mask display device, higher resolution can be achieved through attention to the following parameters:

- Finer pitch
- Smaller-diameter apertures and screen dots
- Smaller *Q*-space
- Thinner mask material

Etching of mask holes becomes more demanding when diameters are smaller than the material thickness. The black matrix tri-dot system has a mask-manufacturing advantage because mask-aperture diameters are larger than in comparable nonmatrix tri-dot tubes or in comparable matrix (or nonmatrix) stripe tubes.

Table 20.2 shows comparative resolution capabilities of three 19-in (48 cm) visible (19V) devices with round-hole masks as screen pitch is reduced to increase resolution. The value $N_r$, a comparative measure of resolution, assumes that resolution is proportional to the square root of the number of trios in a square area with sides equal to screen height. To achieve higher resolution than that of the third design shown in the table (or for smaller tube sizes), it is necessary to use thinner mask material and smaller holes. The *Q*-space also reduces proportionally to pitch, becoming critically small for manufacturing.

The construction of large, wide screen glass tubes presents several design and fabrication problems. The tensile stresses on the bulb must be minimized to avoid cracks resulting from temperature imbalance during and after manufacture, and to prevent subsequent implosion. Extensive study of the stress points in the design stages through computer simulation has reduced the nominal physical stress to about the same level as in conventional tubes of 4:3 aspect ratio.

Reduction of moiré to an unnoticeable level requires an increase in the frequency of the pattern or a decrease in the amplitude. The pattern can be minimized through the use of certain specific values of triad pitch. However, the selection of the pitch is limited not only by the thinner shadow-mask for a finer pitch, but also by a sacrifice in resolution for a more coarse pitch. For a 40-in wide screen tube, a pitch of 0.46 mm provides a resolution of 1000 television lines and satisfies the other requirements of shadow-mask strength and moiré-pattern reduction.

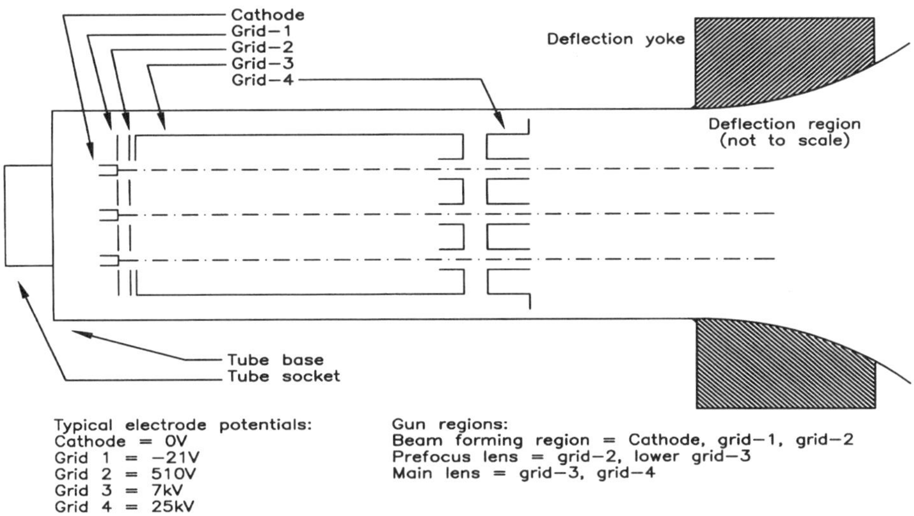


**Figure 20.10** Simplified mechanical structure of a bipotential color electron gun.

## 20.3 Electron Gun

Figure 20.10 illustrates the general electrode configuration for a shadow-mask color electron gun. The device can be subdivided into three major regions:

- Beam-forming region, which consists of the cathode, grid-1, and grid-2 electrodes.
- Prefocus lens region, which consists of the grid-2 and lower grid-3 electrodes.
- Main lens region, which consists of the grid-3 and grid-4 electrodes. These elements create a focusing field for the electron beam.

In more complicated lens systems, additional elements follow grid 4.

Electron guns for color tubes can be classified according to the main lens configuration, which include:

- Unipotential
- Bipotential
- Tripotential
- Hybrid lenses

The unipotential gun is the simplest of all designs. This type of gun rarely is used for color applications, except for small screen sizes. The system suffers from a tendency to arc at high anode voltage and from relatively large low-current spots.

The bipotential lens is the most commonly used gun in shadow-mask color tubes. The main lens of the gun is formed in the gap between grids 3 and 4. When grid 3 operates at 18 to 22 percent of the grid-4 voltage, the lens is referred to as a *low-bipotential* configuration, often called *LoBi* for short. When grid 3 operates at 26 to 30 percent of the grid-4 voltage, the lens is referred to as a *high-bipotential* or *HiBi* configuration.

The LoBi configuration has the advantages of a short grid 3 and shorter overall length, with parts assembly generally less critical than the HiBi configuration. However, with its shorter object distance (grid-3 length), the lens suffers from somewhat larger spot size than the HiBi. The HiBi, on the other hand, with a longer grid-3 object distance, has improved spot size and resolution. The focus voltage supply for the LoBi also can be less expensive.

Further improvement in focus characteristics can been achieved with a tripotential lens. The lens region has more than one gap and requires two focus supplies, one typically at 40 percent and the other at 24 percent of the anode potential. With this refinement, the lens has lower spherical aberration. These features, together with a longer object distance (grids 3 to 5), yield a smaller spot size at the screen than is achieved with the bipotential design. Among the drawbacks of the tripotential gun are:

- The assembly is physically longer.
- It requires two focus supplies.
- It requires a special base to deliver the high focus voltage through the stem of the tube.

Improved performance may be realized by combining elements of the unipotential and bipotential lenses in series. The two more common configurations are known as *UniBi* and *BiUni*. The UniBi structure (sometimes referred to as *quadripotential focus*) combines the HiBi main lens gap with the unipotential type of lens structure to collimate the beam. As shown in Figure 20.11, the grid-2 voltage is tied to grid 4 (inserted in the object region of the gun), causing the beam bundle to collimate to a smaller diameter in the main lens. With this added focusing, the gun is slightly shorter than a bipotential gun having an equal focus voltage.

The BiUni structure, illustrated in Figure 20.12, achieves a similar beam collimation by tying the added element to the anode, rather than to grid 2. The gun structure is shorter because of the added focusing early in the device. With three high-gradient gaps, arcing can be a problem in the BiUni configuration.

The Trinitron gun consists of three beams focused through the use of a single large main-focus lens of unipotential design. Figure 20.13 shows the three in-line beams mechanically tilted to pass through the center of the lens, then reconverged toward a common point on the screen. The gun is somewhat longer than other color guns, and the mechanical structuring of the device requires unusual care and accuracy in assembly.

### 20.3a Guns for High-Resolution Application

Improved versions of both the delta and in-line guns previously noted are used in high-resolution display applications. In both cases, the guns are adjusted for the lower beam current and higher resolution needed in data and/or graphics display. The advantages and disadvantages of delta and in-line guns also apply here. For example, the use of a delta-type cylinder, or *barrel-type* gun, requires as many as 20 carefully tailored convergence waveforms to obtain near-perfect convergence over the full face of the tube. The in-line gun, with a self-converging yoke, avoids the need

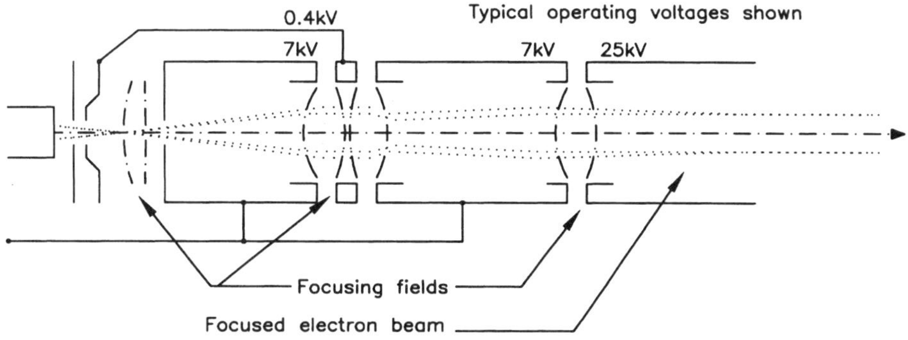


**Figure 20.11** Electrode arrangement of a UniBi gun.

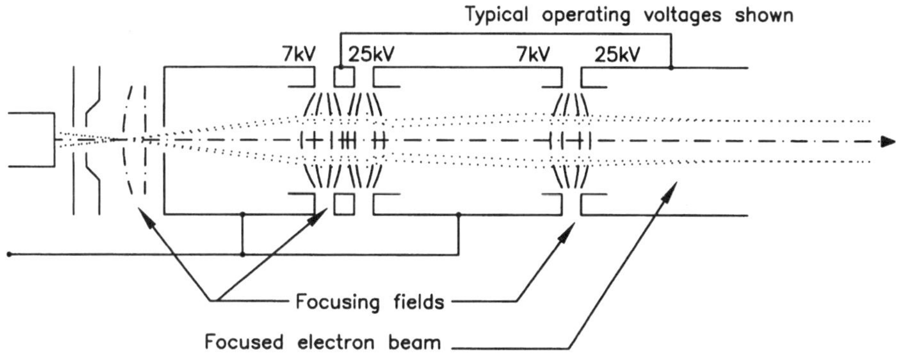


**Figure 20.12** Electrode arrangement of a BiUni gun.

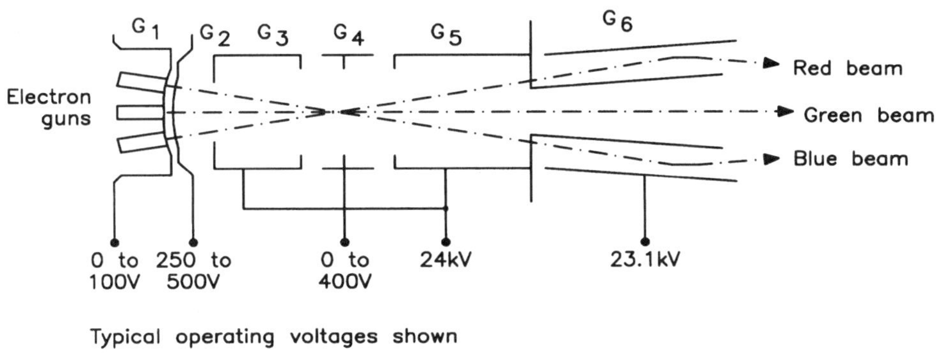


**Figure 20.13** Electrode arrangement of the Trinitron gun. (*After* [6].)

(*a*)

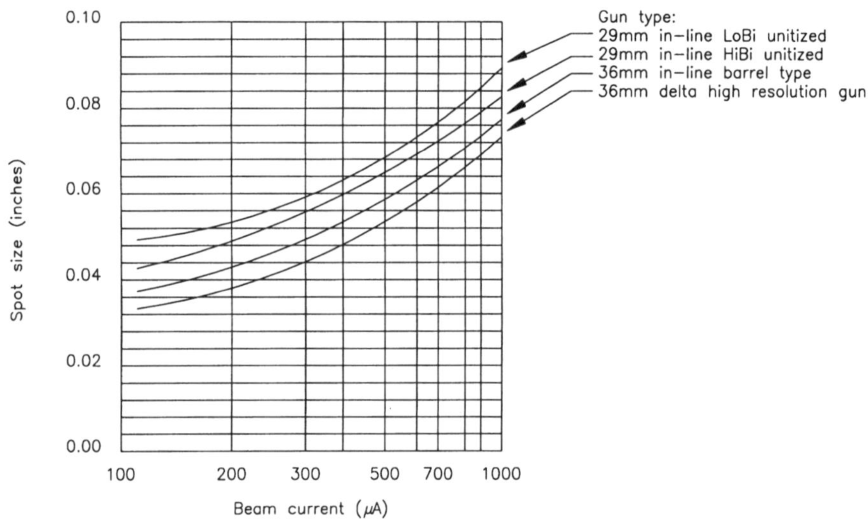


(*b*)

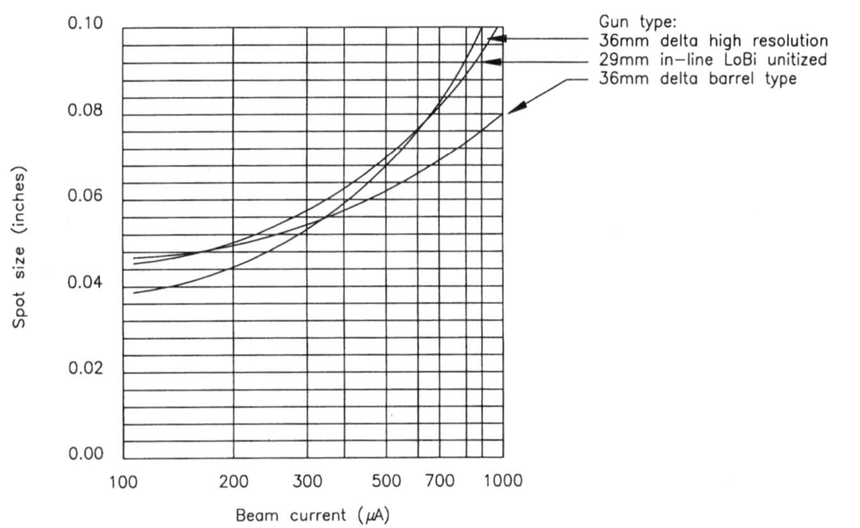


**Figure 20.14** Spot-size comparison of high-resolution guns vs. commercial television guns: (*a*) 13-in vertical display, (*b*) 19-in vertical display.

for these waveforms, but at the expense of slightly larger spots, particularly in the corners where overfocused haze tails can cause problems.

Figure 20.14 compares spot sizes (at up to 1 mA of beam current) for high-resolution designs with those of commercial receiver-type devices. Both delta and in-line 13- and 19-in vertical

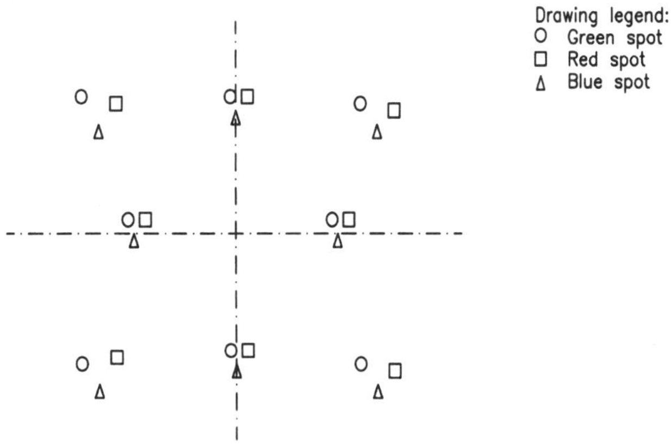


**Figure 20.15** Astigmatism errors in a color CRT. (*After* [7].)

(13V and 19V) devices are shown. Note the marked improvement in spot size at current levels below 500 μA for the high-resolution devices.

## 20.4 Deflecting Multiple Electron Beams

Deflection of the electron beams in a color CRT is a difficult technical exercise. The main problems that occur when the three beams are deflected by a common deflection system are associated with spot distortions that occur in single-beam tubes. The effect is intensified, however, by the need for the three beams to cross over and combine as a spot on the shadow-mask. The two most significant effects are:

- Curvature of the field
- Astigmatism

Misalignment and misregistration of the three beams of the color CRT will lead to loss of purity for colors produced by combinations of the primary colors. Such reproduction distortions also may result in a reduction in luminance output because a smaller part of the beams are passing through the apertures. Additional errors that must be considered include:

- Deflection-angle changes in the yoke-deflection center
- Stray electromagnetic fields

Curvature of field and astigmatism distortions result in a misconvergence of the beam. Figure 20.15 illustrates distortion resulting from astigmatism (for a delta-gun tube). Figure 20.16 illustrates misconvergence of the beam resulting from astigmatism and coma. The misconvergence that occurs in the four corners of the raster is shown in Figure 20.17. The result is that color rendition will not be true, particularly at the edges of the screen. This can be partially compensated by introducing quadripole fields, which cause the beams to be twisted, restoring the equilateral nature of the triangle. The shapes these fields can take are illustrated in Figure 20.18 along with the currents required to produce the fields.

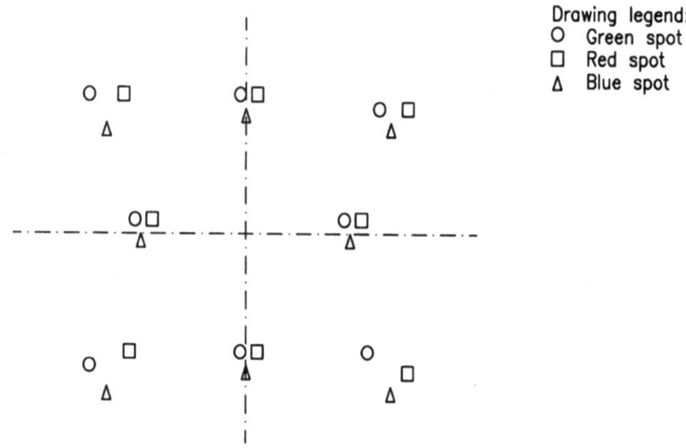


**Figure 20.16** Astigmatism and coma errors. (*After* [7].)

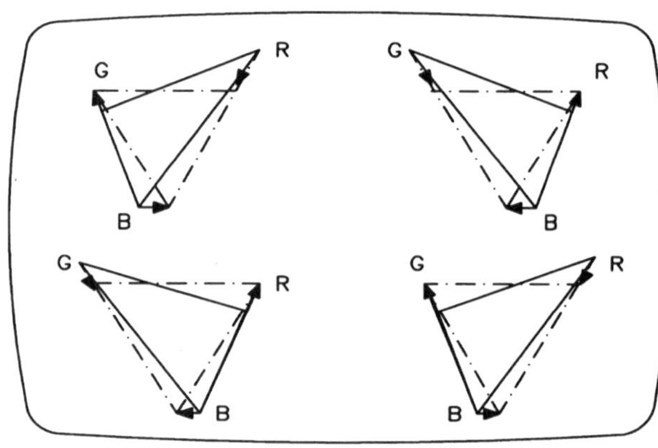


**Figure 20.17** Misconvergence in the four corners of the raster in a color CRT. (*After* [7].)

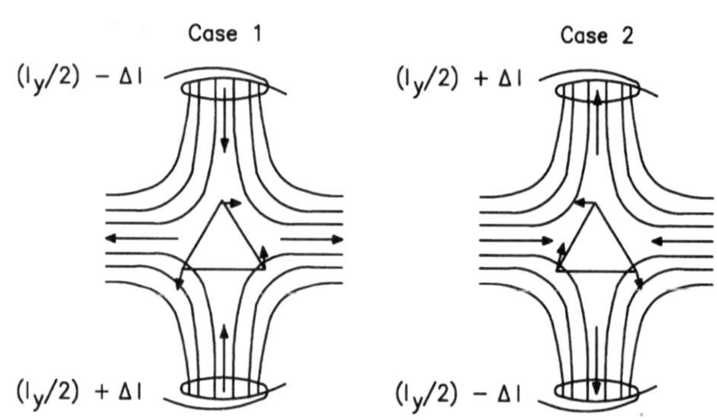


**Figure 20.18** Field configurations suitable for correcting misconvergence in a color CRT. (*After* [7].)

The registration challenge is less severe in the case of the in-line gun; the three beams need only be converged into a vertical line, rather than the round spot required by the delta gun. For

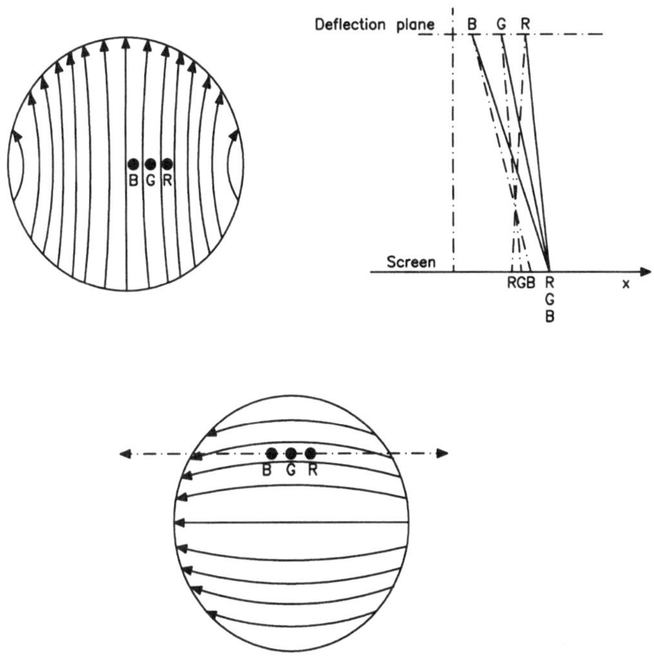


**Figure 20.19** Self-converging deflection field in an in-line gun device: (*a*) horizontal deflection field, (*b*) vertical deflection field. (*After* [8].)

the in-line gun, a precision self-converging system is used where the yoke is designed to operate with one specific tube. This self-converging yoke causes the beams to diverge horizontally in the yoke, resulting in nonuniform fields that counteract the overconvergence. The shape of the fields for horizontal and vertical deflection are shown in Figure 20.19. The horizontal yoke generates a pincushion field, and the vertical yoke generates a barrel-shaped field to accomplish these ends. However, complete self-convergence without the need for compensating adjustments is possible only with narrow-angle tubes; for 110° tubes, it is necessary to add some convergence adjustments, although much less than is required for the delta gun. The drawback is that the yoke usually must be tailored to the specific tube with which it is used.

## 20.5 New Consumer Devices

Progress continues to be made in CRT technology. Numerous improvements have been made over the years, and considerable new efforts have accompanied the move to DTV. A case in point is the FD Trinitron Wega (Sony), the first consumer TV with a flat face CRT. The purpose of making a flat face display was to eliminate distortion, reduce reflections, and improve the picture quality [9].

Conventional curved CRT screens cause geometric distortion in the displayed images and a loss of part of the image when viewed from a side angle. Furthermore, the spherical shape of the screen causes ambient room lighting to be reflected to the viewer. A flat display screen eliminates the geometric distortion and reduces reflections.

Glass is stronger when it is curved. For this reason, detailed modeling and strength analysis were necessary to make the face of the tube capable of withstanding the applied atmospheric pressure without failing or deforming. Because of the pressure gradient on the glass, it was necessary to develop a new tempered glass based upon the principles of automobile windshield glass. For additional implosion protection, a 188 μm thick film was laminated on the panel surface. The film also serves as an anti-reflective coating to improve display contrast, a common practice in computer monitors.

To realize a flat *aperture grille* (AG) with high tension, there were two major problems to overcome: mask wrinkle and AG vibration. Mask wrinkle is caused by localized heating during the welding process. It was found that the AG wires and even the frame itself were susceptible to sympathetic vibrations from the receiver speakers. The answer was to double the springs that attach the frame to the CRT panel.

As the faceplate gets flatter, it becomes more difficult for the electron gun to make a round spot towards the edges of the screen. As a solution to this problem, the focal length of the electron gun was elongated by 20 percent compared with a conventional Trinitron gun. The result was improved focus at the edges of the flat CRT. In order to keep the overall length of the TV set the same, other parts of the gun were widened and shortened.

A new deflection yoke was developed using a horizontal coil with a large square-front-bend to eliminate the usual pincushion distortion introduced by the flat CRT. A modulation coil was added to the vertical winding that used the horizontal frequency to correct vertical misconvergence.

Additional improvements included the following:

- Dynamic focus to vary the voltage across scan lines for sharp focus from edge-to-edge
- Velocity modulation to change the speed of the beam as it scans horizontally across the screen and sharpens the focus

This device is representative of the new generation of CRT displays now finding their way to consumers. The trend to higher quality, lower priced displays will likely accelerate as DTV penetration in the home increases.

## 20.6 References

1. Tong, Hua-Sou: "HDTV Display—A CRT Approach," *Display Technologies*, Shu-Hsia Chen and Shin-Tson Wu (eds.), Proc. SPIE 1815, SPIE, Bellingham, Wash., pp. 2–8, 1992.
2. Hockenbrock, Richard: "New Technology Advances for Brighter Color CRT Displays," *Display System Optics II*, Harry M. Assenheim (ed.), Proc. SPIE 1117, SPIE, Bellingham, Wash., pp. 219-226, 1989.

3. Robinder, R., D. Bates, P. Green: "A High Brightness Shadow Mask Color CRT for Cockpit Displays," *SID Digest*, Society for Information Display, vol. 14, pp. 72-73, 1983.

4. Mitsuhashi, Tetsuo: "HDTV and Large Screen Display," *Large-Screen Projection Displays II*, William P. Bleha, Jr. (ed.), Proc. SPIE 1255, SPIE, Bellingham, Wash., pp. 2-12, 1990.

5. Benson, K. B., and D. G. Fink: *HDTV: Advanced Television for the 1990s*, McGraw-Hill, New York, 1990.

6. Morrell, A., et al.: *Color Picture Tubes*, Academic Press, New York, pp. 91-98, 1974.

7. Hutter, Rudolph G. E.: "The Deflection of Electron Beams," *Advances in Image Pickup and Display*, B. Kazen (ed.), vol. 1, Academic Press, New York, pp. 212-215, 1974.

8. Barkow, W. H., and J. Gross: "The RCA Large Screen 110° Precision In-Line System," ST-5015, RCA Entertainment, Lancaster, Pa.

9. Eccles, D. A., and Y. Zhang: "Digital-Television Signal Processing and Display Technology," *SID 99 Digest*, Society for Information Display, San Jose, Calif., pp. 108–111, 1999.

## 20.7 Bibliography

Aiken, J. A.: "A Thin Cathode Ray Tube," *Proc. IRE*, vol. 45, pg. 1599, 1957.

Ashizaki, S., Y. Suzuki, K. Mitsuda, and H. Omae: "Direct-View and Projection CRTs for HDTV," *IEEE Transactions on Consumer Electronics*, vol. 34, no. 1, pp. 91–98, February 1988.

Barbin, R., and R. Hughes: "New Color Picture Tube System for Portable TV Receivers," *IEEE Trans. Broadcast TV Receivers*, vol. BTR-18, no. 3, pp. 193–200, August 1972.

Blacker, A., et al.: "A New Form of Extended Field Lens for Use in Color Television Picture Tube Guns," *IEEE Trans. Consumer Electronics*, pp. 238–246, August 1966.

Blaha, R.: "Degaussing Circuits for Color TV Receivers," *IEEE Trans. Broadcast TV Receivers*, vol. BTR-18, no. 1, pp. 7–10, February 1972.

Carpenter, C.: et al., "An Analysis of Focusing and Deflection in the Post-Deflection Focus Color Kinescope," *IRE Trans. Electron Devices*, vol. 2, pp. 1–7, 1955.

Casteloano, Joseph A.: *Handbook of Display Technology*, Academic, New York, N.Y., 1992.

Chang, I.: "Recent Advances in Display Technologies," *Proc. SID*, Society for Information Display, San Jose, Calif., vol. 21, no. 2, pg. 45, 1980.

Chen, H., and R. Hughes: "A High Performance Color CRT Gun with an Asymmetrical Beam Forming Region," *IEEE Trans. Consumer Electronics*, vol. CE-26, pp. 459–465, August, 1980.

Chen, K. C., W. Y. Ho and C. H. Tseng: "Invar Mask for Color Cathode Ray Tubes," in *Display Technologies*, Shu-Hsia Chen and Shin-Tson Wu (eds.), Proc. SPIE 1815, SPIE, Bellingham, Wash., pp.42–48, 1992.

Clapp, R., et al.: "A New Beam Indexing Color Television Display System," *Proc. IRE*, vol. 44, no. 9, pp. 1108–1114, September 1956.

Cohen, C.: "Sony's Pocket TV Slims Down CRT Technology," *Electronics*, pg. 81, February 10, 1982.

Credelle, T. L., et al.: *Japan Display '83*, pg. 26, 1983.

Credelle, T. L.: "Modular Flat Display Device with Beam Convergence," U.S. Patent 4,131,823.

Credelle, T. L., et al.: "Cathodoluminescent Flat Panel TV Using Electron Beam Guides," *SID Int. Symp. Digest*, Society for Information Display, San Jose, Calif., pg. 26, 1980.

"CRT Control Grid Having Orthogonal Openings on Opposite Sides," U.S. Patent 4,242,613, Dec. 30, 1980.

"CRTs: Glossary of Terms and Definitions," Publication TEP92, Electronic Industries Association, Washington, 1975.

Davis, C., and D. Say: "High Performance Guns for Color TV—A Comparison of Recent Designs," *IEEE Trans. Consumer Electronics*, vol. CE-25, August 1979.

Donofrio, R.: "Image Sharpness of a Color Picture Tube by MTF Techniques," *IEEE Trans. Broadcast TV Receivers*, vol. BTR-18, no. 1, pp. 1–6, February 1972.

Dressler, R.: "The PDF Chromatron—A Single or Multi-Gun CRT," *Proc. IRE*, vol. 41, no. 7, July 1953.

"Electron Gun with Astigmatic Flare-Reducing Beam Forming Region," U.S. Patent 4,234,814, Nov. 18, 1980.

Fink, Donald, (ed.): *Television Engineering Handbook*, McGraw-Hill, New York, N.Y., 1957.

Fink, Donald, and Donald Christiansen (eds.): *Electronics Engineers Handbook*, 3rd ed., McGraw-Hill, New York, N.Y., 1989.

Fiore, J., and S. Kaplin: "A Second Generation Color Tube Providing More Than Twice the Brightness and Improved Contrast," *IEEE Trans. Consumer Electronics*, vol. CE-28, no. 1, pp. 65–73, February 1982.

Flechsig, W.: "CRT for the Production of Multicolored Pictures on a Luminescent Screen," French Patent 866,065, 1939.

Godfrey, R., et al.: "Development of the Permachrome Color Picture Tube," *IEEE Trans. Broadcast TV Receivers*, vol. BTR-14, no. 1, 1968.

Gow, J., and R. Door: "Compatible Color Picture Presentation with the Single-Gun Tri Color Chromatron," *Proc. IRE*, vol. 42, no. 1, pp. 308–314, January 1954.

Hasker, J.: "Astigmatic Electron Gun for the Beam Indexing Color TV Display," *IEEE Trans. Electron Devices*, vol. ED-18, no. 9, pg. 703, September 1971.

Herold, E.: "A History of Color TV Displays," *Proc. IEEE*, vol. 64, no. 9, pp. 1331–1337, September 1976.

Hoskoshi, K., et al.: "A New Approach to a High Performance Electron Gun Design for Color Picture Tubes," 1980 IEEE Chicago Spring Conf. Consumer Electronics.

Hu, C., Y. Yu and K. Wang: "Antiglare/Antistatic Coatings for Cathode Ray Tube Based on Polymer System," in *Display Technologies*, Shu-Hsia Chen and Shin-Tson Wu (eds)., Proc. SPIE 1815, SPIE, Bellingham, Wash., pp.42–48, 1992.

Johnson, A.: "Color Tubes for Data Display—A System Study," Philips ECG, Electronic Tube Division.

Law, H.: "A Three-Gun Shadowmask Color Kinescope," *Proc. IRE*, vol. 39, pp. 1186–1194, October 1951.

Lucchesi, B., and M. Carpenter: "Pictures of Deflected Electron Spots from a Computer," *IEEE Trans. Consumer Electronics*, vol. CE-25, no. 4, pp. 468–474, 1979.

Maeda, M.: *Japan Display '83*, pg. 2, 1971.

Masterson, W., and R. Barbin: "Designing Out the Problems of Wide-Angle Color TV Tube," *Electronics*, pp. 60–63, April 26, 1971.

Mears, N., "Method and Apparatus for Producing Perforated Metal Webs," U.S. Patent 2,762,149, 1956.

Mokhoff, N.: "A Step Toward Perfect Resolution," *IEEE Spectrum*, IEEE, New York, N.Y., vol. 18, no. 7, pp. 56–58, July 1981.

Morrell, A.: "Color Picture Tube Design Trends," *Proc. SID*, Society for Information Display, San Jose, Calif., vol. 22, no. 1, pp. 3–9, 1981.

Moss, H.: *Narrow Angle Electron Guns and Cathode Ray Tubes*, Academic, New York, N.Y., 1968.

Oess, F.: "CRT Considerations for Raster Dot Alpha Numeric Presentations," *Proc. SID*, Society for Information Display, San Jose, Calif., vol. 20, no. 2, pp. 81–88, second quarter, 1979.

Ohkoshi, A., et al.: "A New 30V-in Beam Index Color Cathode Ray Tube," *IEEE Trans. Consumer Electronics*, vol. CE-27, p. 433, August 1981.

Palac, K.: Method for Manufacturing a Color CRT Using Mask and Screen Masters, U.S. Patent 3,989,524, 1976.

Pitts, K., and N. Hurst: "How Much do People Prefer Widescreen (16 × 9) to Standard NTSC (4 × 3)?," *IEEE Transactions on Consumer Electronics*, vol. 35, no. 3, pp. 160–169, August 1989.

"Recommended Practice for Measurement of X-Radiation from Direct View TV Picture Tubes," Publication TEP 164, Electronics Industries Association, Washington, D.C., 1981.

Robbins, J., and D. Mackey: "Moire Pattern in Color TV," *IEEE Trans. Consumer Electronics*, vol. CE-28, no. 1, pp. 44–55, February 1982.

Rublack, W.: "In-Line Plural Beam CRT with an Aspherical Mask," U.S. Patent 3,435,668, 1969.

Sakamoto, Y.: and E. Miyazaki, *Japan Display '83*, pg. 30, 1983.

Say, D.: "Picture Tube Spot Analysis Using Direct Photography," *IEEE Trans. Consumer Electronics*, vol. CE-23, pp. 32–37, February 1977.

Say, D.: "The High Voltage Bipotential Approach to Enhanced Color Tube Performance," *IEEE Trans. Consumer Electronics*, vol. CE-24, no. 1, pg. 75, February 1978.

Schwartz, J.: "Electron Beam Cathodoluminescent Panel Display," U.S. Patent 4,137,486.

Sherr, S.: *Electronic Displays*, Wiley, New York, N.Y., 1979.

Sinclair, C.: "Small Flat Cathode Ray Tube," *SID Digest*, Society for Information Display, San Jose, Calif., pg. 138, 1981.

Stanley, T.: "Flat Cathode Ray Tube," U.S. Patent 4,031,427.

Swartz, J.: "Beam Index Tube Technology," *SID Proceedings*, Society for Information Display, San Jose, Calif., vol. 20, no. 2, p. 45, 1979.

Uba, T., K. Omae, R. Ashiya, and K. Saita: "16:9 Aspect Ratio 38V-High Resolution Trinitron for HDTV," *IEEE Transactions on Consumer Electronics*, vol. 34, no. 1., pp. 85–89, February 1988.

Woodhead, A., et al.: *1982 SID Digest*, Society for Information Display, San Jose, Calif., pg. 206, 1982.

Yoshida, S., et al.: "25-V Inch 114-Degree Trinitron Color Picture Tube and Associated New Development," *Trans. BTR*, pp. 193-200, August 1974.

Yoshida, S., et al.: "A Wide Deflection Angle (114°) Trinitron Color Picture Tube," *IEEE Trans. Electron Devices*, vol. 19, no. 4, pp. 231–238, 1973.

Yoshida. S.: et al., "The Trinitron—A New Color Tube," *IEEE Trans. Consumer Electronics*, vol. CE-28, no. 1, pp. 56–64, February 1982.

Chapter

# 21

# Projection Display Systems

**Donald L. Say, R. A. Hedler, L. L. Maninger, R. A. Momberger, J. D. Robbins, K. Blair Benson, Donald G. Fink**

**Jerry C. Whitaker, Editor**

## 21.1 Introduction

As the need to present high-resolution video and graphics information steadily increases, the use of large-screen projection displays rapidly expands. High-definition television will require large screens (greater than 40-in diagonal) to provide effective presentation. Given the physical limitations of CRTs, some form of projection is the only practical solution. The role of HDTV and film in theaters of the future also is being explored, along with performance criteria for effective large-screen video presentations.

New projection-system hardware is taking advantage of *liquid crystal* (LC) and *thin-film transistor* (TFT) technology originally developed for direct-view flat-panel displays. Also, new deformable-membrane light valves used in Schlieren systems are being developed to update oil film light-valve projectors, which have been the mainstay of large-screen projection technology for many years.

Extensive developmental efforts are being directed toward large-screen display systems. Much of this research is aimed at advancing new technologies such as plasma, electroluminescent and LCD, lasers, and new varieties of CRTs. Currently, CRT systems lead the way for applications requiring full color and high resolution. LCD systems, which are advancing rapidly, may capture a sizable portion of the video-only marketplace.

Large-screen projectors fall into four broad classes, or grades:

- **Graphics**. Graphics projectors are the highest-quality—and generally the highest-priced—projectors. These systems are capable of the highest operating frequency and resolution. They can gen-lock to almost any computer or image source, and they offer resolutions of 2000 × 2000 pixels (or more) with horizontal sweep rates to 89 kHz and higher.
- **Data**. Data projectors are less expensive than graphics projectors and are suitable for use with common computer image generators, such as PCs equipped with VGA graphics cards. Data

projectors typically offer resolutions of 1024 × 768 pixels and horizontal sweep rates of approximately 49 kHz.

- **HDTV**. HDTV projectors provide the quality level necessary to take full advantage of high-definition imaging systems. Resolution in excess of 1000 TV lines is provided at an aspect ratio of 16:9.
- **Video**. Video projectors provide resolution suitable for NTSC-level images. Display performance of 380 to 480 TV lines is typical.

Computer signal sources can follow many different—and sometimes incompatible—standards. It is not a trivial matter to connect any given projector to any given source, although multisync projectors are becoming the norm. Generally speaking, projectors are *downward compatible*. In other words, most graphics projectors can function as data projectors, HDTV projectors, and video projectors; most data projectors also can display HDTV and video, and so on.

## 21.1a Displays for HDTV Applications

The most significant differences between a conventional display and one designed for high-definition video are the increased resolution and wider aspect ratio of HDTV. Four technologies have emerged as practical for viewing high-definition images of 40-in diagonal screens and larger:

- **Light-valve projection display**. Capable of modulating high-power external light sources, light valves are used mainly for large-screen displays measuring greater than 200 in.
- **CRT projection displays**. Widely used for midsize-screen displays of 45 to 200 in, CRT projection systems are popular because of their relative ease of manufacturing (hence, competitive cost) and good performance. These displays may be the mainstay technology for HDTV in the near future.
- **Flat-panel plasma display panel (PDP) and LCD (liquid crystal display)**. Flat-panel PDP displays have been produced in 40-in and larger displays. Similar LCD systems also have been shown. Both hold great promise for future HDTV applications.
- **Digital micromirror device**. The DMD (Texas Instruments) is a spatial light modulator that combines electronic, mechanical, and optical functions to yield high-resolution projection images.

In addition, several hybrid technologies that are under study combine various elements from the mainstay projection techniques.

## 21.1b Projection-System Fundamentals

Video projection systems provide a method of displaying a much larger image than can be generated on a direct-view cathode ray picture tube. Optical magnification and other techniques are employed to throw an expanded image on a passive viewing surface that may have a diagonal dimension of 75 in or more.

The basic elements of a projection system, illustrated in Figure 21.1, include:

- Viewing screen
- Optical elements

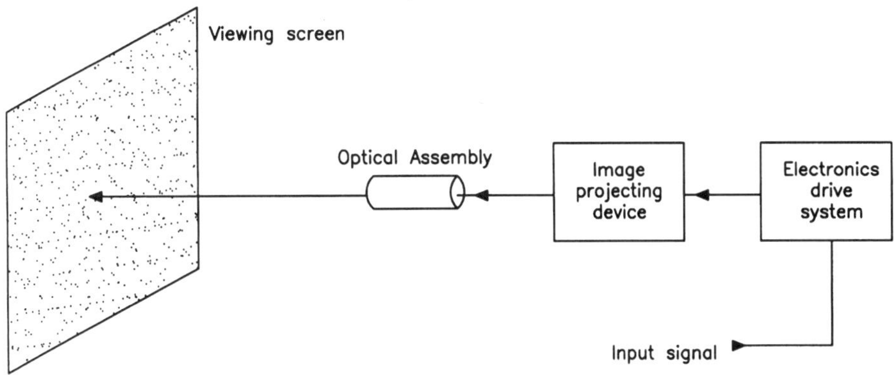


**Figure 21.1** Principal elements of a video projection system.

- Image source
- Drive electronics

The major differences between projection systems and direct-view displays are embodied in the first three elements, but the electronics assembly of a typical projection system is essentially the same as that of a direct-view system.

To provide an acceptable image, a projection system must approach or equal the performance of a direct-view device in terms of brightness, contrast, and resolution. Whereas brightness and contrast may be compromised to some extent, large displays must excel in resolution because of the tendency of viewers to be positioned at less than the normal standard-definition-image relative distance of 4 to 8 times the picture height from the viewing surface. Table 21.1 provides performance levels achieved by direct-view video displays and conventional film theater equipment.

Evaluation of overall projection system brightness *B*, as a function of its optical components, can be calculated using the following equation:

$$B = \frac{L_G \times G \times T \times R^M \times D}{4 \times W_G(F/\ N)^2 \times (1 + m)^2} \tag{21.1}$$

Where:
$L_G$ = luminance of the green source (CRT or other device)
$G$ = screen gain
$T$ = lens transmission
$R$ = mirror reflectance
$M$ = number of mirrors
$D$ = dichroic efficiency
$W_G$ = green contribution to desired white output (percent)

**Table 21.1 Performance Levels of Video and Theater Displays**

| Display System | Luminous Output (Brightness), nits (ft-L) | Contrast Ratio at Ambient Illumination (fc) | Resolution (TVL) |
|---|---|---|---|
| Television receiver | 200–400, 60–120 | 30:1 at 5 | 275 |
| Theater (film projector) | 34–69, 10–20[1] | 100:1 at 0.1[2] | 1000 and up |

[1] U. S. standard (PH-22.124-1961)
[2] Limited by lens flare.

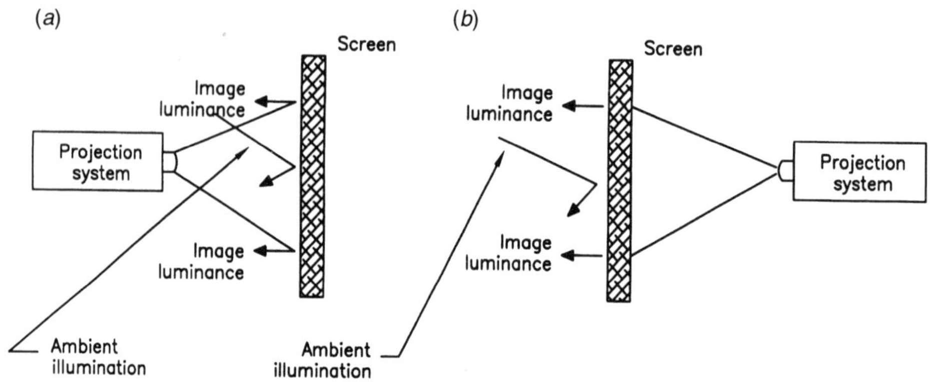


**Figure 21.2** Projection-system screen characteristics: (*a*) front projection, (*b*) rear projection.

$f/N$ = lens *f*-number
$m$ = magnification

For systems in which dichroics or mirrors are not employed, those terms drop out.

Two basic categories of viewing screens are employed for projection video displays. As illustrated in Figure 21.2, the systems are:

- Front projection, where the image is viewed from the same side of the screen onto which it is projected.
- Rear projection, where the image is viewed from the opposite side of the screen onto which it is projected.

Front projection depends upon reflectivity to provide a bright image, while rear projection requires high transmission to achieve that same characteristic. In either case, screen size influences display brightness inversely as follows:

$$B = \frac{L}{A} \tag{21.2}$$

**Table 21.2 Screen Gain Characteristics for Various Materials** (*After* [1].)

| Screen Type | Gain |
|---|---|
| Lambertian (flat-white paint, magnesium oxide) | 1.0 |
| White semigloss | 1.5 |
| White pearlescent | 1.5–2.5 |
| Aluminized | 1–12 |
| Lenticular | 1.5–2 |
| Beaded | 1.5–3 |
| Ektalite (Kodak) | 10–15 |
| Scotch-light (3M) | Up to 200 |

Where:
$B$ = apparent brightness (cd/m$^2$)
$L$ = projector light output (lm)
$A$ = screen viewing area (m$^2$)

Thus, for a given projector luminance output, viewed brightness varies in proportion to the reciprocal of the square of any screen linear dimension (width, height, or diagonal). An increase in screen width from the conventional aspect ratio of 4:3 (1.33) to an HDTV ratio of 16:9 (1.777) requires an increase in projector light output of approximately 33 percent for the same screen brightness.

To improve apparent brightness, directional characteristics can be designed into viewing screens. This property is termed *screen gain G*, and the previous equation becomes:

$$B = G \times \frac{L}{A} \tag{21.3}$$

Gain is expressed as screen brightness relative to a *lambertian* surface. Table 21.2 lists some typical front-projection screens and their associated gains.

Screen contrast is a function of the manner in which ambient illumination is treated. Figure 21.2 illustrates that a highly reflective screen (used in front projection) reflects ambient illumination, as well as the projected illumination (image). Thus, the reflected light tends to dilute contrast, although highly directional screens diminish this effect. A rear-projection screen depends upon high transmission for brightness but can capitalize on low reflectance to improve contrast. A scheme for achieving this is equivalent to the black matrix utilized in tricolor CRTs. Illustrated in Figure 21.3, the technique focuses projected light through lenticular lens segments onto strips of the viewing surface, allowing intervening areas to be coated with a black (nonreflective) material. The lenticular segments and black stripes normally are oriented in the vertical dimension to broaden the horizontal viewing angle. The overall result is a screen that transmits most of the light (typically 60 percent) incident from the rear, while absorbing a large percentage of the light (typically 90 percent) incident from the viewing side, thus providing greater contrast.

Rear-projection screens usually employ extra elements, including diffusers and directional correctors, to maximize brightness and contrast in the viewing area.

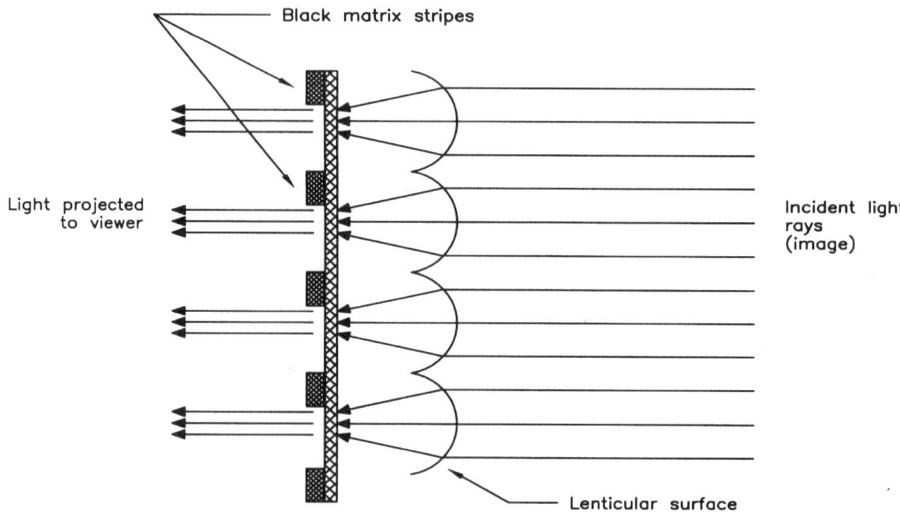


**Figure 21.3** High-contrast rear-projection system.

As with direct-view CRT screens, resolution can be affected by screen construction. This is not usually a problem with front-projection screens, although granularity or lenticular patterns can limit image detail. In general, any screen element (such as the matrix arrangement previously described) that quantizes the image (breaks it into discrete segments) limits resolution. For 525- and 625-line video, this factor does not provide the limiting aperture. High-resolution applications, however, require attention to this parameter.

## 21.2 Optical Projection Systems

Both refractive and reflective lens configurations have been used for the display of a CRT raster on screens that are 40-in diagonal or larger. The first attempts merely placed a lenticular Fresnel lens, or an inefficient *f*/1.6 projection lens, in front of a shadow-mask direct view tube, as shown in Figure 21.4*a*. The resulting brightness of no greater than 2 or 2 ft-L was suitable for viewing only in a darkened room. Figure 21.4*b* shows a variation on this basic theme. Three individual CRTs are combined with cross-reflecting mirrors and focused onto the screen. The in-line projection layout is shown in Figure 21.4*c* using three tubes, each with its own lens. This is the most common system used for multitube displays. Typical packaging to reduce cabinet size for front or rear projection is shown in Figures 21.4*d* and 21.4*e*, respectively.

Because of off-center positioning of the outboard color channels, the optical paths differ from the center channel, and keystone scanning height modulation is necessary to correct for differ-

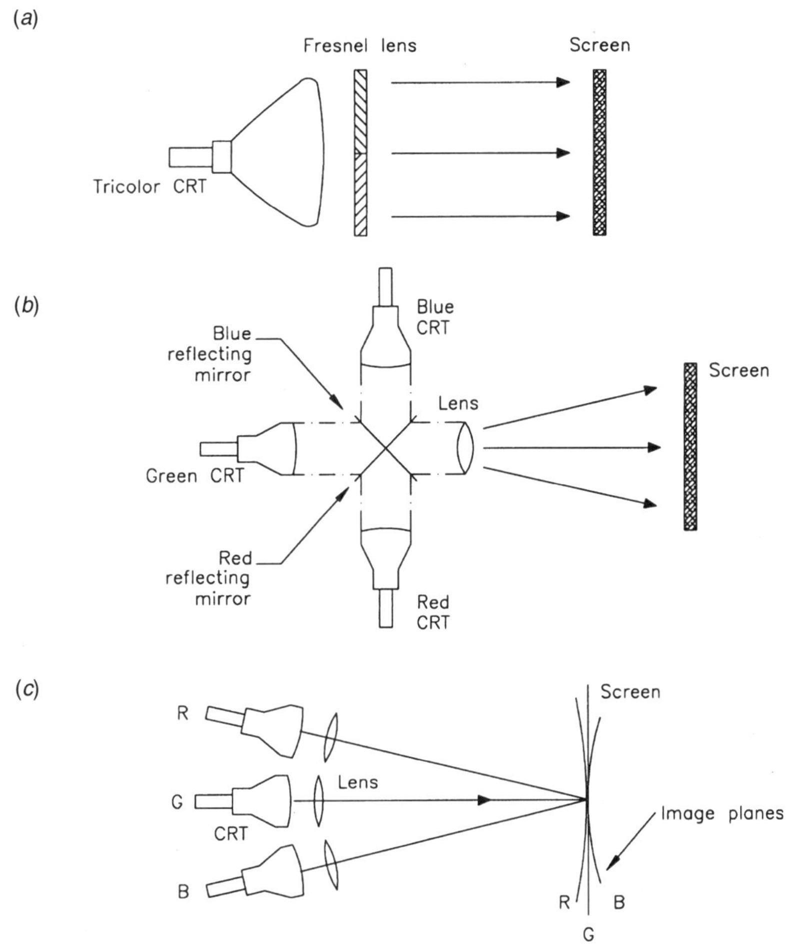


**Figure 21.4** Optical projection configurations: (*a*) single-tube, single-lens rear-projection system; (*b*) crossed-mirror trinescope rear projection; (*c*) 3-tube, 3-lens rear-projection system; (*d*, next page) folded optics, front projection with three tubes in-line and a dichroic mirror; (*e*, next page) folded optics, rear projection.

ences in optical throw from left to right. The problem, illustrated in Figure 21.5, is more severe for wide-screen formats.

Variables to be evaluated in choosing from the many available projection schemes include the following:

- Source luminance
- Source area

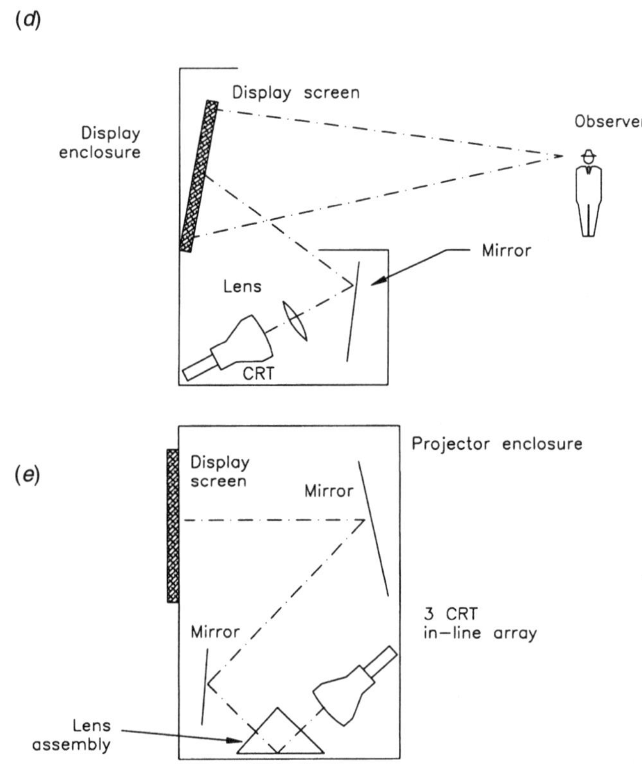


**Figure 21.4** continued

- Image magnification (screen size required)
- Optical-path transmission efficiency
- Light-collection efficiency (of the lens)
- Cost, weight, and complexity of components and corrective circuitry

The lens package is a critical factor in rendering a projection system cost-effective. The package must offer good luminance-collection efficiency (small *f*-number), high transmission, good *modulation transfer function* (MTF), light weight, and low cost.

The total light incident upon a projection screen is equal to the total light emerging from the projection optical system, neglecting losses in the intervening medium. Distribution of this light generally is not uniform. Its intensity is less at screen edges than at the center in most projection systems as a result of light-ray obliquity through the lens ($cos^4\ \theta$ *law*) and *vignetting effects* [1].

Light output from a lens is determined by collection efficiency and transmittance, as well as source luminance. Typical figures for these characteristics are:

- Collection efficiency: 15 to 25 percent
- Transmittance: 75 to 90 percent

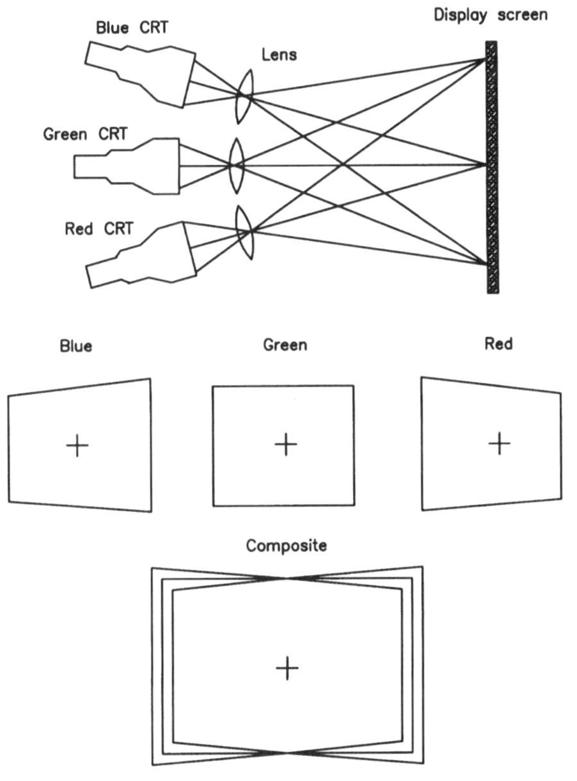


**Figure 21.5** A 3-tube in-line array. The optical axis of the outboard (red and blue) tubes intersect the axis of the green tube at screen center. Red and blue rasters show trapezoidal (keystone) distortion and result in misconvergence when superimposed on the green raster, thus requiring electrical correction of scanning geometry and focus.

Collection efficiency is partially a function of the light source, and the figure given is typical for a lambertian source (CRT) and lens having a half-field angle of approximately 25°.

## 21.2a Optical Distortions

Optical distortions are important to image geometry and resolution. Geometry generally is corrected electronically, for both pincushion/barrel effects and keystoning, which result from the fact that the three image sources are not coaxially disposed in the common in-line array.

Resolution, however, is affected by lens astigmatism, coma, spherical aberration, and chromatic aberration. The first three factors are dependent upon the excellence of the lens, but chromatic aberration can be minimized by using line emitters (monochromatic) or narrowband emitters for each of the three image sources. Because a specific lens design possesses different magnification for each of the three primary colors (the index of refraction varies with wavelength), throw distance for each also must be adjusted independently to attain proper registration.

To determine the ultimate luminance performance of the device, transmission, reflectance, and scattering by additional optical elements such as dichroic filters, optical-path folding mirrors, or corrective lenses must also be accounted for. Dichroics exhibit light attenuations of 5 to 30 percent, and mirrors can reduce light transmission by as much as 5 percent each. Front-surface mirrors exhibit minimum absorption and scattering, but they are susceptible to damage during cleaning.

Contrast also is affected by the number and nature of optical elements employed in the projection system. Each optical interface generates internal reflections and scattering, which dilute contrast and reduce MTF amplitude response. Optical coatings may be utilized to minimize these effects, but their contribution must be balanced against their cost.

## 21.2b Image Devices

CRTs and light valves are the two most common devices for creating images to be optically projected. Each is available in a multitude of variations. Projection CRTs have historically ranged in size from 1-in (2.5 cm) to 13-in (33 cm) diagonal (diameter for round envelope types). Because screen power must increase in proportion to the square of the magnification ratio, it is clear that faceplate dissipation for CRTs used in projection systems must be extremely high. Electrical-to-luminance conversion efficiency for common video phosphors is on the order of 15 percent [2]. A 50-in (1.3 m) diagonal measure screen at 60 ft-L requires a 5-in (12.7 cm) CRT to emit 6000 ft-L, exclusive of system optical losses, resulting in a faceplate dissipation of approximately 20 W in a 3-in (7.6 cm) by 4-in (10.2 cm) raster. A practical limitation for ambient air-cooled glass envelopes (to minimize thermal breakage) is 1 mW/mm$^2$ or 7.74 W for this size display. This incompatibility must be accommodated through either improved phosphor efficiency or reduced strain on the envelope via cooling. Because phosphor development is a mature science, maximum benefits are found in the latter course of action, with liquid-cooling assemblies employed to equalize differential strain on the CRT faceplate. Such an implementation produces an added benefit through reduction of phosphor *thermal quenching*, thereby supplying up to 25 percent more luminance output than is attainable in an uncooled device at equal screen dissipation [3].

A liquid-cooled CRT assembly, shown in Figure 21.6, depends upon a large heat sink to carry away and dissipate a substantial portion of the heat generated in the CRT. Large-screen projectors using such assemblies commonly operate CRTs at 4 to 5 times their rated thermal capacities. Economic constraints must be weighed against the added cost of cooling assemblies, however, and methods to improve phosphor conversion efficiency and optical coupling/transmission efficiencies continue to be investigated.

Concomitant to high power is high voltage and/or high beam current. Each has its benefits and penalties. Resolution, dependent on spot diameter, is improved by increased anode voltage and reduced beam current. For a 525- or 625-line display, spot diameter should be 0.006-in (0.16 mm) on the 3 × 4-in (7.6 × 10.2 cm) raster previously discussed. Higher-resolving-power displays require yet smaller spot diameters, but a practical maximum anode-voltage limit is 30 to 32 kV when x-radiation, arcing, and stray emission are considered. Exceeding 32 kV typically requires special shielding and CRT processing during assembly.

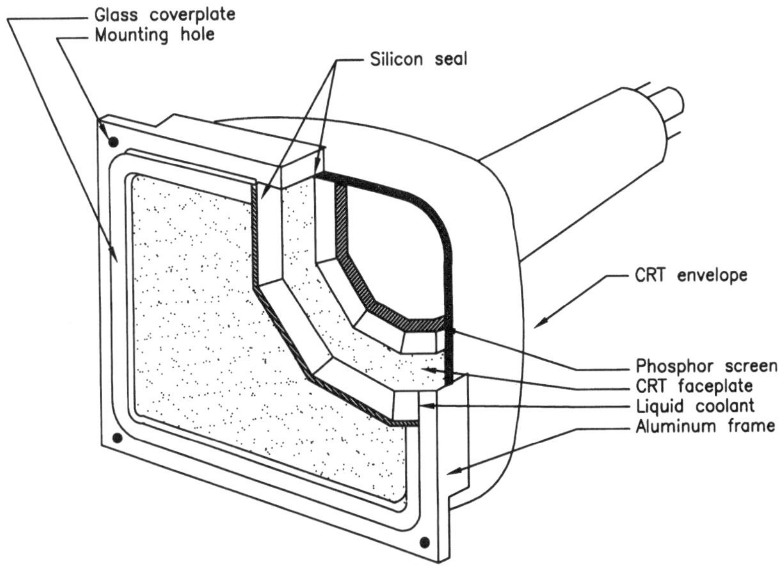


**Figure 21.6** Mechanical construction of a liquid-cooled projection CRT.

One form of the in-line array benefits from the relatively large aperture and light transmission efficiency of Schmidt reflective optics by combining the electron optics and phosphor screen with the projection optics in a single tube. The principal components of an integral Schmidt system are shown in Figure 21.7. Electrons emitted from the electron gun pass through the center opening in the spherical mirror of the reflective optical system to scan a metal-backed phosphor screen. Light from the phosphor (red, green, or blue, depending on the color channel) is reflected from the spherical mirror through an *aspheric corrector lens*, which serves as the face of the projection tube. Schmidt reflective optical systems are significantly more efficient than refractive systems because of the lower *f* characteristic and the reduced attenuation by glass in the optical path.

## Advanced CRTs for Projection Display

Among the promising technologies for improved CRT devices intended for HDTV applications is the YAG (*yttrium aluminum garnet*) luminescent screen. Prepared using a *liquid-phase epitaxy* (LPE) process, such devices can obtain high luminance and high definition, but the tube envelope—made from high-density sintered alumina—is costly, and the sealing process between the YAG faceplate and tube envelope is complicated. Screening is likewise difficult, and efficiency is a concern [4].

Further work has resulted in a 2–3 in. diagonal YAG phosphor screen for HDTV [5, 6]. The substrate material (i.e. the CRT faceplate—YAG), replaces conventional glass, offering high

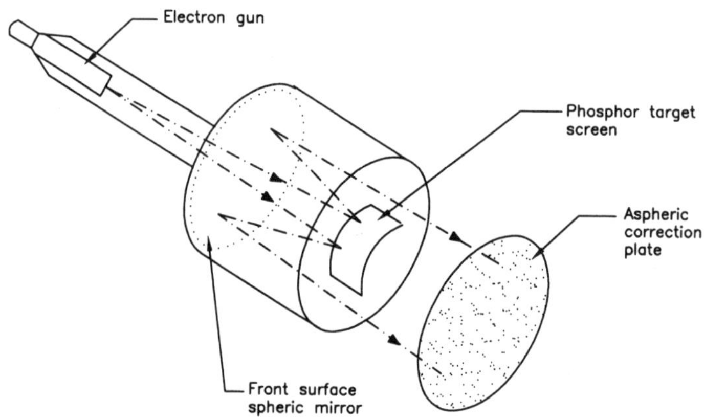


**Figure 21.7** Projection CRT with integral Schmidt optics.

thermal conductivity and optical clarity. The phosphor screen is prepared by depositing phosphor on YAG using improved centrifugal sendimentation screening techniques.

Additional development has resulted in a practical device [7]. The YAG phosphor screen faceplate has significantly higher thermal conductivity ($\lambda = 0.12$ W-cm$^{-1}$-K$^{-1}$ at room temperature) than a conventional glass faceplate ($\lambda = 0.01$ W-cm$^{-1}$-K$^{-1}$ at room temperature). Moreover, the new device has high mechanical strength, excellent insulation characteristics, and is much thinner ($D = 2$ mm) than glass ($D = 10$ mm, approximately).

Common rules of thermodynamics dictate that when the phosphor screen is in thermal equilibrium, the conducted thermal energy $Q$ per unit area of phosphor screen faceplate per unit time is equal to:

$$Q = \frac{\lambda \Delta T}{D} \tag{21.4}$$

Where:
$\lambda$ = thermal conductivity of the faceplate
$\Delta T$ = temperature difference between the inside and outside of faceplate
$D$ = thickness of faceplate

Using this equation, it can be determined that for the same electron excitation power density, the YAG phosphor screen temperature difference is only 1/60 that of the glass phosphor screen. By means of calculation and experiment, the temperature difference between the YAG phosphor screen inside and outside surfaces is only 1.8°C, compared to 110°C for a typical glass phosphor screen at 1-W/cm$^2$ electron excitation power density [7].

Through the use of liquid cooling, the YAG phosphor screen avoids to a great extent phosphor thermal quenching and burn out caused by high-density electron bombardment, which is a serious problem with conventional glass projection phosphor screens. Therefore, a device based on the YAG phosphor screen can obtain higher luminance and longer life than one based on a glass phosphor screen.

In order to further improve the luminance, chromaticity, and resolution on the YAG projection system, a multilayer interference filter an be applied between the phosphor layer and the YAG faceplate, comprising a number of layers that are alternately of high and low refractive index [8]. This filter has the transmittance characteristics of a shortwave pass for the central wavelength of light emitted by the phosphor. Therefore, large-angle light is reflected by the filter to the phosphor and is scattered with an angular redistribution, resulting in part of the light being re-emitted at small angles. This changes the angular distribution of the emitted light from Lambertian to a more peaked emission in the forward direction with a gain at the smaller angles. Using this approach, significantly improved collection of light by the lens is achieved, which increases the luminance on the projection screen. Furthermore, because the filter is a narrow-bandpass type, it changes the spectral distribution of the emitted light and results in an attenuation of unwanted long-wave components. This improves the chromaticity and reduces chromatic aberration.

## 21.2c Application Example

A CRT-based 61-in high-definition projection television conforming to the ATSC standard and intended for consumer applications has been developed (Hitachi) that provides 900 TV lines resolution and 550 Cd/m$^2$ peak brightness with a contrast ratio of 100. The key elements of the new display device include the following [8]:

- **Projection lens system**. The design requirements included minimum distortion, necessary because of the high-resolution displayed image, and high optical efficiency for maximum brightness. The system consists of a spherical glass lens and five aspherical plastic lenses, which are used primarily to control monochromatic aberrations.
- **Large diameter CRT gun**. A 36.5-mm diameter CRT neck was used to house the double-cylinder type *high focus voltage unipotential focusing* (Hi-UPF) electron gun. The gun has three lenses to finely control and shape the emitted beam with a minimum of distortion.
- **Deflection system**. In general, the deflection sensitivity of the deflection yoke (DY) and convergence yoke (CY) is reduced as a result of the larger neck diameter of the tube. Thus, more power must be delivered to the DY and CY devices. Coupled with this requirement was the higher horizontal scanning frequency of 33.75 kHz, rather than the conventional 15.73 kHz for a consumer television receiver. The higher frequency operation brought into play the *skin effect*, which required considerable changes in the windings to achieve the required performance.
- **Video driver circuit**. The requirements of high-definition operation demanded considerable improvements in the video driver, relative to a conventional display. To achieve high brightness, the circuit must operate up to approximately 150 V p-p at a bandwidth of 30 MHz. These performance levels are not unusual in themselves—high-end computer monitors achieve these levels and greater—but in the case of a consumer receiver, the cost also must be reasonable for commercial success.

- **Digital convergence system**. A two-mode convergence system—a *full scan mode* and a so-called *smooth wide mode*—was implemented to maintain the accuracy of beam landing across the face of the device. The convergence accuracy is 12 bits for vertical R, G, and B components. Horizontally, G resolution is 10 bits and both B and R are 9 bits. The resolution of the green channel is higher than that of the red and blue channels because the eye is more sensitive to green. Eight photo sensors are mounted on the screen frame to facilitate the auto convergence function.

The display is packaged as a rear-projection system. The 61-in diagonal picture size is set in a 16:9 aspect ratio. The effective optical resolution of the display is 900 × 1600 pixels. The total resolution in pixels is 1400 and the screen pitch is 0.72.

## 21.3 Light-Valve Systems

Light valves may be defined as devices that, like film projectors, employ a fixed light source modulated by an optical-valve intervening source and projection optics. Although light-valve displays have been commercially available for some time, it is still a rapidly developing discipline. Light-valve systems offer high brightness, variable image size, and high resolution. Progress in light-valve technology for HDTV depends upon developments in two key areas:

- Materials and technologies for light control
- Integrated electronic driving circuits for addressing picture elements

### 21.3a Eidophor Reflective Optical System

Light-valve systems are capable of producing images of substantially higher resolution than are required for 525-/625-line systems. They are ideally suited to large-screen theater displays of HDTV. The *Eidophor* system (Gretag) is in common usage.

In a manner similar to the operation of a film projector, a fixed light source is modulated by an optical valve system (Schlieren optics) located between the light source and the projection optics (see Figure 21.8). In the basic Eidophor system, collimated light—typically from a 2 kW xenon source (component 1 in the figure)—is directed by a mirror to a viscous oil surface in a vacuum by a grill of mirrored slits (component 3).

The slits are positioned relative to the oil-coated reflective surface so that when the surface is flat, no light is reflected back through the slits. An electron beam scanning the surface of the oil with a video picture raster (components 4 to 6) deforms the surface in varying amounts, depending upon the video modulation of the scanning beam. Where the oil is deformed by the modulated electron scanning beam, light rays from the mirrored slits are reflected at an angle that permits them to pass through the slits to the projection lens. The viscosity of the liquid is high enough to retain the deformation over a period slightly greater than a television field.

Projection of color signals is accomplished through the use of three units, one for each of the red, green, and blue primary colors converged on a screen.

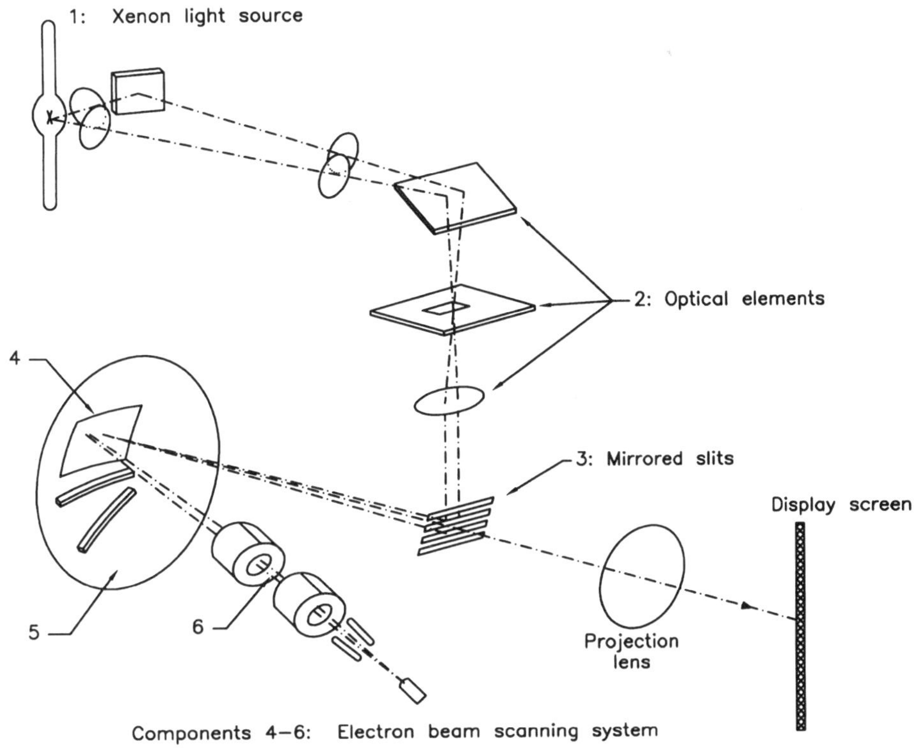


**Figure 21.8** Mechanical configuration of the Eidophor projector optical system.

## 21.3b Talaria Transmissive Color System

The *Talaria* system (General Electric) also uses the principle of deformation of an oil film to modulate light rays with video information. The oil film, however, is transmissive rather than reflective. In addition, for full-color displays, only one gun is used to produce red, green, and blue colors. This is accomplished in a single light valve by the more complex Schlieren optical system shown in Figure 21.9.

Colors are created by writing *diffraction grating*, or grooves, for each pixel on the fluid by modulating the electron beam with video information. These gratings break up the transmitted light into its spectral colors, which appear at the output bars where they are spatially filtered to permit only the desired color to be projected onto the screen.

Green light is passed through the horizontal slots and is controlled by modulating the width of the raster scan lines. This is accomplished by means of a high-frequency carrier, modulated by the green information applied to the vertical deflection plates. Magenta light, composed of red and blue primaries, is passed through the vertical slots and is modulated by diffraction gratings created at right angles (*orthogonal diffraction*) to the raster lines by velocity-modulating the elec-

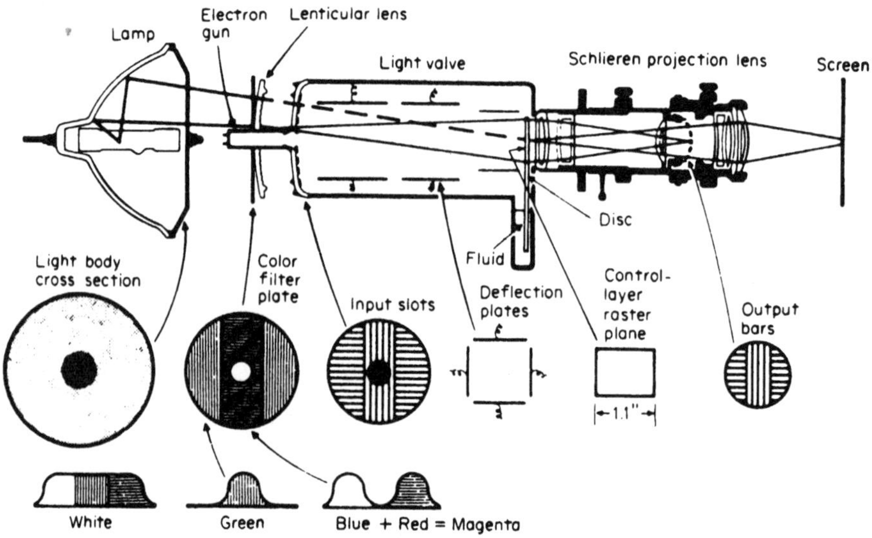


**Figure 21.9** Functional operation of the General Electric single-gun light-valve system.

tron beam in the horizontal direction. This typically is achieved by applying 16 and 12 MHz carrier signals for red and blue, respectively, to the horizontal deflection plates and modulating them with the red and blue video signals. The grooves created by the 16 MHz carrier have the proper spacing to diffract the red portion of the spectrum through the output slots while the blue light is blocked. For the 12 MHz carrier, the blue light is diffracted onto the screen while the red light is blocked. The three primary colors are projected simultaneously onto the screen in register as a full-color picture.

To meet the requirements of HDTV, the basic Talaria system can be modified as shown in Figure 21.10. In the system (Talaria MLV-HDTV), one monochromatic unit with green dichroic filters produces the green spectrum. Because of the high scan rate for HDTV, the green video is modulated onto a 30 MHz carrier instead of the 12 or 15 MHz used for 525- or 625-line displays. Adequate brightness levels are produced using a 700 W xenon lamp for the green light valve and a 1.3 kW lamp for the magenta (red and blue) light valve.

A second light valve with red and blue dichroic filters produces the red and blue primary colors. The red and blue colors are separated through the use of orthogonal diffraction axes. Red is produced when the writing surface diffracts light vertically. This is accomplished by negative amplitude modulation of a 120 MHz carrier, which is applied to the vertical diffraction plates of the light valve. Blue is produced when the writing surface diffracts light horizontally. This is accomplished by modulating a 30 MHz carrier with the blue video signal and applying it to the horizontal plate, as is done in the green light valve.

The input slots and the output bar system of the conventional light valve are used, but with wider spacing of the bars. Therefore, the resolution limit is increased. The wider bar spacing is

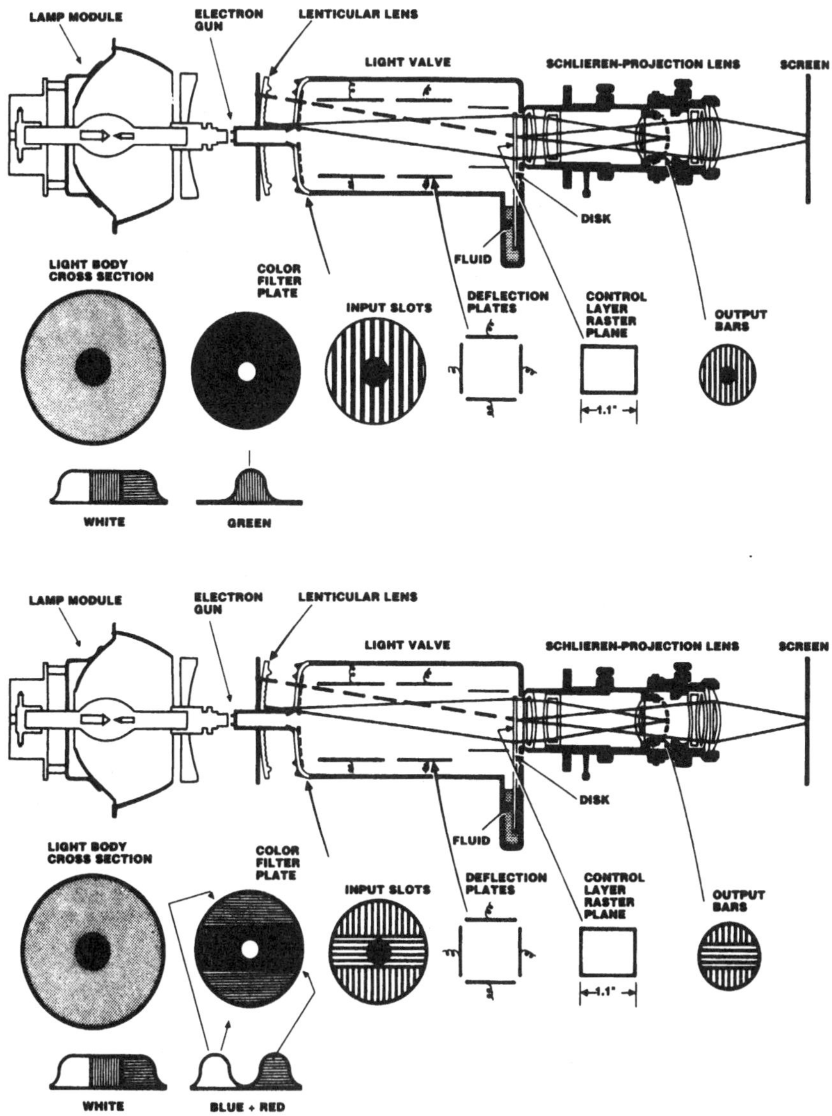


**Figure 21.10** Functional operation of the 2-channel HDTV light-valve system.

achievable because the red and blue colors do not have to be separated on the same diffraction axis as in the single light-valve system. This arrangement eliminates the cross-color artifact present with the single light-valve system, and therefore improves the overall colorimetric characteristics.

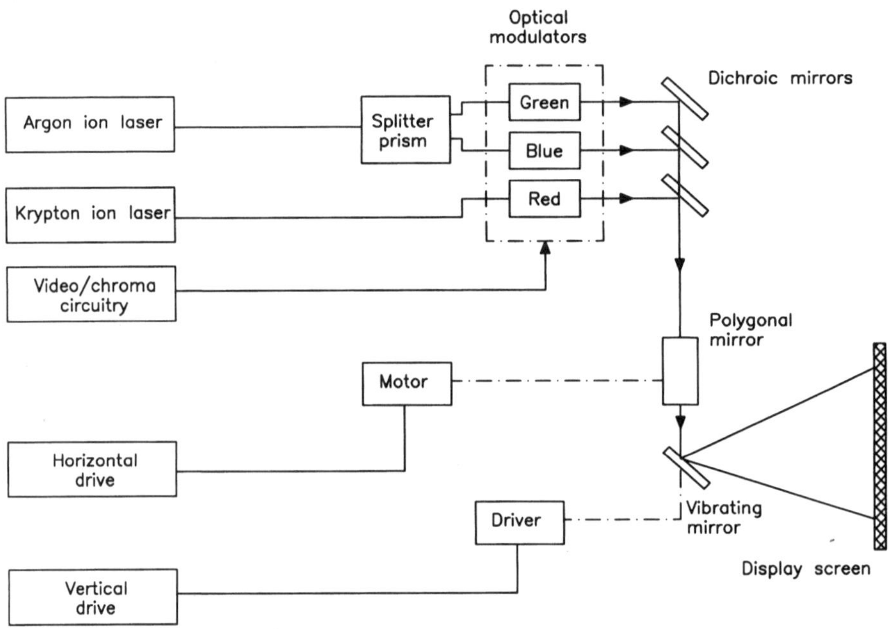


**Figure 21.11** Block diagram of a laser-scanning projection display.

High-performance electron guns help provide the required resolution and modulation efficiency for HDTV systems. The video carriers are optimized to increase the signal bandwidth capability to approximately 30 MHz.

In the 3-element system, all three devices are monochrome light valves with red, blue, and green dichroic filters. The use of three independent light valves improves color brightness, resolution, and colorimetry. Typically, the three light valves are individually illuminated by xenon arc lamps operating at 1 kW for the green and at 1.3 kW for the red and blue light valves.

The contrast ratio is an important parameter in light-valve operation. The amount of light available from the arc lamp is basically constant; the oil film modulates the light in response to picture information. The key parameter is the amount of light that is blocked during picture conditions when totally dark scenes are being displayed (the *darkfield* performance). Another important factor is the capability of the display device to maintain a linear relationship between the original scene luminance and the reproduced picture. The amount of predistortion introduced by the camera must be compensated for by an opposite amount of distortion at the display device.

## 21.3c Laser-Beam Projection Scanning System

Several approaches to laser projection displays have been implemented. The most successful employs three optical laser light sources whose coherent beams are modulated electro-optically

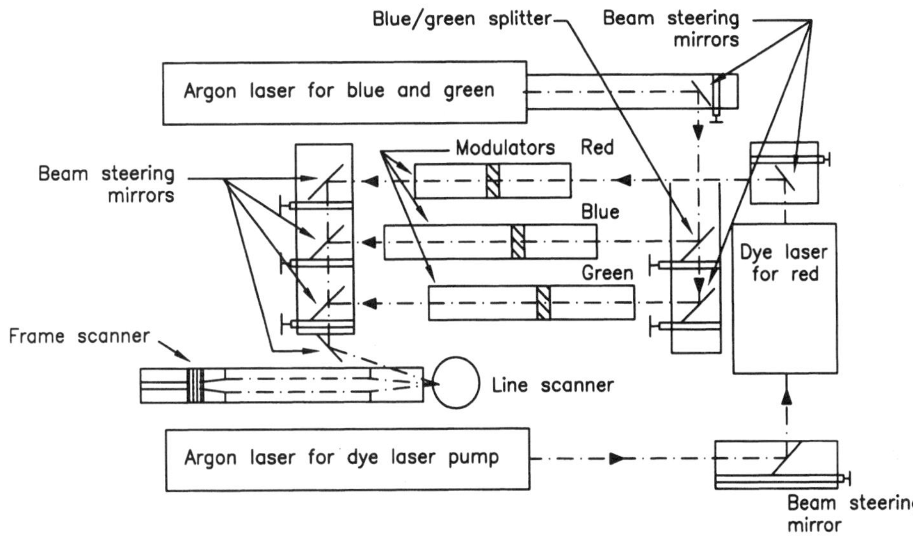


**Figure 21.12** Configuration of a color laser projector. (*After* [9].)

and deflected by electromechanical means to project a raster display on a screen. The scanning functions typically are provided by a rotating polygon mirror and a separate vibrating mirror. A block diagram of the basic system is shown in Figure 21.11.

The flying spots of light used in this approach are one scan line (or less) in height and a small number of pixels wide. This means that any part of the screen may be illuminated for only a few nanoseconds. The scanned laser light projector is capable of high contrast ratios (as high as 1000:1) in a darkened environment. A laser projector, however, may be subject to a brightness variable referred to as *speckle*. Speckle is a sparkling effect resulting from interference patterns in coherent light. The effect causes a flat, dull projection surface to look as if it has a beaded texture. This tends to increase the perception of brightness, at the expense of image quality.

Figure 21.12 shows the configuration of a laser projector using continuous-wave lasers and mechanical scanners. The requirements for the light wavelengths of the lasers are critical. The blue wavelength must be shorter than 477 nm, but as long as possible. The red wavelength should be longer than 595 nm, but as short as possible. Greens having wavelengths of 510, 514, and 532 nm have been used with success. Because laser projectors display intense colors, small errors in color balance can result in significant distortions in gray scale or skin tone. The requirement for several watts of continuous-wave power further limits the usable laser devices. Although several alternatives have been considered, color laser projectors typically use argon ion lasers for blue and green, and an argon ion pumped-dye laser for red.

Production of a conventional video signal requires a modulator with a minimum operating frequency of 6 MHz. Several approaches are available. The bulk acousto-optic modulator is well suited to this task in high-power laser projection. A directly modulated laser diode is used for

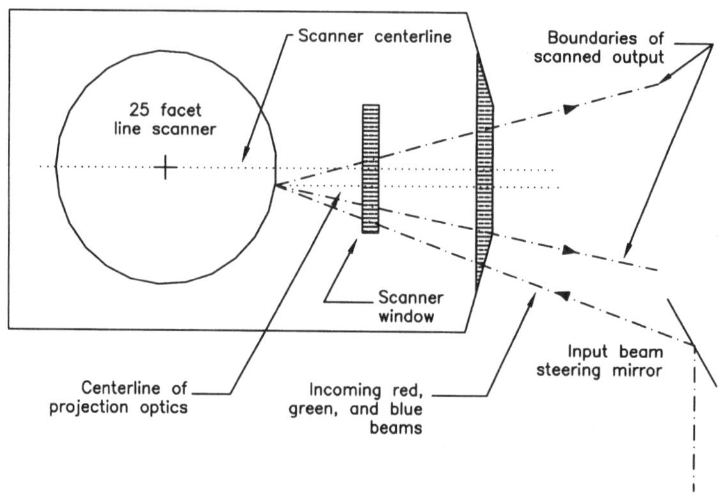


**Figure 21.13** Rotating polygon line scanner for a laser projector. (*After* [9].)

some low-power operations. The modulator does not absorb the laser beam; instead, it deflects it as needed. The angle of deflection is determined by the frequency of the acoustic wave.

The scanner is the component of the laser projector that moves the point of modulated light across and down the image plane. Several types of scanners may be used, including mechanical, acousto-optic, and electro-optic. Two categories of scanning devices are used in the system:

- *Line scanner*, which scans the beam in horizontal lines across the screen. Lines are traced at 15,000 to 70,000 times per second. The rotating polygon scanner is commonly used, featuring 24 to 60 (or more) mirrored facets. Figure 21.13 illustrates a mechanical rotating polygon scanner.
- *Frame scanner*, which scans the beam vertically and forms frames of horizontal scan lines. The frame scanner cycles 50 to 120 (or more) times per second. A galvanometer-based scanner typically is used in conjunction with a mirrored surface. Because the mirror must fly back to the top of the screen in less than a millisecond, the device must be small and light. Optical elements are used to force the horizontally scanned beam into a small spot on the frame scanner mirror. The operating speed of the deflection components may be controlled from an internal clock, or it may be locked to an external time base.

The laser projector is said to have *infinite focus*. To accomplish this, optics are necessary to keep the scanned laser beam thin. Such optics allow a video image to remain in focus from approximately 4 ft to infinity. For high-resolution applications, the focus is more critical.

Heat is an undesirable by-product of laser operation. Most devices are water-cooled and use a heat exchanger to dump the waste heat.

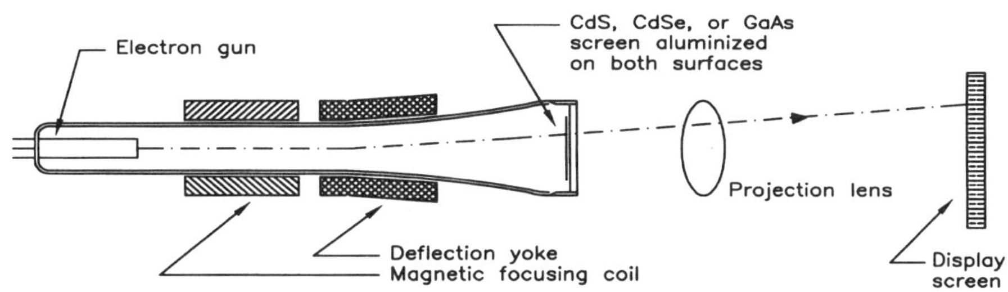


**Figure 21.14** Laser-screen projection CRT.

## 21.3d Electron-Beam Projection System

Another approach to large-screen display employs an electron-beam pumped monocrystalline screen in a CRT to produce a 1-in raster. The raster screen image is projected by conventional optics, as shown in Figure 21.14. This technology promises three optical benefits:

- High luminance in the image plane
- Highly directional luminance output for efficient optical coupling
- Compact, lightweight, and inexpensive projection optics

Full-color operation is accomplished using three such devices, one for each primary color.

# 21.4 LCD Projection Systems

Liquid crystal displays have been widely employed in high-information content systems and high-density projection devices. The steady progress in active-matrix liquid crystal displays, in which each pixel is addressed by means of a thin-film transistor (TFT), has led to the development of full-color video TFT-addressed *liquid crystal light valve* (LCLV) projectors. Compared with conventional CRT-based projectors, LCLV systems have a number of advantages, including:

- Compact size
- Light weight
- Low cost
- Accurate color registration

Total light output (brightness) and contrast ratio, however, are issues of concern. Improvements in the transmittance of polarizer elements have helped increase display brightness. By arrangement of the direction of the polarizers, the LCD can be placed in either the *normally black* (NB) or the *normally white* (NW) mode. In the NW mode, light will be transmitted through the cell at $V_{off}$ and will be blocked at $V_{on}$. The opposite situation applies for the NB mode. A high-contrast display cannot be obtained without a satisfactory dark state.

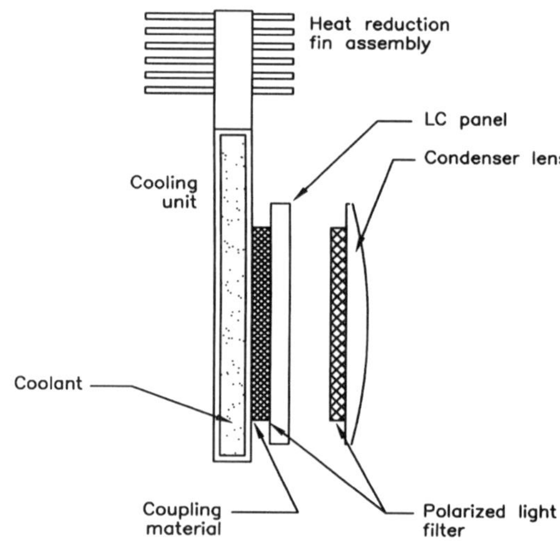


**Figure 21.15** Structure of a liquid-cooled LC panel. (*After* [10].)

Cooling the LC panels represents a significant technical challenge as the light output of projection systems increases. The contrast ratio of the displayed image will decrease as the temperature of the LC panels rises. Furthermore, long-term operation under conditions of elevated temperature will result in shortened life for the panel elements.

The conventional approach to cooling has been to circulate forced air over the LC panels. This approach is simple and works well for low- to medium-light output systems. Higher-power operation, however, may require liquid-cooling, as illustrated in Figure 21.15. The cooling unit, mounted behind the LC panel, is made of frame glass and two glass panels. The coolant is a mixture of water and ethylene glycol, which prevents the substance from freezing at low temperatures.

Figure 21.16 compares the temperature in the center of the panel for conventional air-cooling and for liquid-cooling. Because LC panel temperature is a function of the brightness distribution of the light source, the highest temperature is at the center of the panel, where most of the light usually is concentrated. As the amount of light from the lamp increases, the cooling activity of the liquid-based system accelerates. To carry waste heat away from the projector, cooling air is directed across the heat-reduction fins.

In addition to the basic LCD projector, a number of variations on the fundamental technology have been developed, including:

- Homeotropic LCLV
- Laser-addressed LCLV
- Image light amplifier

The applications of these technologies are discussed in the following sections.

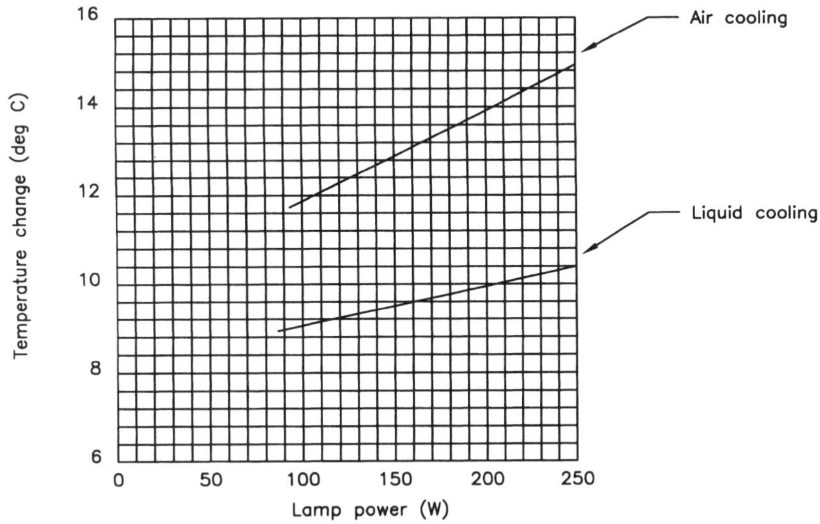


**Figure 21.16** LC panel temperature dependence on lamp power in a liquid crystal projector. (*After* [10].)

## 21.4a Homeotropic LCLV

The *homeotropic* (perpendicular alignment) LCLV (Hughes) is based on the optical switching element shown in Figure 21.17 [11]. Sandwiched between two transparent idium-tin oxide electrodes are a layer of cadmium sulfide photoconductor, a cadmium telluride light-blocking layer, a dielectric mirror, and a 6-μm-thick layer of liquid crystal. An ac bias voltage connects the transparent electrodes. The light-blocking layer and the dielectric mirror are thin and have high dielectric constants, so the ac field is primarily across the photoconductor and the liquid crystal layer. When there is no writing light impinging on the photoconductor, the field or voltage drop is primarily across the photoconductor and not across the liquid crystal. When a point on the photoconductor is activated by a point of light from an external writing CRT, the impedance at that point drops and the ac field is applied to the corresponding point in the liquid crystal layer. The lateral impedances of the layers are high, so the photocarriers generated at the point of exposure do not spread to adjacent points. Thus, the image from the CRT that is exposing the photoconductor is reproduced as a voltage image across the liquid crystal. The electric field causes the liquid crystal molecules to rotate the plane of polarization of the projected light that passes through the liquid crystal layer.

The homeotropic LCLV provides a perpendicular alignment in the *off* state (no voltage across the liquid crystal) and a pure optical birefringence effect of the liquid crystal in the *on* state (voltage applied across the liquid crystal). To implement this electro-optical effect, the liquid crystal layer is fabricated in a perpendicular alignment configuration; the liquid crystal molecules at the electrodes are aligned with their long axes perpendicular to the electrode surfaces. In addition,

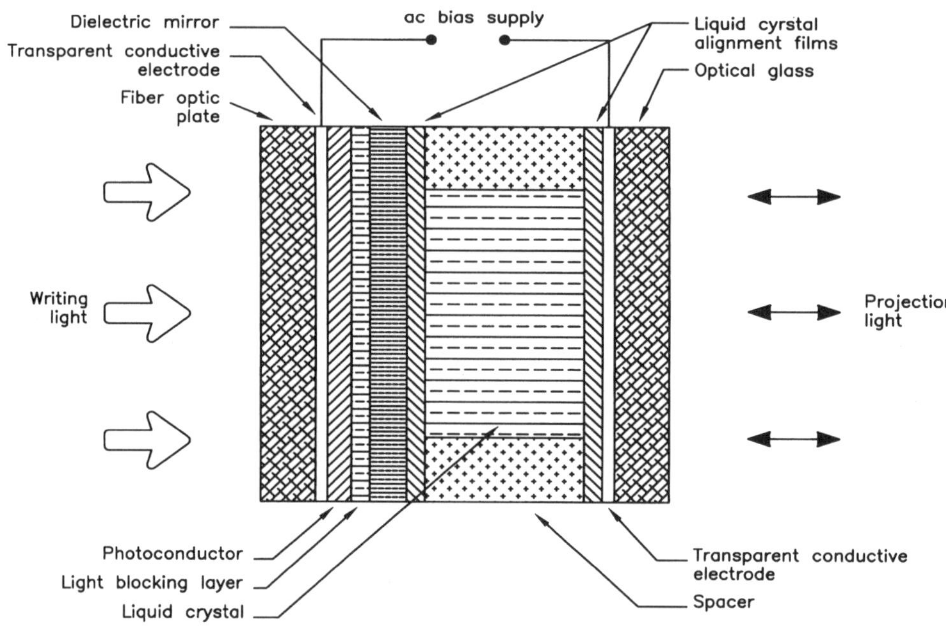


**Figure 21.17** Layout of a homeotropic liquid crystal light valve. (*After* [11].)

they are aligned with their long axes parallel to each other along a preferred direction that is fabricated into the surface of the electrodes.

A conceptual drawing of an LCLV projector using this technology is shown in Figure 21.18. Light from a xenon arc lamp is polarized by a polarizing beam-splitting prism. The prism is designed to reflect *S* polarized light (the polarization axis, the E-field vector, is perpendicular to the plane of incidence) and transmit *P* polarized light (the polarization axis, the E-field vector, is parallel to the plane of incidence). The *S* polarization component of the light is reflected to the light valve. When the light valve is activated by an image from the CRT, the reflecting polarized light is rotated 90° and becomes *P* polarized. The *P* polarization component transmits through the prism to the projection lens. In this type of system, the prism functions as both a polarizer and an analyzer. Both monochrome and full-color versions of this design have been implemented. Figure 21.19 shows a block diagram of the color system.

The light valve is coupled to the CRT assembly by a fiber optic backplate on the light valve and a fiber optic faceplate on the CRT.

The heart of this system is the fluid-filled prism assembly. This device contains all of the filters and beam-splitter elements that separate the light into different polarizations and colors (for the full-color model). Figure 21.20 shows the position of the beam-splitter elements inside the prism assembly.

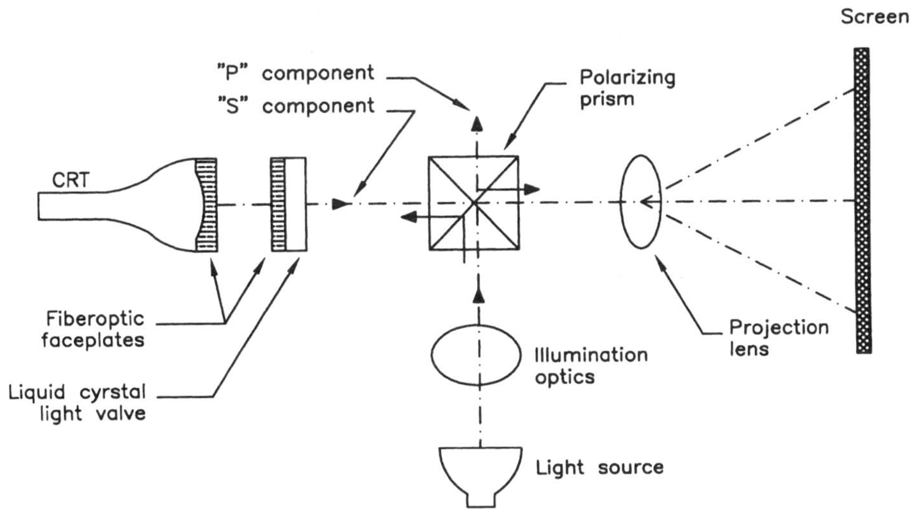


**Figure 21.18** Block diagram of an LCLV system. (*After* [11].)

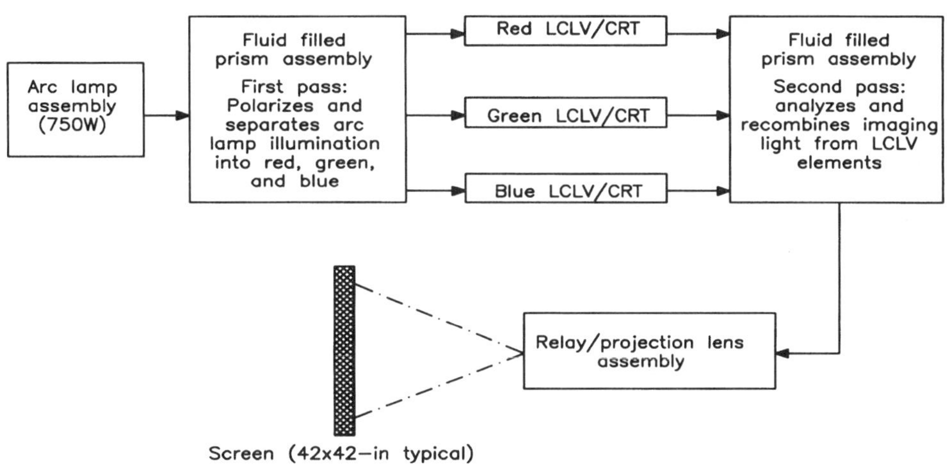


**Figure 21.19** Functional optical system diagram of an LCLV projector. (*After* [11].)

The light that exits the arc lamp housing enters the prism through the ultraviolet (UV) filter. The UV filter is a combination of a UV reflectance dichroic on a UV absorption filter. The UV energy below 400 nm can be destructive to the life of the LCLV. Two separate beam-splitters *pre-*

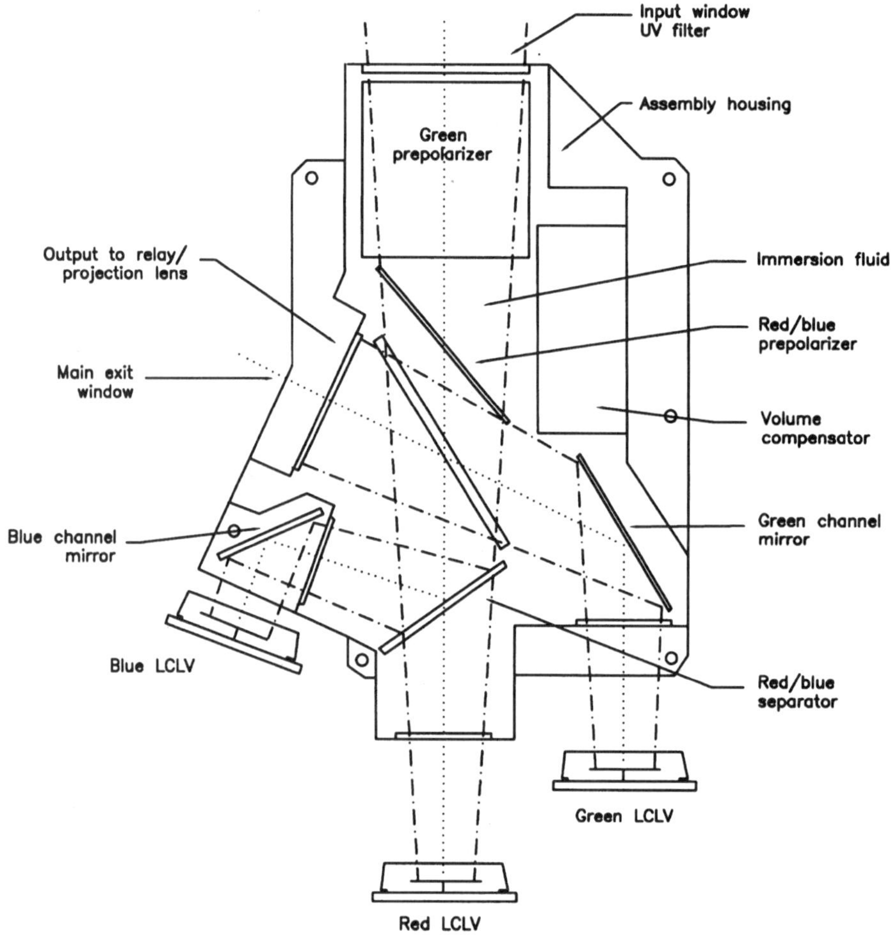


**Figure 21.20** Layout of the fluid-filled prism assembly for an LCLV projector. (*After* [11].)

*polarize* the light: the green prepolarizer and the red/blue prepolarizer. Prepolarizers are used to increase the extinction ratio of the two polarization states. The green prepolarizer is designed to reflect only the *S* polarization component of green light. The *P* polarization component of green and both the *S* and *P* polarizations of red and blue are transmitted through the green prepolarizer. The green prepolarizer is oriented 90° with respect to the other beam-splitters in the prism. This change of orientation means that the transmitted *P* component of the green light appears to be *S* polarized relative to the other beam-splitters in the prism assembly. The red/blue prepolarizer is designed to reflect the *S* polarization component of red and blue. The remaining green that is now *S* polarized and the red and blue *P* polarized components are transmitted to the main beam-splitter.

The main beam-splitter is a broadband polarizer designed to reflect all visible *S* polarized light and transmit all visible *P* polarized light. The *S* polarized green light is reflected by the main beam-splitter to the green mirror. The green mirror reflects the light to the green LCLV. The *P* polarized component of red and blue transmits through the main polarizer. The red/blue separator reflects the blue light and transmits the red to the respective LCLVs.

Image light is rotated by the light valve and reflected to the prism. The rotation of the light or phase change (*P* polarization to *S* polarization) allows the image light from the red and blue LCLVs to reflect off the main beam-splitter toward the projection lens. When the green LCLV is activated, the green image light is transmitted through the main beam-splitter.

The fluid-filled prism assembly is constructed with thin-film coatings deposited on quartz plates and immersed in an index-matching fluid. Because polarized light is used, it is necessary to eliminate stress birefringence in the prism, which, if present, would cause contrast degradation and background nonuniformity in the projected image. Using a fluid-filled prism effectively eliminates mechanically and thermally induced stress birefringence, as well as residual stress birefringence. Expansion of the fluid with temperature is compensated for by the use of a bellows arrangement.

## 21.4b Laser-Addressed LCLV

The laser-addressed liquid crystal light valve is a variation on the CRT-addressed LCLV that was discussed in the previous section. Instead of using a CRT to write the video information on the LC element, however, a laser is used [12]. The principal advantage of this approach is the small spot size possible with laser technology, a factor that increases the resolution of the displayed image. In addition, a single beam from a laser scanner can be directed to three separate RGB light valves; the use of only one source for the image reduces the potential for convergence problems. A block diagram of the scheme is shown in Figure 21.21. Major components of the system include:

- Laser
- Video raster scanner
- Polarizing multiplexing switches
- LCLV assemblies
- Projection optics

Projector operation can be divided into two basic subsystems:

- Input subsystem, which includes the laser, scanning mechanism, and polarizing beam-splitters
- Output subsystem, which includes the LCLV assemblies, projection optics, and light source

The output subsystem is similar in nature to the homeotropic LCLV discussed previously.

The heart of the input subsystem is the *laser raster scanner* (LRS). The X-Y deflection system incorporated into the LRS uses four acousto-optic devices to achieve all solid-state modulation and scanning. A diode-pumped Nd:YAG, doubled to 535 nm (18 mW at 40:1 polarization ratio), is used as the laser source for the basic raster scan system.

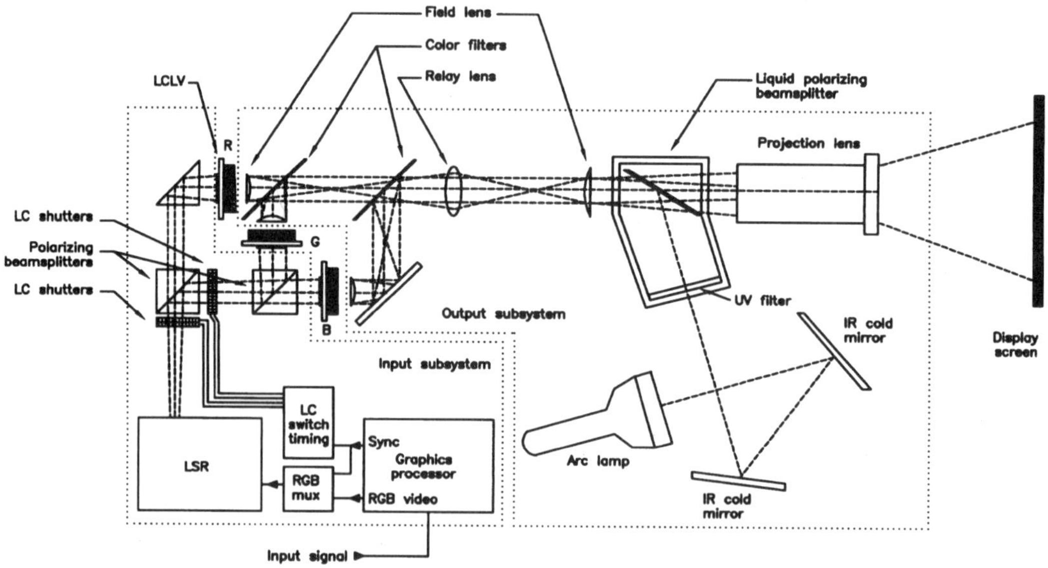


**Figure 21.21** Block diagram of a color laser-addressed light-valve display system. (*After* [12].)

Because only a single-channel scanner is used to drive the LCLV assemblies, a method is required for sequencing the RGB video fields to their respective light valves. This sequencing is accomplished through the use of LC polarization switches. These devices act in a manner similar to the LCLV in that they change the polarization state of incident light. In this case, the intent is to rotate the polarization by 0° or 90° and exploit the beam-steering capabilities of polarizing beam-splitters to direct the image field to the proper light valve at the proper time.

As shown in Figure 21.21, each switch location actually consists of a pair of individual switches. This feature was incorporated to achieve the necessary fall time required by the LRS to avoid overlapping of the RGB fields. Although these devices have a short rise time (approximately 100 μs), their decay time is relatively long (approximately 1 ms). By using two switches in tandem, compensation for the relatively slow relaxation time is accomplished by biasing the second switch to relax in an equal, but opposite, polarization sense. This causes the intermediate polarization state that is present during the switch-off cycle to be nulled out, allowing light to pass through the LC switch during this phase of the switching operation.

## 21.4c Image Light Amplifier (ILA)

The Image Light Amplifier (Hughes, JVC) combines the advantages of a CRT-type projector with the high-brightness capability of a xenon arc lamp projection source. The enabling technology of the projector is the ILA modulator, which accepts a low-intensity image from a CRT and replicates the image on a high-intensity white xenon arc lamp beam. The ILA uses solid-state

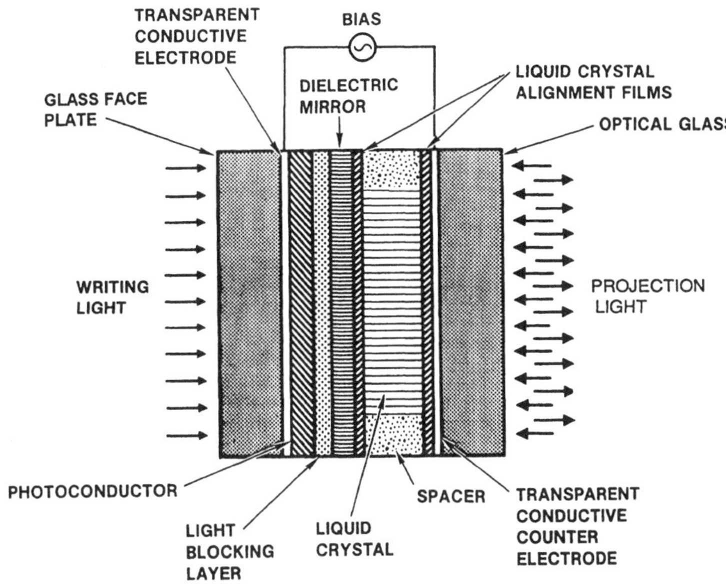


**Figure 21.22** Cross-sectional diagram of the Image Light Amplifier light valve. (*After* [13]. *Courtesy of Hughes-JVC Technology.*)

thin-film and liquid crystal technologies to produce a system with up to 12,000 lm output, at a contrast ratio of 1000:1. Resolutions of 2800 × 1500 TV lines (and higher) have been produced [13]. The ILA projector incorporates three modulators for the RGB channels of a color system.

The ILA light valve (ILA-LV) is a spatial light modulator that accepts a low-intensity input image and converts it, in real time, into an output image with light from another source. [14]. In the system, the image light sources are high-resolution CRTs, and the output light source is a xenon arc lamp. The ILA-LV is designed to operate in a reflective mode so that the input CRT and output xenon light beam are incident on opposite faces of the device. A cross-sectional diagram of the ILA-LV is shown in Figure 21.22. Note the similarities to the homeotropic LCLV shown in Figure 21.17.

The basic ILA projector optical channel is shown in Figure 21.23. This is the building block for the full-color system, which is shown in Figure 21.24. The input image, provided by the high-resolution CRT, is imaged on the ILA-LV through a relay lens. The xenon arc lamp and condensing optics provide the output projection light beam, which is linearly polarized by a McNeille-type *polarizing beam-splitter* (PBS) before reaching the ILA-LV. The PBS polarizes the light to a high extinction ratio without absorption. The projected beam passes through the liquid crystal, reflects from the dielectric mirror, and passes through the liquid crystal again before returning to the polarizing beam-splitter.

As the beam passes through the liquid crystal, the direction of the linearly polarized light is rotated in direct response to the level of input image modulation of the liquid crystal birefringence. The PBS then operates on the output image from the ILA-LV, passing rotated polarized

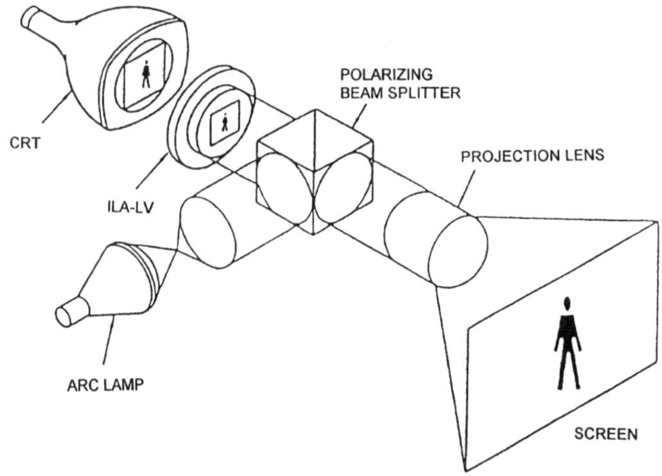


**Figure 21.23** Optical channel of the basic ILA projector system. (*After* [13]. *Courtesy of Hughes-JVC Technology.*)

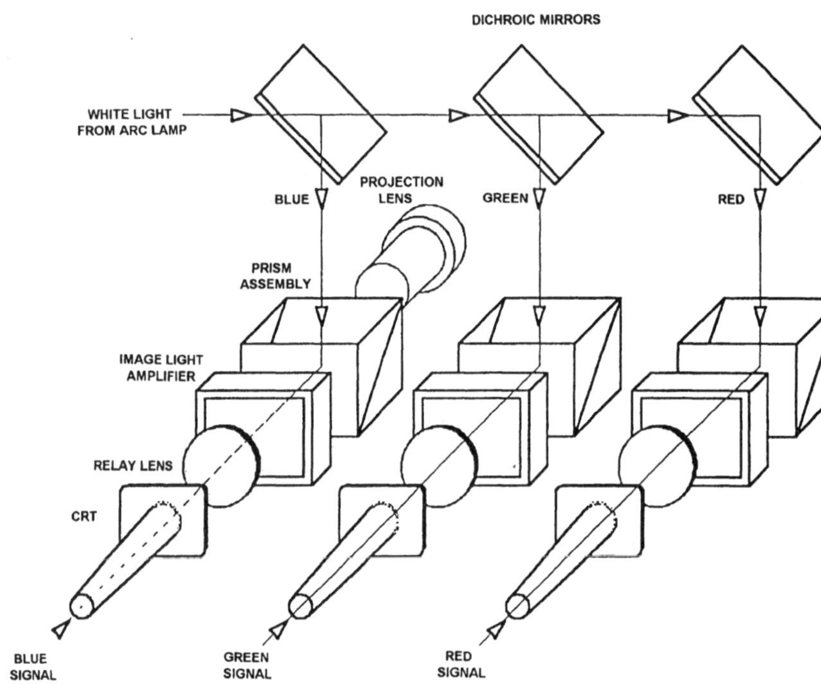


**Figure 21.24** Full-color optical system for the ILA projector (Series 300/400). (*After* [13]. *Courtesy of Hughes-JVC Technology.*)

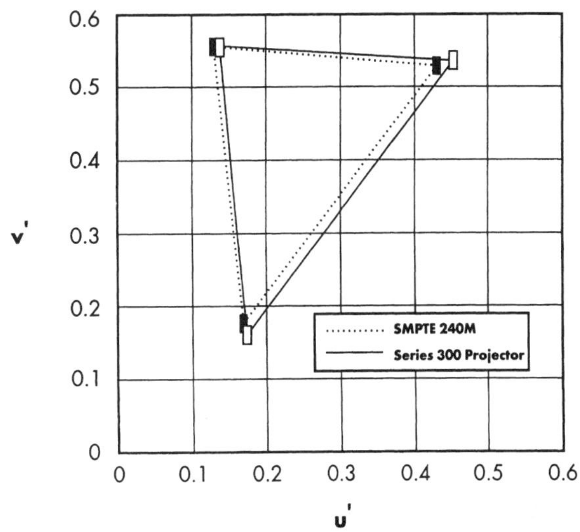


**Figure 21.25** Color gamut of the ILA projector (Series 300/400) compared with SMPTE 240M. (*After* [13]. *Courtesy of Hughes-JVC Technology.*)

light to the projection lens and returning nonrotated light toward the lamp. Finally, the projection lens focuses and magnifies the ILA-LV image onto a screen.

Among the benefits of this system are:

- High brightness
- Undefined pixel structure (which permits a variety of input signal types and formats)
- High resolution
- Good contrast ratio
- Excellent color rendition, as illustrated in Figure 21.25.

## 21.5 Mechanical Switching Systems

A class of projection display systems has emerged based on an electromechanical architecture. Such devices, a subset of *microelectromechanical systems* (MEMS), offer a number of benefits, not the least of which is intrinsic simplicity.

### 21.5a Digital Micromirror Device

DMD is a semiconductor-based array of fast, reflective digital light switches that precisely control a light source using a binary pulse-width modulation (PWM) technique [14]. Individual DMD elements can be combined with image processing, memory, a light source, and optics to form a *Digital Light Processing* (DLP, Texas Instruments) system capable of projecting high-resolution color images. DLP-based projection displays are well suited to high-brightness and high-resolution applications. Attributes of the technology include the following:

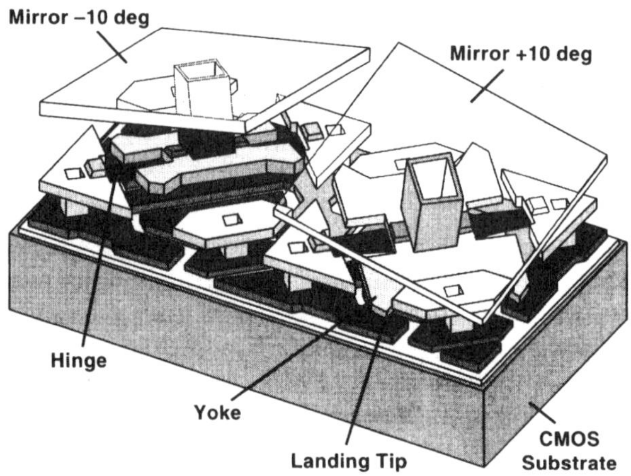


**Figure 21.26** A pair of DMD pixels with one mirror shown at –10° and the other at +10°. (After [14]. Courtesy of Texas Instruments.)

- The digital light switch is reflective and has a high *fill factor*, resulting in high optical efficiency at the pixel level and low pixelation effects in the projected image.
- As the resolution and size of the DMD array increases, the overall system optical efficiency grows because of higher lamp-coupling efficiency.
- The DMD operates with conventional CMOS voltage levels (5 V), so integrated row and column drivers are readily employed to minimize the complexity and cost of scaling to higher resolutions.
- Because the DMD is a reflective technology, the DMD chip can be effectively cooled through the chip substrate, thus facilitating the use of high-power projection lamps without thermal degradation of the DMD.

The DMD light switch, shown in Figure 21.26, is a member of a class of devices known as microelectromechanical systems. Other MEMS devices include pressure sensors, accelerometers, and microactuators. The DMD is monolithically fabricated by CMOS-like processes over a CMOS memory element. Each light switch has a 16-μm-square aluminum mirror that can reflect light in one of two directions, depending on the state of the underlying memory cell. Rotation of the mirror is accomplished through electrostatic attraction produced by voltage differences developed between the mirror and the underlying memory. With the memory cell in the *on* (1) state, the mirror rotates to +10°; with the memory cell in the *off* (0) state, the mirror rotates to –10°.

When the DMD is combined with a suitable light source and projection optics, as illustrated in Figure 21.27, the mirror reflects incident light either into or out of the pupil of the projection lens by a simple beam-steering process. As a result, the (1) state of the mirror appears bright, and

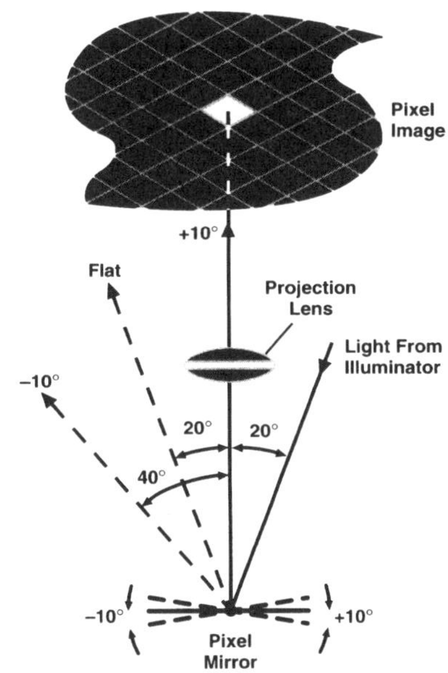


**Figure 21.27** The basics of DMD optical switching. (*After* [14]. *Courtesy of Texas Instruments.*)

the (0) state of the mirror appears dark. Compared with diffraction-based light switches, the beam-steering action of the DMD light switch provides a favorable tradeoff between contrast ratio and the overall brightness efficiency of the system.

Image gray scale is achieved through pulse-width modulation of the incident light. Color is achieved by using color filters, either stationary or rotating, in combination with one, two, or three DMD chips. A detailed photo of a DMD element is shown in Figure 21.28.

The simultaneous update of all mirrors produces an inherently low flicker display; there is no line-to-line temporal phase shift. Furthermore, a bit-splitting PWM algorithm produces short-duration light pulses that are distributed uniformly throughout the video field time, eliminating a temporal decay in brightness.

DLP optical systems have been designed in a variety of configurations, distinguished by the number of DMD chip arrays in the system [15]. The 1- and 2-chip schemes rely on a rotating color disk to time-multiplex the colors. The 1-chip configuration is used for lower brightness applications and is the most compact. The 2-chip systems yield higher brightness performance, but are primarily intended to compensate for the color deficiencies resulting from spectrally imbalanced lamps (e.g., the red deficiency in many metal halide lamps). For applications requiring the highest brightness, 3-chip systems are used.

A 3-chip DLP optical system is shown in Figure 21.29. Because the DMD is a simple array of reflective light switches, no polarizers are required. Light from a metal halide or xenon lamp is collected by a condenser lens. For proper operation of the DMD light switch, this light must be directed at 20° relative to the normal of the DMD chip. To accomplish this in a method that eliminates mechanical interference between the illuminating and projecting optics, a *total internal reflection* (TIR) prism is interposed between the projection lens and the DMD color-splitting/color-combining prisms.

The color-splitting/color-combining prisms use dichroic interference filters deposited on their surfaces to split the light by reflection and transmission into red, green, and blue compo-

**Figure 21.28** A scanning electron microscope view of the DMD yoke and spring tips (the mirror has been removed). (*After* [14]. *Courtesy of Texas Instruments.*)

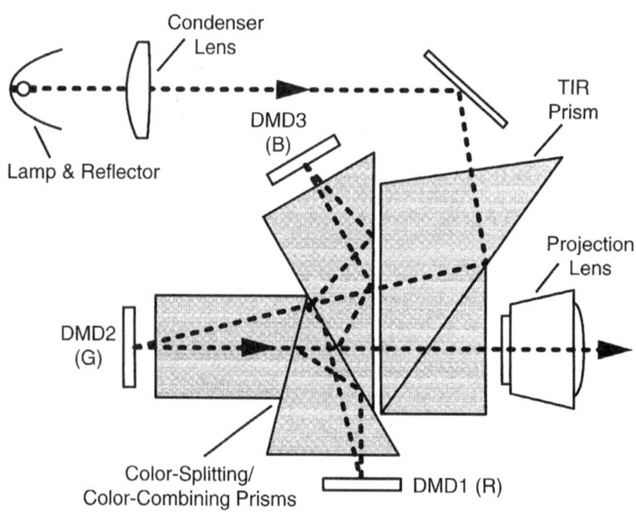


**Figure 21.29** Optical system of the DLP 3-chip implementation. (*After* [14]. *Courtesy of Texas Instruments.*)

nents. The red and blue prisms require an additional reflection from a TIR surface of the prism in order to direct the light at the correct angle to the red and blue DMDs. Light reflected from the on-state mirrors of the three DMDs is directed back through the prisms, and the color compo-

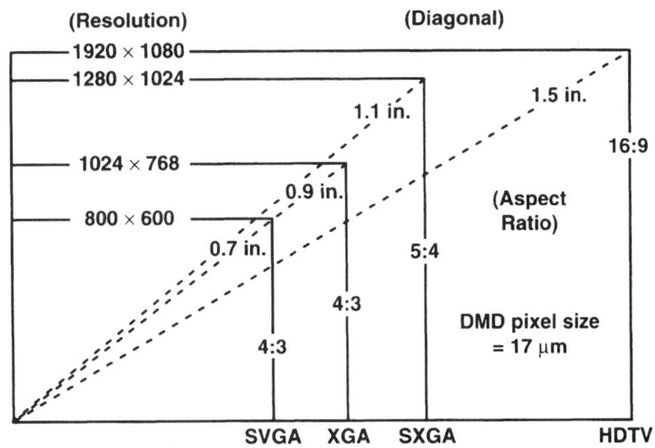


**Figure 21.30** DMD display resolution as a function of chip diagonal. (*After* [14]. *Courtesy of Texas Instruments.*)

nents are recombined. The combined light then passes through the TIR prism and into the projection lens because its angle has been reduced to below the critical angle for total internal reflection in the prism air gap.

As the DMD resolution is increased, the pixel pitch is held constant and the chip diagonal is allowed to increase, as detailed in Figure 21.30. This approach to display design has several advantages:

- The high optical efficiency and contrast ratio of the pixel are maintained at all resolutions.
- Pixel timing is common to all designs, and high address margins are maintained.
- The chip diagonal increases with resolution, which improves the DMD system optical efficiency.

DLP projection systems have been demonstrated at a variety of resolutions (and aspect ratios), including a 16:9 aspect ratio high-definition (1920 × 1080) system [16, 17, 18].

## 21.5b Grating Light Valve Display

The Grating Light Valve (GLV) technology is a means for creating a high-performance spatial light modulator on the surface of a silicon chip [19]. It is based on simple optical principles that leverage the wavelike behavior of light, using diffractive interference as the basis for discriminating between on and off pixel states. A GLV array is fabricated using conventional CMOS materials and equipment, adopting techniques from the emerging field of microelectromechanical systems. Pixels are comprised of a series of identical mechanical structures, fabricated using a relatively small number of masks and processing steps.

A GLV pixel is an addressable diffraction grating created by moving microscopic planar structures. A typical GLV pixel consists of an even number of parallel, dual-supported "ribbons"

**Figure 21.31** A typical GLV pixel showing alternate ribbons being addressed. (*From* [19]. *Courtesy of Silicon Light Machines.*)

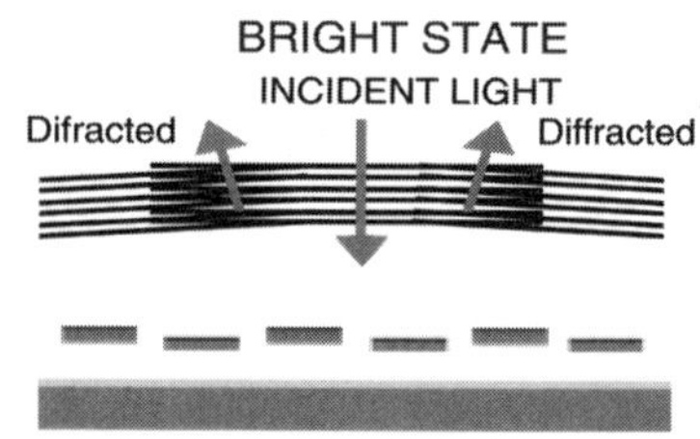


**Figure 21.32** The diffractive (bright) state of a GLV pixel. (*After* [19]. *Courtesy of Silicon Light Machines.*)

formed of silicon nitride and coated with a reflective aluminum top-layer, as illustrated in Figure 21.31. These ribbons are suspended above a thin air gap, allowing them to move vertically relative to the plane of the surface. The ribbons are held in tension, such that in their unaddressed state, the surfaces of the ribbons collectively function as a mirror. A GLV pixel is addressed by inducing a voltage potential between the top of the ribbons and the substrate, thereby deflecting alternate ribbons. Viewed in cross-section (as in Figure 21.32), the up/down pattern of reflective surfaces creates a *square-well diffraction grating* [20]. The GLV device uses digitally-generated signals to produce specific drive voltages for each pixel, achieving precise variable control over the proportion of light that is reflected or diffracted.

## Scanned Linear GLV Architecture

In contrast to a more conventional 2-D array made up of physical pixels that correspond to each bright or dark spot in the final image, the Scanned Linear GLV architecture consists of a linear array of physical GLV pixels oriented along a vertical column of image data [21]. Once per image refresh, this linear array is optically scanned across the screen to produce a complete two-dimensional image. During one scan, each pixel writes successive values corresponding to one entire row of image data. In this way, a single scan of the linear GLV array creates the complete image.

The GLV architecture imposes considerable performance requirements on the spatial light modulator used in the system, specifically:

- It must be capable of extremely fast switching speeds
- Capable of withstanding very high optical power density
- Capable of rendering continuous tone gray scale

## GLV Video Processing Architecture

A 1080P projection display prototype based on the Scanned Linear GLV architecture has been developed [19]. (See Figure 21.33.) The system accepts 1080P video data at 24 or 30 f/s via a

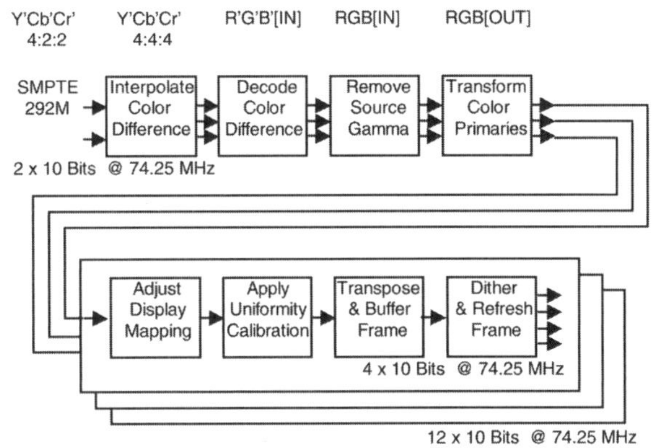


**Figure 21.33** Video processing architecture for a Scanned Linear GLV system. (*After* [19]. *Courtesy of Silicon Light Machines.*)

standard SMPTE 292M serial digital interface. The electronics architecture supports the following system parameters:

- 1920 × 1080 resolution
- Up to 120 Hz refresh, progressive scan
- 10 bits/channel R, G, B

The SMPTE 292M serial digital input contains luminance for all pixels and chrominance for odd pixels. The even pixel chroma values are generated by FIR filtering the red and blue chroma inputs. The luma and chroma are decoded into red, green, and blue with gamma correction using multipliers and adders. The decoder can support any 10-bit standard such as ITU-Rec. 601 or SMPTE 240M.

Gamma removal and color transformation are combined to allow a 9K × 16-bit table to take the place of three nonlinear mappings and nine multiply steps. This approach supports an arbitrary change of color primaries and any 10-bit gamma standard, such as ITU-Rec. BT.709 or SMPTE 240M. The table outputs are added and rounded to yield linear RGB for the system primaries. The prototype system uses solid-state RGB lasers as illumination sources and is capable of displaying a considerably wider color gamut than SMPTE RP 145. The table entries can be modified to change the whitepoint and/or to interpret highlights, depending on whether the input is standard video for HDTV or a custom transfer intended to produce a more film-like appearance.

The display mapping/gamma adjustment step maps RGB intensity to the GLV intensity-voltage characteristic. Conventional spatial light modulators that create grayscale values through pulse width modulation have an inherently linear optical response. However, the inherent GLV electro-optic response creates a natural, continuous grayscale with wide dynamic range that is well matched to the human visual system. (See Figure 21.34).

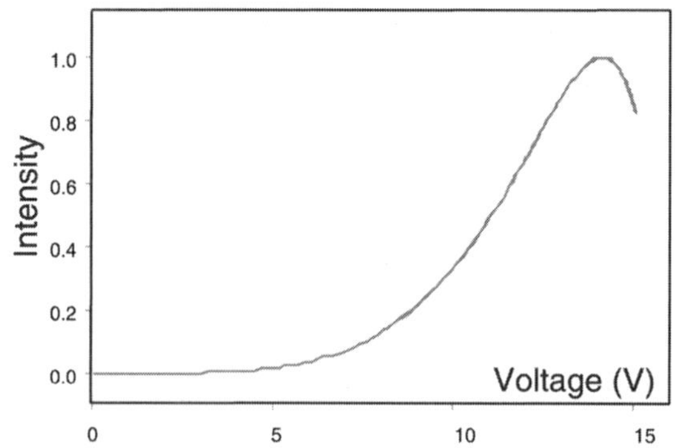


**Figure 21.34** GLV electro-optic response. (*After* [19]. *Courtesy of Silicon Light Machines.*)

The electro-optic response of a GLV pixel is quite uniform because it is based on a simple and repeatable system of electrostatic (attractive) and mechanical spring (restorative) forces. Because of this mechanical simplicity, the GLV response is highly predictable and can be mathematically calculated from relatively simple models.

The SMPTE 292M input is *row-centric* (i.e., the video data is presented sequentially by row). Because the scanned linear GLV system as currently implemented scans left to right by column, a frame buffer is used to store data by rows and transpose it into column data for display. Because higher refresh rates produce better image quality, the frame buffer accepts progressive data at the source rate and sends it out at a faster rate for display. The frame buffer in the prototype system typically reads data in at 24 or 30 f/s and refreshes the display up to four times the input rate.

Scanning horizontally has several benefits:

- It requires a smaller and less expensive linear GLV array (1080 pixels vs. 1920 pixels, a 44 percent pixel count reduction).
- The smaller modulator allows additional system cost savings, such as smaller recombination and projection optics, smaller look-up tables, and so on.
- A horizontal scan enables electronic support for variable aspect ratios (Figure 21.35). For example, a horizontal scan system can change from 4:3 to 16:9 for HDTV or to cinemascope (2.35) for electronic cinema, without requiring anamorphic lenses or complex scaling algorithms that tend to degrade image quality.

By refreshing the display 3 or 4 times per frame, it is possible to achieve 1.6 or 2 additional effective bits of grayscale through dithering [19]. The prototype system uses conventional drivers, similar to those that might be used as LCD column drivers. Through temporal dithering, the system exploits the inherent speed and the scanned line approach of the GLV device to achieve 10-bit gray scale using simple, low cost 8-bit drivers.

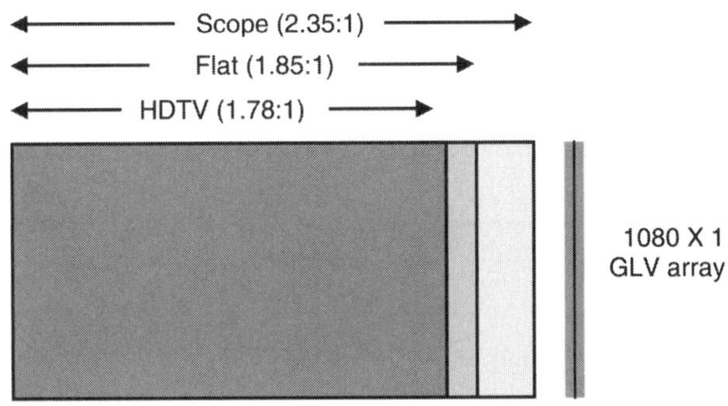


**Figure 21.35** The GLV horizontally scanned linear system supports variable aspect ratios. (*After* [19]. *Courtesy of Silicon Light Machines.*)

## 21.6 Projection Requirements for Cinema Applications

Screen brightness is a critical element in providing an acceptable HDTV large-screen display. Without adequate brightness, the impact on the audience is reduced. Theaters have historically used front-projection systems for 35 mm film. This arrangement provides for efficient and flexible theater seating. Large-screen HDTV is likely to maintain the same arrangement. Typical motion picture theater projection-system specifications are as follows:

- Screen width: 30 ft
- Aspect ratio: 2.35:1
- Contrast ratio: 300:1
- Screen luminance: 15 ft-L
- Center-to-edge brightness uniformity: 85 percent

These specifications meet the expectations of motion picture viewers. It follows that for HDTV projectors to be competitive in large-screen theatrical display applications, they must meet similar specifications. The major hurdle for making the electronic cinema a commercial reality is the availability—actually, the lack—of high-performance HDTV projectors with theater brightness at a reasonable cost [22]. Illuminating a 20- × 36-ft theater screen with 20 ft-L (assuming a screen gain of 1.5) requires 9600 lm output from the projector. Table 21.3 compares key performance requirements for theater HDTV systems and consumer displays.

### 21.6a Operational Considerations

Accurate convergence is critical to display resolution. Figure 21.36 shows the degradation in resolution resulting from convergence error. It is necessary to keep convergence errors to less than half the distance between the scanning lines to hold resolution loss below 3 dB. Errors in convergence also result in color contours in a displayed image. Estimates have put the detectable thresh-

**Table 21.3 Minimum Projector Specifications for Consumer and Theater Displays** (*After* [22].)

| Parameter | Consumer Display | Theater Display |
|---|---|---|
| Resolution | Greater than 750 TV lines | Greater than 750 TV lines |
| Light output | 1000 lm | 10,000 lm |
| Cost | Less than $2000 | Less than $50,000 |
| Response time | Less than 10 ms | Less than 10 ms |
| Power consumption | Less than 300 W | Less than 3000 W |
| Small area uniformity | ±0.25 percent | ±0.25 percent |
| Contrast ratio | Greater than 90:1 | Greater than 90:1 |
| Flicker | Undetectable | Undetectable |

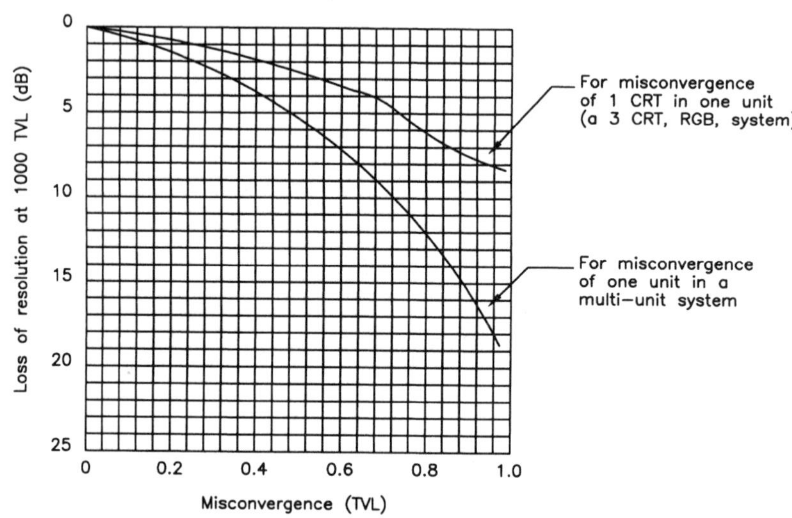


**Figure 21.36** The relationship between misconvergence and display resolution. (*After* [23].)

old of color contours at 0.75 to 0.5 minutes of arc. This figure also indicates that convergence error must be held under 0.5 scanning lines.

Raster stability influences the short-term stability of the projected display. The signal-to-noise ratio (S/N) and raster stability relationships for deflection circuits are shown in Table 21.4. The S/N equivalent of one-fifth of a scanning line is necessary to obtain sufficient raster stability. In HDTV applications, this translates to approximately 80 dB S/N. Other important factors are high speed and improved efficiency of the deflection circuit.

**Table 21.4 Appearance of Scan Line Irregularities in a Projection Display** (*After* [23].)

| Condition of Adjacent Lines | S/N (p-p) |
|---|---|
| Clearly overlapped | Less than 69 dB |
| Just before overlapped | 75 dB |
| No irregularities | Greater than 86 dB |

Some manufacturers have begun to incorporate automatic convergence systems into their products. These systems usually take the form of a CCD camera sensor that scans various portions of the screen as test patterns are displayed.

## 21.7 References


1. Luxenberg, H., and R. Kuehn: *Display Systems Engineering*, McGraw-Hill, New York, 1968.
2. McKechnie, S.: Philips Laboratories (NA) report, 1981, unpublished.
3. Kikuchi, M., et al.: "A New Coolant-Sealed CRT for Projection Color TV," *IEEE Trans.*, vol. CE-27, IEEE, New York, no. 3, pp. 478-485, August 1981.
4. Van, M. W., and J. Von Esdonk: "A High Luminance High-Resolution Cathode-Ray Tube for Special Purposes," *IEEE Trans Electron Dev.*, IEEE, New York, N.Y., ED-30, pg. 193, 1983.
5. Cheng, J. B., and Q. H. Wang: "Studies on YAG Phosphor Screen for HDTV Projector," *Proc SPIE 2892*, SPIE, Bellingham, Wash., pg. 36, 1996.
6. Wang, Q. H., J. B. Cheng, and Z. L. Lin: "A New YAG Phosphor Screen for Projection CRT," *Electron Lett.*, vol. 34, no. 14, pg. 1420, 1998.
7. Wang, Qionghua, Jianbo Cheng, Zulun Lin, and Gang Yang: "A High-Luminance and High-Resolution CRT for Projection HDTV Display," *Journal of the SID*, Society for Information Display, San Jose, Calif., vol. 7, no. 3, pg 183–186, 1999.
8. Yoshizawa, T., S. Hatakeyama, A. Ueno, M. Tsukahara, K. Matsumi, K. Hirota: "A 61-in High-Definition Projection TV for the ATSC Standard," *SID 99 Digest*, Society for Information Display, San Jose, Calif., pp. 112–115, 1999.
9. Pease, Richard W.: "An Overview of Technology for Large Wall Screen Projection Using Lasers as a Light Source," *Large Screen Projection Displays II*, William P. Bleha, Jr. (ed.), Proc. SPIE 1255, SPIE, Bellingham, Wash., pp. 93-103, 1990.
10. Takeuchi, Kazuhiko, et. al.: "A 750-TV-Line Resolution Projector Using 1.5 Megapixel a-Si TFT LC Modules," *SID 91 Digest*, Society for Information Display, pp. 415–418, 1991.

11. Fritz, Victor J.: "Full-Color Liquid Crystal Light Valve Projector for Shipboard Use," *Large Screen Projection Displays II*, William P. Bleha, Jr. (ed.), Proc. SPIE 1255, SPIE, Bellingham, Wash., pp. 59–68, 1990.

12. Phillips, Thomas E., et. al.: "1280 × 1024 Video Rate Laser-Addressed Liquid Crystal Light Valve Color Projection Display," *Optical Engineering*, Society of Photo-Optical Instrumentation Engineers, vol. 31, no. 11, pp. 2300–2312, November 1992.

13. Bleha, W. P.: "Image Light Amplifier (ILA) Technology for Large-Screen Projection," *SMPTE Journal*, SMPTE, White Plains, N.Y., pp. 710–717, October 1997.

14. Hornbeck, Larry J.: "Digital Light Processing for High-Brightness, High-Resolution Applications," *Projection Displays III*, Electronic Imaging '97 Conference, SPIE, Bellingham, Wash., February 1997.

15. Florence, J., and L. Yoder: "Display System Architectures for Digital Micromirror Device (DMD) Based Projectors," *Proc. SPIE*, SPIE, Bellingham, Wash., vol. 2650, *Projection Displays II*, pp. 193–208, 1996.

16. Gove, R. J., V. Markandey, S. Marshall, D. Doherty, G. Sextro, and M. DuVal: "High-Definition Display System Based on Digital Micromirror Device," International Workshop on HDTV (HDTV '94), International Institute for Communications, Turin, Italy (October 1994).

17. Sextro, G., I. Ballew, and J. Lwai: "High-Definition Projection System Using DMD Display Technology," *SID 95 Digest*, Society for Information Display, San Jose, Calif., pp. 70–73, 1995.

18. Younse, J. M.: "Projection Display Systems Based on the Digital Micromirror Device (DMD)," SPIE Conference on Microelectronic Structures and Microelectromechanical Devices for Optical Processing and Multimedia Applications, Austin, Tex., *SPIE Proceedings*, SPIE, Bellingham, Wash., vol. 2641, pp. 64–75, Oct. 24, 1995.

19. Corrigan, R. W., B. R. Lang, D. A. LeHoty, and P. A. Alioshin: "An Alternative Architecture for High Performance Display," Silicon Light Machines, Sunnyvale, Calif., 1999. Presented at the 141st SMPTE Technical Conference (paper 141-25).

20. Amm, D. T., and R. W. Corrigan: "Optical Performance of the Grating Light Valve Technology," Projection Displays V Symposium, SPIE Proceedings, SPIE, San Jose, Calif., vol. EI 3634-10, February 1999.

21. Amm, D. T., and R.W. Corrigan: "Grating Light Valve Technology: Update and Novel Applications," SID Symposium—Anaheim, SID, San Jose, Calif., May 1998.

22. Glenn, W. E., C. E. Holton, G. J. Dixon, and P. J. Bos: "High-Efficiency Light Valve Projectors and High-Efficiency Laser Light Sources," *SMPTE Journal*, SMPTE, White Plains, N.Y., pp. 210–216, April 1997.

23. Mitsuhashi, Tetsuo: "HDTV and Large Screen Display," *Large-Screen Projection Displays II*, William P. Bleha, Jr. (ed.), Proc. SPIE 1255, SPIE, Bellingham, Wash., pp. 2–12, 1990.

## 21.8 Bibliography


Ashizaki, S., Y. Suzuki, K. Mitsuda, and H. Omae: "Direct-View and Projection CRTs for HDTV," *IEEE Transactions on Consumer Electronics*, IEEE, New York, N.Y., 1988.

Bates, W., P. Gelinas, and P. Recuay: "Light Valve Projection System for HDTV," *Proceedings of the ITS*, International Television Symposium, Montreux, Switzerland, 1991.

Bates, W., P. Gelinas, and P. Recuay: "Light Valve Projection System for HDTV," *Proceedings of the International Television Symposium*, ITS, Montreux, Switzerland, 1991.

Bauman, E.: "The Fischer Large-Screen Projection System," *SMPTE Journal*, SMPTE, White Plains, N.Y., vol. 60, pg. 351, 1953.

Benson, K. B., and D. G. Fink: *HDTV: Advanced Television for the 1990s*, McGraw-Hill, New York, N.Y., 1990.

Blaha, Richard J.: "Large Screen Display Technology Assessment for Military Applications," *Large-Screen Projection Displays II*, William P. Bleha, Jr., (ed.), Proc. SPIE 1255, SPIE, Bellingham, Wash., pp. 80–92, 1990.

Bleha, William P., Jr., (ed.): *Large-Screen Projection Displays II*, Proc. SPIE 1255, SPIE, Bellingham, Wash., 1990.

Cheng, Jia-Shyong, et. al.: "The Optimum Design of LCD Parameters in Projection and Direct View Applications," *Display Technologies*, Shu-Hsia Chen and Shin-Tson Wu (eds.), Proc. SPIE 1815, SPIE, Bellingham, Wash., pp. 69–80, 1992.

Fink, D. G. (ed.): *Television Engineering Handbook*, McGraw-Hill, New York, N.Y., 1957.

Fink, D. G.: *Color Television Standards*, McGraw-Hill, New York, N.Y., 1986.

Fink, D. G., et al.: "The Future of High-Definition Television," *SMPTE Journal*, SMPTE, White Plains, N.Y., vol. 89, February/March 1980.

Fritz, Victor J.: "Full-Color Liquid Crystal Light Valve Projector for Shipboard Use," *Large Screen Projection Displays II*, William P. Bleha, Jr., (ed.), Proc. SPIE 1255, SPIE, Bellingham, Wash., pp. 59–68, 1990.

Fujio, T., J. Ishida, T. Komoto, and T. Nishizawa: "High-Definition Television Systems—Signal Standards and Transmission," *SMPTE Journal*, SMPTE, White Plains, N.Y., vol. 89, August 1980.

Gerhard-Multhaupt, R.: "Light Valve Technologies for HDTV Projection Displays: A Summary," *Proceedings of the ITS*, International Television Symposium, Montreux, Switzerland, 1991.

Glenn, William. E.: "Principles of Simultaneous Color Projection Using Fluid Deformation," *SMPTE Journal*, SMPTE, White Plains, N.Y., vol. 79, pg. 788, 1970.

Glenn, William E.: "Large Screen Displays for Consumer and Theater Use," *Large Screen Projection Displays II*, William P. Bleha, Jr., (ed.), Proc. SPIE 1255, SPIE, Bellingham, Wash., pp. 36–43, 1990.

Good, W.: "Recent Advances in the Single-Gun Color Television Light-Valve Projector," *Soc. Photo-Optical Instrumentation Engrs.*, vol. 59, 1975.

Good, W.: "Projection Television," *IEEE Trans.*, vol. CE-21, no. 3, pp. 206–212, August 1975.

Gretag AG: "What You May Want to Know about the Technique of Eidophor," Regensdorf, Switzerland.

Grinberg, J. et al.: "Photoactivated Birefringent Liquid-Crystal Light Valve for Color Symbology Display," *IEEE Trans. Electron Devices*, vol. ED-22, no. 9, pp. 775–783, September 1975.

Hardy, A. C., and F. H. Perrin: *The Principles of Optics*, McGraw-Hill, New York, N.Y., 1932.

Howe, R., and B. Welham: "Developments in Plastic Optics for Projection Television Systems," *IEEE Trans.*, vol. CE-26, no. 1, pp. 44–53, February 1980.

Hubel, David H.: *Eye, Brain and Vision*, Scientific American Library, New York, N.Y., 1988.

Itah, N., et al.: "New Color Video Projection System with Glass Optics and Three Primary Color Tubes for Consumer Use," *IEEE Trans. Consumer Electronics*, vol. CE-25, no. 4, pp. 497–503, August 1979.

Judd, D. B.: "The 1931 C.I.E. Standard Observer and Coordinate System for Colorimetry," *Journal of the Optical Society of America*, vol. 23, 1933.

Kikuchi, M., et al.: "A New Coolant-Sealed CRT for Projection Color TV," *IEEE Trans.*, vol. CE-27, no. 3, pp. 478–485, August 1981.

Kingslake, Rudolf (ed.): *Applied Optics and Optical Engineering*, vol. 1, Chap. 6, Academic, New York, N.Y., 1965.

"Kodak Filters for Scientific and Technical Uses," Eastman Kodak Co., Rochester, N.Y.

Kurahashi, K., et al.: "An Outdoor Screen Color Display System," *SID Int. Symp. Digest 7*, Technical Papers, vol. XII, Society for Information Display, San Jose, Calif., pp. 132–133, April 1981.

Lakatos, A. I., and R. F. Bergen: "Projection Display Using an Amorphous-Se-Type Ruticon Light Valve," *IEEE Trans. Electron Devices*, vol. ED-24, no. 7, pp. 930–934, July 1977.

Luxenberg, H., and R. Kuehn: *Display Systems Engineering*, McGraw-Hill, New York, N.Y., 1968.

Maseo, Imai, et. al.: "High-Brightness Liquid Crystal Light Valve Projector Using a New Polarization Converter," *Large Screen Projection Displays II*, William P. Bleha, Jr., (ed.), Proc. SPIE 1255, SPIE, Bellingham, Wash., pp. 52–58, 1990.

Mitsuhashi, Tetsuo: "HDTV and Large Screen Display," *Large-Screen Projection Displays II*, William P. Bleha, Jr., (ed.), Proc. SPIE 1255, SPIE, Bellingham, Wash., pp. 2–12, 1990.

Morizono, M.: "Technological Trends in High Resolution Displays Including HDTV," *SID International Symposium Digest*, paper 3.1, Society for Information Display, San Jose, Calif., May 1990.

Nasibov, A., et al.: "Electron-Beam Tube with a Laser Screen," *Sov. J. Quant. Electron.*, vol. 4, no. 3, pp. 296–300, September 1974.

Pease, Richard W.: "An Overview of Technology for Large Wall Screen Projection Using Lasers as a Light Source," *Large Screen Projection Displays II*, William P. Bleha, Jr., (ed.), Proc. SPIE 1255, SPIE, Bellingham, Wash., pp. 93–103, 1990.

Pfahnl, A.: "Aging of Electronic Phosphors in Cathode Ray Tubes," *Advances in Electron Tube Techniques*, Pergamon, New York, N.Y., pp. 204–208.

Phillips, Thomas E., et. al.: "1280 × 1024 Video Rate Laser-Addressed Liquid Crystal Light Valve Color Projection Display," *Optical Engineering*, Society of Photo-Optical Instrumentation Engineers, vol. 31, no. 11, pp. 2300–2312, November 1992.

Pitts, K., and N. Hurst: "How Much Do People Prefer Widescreen (16 × 9) to Standard NTSC (4 × 3)?," *IEEE Transactions on Consumer Electronics*, IEEE, New York, N.Y., August 1989.

Pointer, R. M.: "The Gamut of Real Surface Colors," *Color Res. App.*, vol. 5, 1945.

Poorter, T., and F. W. deVrijer: "The Projection of Color Television Pictures," *SMPTE Journal*, SMPTE, White Plains, N.Y., vol. 68, pg. 141, 1959.

Robertson, A.: "Projection Television—1 Review of Practice," *Wireless World*, vol. 82, no. 1489, pp. 47–52, September 1976.

Schiecke, K.: "Projection Television: Correcting Distortions," *IEEE Spectrum*, IEEE, New York, N.Y., vol. 18, no. 11, pp. 40–45, November 1981.

Sears, F. W.: *Principles of Physics, III*, Optics, Addison-Wesley, Cambridge, Mass., 1946.

Sherr, S.: *Fundamentals of Display System Design*, Wiley-Interscience, New York, N.Y., 1970.

Takeuchi, Kazuhiko, et. al.: "A 750-TV-Line Resolution Projector using 1.5 Megapixel a-Si TFT LC Modules," *SID 91 Digest*, Society for Information Display, San Jose, Calif., pp. 415–418, 1991.

Taneda, T., et al.: "A 1125-Scanning Line Laser Color TV Display," *SID 1973 Symp. Digest Technical Papers*, Society for Information Display, San Jose, Calif., vol. IV, pp. 86–87, May 1973.

Tomioka, M., and Y. Hayshi: "Liquid Crystal Projection Display for HDTV," *Proceedings of the International Television Symposium*, ITS, Montreux, Switzerland, 1991.

Tsuruta, Masahiko, and Neil Neubert: "An Advanced High Resolution, High Brightness LCD Color Video Projector," *SMPTE Journal*, SMPTE, White Plains, N.Y., pp. 399–403, June 1992.

Wang, S., et al.: "Spectral and Spatial Distribution of X-Rays from Color Television Receivers," *Proc. Conf. Detection and Measurement of X-radiation from Color Television Receivers*, Washington, D.C., pp. 53–72, March 28–29, 1968.

Wedding, Donald K., Sr.: "Large Area Full Color ac Plasma Display Monitor," *Large Screen Projection Displays II*, William P. Bleha, Jr., (ed.), Proc. SPIE 1255, SPIE, Bellingham, Wash., pp. 29–35, 1990.

Williams, Charles S., and Becklund, Orville A.: *Optics: A Short Course for Engineers and Scientists*, Wiley Interscience, New York, N.Y., 1972.

"X-Radiation Measurement Procedures for Projection Tubes," TEPAC Publication 102, Electronic Industries Association, Washington, D. C.

Yamamoto, Y., Y. Nagaoka, Y. Nakajima, and T. Murao: "Super-compact Projection Lenses for Projection Television," *IEEE Transactions on Consumer Electronics*, IEEE, New York, N.Y., August 1986.

Chapter

# 22

# Flat Panel Displays

**Jerry C. Whitaker, Editor**

## 22.1 Introduction

A wide variety of flat panel devices are available in the marketplace. For each basic technology, several variations typically exist. This chapter will concentrate on those flat panel display technologies that are applicable for video purposes, specifically:

- Liquid crystal display (LCD), including *active-matrix liquid crystal display* (AMLCD) and *plasma-addressed liquid crystal* (PALC)
- Plasma display panel (PDP)
- Field emission display (FED)

It is fair to point out that many of the display technologies that do not now support full-color video may do so in the near future. Flat panel development is racing along at an unprecedented pace, driven largely by the demand of consumers for display devices that are more compact than CRTs (for a given screen size), lighter, and more power-efficient.

It is commonplace to regard flat panel displays and solid-state displays as essentially the same product category. While this is—generally speaking—true, it should be remembered that some flat panel displays have as much in common with vacuum devices as they do with solid-state devices (the FED is a convenient example).

## 22.2 Liquid Crystal Displays

With the exception of the *light-emitting diode* (LED), the liquid crystal display (LCD) is the best known and most highly developed flat panel display device. Applications range from simple monochrome displays for cellular telephones to full-color large-screen video systems.

### 22.2a Principles of Operation

Materials classified as *liquid crystals* usually exist in a liquid form at high temperatures and in a solid form at low temperatures. In between these two extremes they exhibit the characteristics of

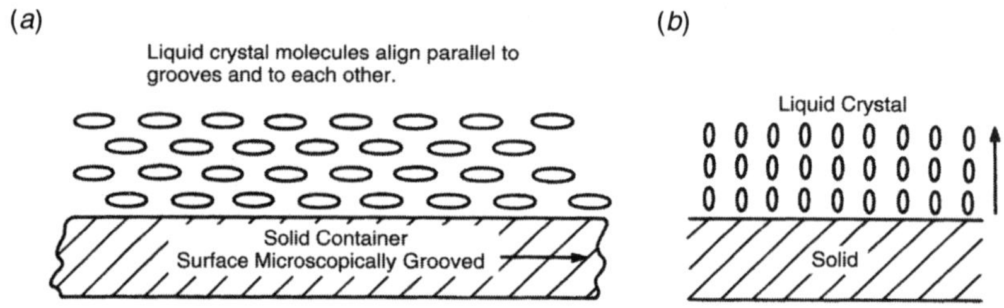


**Figure 22.1** The liquid-crystal/grooved interface: (*a*) no field applied, (*b*) an electric field greater than the critical value applied. (*After* [2].)

both states [1]. There are a number of different types of liquid crystals, although *nematic crystals* are by far the most common found in display devices.

The essential feature of a liquid crystal is the long rod-like molecule physical shape. In a nematic crystal, the molecules align in a predictable manner according to certain rules (Figure 22.1). If a surface—such as a display device panel—is microscopically groved, the interface molecules will be aligned by the groves and intermolecule forces will maintain the orientation across the liquid crystal, as illustrated in Figure 22.1.*a*. The molecules will align in an electric field, but beyond a certain *critical value*, the field will be sufficient to overcome the alignment, as shown in Figure 22.1*b*. In the practical application of liquid crystal materials, the transition is not so abrupt; groove alignment persists at the interface itself (Figure 22.2). The process of alignment in the electric field results from the *anisotropic dielectric constant* characteristics of the liquid crystal material.

The principle of the *twisted nematic cell* is a bit more complicated. As illustrated in Figure 22.3, confining plates, typically spaced 10 μm apart, are grooved orthogonally, forcing the molecular orientation to spiral through 90°. Two polarizers and a mirror are added, as shown in the figure; incident ambient light is polarized and enters the liquid-crystal cell with the plane of polarization parallel to the molecular orientation. As the light traverses the cell, the plane of polarization is rotated by the twist in the liquid crystal so that it reaches the opposite face with a polarization 90° to the original direction, but now parallel to the direction of the second polarizer, through which it can now pass. The light is next reflected from the mirror and passed back through the cell, reversing the prior sequence.

When an electric field of a potential greater than the critical field is applied between the transparent electrodes, the 90° twist in the crystal is disrupted; the molecules align parallel to the field so that the rotation of the light's plane of polarization cannot be sustained. The *crossed polarizers*, therefore, effectively block reflection of the incident light from the backing mirror. The surface thus appears to be dark. The contrast ratio can be further improved through the use of a so-called *super twisted nematic crystal*, where the molecular orientation is rotated through 270° rather than 90°.

The LCD device thus described uses ambient light as the illumination source in conjunction with the mirror element. The LCD can also be oriented to serve a transmissive rather than reflective function.

The transmission LCD functions in a manner similar to the reflective LCD except that the mirror is eliminated, replaced by a powered backlight source. Although the energy-efficient

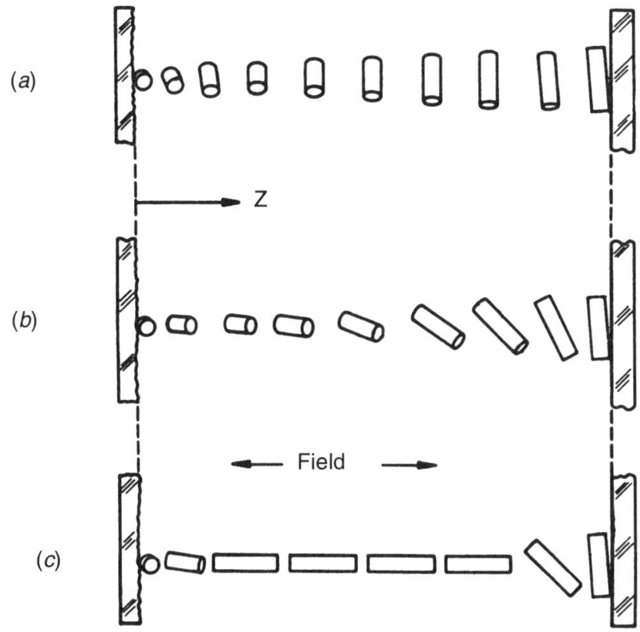


**Figure 22.2** The orientation of the liquid-crystal axis in a cell: (*a*) no field applied, (*b*) a field of approximately twice the critical field applied, (*c*) a field several times the critical field applied. (*After* [3].)

advantage of the passive device is lost in this arrangement, monochromatic backlighting provides one simple means of constructing displays with varying (albeit limited) background colors.

A more sophisticated method of achieving color in an LCD is through the use of *cholesteric crystals* in the light-controlling element. The three primary types of liquid crystals—nematic, cholesteric, and *smectic*—are distinguished by the different types of molecular ordering that they display. In the cholesteric crystal, the direction of molecular alignment rotates in each successive parallel plane, illustrated in Figure 22.4. As shown, the spatial period of the rotation is known as the *pitch*. Braggs reflections occur when the wavelength of the incident light meets the condition $\lambda = p/n$, where $p$ is the pitch and $n$ is an integer. The liquid-crystal can thus appear to be colored in incident white light, although it is important to point out that in practice the color is strongly depending upon the temperature of the cell.

## Display Addressing

Addressing techniques for liquid-crystal devices can be classified into three basic types [5]:

- Direct (static) addressing
- Passive matrix addressing
- Active matrix addressing.

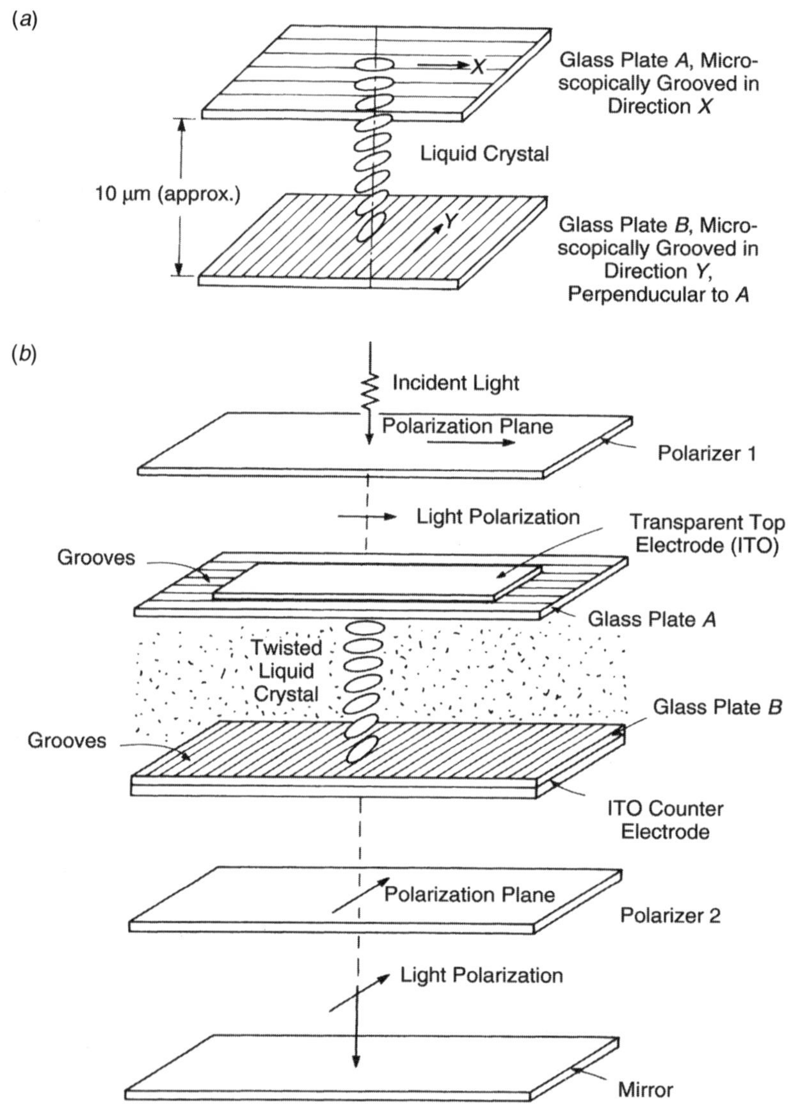


**Figure 22.3** LCD architecture: (*a*) twisted nematic cell, (*b*) LCD element. (*After* [2].)

Direct addressing is used for displays of low information content, such as numeric, bar graph, and other fixed pattern displays. For high information content applications, this approach is impractical because of the huge number of interconnects required.

Passive matrix addressing is the simplest and least expensive approach for medium-to-high information content displays. Each liquid-crystal cell is addressed through a row and

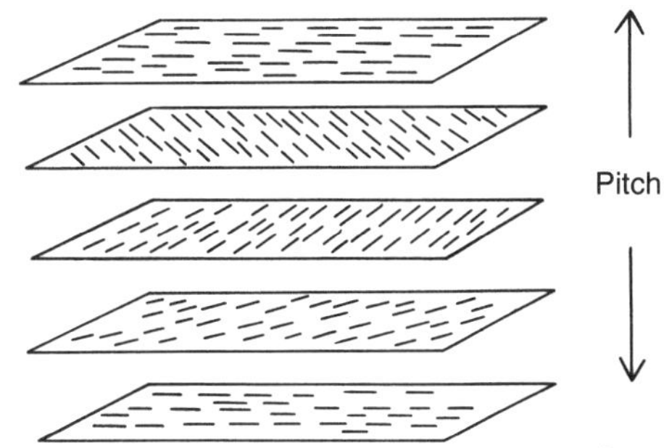


**Figure 22.4** Cholesteric ordering: a large number of planes of nematic ordering are formed where the directors rotate as light moves along a direction perpendicular to the planes. (*After* [4].)

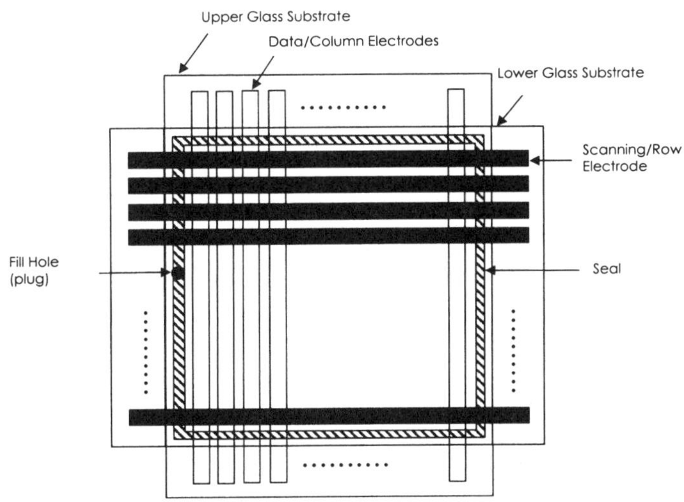


**Figure 22.5** Passive matrix addressing. (*From* [5]. *Used with permission.*)

column matrix, as shown in Figure 22.5. Although simple and intuitive, this approach quickly gets out of hand in a display of large dimensions. Furthermore, issues relating to cell turn-on and turn-off times come in to play, making the timing of the multiplexing system critical in many practical applications.

Active matrix addressing removes the limitations inherent in the multiplexing approach by incorporating a nonlinear control element in series with each pixel. This provides a 100 percent duty cycle for the pixel using the charge stored at the pixel during the row addressing time. The active matrix pixel array, with a control element at each pixel, is illustrated in Figure 22.6.

The active-matrix LCD (AMLCD) typically uses a three terminal *thin film transistor* (TFT) as the active matrix switching element [5]. This provides superior image quality as a result of the complete isolation of each pixel from its neighboring pixel. A common active matrix array employing three terminal TFT devices is shown in Figure 22.7. A storage capacitor ($C_s$ in the

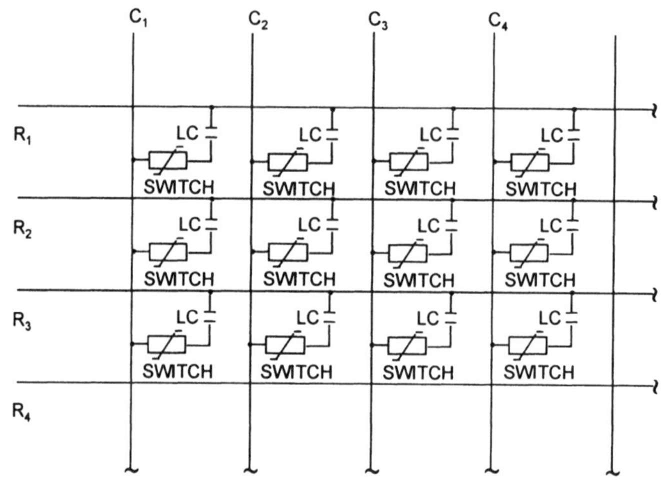


**Figure 22.6** Active matrix addressing scheme. (*From* [5]. *Used with permission.*)

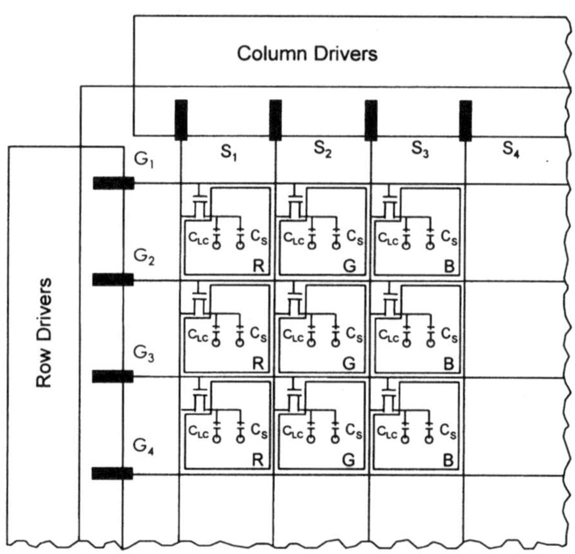


**Figure 22.7** Addressing topology using three-terminal TFT devices. (*From* [5]. *Used with permission.*)

figure) is usually incorporated at each pixel to reduce the pixel voltage offset, improve display uniformity, and broaden the acceptable operating temperature range.

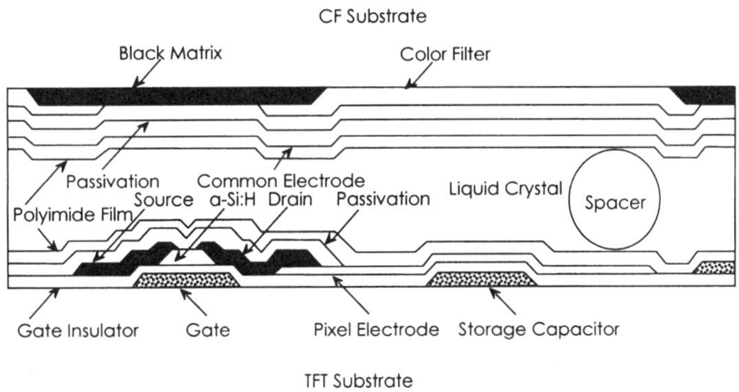


**Figure 22.8** Cross-section of an AMLCD display panel. (*From* [5]. *Used with permission.*)

Figure 22.8 shows a cross-section of a TFT-LCD pixel illustrating the various layers of the TFT, specifically:

- The pixel electrode
- Storage capacitor
- Liquid-crystal cell with polyimide alignment layers, spacers, color filters, and black matrix

## 22.2b AMLCD Performance Parameters

The performance of an AMLCD can be characterized by the following parameters [5]:

- Viewing angle
- Contrast ratio
- Gray-level luminance stability
- Chromaticity stability
- Color gamut
- Response time
- Reflectance
- Power consumption
- Luminance
- Imaging "sticking"
- Crosstalk among pixels

It is intuitive that certain performance aspects are more important in some applications than in others, i.e. portable applications require the lowest possible power consumption, while at the

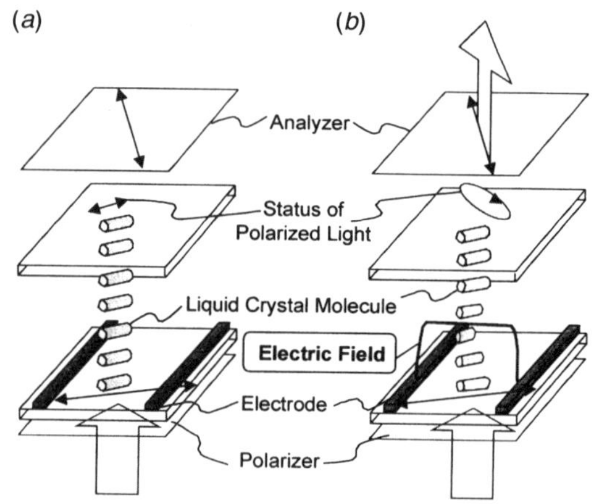


**Figure 22.9** The principle of in-plane switching for the AMLCD: (*a*) off state, (*b*) on state. (*From* [5]. *Used with permission.*)

other end of the applications scale, video requires high resolution, high luminance, and wide color gamut (among other things).

One parameter that has seen considerable improvement in recent years is the viewing angle. This specification, the range of horizontal and vertical viewing angles (the *viewing envelope*), has improved the point that AMLCD displays are practical for instrumentation applications such as airborne avionics.

One of the techniques used to accomplish improved viewing angles is *in-plane switching* (IPS) [5]. In the IPS scheme, illustrated in Figure 22.9, the bottom TFT substrate contains both the pixel electrode and the common electrodes in an interdigitated configuration. The top color filter substrate does not contain any electrode. The liquid crystal is aligned homogeneously, and the input polarizer axis is parallel to the liquid-crystal director orientation. The output polarizer is a 90° cross-polarizer configuration. For the field-off condition, the input polarizer light travels unaltered through the cell, and then is blocked by the crossed exit polarizer. This provides the normally black state. The field-on state causes in-plane rotation of the liquid-crystal modules (to align with the field), and the resulting optical retardation causes rotation of the polarization axis of the input light, which then emerges through the output polarizer.

### AMLCD Applications

High-performance AMLCDs have successfully replaced CRTs in demanding applications ranging from airline cockpit displays to standard- and high-definition video. Current AMLCD work is focusing on direct view devices with sizes of 25 to 30 in diagonal.

AMLCDs have become the dominant technology for portable display applications, such as laptop computers because they offer a unique combination of low-voltage drive, low-power consumption, low cost, and acceptable image quality.

## 22.2c Plasma-Address Liquid Crystal Display

Building upon the inherent simplicity and scalability of the AMLCD addressing scheme, *plasma-addressed liquid crystal display* (PALC) technology has shown considerable progress

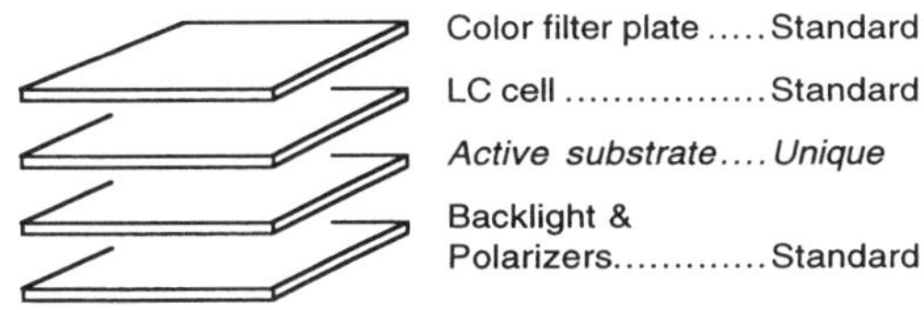


**Figure 22.10** Simplified illustration of the PALC and AMLCD stack. Compared to conventional LCDs, only the active substrate is unique. *(From* [6]. *Used with permission.*)

within the past few years. PALC is an active-matrix liquid crystal display that uses an alternative matrix technique that eliminates the silicon-based integrated circuit processing used with TFT designs [6]. PALC instead relies on the electrical properties of plasma to provide large-sized AMLCDs suitable for high-definition applications. Screen sizes of 42-in are practical at this writing.

Although structurally different, the PALC and TFT active substrates are functionally equivalent. Figure 22.10 illustrates additional similarities.

The color filter plate contains transparent conductors, patterned in a manner identical to those used in passively addressed LCDs, and color filters identical to those used in passive and active matrix devices. These color filters provide the excellent color rendition common to the AMLCD family of devices. In addition, color filters—in conjunction with the polarizers—are responsible for the good contrast performance of AMLCDs.

The liquid crystal cell used in the PALC device is typically a standard TN cell. The use of nematic liquid crystals results in continuous grayscale performance, i.e. the number of displayed gray levels can be as large as desired; it is limited only by the number of bits supported by the drive circuitry.

Backlighting requirements for the AMLCD separates the image generation process from light production, which allows each parameter to be optimized independently.

## Application Considerations

Progressively updating a 1000 line display at a 60 Hz rate requires a line-addressing rate of approximately 16 μs. The plasma therefore must capture data and then decay to a non-active state within the 16 μs time-line. In a PALC display, the decay time of the plasma is strongly influenced by gas decomposition within the channels. Available compositions exhibit decay rates as short as 2.5 μs, well within the requirements for HD imaging [7].

The resolution performance of a large PALC display is, at this writing, rather modest by AMLCD standards [6]. For example, a full-color pixel pitch of 0.5 mm is required for a 42-in. HD display.

The measured performance lifetime of common PALC panels has exceeded 30,000 hrs. The PALC utilizes a dc plasma to generate the charge carriers necessary to address the liquid crystal pixels. Because plasma light generation is not required, power levels are quite low. Furthermore, pixel addressing requires only a few microseconds, and so the plasma duty cycle is likewise short.

Another important consideration in panel life is the distribution of power over the device. PALC dissipation is independent of the displayed image, eliminating the problem of image burn-in.

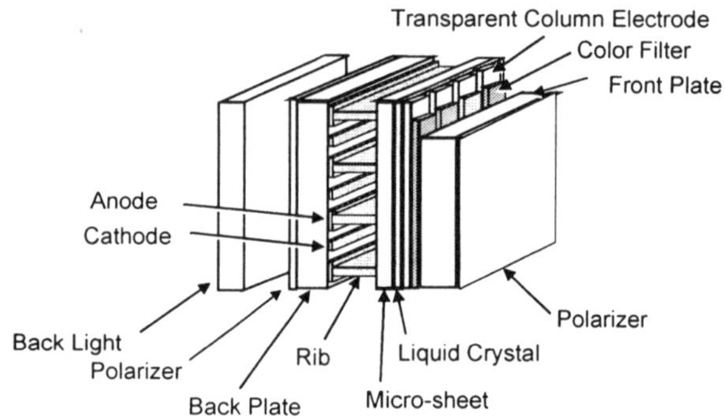


**Figure 22.11** Physical structure of the PALC. (*From* [9]. *Used with permission.*)

Many of the advancements made in AMLCD viewing angles are applicable to PALC displays. One approach has extended the horizontal and vertical viewing angle to 140º [8].

### HD Display Device

The demands of high-definition imaging place considerable requirements on a display device. Despite the challenges, a 42-in. HD PALC has been developed into a practical product [9]. The display incorporates 960 × 1920 color pixels. Twice the vertical resolution was realized while maintaining the VGA-pitch plasma structure. The key feature of this technology is the precise control of local plasma conductivity.

To make a fine-pitch PALC with conventional technologies, two fine-pitch substrates are prepared: one is a substrate with fine-pitch color filter and transparent column electrode elements (ITO), and the other is a fine-pitch plasma substrate. (See Figure 22.11.) For the color filter and ITO substrate, the structure for a PALC is basically identical to a passive matrix LCD. The issue, therefore, lies in making a fine pattern on a large panel.

For the plasma substrate, however, a fine pitch structure reduces the aperture ratio, which results in lower brightness.

In this device, twice the vertical resolution is achieved by precisely controlling the spread of the plasma switch. (See Figure 22.12.) The plasma structure itself is the same as for a VGA panel. As shown in the figure, an electrode that acts as the cathode during regular operation is formed at the center of the channel. The other electrode, which acts as the anode during regular operation, is formed beneath the rib. During regular operation, one channel that forms between two ribs corresponds to one scan line. In the high-definition device, one channel is divided into two scan lines and through manipulation of the addressing and driving systems, the resolution is effectively doubled.

## 22.3 Plasma Displays

Plasma display panel (PDP) devices have emerged as one of the dominant flat-panel display technologies [10].[1] The major market success has come in portable instrumentation and laptop computers. PDP devices can be grouped into three fundamental technologies:

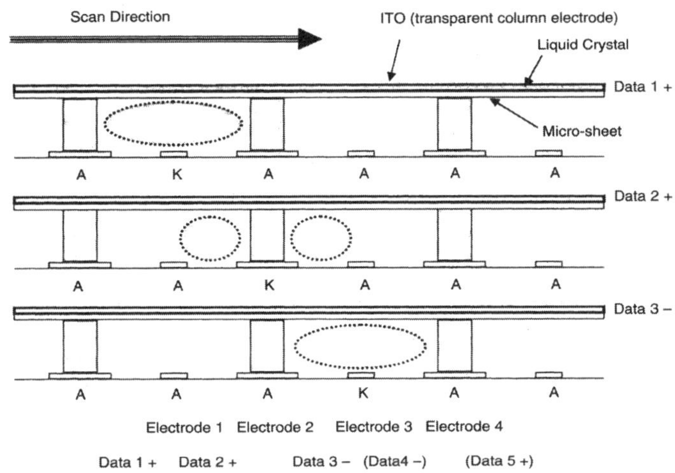


**Figure 22.12** Addressing scheme for a high-definition PALC device. (*After* [9]. *Used with permission.*)

- **dc displays**. Subdivisions include those devices with memory and those using dc refresh.
- **ac displays**. Subdivisions include those devices with memory and those using ac refresh.
- **ac-dc hybrids**.

In its basic design, the plasma display panel has the following attributes:

- Large nonlinearity resulting from the electrical characteristics of the gas discharge used in the device; that is, below a certain threshold voltage, the gas discharge will emit no light. This sharp nonlinearity allows the plasma display to be multiplexed virtually without limit, which means that very large plasma displays are practical.
- Memory-based display cells are practical. Stored directly in the glass plasma panel, memory is a desirable feature for a flat panel display because it facilitates high brightness, eliminates flicker, and can simplify the interface scanning system. For a memory-based display cell, the duty cycle is unity; the cells that should be illuminated are always on.
- Discharge switching is practical. In a PDP, the gas discharge itself can perform high-voltage switching operations that allow certain silicon-based switching devices to be eliminated, and others to be reduced in number. Because the circuits that drive a plasma panel constitute a large portion of the final cost of a PDP product, reducing the number of required drivers can have a dramatic effect on product price and, therefore, market penetration.

---

1. Portions of this section were adapted from: Weber, Larry F.: "Plasma Displays," in *The Electrical Engineering Handbook*, Richard C. Dorf (ed.), CRC Press, Boca Raton, Fla., 1786-1798, 1993. Used with permission.

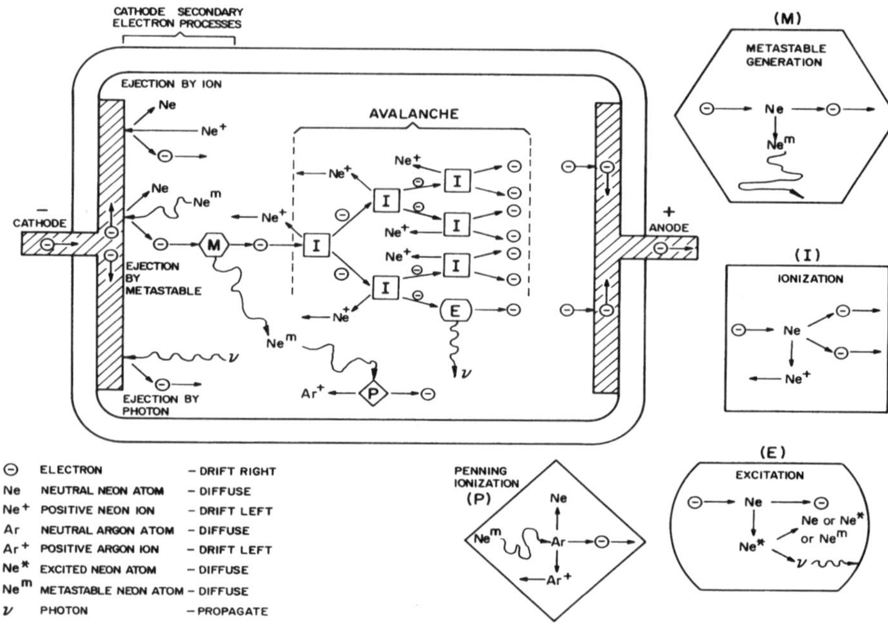


**Figure 22.13** Gas discharge reactions model for a plasma display panel. (*From* [10]. *Used with permission.*)

- Long operating life. Lifetime projections for ac-based PDPs range up to 350,000 hrs. Somewhat lower numbers are assigned for dc-based devices. Whether these very long lifetimes can ever be realized (350,000 hrs ≈ 40 years), it is safe to say that plasma displays will last considerably longer than other emissive devices.
- Performance parameters, including brightness, luminance efficiency, and viewing angle are good.

## 22.3a Gas Discharge Characteristics

Figure 22.13 illustrates the important reactions that occur in a gas discharge [10]. The reactions in the gas volume include the following:

- Ionization (I)
- Excitation (E)
- Metastable generation (M)
- Penning ionization (P)

The three surface reactions that occur at the cathode cause electrons to be ejected from the cathode by a bombarding neon ion, a neon metastable atom, or a high-energy photon. The most

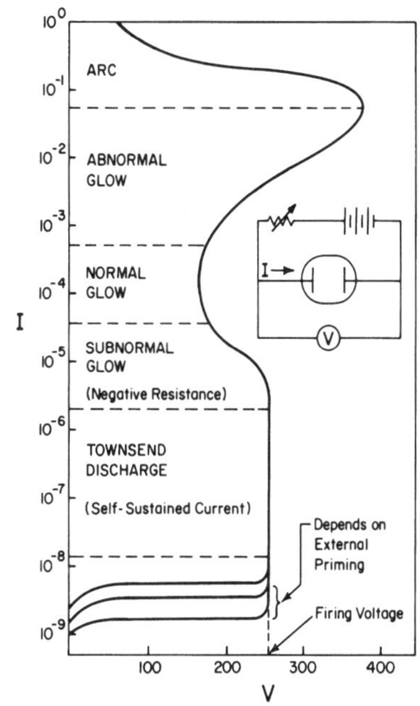


**Figure 22.14** Typical voltage-current characteristics of a gas discharge. (*From* [10]. *Used with permission.*)

important volume reaction is ionization (I), which can cause the generation of an avalanche in the gas volume, as shown in the figure. This avalanche is started by an electron near the cathode and, as it grows toward the anode, the avalanche generates a large number of election-ion pairs. The number of electron-ion pairs increases with increasing applied voltage across the gas. Ions, photos, or metastable atoms that are transported to the cathode can then eject electrons with a cathode surface-dependent probability, and these ejected electrons then initiate further avalanches. These mechanisms act as a positive feedback system that becomes unstable when the loop gain is greater than unity. The onset of the unstable condition is defined as the *gas firing voltage*. Above this firing voltage, the discharge current will continue to grow without bounds if the initial avalanche is primed with at least a single electron.

Figure 22.14 shows the voltage-current characteristics for a typical gas discharge found in a plasma display. Note the strong nonlinearity at the firing voltage. This is a major attribute of gas discharge that allows matrix addressing. When the discharge current has sufficient magnitude, space charge distortion sets in and the characteristic achieves a *negative resistance region*. Most plasma displays operate near the junction of the normal and the abnormal glow regions of the characteristic.

Figure 22.15 shows the characteristics of the glow discharge commonly found in a plasma display. The light comes from two luminous regions:

- The *negative glow region*
- Positive column region

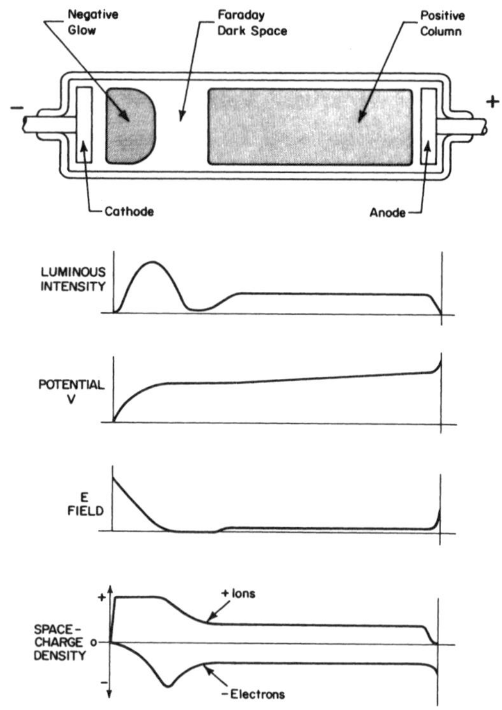


**Figure 22.15** The luminous regions of a gas discharge. (*From* [10]. *Used with permission.*)

These regions are caused by the space charge distribution of the electrons and ions that distort the electric field and voltage distribution. Figure 22.16 shows the wavelength distribution found in plasma displays. It is the classic line spectrum of neon gas.

To avoid device failure resulting from the arc regions shown in Figure 22.14, some form of current limiting must be provided. The two most common basic approaches are illustrated in Figure 22.17. As shown, for the dc plasma display, a series resistor (usually external to the device) is used for current limiting. For the ac plasma display, current limiting is accomplished using an internal glass dielectric that couples the electrodes capacitively to the gas discharge.

Memory capability for an ac plasma display can be implemented with relative simplicity. During fabrication, a dielectric layer is constructed with a specified impedance and a current-limiting capacitor is placed in series with each pixel. When a voltage pulse is applied to the ac panel, the discharge deposits a charge on the wall that reduces the voltage across the gas. After a short time, the discharge will extinguish and the light output will end until the applied voltage reverses polarity and a new discharge pulse occurs. This charge allows the ac plasma display to operate in a memory mode, which greatly increases the brightness of large area displays.

## 22.3b Memory-Type ac Plasma Display

The basic ac PDP topology is show in Figure 22.18. This type of device is made by depositing thin-film electrodes on the front and back substrates and then covering those electrodes with a

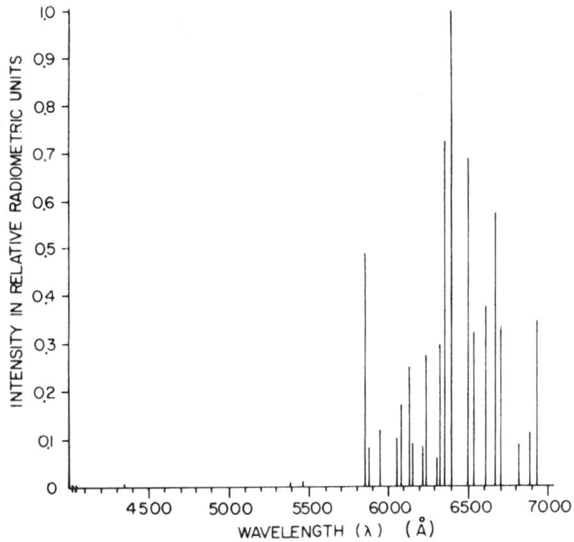


**Figure 22.16** Light wavelength distribution for a neon gas plasma display. (*From* [10]. *Used with permission.*)

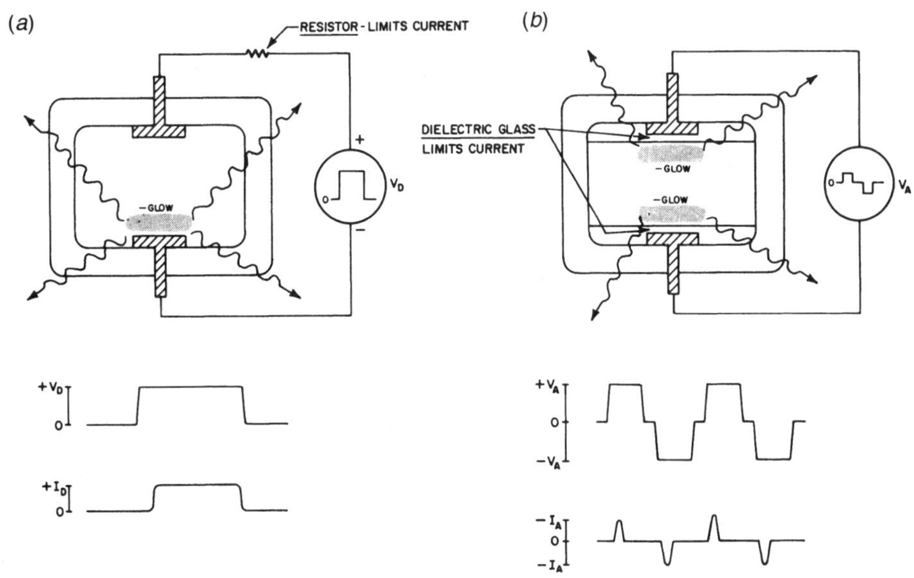


**Figure 22.17** Common current-limiting techniques for PDP devices: (*a*) dc display, (*b*) ac display. (*From* [10]. *Used with permission.*)

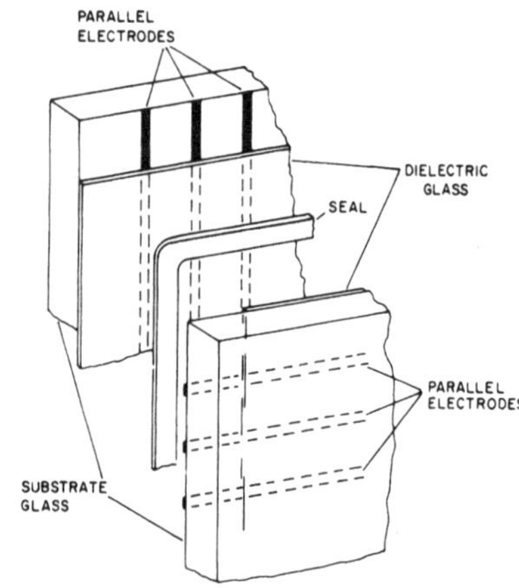


**Figure 22.18** Structure of an ac PDP display cell with memory. (*From* [10]. *Used with permission.*)

thin dielectric glass [10]. As mentioned in the previous section, the dielectric glass serves as a capacitor that limits current discharge. This dielectric also is used to store the charge, which gives this type of panel its inherent memory. The two substrates are then sealed around the perimeter and filled with neon gas. There are no barrier ribs between the discharge areas; pixels are isolated simply through the action of electric fields.

The inner surface of the dielectric, which is coated with magnesium oxide, is in contact with the neon gas. This MgO layer facilitates low operating voltages and long life.

The ac PDP requires an ac signal, referred to as the *sustaining voltage*, applied during operation. When a pixel is discharging, charge collects on the dielectric glass walls and influences the voltage across the gas. The component of voltage resulting from this charge is known as the *wall voltage*. When a pixel is in the on state, the wall voltage changes for each polarity reversal of the sustain voltage. This change in wall voltage coincides with a pulse of light because of the gas discharge. When the pixel is off, there are no light pulses, and the wall voltage remains zero.

Pixel addressing is achieved through a partial discharge introduced by an address pulse, timed between the sustain voltage pulses. A write pulse causes the wall voltage to transit from zero volts to the final equilibrium wall voltage. In the same manner, an erase pulse causes the wall voltage to return to zero.

## 22.3c Hybrid ac-dc Plasma Display

As show in Figure 22.19, the hybrid PDP is basically a dc plasma panel with the normal anode and cathode dc electrodes and standard barrier ribs to isolate the discharges. The distinction of this panel is the *trigger electrode* that is buried under the dc cathodes. This ac electrode is not used for display purposes but rather to create *priming particles* for the dc discharge.

The addressing sequence is initiated with a negative pulse applied to the ac trigger electrode. This creates the priming particles that eventually are used to ignite the normal dc discharge that is used for the display. These priming particles have two beneficial effects:

- They allow the use of lower voltages to drive the normal dc address electrodes

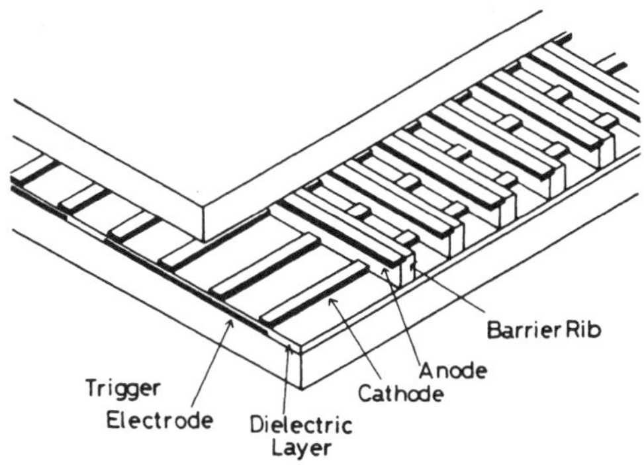


**Figure 22.19** Structure of a hybrid ac-dc plasma panel. (*Courtesy of Sony. From* [10]. *Used with permission.*)

- They permit a shorter time delay between the start of the address pulse and the ignition of the discharge

## 22.3d Color Plasma Displays

Full-color operation is possible with a PDP by introducing *vacuum ultraviolet* (VUV) excitation phosphors into the discharge cell [11]. (See Figure 22.20.) Because the plasma display controls electron motions, a response time of less than 1 μs is possible, allowing the device to be used for displaying HDTV images with 8-bit (256) intensity levels.

The primary color red, green, and blue phosphors are deposited onto walls of each discharge cell. In order to retain the same resolution of an equivalent monochrome display, the discharge *cell pitch* (*sub-pixel pitch*) must be reduced by a factor of three. The deposition process on the bottom and side walls of the discharge cells is somewhat complicated. Practical methods include:

- Thick-film printing. This well-proved technique requires careful adjustments of the binder contents and printing conditions.
- Photo-sensitive phosphor paste. In this method, a layer having special optical properties is first deposited on the rib surface to which the phosphor is to be deposited. Through irradiation with UV light, the layer exhibits tackiness, which causes the phosphor powder to adhere to the layer.
- Lamination. Phosphor in the form of thin layers is laminated along the rib surface and then etched using conventional photo processing techniques.

Display devices for video applications require at least 8-bit gray levels (256). In neon-red monochrome displays, gray scale expression is achieved by intensity modulation of the discharge current. This technique, however, is not applicable to color plasma devices whose output luminance has a strong saturation characteristic with respect to the discharge current.

Gray scales for color PDPs, rather, are obtained by using a *pulse-number-modulation* scheme. One video field is divided into eight sub-fields, each of which consists of an *address period* and

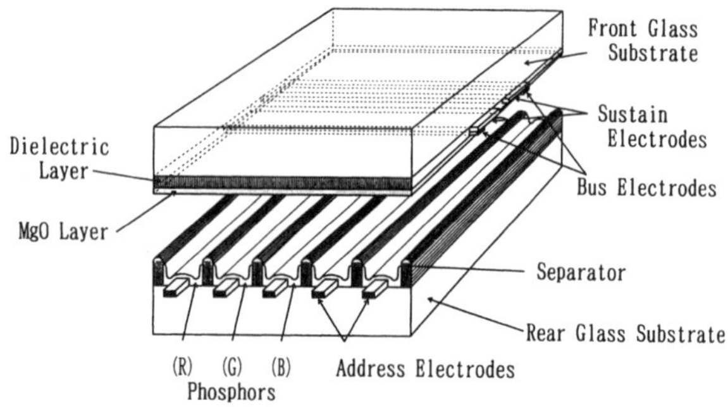


**Figure 22.20** Full-color ac memory plasma display cell. (*Courtesy of Fujitsu. From* [10]. *Used with permission.*)

a *display period.* The lengths of the display periods are arranged according to the binary sequence, 1:2:4:8:16:32:64:128. The display periods are filled with trains of constant width and constant period pulses. With the 8 sub-field technique, gray levels of $2^8$ can be achieved by combining appropriate sub-fields.

The technique of gray scale expression using sub-fields is quite adequate for still images, however, disturbances of gray scales are often observed when displaying moving images [12]. When a gradation pattern is stationary, there is no disturbance, but as soon as the pattern starts moving and the eyes follow it, disturbances are perceived. When the eyes stop following the image, the disturbances immediately disappear. Given this pattern movement, color disturbances occur as well.

Various methods of reducing these motional artifacts have been proposed [11], with varying degrees of success. One method, referred to as *distributed-address and sustain* (DAS), has shown promise [13].

In the DAS approach, the address period and the display period are not separated. In this manner, the problem of reduced light-emitting time in conventional approaches is eliminated. DAS is implemented using revised driver timing and modified driver waveforms.

## 22.3e Performance Issues for Video Applications

Although the performance of PDP devices is acceptable for home video (consumer) use, several issues remain [11]. Prime among these is image quality. While high luminance is necessary for attractive images, it is more important to achieve high contrast under high ambient light environments, particularly in consumer applications. Contrast performance can be improved through the use of one or more of the following techniques:

- Increasing emission intensity from the phosphor
- Minimizing background lights by reducing emission from control discharges or by hiding auxiliary discharge emissions

- Reducing ambient light reflections at the front glass and at the phosphor surface through the use of a neutral density filter or a circular polarizing filter

Of course, any filter placed in front of an image-producing surface also reduces the output luminance, requiring higher luminance and luminous efficiency. In this regard there is, naturally, the necessity for compromise among operating parameters.

Reflected light can also be reduced with color filters (a red-transmitting filter in front of each red cell, and so on). With ideal color filters, the contrast ratio can be improved by a factor of three without a reduction in output luminance.

For video applications, certain color gamut issues—principally involving the red phosphor—also remain to be solved for PDP devices [11].

Despite these challenges, PDPs hold great promise and are expected to claim a portion of the small- to medium-sized projection display market. The cost economies of the display elements used in building PDPs (standardization of glass substrate, use of non-rare-earth phosphors, and combined filters) permit large displays to be constructed at prices competitive with projection systems. On the negative side insofar as cost is concerned, however, is the large number of high voltage circuit drivers required. The cost of the drive electronics in a plasma display usually exceeds that of the panel itself.

Work continues to extend PDP performance, largely concentrating on the following parameters:

- Large screen sizes, in the range of 50 to 60 in.
- Improved contrast ratio of as high as 100:1 in normal ambient light
- Increased peak white luminance to as high as 700 cd/m$^2$ to compete with CRT devices
- Improved luminous efficiency to 5 lm/W or higher

At this writing the technologies necessary to achieve these goals were believed to exist. Implementation remained a challenge, however.

### 22.3f Advanced XGA Device

Among the products on the cutting edge of PDP device performance is a high-luminance 50-in. diagonal color display (Pioneer, [14]). Key performance parameters include:

- Luminous efficiency of 1.0 lm/W in a 50-in. diagonal XGA/wide panel display
- Sub-pixel pitch of 0.286 mm
- 768(V) × 1260(H) pixel configuration
- The display cell, utilizing "T"-shaped electrodes, is shown in Figure 22.21*a*. This configuration is contrasted with a more conventional rectangular cell, shown in *b*.

## 22.4 Field Emission Displays

A field emission display (FED) is—in essence—an offspring of the CRT. As shown in Figure 22.22, the FED is a vacuum tube display device in which electrons from millions of individual

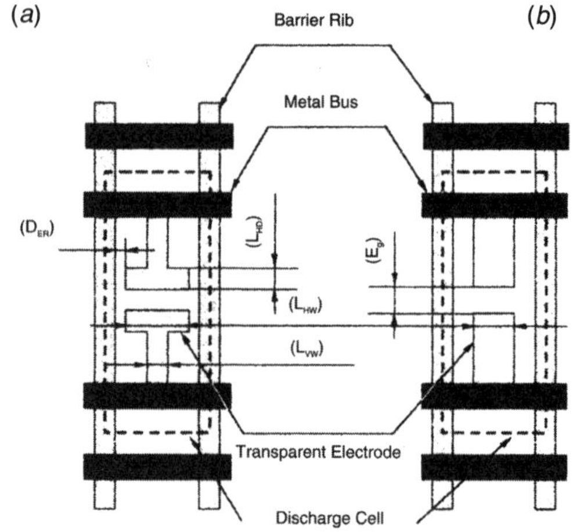


**Figure 22.21** Co-planar PDP discharge cell: (*a*) incorporating a "T"-shaped electrode, (*b*) incorporating a rectangular electrode. (*After* [14].)

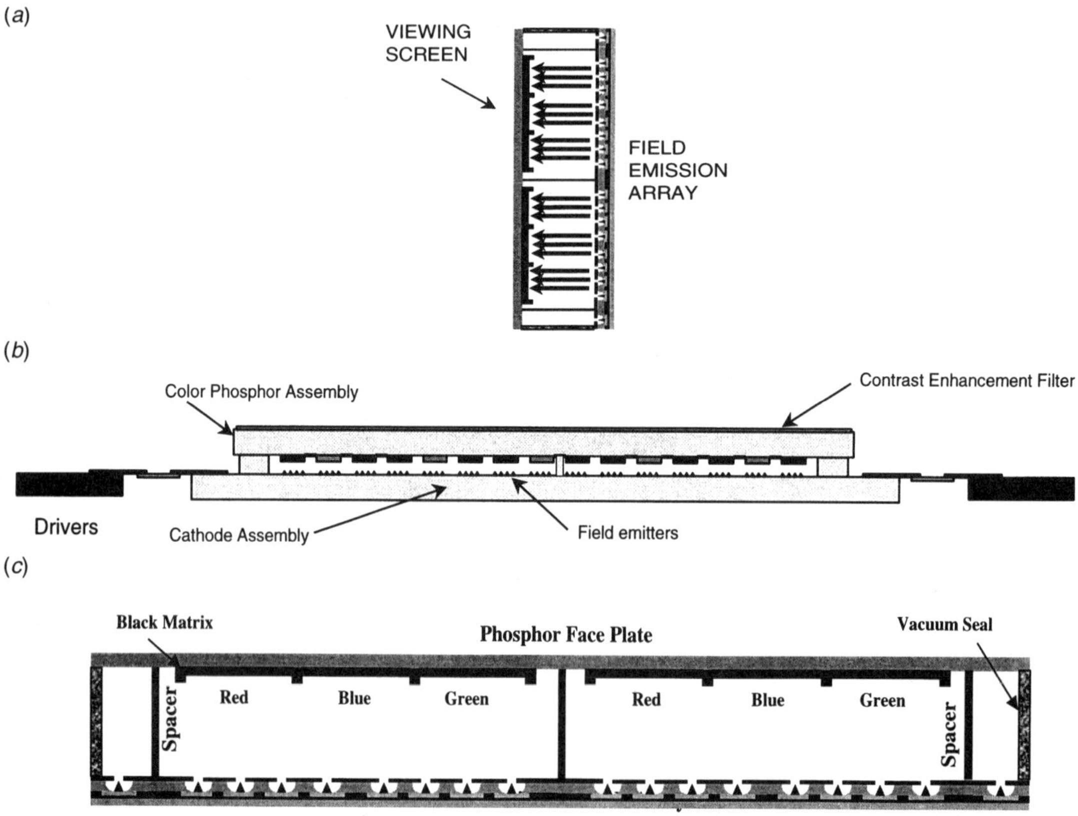


**Figure 22.22** Field emission display device: (*a*) basic configuration, (*b*) emission element, (*c*) emission element detail. (*From* [15]. *Used with permission.*)

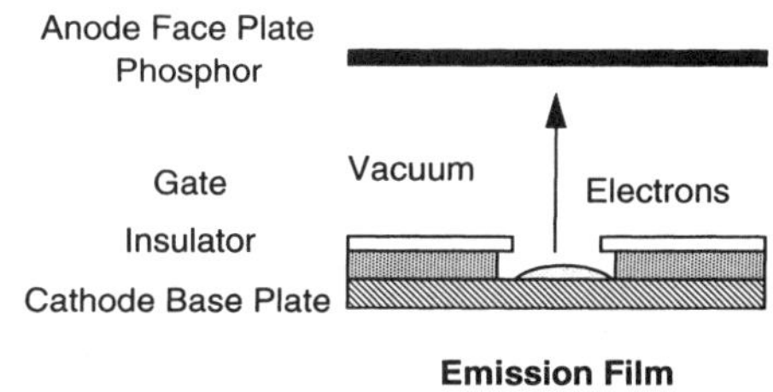


**Figure 22.23** A planar carbon emitter. (*From* [15]. *Used with permission.*)

(and very small) cathodes travel to a multicolor phosphor viewing screen to create an image [15]. A matrix of emitters makes up the displayed raster.

Owing to its CRT heritage, the FED enjoys the benefits of the CRT over other display technologies, while foregoing the physical size penalties (depth) of the CRT.

## 22.4a Operational Elements

As with the CRT, the ultimate performance of the display is determined to a large extent by the emissive surfaces—*microtips* for the FED. Figure 22.23 shows a simple planar carbon emitter. Advantages of this approach include simplicity of the structure and the resulting low manufacturing costs.

A variety of other emitter shapes are possible, ranging from cylindrical to pyramid. The shape chosen for a given device involves a trade-off between maximum emitter efficiency, long term performance, and manufacturing constraints. A microphotograph of a high-performance emitter is shown in Figure 22.24. Emission for this configuration—a *Spindt array*—strongly depends on the tip height and sharpness, and the position of the tip within the well.

Whatever form the emitter takes, preserving the life of the cathode is an important design criteria. Conditions that can lead to cathode destruction include the following:

- Emitter damage resulting from uncontrolled emission current.
- Undesirable vacuum levels, which can lead to impact ionization of the residual gases, followed by ion bombardment damage to the emitter.
- Electron-stimulated desorption from surfaces and device degradation resulting from the gas-emitter surface interactions.
- Electron beam induced desorption of gas from the phosphor and subsequent absorption on emitter tips.

Addressing of the FED is based on a matrix where rows are addressed sequentially in time. During each row address time, all columns are addressed simultaneously. The row pattern repeats at the frame rate and is content independent.

Gray scale can be achieved by pulse width modulation of column voltages, amplitude modulation of column voltages, or a combination of the two. Among the challenges of gray scale performance are that emission uniformity is difficult to control across the dimensions of a display screen. Using voltage modulation to control gray scale can, therefore, be problematic [16]. Pulse modulation, on the other hand, provides uniform gray levels compared to voltage modulation,

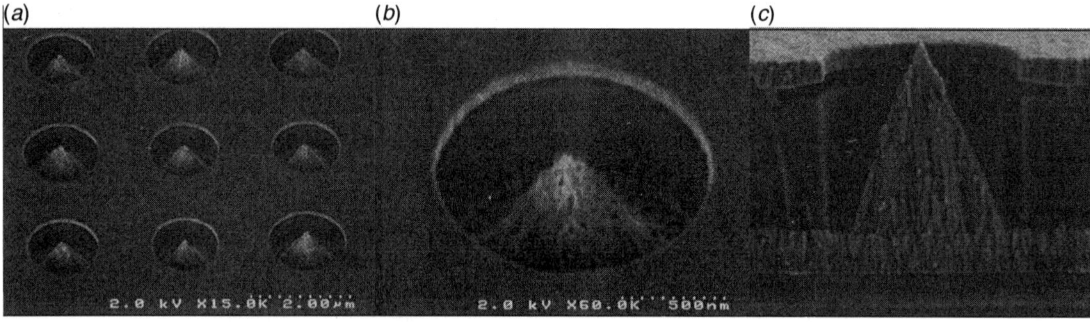


**Figure 22.24** Electron microphotograph of a Spindt array: (*a*) group of nine devices, (*b*) closeup of a single emitter, (*c*) emitter cross-section detail. (*From* [15]. *Used with permission.*)

but requires resistive layers in the FED panel to get an averaging effect and enhanced emission current uniformity.

As with most other vacuum devices, getters are used to maintain a good vacuum in the package. Vacuum levels in range of $10^{-6}$ to $10^{-7}$ Torr are required for long life.

## 22.5 Fixed Pixel Pattern

The flat panel devices discussed in this chapter are on the cutting edge of display technology. All, however, share one common drawback: a fixed pixel pattern. It is quite practical—even commonplace—for CRT-based displays to offer multisync operation, where the scanning parameters of the display conform to the input signal. With a fixed pattern device, however, the input signal must conform to the display. In many applications this issue is not important, but in others it can be a serious limiting factor.

It is important also to point out that several of the projection devices discussed in the previous chapter share this limitation.

## 22.6 References

1. Morris, James E.: "Liquid-Crystal Displays," in *The Electrical Engineering Handbook*, Richard C. Dorf (ed.), CRC Press, Boca Raton, Fla., 1993.

2. Allison, J.: *Electronic Engineering Semiconductors and Devices*, 2nd ed., McGraw-Hill, London, pg. 308–309, 1990.

3. Baur, G.: *The Physics and Chemistry of Liquid Crystal Devices*, G. J. Sprokel (ed.), Plenum, New York, N.Y., pg. 62, 1980.

4. Wilson, J., and J. F. B. Hawkes: *Optoelectronics: An Introduction*, Prentice-Hall, London, pg. 145, 1989.

5. Sarma, Kalluri R.: "Active-Matrix LCDs," in *Seminar Lecture Notes*, Society for Information Display, San Jose, Calif., vol. 1, pp. M3/3–M3/45, 1999.
6. Buzak, Thomas S.: "Recent Advances in PALC Technology," in *Proceedings of the 18th International Display Research Conference*, Society for Information Display, San Jose, Calif., Asia Display '98, pp. 273–276, 1998.
7. Ilcisen, K. J., et. al.: *Eurodisplay '96*, pg. 595, 1996.
8. Burgmans, A., et. al.: *Information Display*, pg. 14, April/May 1998.
9. Hayashi, M., N. Yamada, and B. Sastra: "Development of a 42-in. High-Definition Plasma-Addressed LCD," in *SID International Symposium Digest of Technical Papers*, Society for Information Display, San Jose, Calif., pp.280–284, 1999.
10. Weber, Larry F.: "Plasma Displays," in *The Electrical Engineering Handbook*, Richard C. Dorf (ed.), CRC Press, Boca Raton, Fla., 1786–1798, 1993.
11. Mikoshiba, Shigeo: *Color Plasma Displays*, Society for Information Display, San Jose, Calif., pp M-4/3–M-4/68, 1999.
12. Masuda, T., et. al: *Conference Record—International Display Resolution Conference*, pg. 357, 1994.
13. Lim, G. S., et. al.: "New Driving Method for Improvement of Picture Quality in 40-in. AC PDP," in *Asia Display '98—Proceedings of the 18th International Display Research Conference*, Society for Information Display, San Jose, Calif., pp. 591–594, 1998.
14. Nishio, T., and K. Amemiya: "High-Luminance and High-Definition 50-in.-Diagonal Co-Planar Color PDPs with T-Shaped Electrodes," in *SID International Symposium Digest of Technical Papers*, Society for Information Display, San Jose, Calif., pp.268–272, 1999.
15. Dworsky, Lawrence N., and Babu R. Chalamala: *Field-Emission Displays*, Society for Information Display, San Jose, Calif., pp. F-1/3–F1/66, 1999.
16. Na, Young-Sun, et. al.: "A New Data Driver Circuit for Field Emission Display," in *Asia Display '98—Proceedings of the 18th International Display Research Conference*, Society for Information Display, San Jose, Calif., pp. 137–140, 1998.

Chapter

23

# Measurement of Color Displays

**Jerry C. Whitaker, Editor**

## 23.1 Introduction

The chromaticity and luminance of a portion of a color display device may be measured in several ways. The most fundamental approach involves a complete spectroradiometric measurement followed by computation using tables of color-matching functions. Portable spectroradiometers with built-in computers are available for this purpose. Another method, somewhat faster but less accurate, involves the use of a photoelectric colorimeter. Because these devices have spectral sensitivities approximately equal to the CIE color-matching functions, they provide direct readings of tristimulus values.

For setting up the reference white, it is often simplest to use a split-field visual comparator and to adjust the display device until it matches the reference field (usually $D_{65}$) of the comparator. However, because a large spectral difference (large *metamerism*) usually exists between the display and the reference, different observers may make different settings by this method. Consequently, settings by one observer—or a group of observers—with normal color vision often are used simply to provide a reference point for subsequent photoelectric measurements.

An alternative method of determining the luminance and chromaticity coordinates of any area of a display involves measuring the output of each phosphor separately, then combining the measurements using the *center of gravity law*, by which the total tristimulus output of each phosphor is considered as an equivalent weight located at the chromaticity coordinates of the phosphor.

Consider the CIE chromaticity diagram shown in Figure 23.1 to be a uniform flat surface positioned in a horizontal plane. For the case illustrated, the center of gravity of the three weights ($T_r$, $T_g$, $T_b$), or the *balance point*, will be at the point $C_o$. This point determines the chromaticity of the mixture color. The luminance of the color $C_o$ will be the linear sum of the luminance outputs of the red, green, and blue phosphors. The chromaticity coordinates of the display primaries may be obtained from the manufacturer. The total tristimulus output of one phosphor may be determined by turning off the other two CRT guns, measuring the luminance of the specified area, and dividing this value by the $y$ chromaticity coordinate of the energized phosphor. This procedure then is repeated for the other two phosphors. From this data, the color resulting from given excitations of the three phosphors may be calculated as follows:

- Chromaticity coordinates of red phosphor = $x_r$, $y_r$
- Chromaticity coordinates of green phosphor = $x_g$, $y_g$

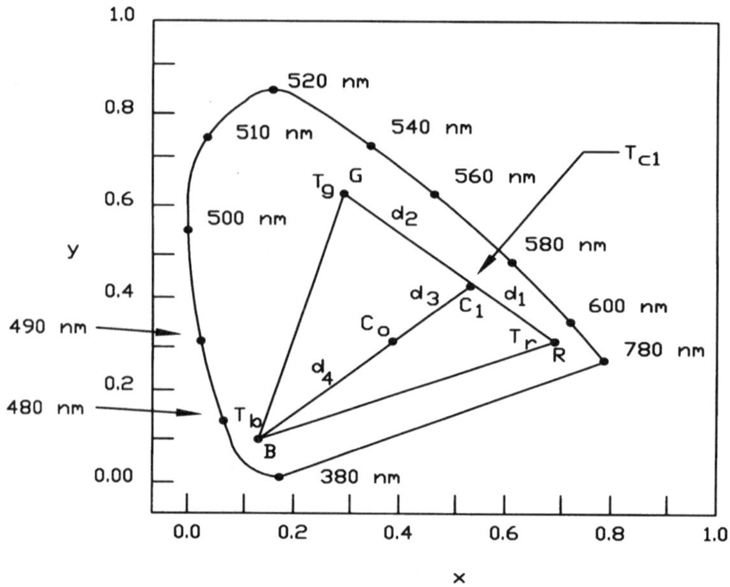


**Figure 23.1** The CIE 1931 chromaticity diagram illustrating use of the *center of gravity law* ($T_r d_1 = T_g d_2$, $T_{c1} = T_r + T_g$, $T_{c1} d_3 = T_b d_4$).

- Chromaticity coordinates of blue phosphor = $x_b, y_b$
- Luminance of red phosphor = $Y_r$
- Luminance of green phosphor = $Y_g$
- Luminance of blue phosphor = $Y_b$

$$\text{Total tristimulus value of red phosphor} = X_r + Y_r + Z_r = \frac{Y_r}{y_r} = T_r \tag{23.1}$$

$$\text{Total tristimulus value of green phosphor} = X_g + Y_g + Z_g = \frac{Y_g}{y_g} = T_g \tag{23.2}$$

$$\text{Total tristimulus value of blue phosphor} = X_b + Y_b + Z_b = \frac{Y_b}{y_b} = T_b \tag{23.3}$$

Consider $T_r$ as a weight located at the chromaticity coordinates of the red phosphor and $T_g$ as a weight located at the chromaticity coordinates of the green phosphor. The location of the chromaticity coordinates of color $C_1$ (blue gun of color CRT turned off) can be determined by taking moments along line RG to determine the center of gravity of weights $T_r$ and $T_g$:

$$T_r \times d_1 = T_g \times d_2 \tag{23.4}$$

The total tristimulus value of $C_1$ is equal to $T_r + T_g = T_{c1}$ (23.5)

Taking moments along line $C_{1B}$ will locate the chromaticity coordinates of the mixture color *Co*:

$$T_{c1} \times d_3 = T_b \times d_4 \tag{23.6}$$

The luminance of the color $C_o$ is equal to $Y_r + Y_g + Y_b$ (23.7)

## 23.1a Assessment of Color Reproduction

A number of factors may contribute to poor color rendition in a display system. To assess the effect of these factors, it is necessary to define system objectives, then establish a method of measuring departures from the objectives. Visual image display may be categorized as follows:

- *Spectral color reproduction*: The exact reproduction of the spectral power distributions of the original stimuli. Clearly, this is not possible in a video system with three primaries.
- *Exact color reproduction*: The exact reproduction of tristimulus values. The reproduction is then a metameric match to the original. Exact color reproduction will result in equality of appearance only if the viewing conditions for the picture and the original scene are identical. These conditions include the angular subtense of the picture, the luminance and chromaticity of the surround, and glare. In practice, exact color reproduction often cannot be achieved because of limitations on the maximum luminance that can be produced on a color monitor.
- *Colorimetric color reproduction*: A variant of exact color reproduction in which the tristimulus values are proportional to those in the original scene. In other words, the chromaticity coordinates are reproduced exactly, but the luminances all are reduced by a constant factor. Traditionally, color video systems have been designed and evaluated for colorimetric color reproduction. If the original and the reproduced reference whites have the same chromaticity, if the viewing conditions are the same, and if the system has an overall gamma of unity, colorimetric color reproduction is indeed a useful criterion. These conditions, however, often do not hold; then, colorimetric color reproduction is inadequate.
- *Equivalent color reproduction*: The reproduction of the original color appearance. This might be considered as the ultimate objective, but it cannot be achieved because of the limited luminance generated in a display system.
- *Corresponding color reproduction*: A compromise by which colors in the reproduction have the same appearance that colors in the original would have had if they had been illuminated to produce the same average luminance level and the same reference white chromaticity as that

of the reproduction. For most purposes, corresponding color reproduction is the most suitable objective of a color video system.

- *Preferred color reproduction*: A departure from the preceding categories that recognizes the preferences of the viewer. It is sometimes argued that corresponding color reproduction is not the ultimate aim for some display systems, such as color television, and that it should be taken into account that people prefer some colors to be different from their actual appearance. For example, suntanned skin color is preferred to average real skin color, and sky is preferred bluer and foliage greener than they really are.

Even if corresponding color reproduction is accepted as the target, some colors are more important than others. For example, flesh tones must be acceptable—not obviously reddish, greenish, purplish, or otherwise incorrectly rendered. Likewise, the sky must be blue and the clouds white, within the viewer's range of acceptance. Similar conditions apply to other well-known colors of common experience.

## 23.1b Chromatic Adaptation and White Balance

With properly adjusted cameras and displays, whites and neutral grays are reproduced with the chromaticity of D65. Tests have shown that such whites (and grays) appear satisfactory in home viewing situations even if the ambient light is of quite different color temperature. Problems occur, however, when the white balance is slightly different from one camera to the next or when the scene shifts from studio to daylight or vice versa. In the first case, unwanted shifts of the displayed white occur, whereas in the other, no shift occurs even though the viewer subconsciously expects a shift.

By always reproducing a white surface with the same chromaticity, the system is mimicking the human visual system, which adapts so that white surfaces always appear the same, whatever the chromaticity of the illuminant (at least within the range of common light sources). The effect on other colors, however, is more complicated. In video cameras, the white balance adjustment usually is made by gain controls on the R, G, and B channels. This is similar to the *von Kries* model of human chromatic adaptation, although the R, G, and B primaries of the model are not the same as the video primaries. It is known that the von Kries model does not accurately account for the appearance of colors after chromatic adaptation, and so it follows that making simple gain changes in a video camera is not the ideal approach. Nevertheless, this approach seems to work well in practice, and the viewer does not object to the fact, for example, that the relative increase in the luminances of reddish objects in tungsten light is lost.

## 23.1c Overall Gamma Requirements

Colorimetric color reproduction requires that the overall gamma of the system—including the camera, the display, and any gamma-adjusting electronics—be unity. This simple criterion is the one most often used in the design of a video color rendition system. However, the more sophisticated criterion of corresponding color reproduction takes into account the effect of the viewing conditions. In particular, several studies have shown that the luminance of the surround is important. For example, a dim surround requires a gamma of about 1.2, and a dark surround requires a gamma of about 1.5 for optimum color reproduction.

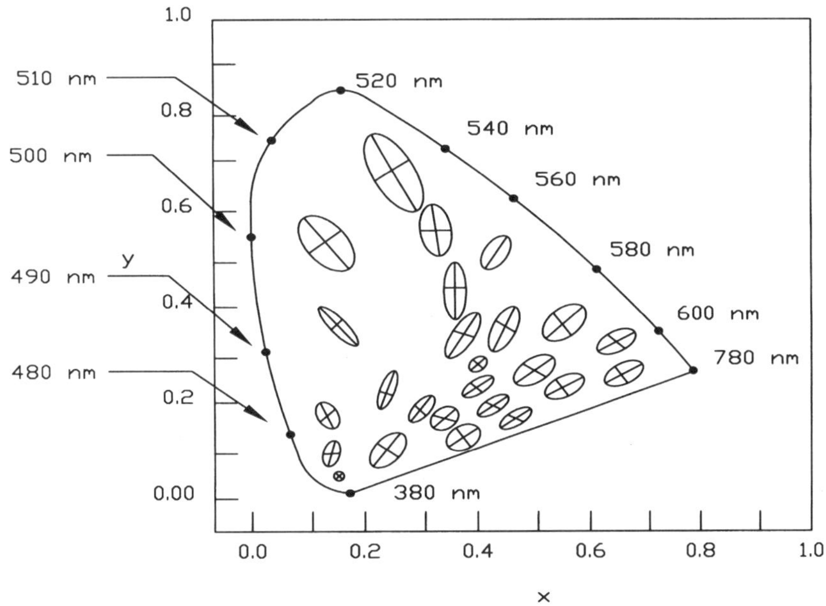


**Figure 23.2** Ellipses of equally perceptible color differences (as identified by MacAdam).

## 23.1d Perception of Color Differences

The CIE 1931 chromaticity diagram does not map chromaticity on a uniform-perceptibility basis. A just-perceptible change of chromaticity is not represented by the same distance in different parts of the diagram. Many investigators have explored the manner in which perceptibility varies over the diagram. The study most often quoted is that of MacAdam, who identified a set of ellipses that are contours of equal perceptibility about a given color, as shown in Figure 23.2 [1].

From this and similar studies it is apparent, for example, that large distances represent relatively small perceptible changes in the green sector of the diagram. In the blue region, much smaller changes in the chromaticity coordinates are readily perceived.

Furthermore, viewing identical images on dissimilar displays can result in observed differences in the appearance of the image [2]. Several factors affect the appearance of the image, including:

- Physical factors, including display gamut, illumination level, and black point
- Psychophysical factors, including chromatic induction and color constancy

Each of these factors interact in such a way that prediction of the appearance of an image on a given display becomes difficult. System designers have experimented with colorimetry to facilitate the sharing of image data among display devices that vary in manufacture, calibration, and location. Of particular interest today is the application of colorimetry to imaging in a networked window system environment, where it is often necessary to ensure that an image displayed remotely looks like the image displayed locally.

Studies have indicated that image context and image content are also factors that affect color appearance. The use of highly chromatic backgrounds in a windowed display system is popular, but it usually will affect the appearance of the colors in the foreground.

## 23.1e Display Resolution and Pixel Format

The pixel represents the smallest resolvable element of a display. The size of the pixel varies from one type of display to another. In a monochrome CRT, pixel size is determined primarily by the following factors:

- Spot size of the electron beam (the current density distribution)
- Phosphor particle size
- Thickness of the phosphor layer

The term *pixel* was developed in the era of monochrome television, and the definition was—at that time—straightforward. With the advent of color-triad-based CRTs and solid-state display systems, the definition is not nearly so clear.

For a color CRT, a single triad of red, green, and blue phosphor dots constitutes a single pixel. This definition assumes that the mechanical and electrical parameters of the CRT will permit each triad to be addressed without illuminating other elements on the face of the tube. Most display systems, however, will not meet this criterion. Depending on the design, a number of triads may constitute a single pixel in a CRT display. A more all-inclusive definition for the pixel is: the smallest spatial-information element as seen by the viewer [3].

*Dot pitch* is one of the principal mechanical criteria of a CRT that determines, to a large extent, the resolution of the display. Dot pitch is defined as the center-to-center distance between adjacent green phosphor dots of the red, green, blue triad.

The *pixel format* is the arrangement of pixels into horizontal rows and vertical columns. For example, an arrangement of 640 horizontal pixels by 480 vertical pixels results in a 640 × 480 pixel format. This description is not a resolution parameter in itself, simply the arrangement of pixel elements on the screen. *Resolution* is the measure of the ability to delineate picture detail; it is the smallest discernible and measurable detail in a visual presentation [4].

*Pixel density* is a parameter closely related to resolution, stated in terms of pixels per linear distance. Pixel density specifies how closely the pixel elements are spaced on a given display. It follows that a display with a given pixel format will not provide the same pixel density, or resolution, on a large-size screen (such as 27-in diagonal) as on a small-size screen (such as 12-in diagonal).

*Television lines* is another term used to describe resolution. The term refers to the number of discernible lines on a standard test chart. As before, the specification of television lines is not, in itself, a description of display resolution. A 525-line display on a 17-in monitor will appear to a viewer to have greater resolution than a 525-line display on a 30-inch monitor. Pixel density is the preferred resolution parameter.

### 23.1f Contrast Ratio

The purpose of a video display is to convey information by controlling the illumination of phosphor dots on a screen, or by controlling the reflectance or transmittance of a light source. The *contrast ratio* specifies the observable difference between a pixel that is switched on and one that is in its corresponding off state:

$$C_r = \frac{L_{on}}{L_{off}} \tag{23.8}$$

Where:
$C_r$ = contrast ratio of the display
$L_{on}$ = luminance of a pixel in the on state
$L_{off}$ = luminance of a pixel in the off state

The area encompassed by the contrast ratio is an important parameter in assessing the performance of a display. Two contrast ratio divisions typically are specified:

- *Small area*: comparison of the on and off states of a pixel-sized area
- *Large area*: comparison of the on and off states of a group of pixels

For most display applications, the small-area contrast ratio is the more critical parameter.

## 23.2 Display Measurement Techniques

A number of different techniques have evolved for measuring the static performance of picture display devices and systems [5]. Most express the measured device performance in a unique figure of merit or metric. Although each approach provides useful information, the lack of standardization in measurement techniques makes it difficult or even impossible to directly compare the performance of a given class of devices.

Regardless of the method used to measure performance, the operating parameters must be set for the anticipated operating environment. Key parameters include:

- Input signal level
- System/display line rate
- Luminance (brightness)
- Contrast
- Image size
- Aspect ratio

If the display is used in more than one environmental condition—such as under day and night conditions—a set of measurements is appropriate for each application.

## 23.2a Subjective CRT Measurements

Three common subjective measurement techniques are used to assess the performance of a CRT [7]:

- Shrinking raster
- Line width
- TV limiting resolution

Predictably, subjective measurements tend to exhibit more variability than objective measurements. Although they generally are not used for acceptance testing or quality control, subjective CRT measurements provide a fast and relatively simple means of performance assessment. Results usually are consistent when performed by the same observer. Results for different observers often vary, however, because different observers use different visual criteria to make their judgments.

The shrinking raster and line-width techniques are used to estimate the vertical dimension of the display (CRT) beam spot size (*footprint*). Several underlying assumptions accompany this approach:

- The spot is assumed to be symmetrical and Gaussian in the vertical and horizontal planes.
- The display modulation transfer function (MTF) calculated from the spot-size measurement results in the best performance envelope that can be expected from the device.
- The modulating electronics are designed with sufficient bandwidth so that spot size is the limiting performance parameter.
- The modulation contrast at low spatial frequencies approaches 100 percent.

Depending on the application, not all of these assumptions are valid:

- **Assumption 1.** Verona [5] has reported that the symmetry assumption is generally not true. The vertical spot profile is only an approximation to the horizontal spot profile; most spot profiles exhibit some degree of astigmatism. However, significant deviations from the symmetry and Gaussian assumptions result in only minor deviations from the projected performance when the assumptions are correct.
- **Assumption 2.** The optimum performance envelope assumption infers that other types of measurements will result in the same or lower modulation contrast at each spatial frequency. The MTF calculations based on a beam footprint in the vertical axis indicate the optimum performance that can be obtained from the display because finer detail (higher spatial frequency information) cannot be written onto the screen smaller than the spot size.
- **Assumption 3.** The modulation circuit bandwidth must be sufficient to pass the full incoming video signal. Typically, the video circuit bandwidth is not a problem with current technology circuits, which usually are designed to provide significantly more bandwidth than the display is capable of reproducing. However, in cases where this assumption is not true, the calculated MTF based purely on the vertical beam profile will be incorrect. The calculated performance will be better than the actual performance of the display.
- **Assumption 4.** The calculated MTF is normalized to 100 percent modulation contrast at zero spatial frequency and ignores the light scatter and other factors that degrade the actual mea-

sured MTF. Independent modulation-contrast measurements at a low spatial frequency can be used to adjust the MTF curve to correct for the normalization effects.

### Shrinking Raster Method

The shrinking raster measurement procedure is relatively simple. Steps include the following [5]:

- The brightness and contrast controls are set for the desired peak luminance with an active raster background luminance (1 percent of peak luminance) using a stair-step video signal.
- While displaying a flat-field video signal input corresponding to the peak luminance, the vertical gain/size is reduced until the raster lines are barely distinguishable.
- The raster height is measured and divided by the number of active scan lines to estimate the average height of each scan line. The number of active scan lines is typically 92 percent of the line rate. (For example, a 525-line display has 480 active lines, an 875-line display has 817, and a 1025-line display has 957 active lines.)

The calculated average line height typically is used as a stand-alone metric of display performance.

The most significant shortcoming of the shrinking raster method is the variability introduced through the determination of when the scan lines are *barely distinct* to the observer. Blinking and other eye movements often enhance the distinctness of the scan lines; lines that were indistinct become distinct again.

### Line-Width Method

The line-width measurement technique requires a microscope with a calibrated graticule [5]. The focused raster is set to a 4:3 aspect ratio, and the brightness and contrast controls are set for the desired peak luminance with an active raster background luminance (1 percent of peak luminance) using a stair-step video signal. A single horizontal line at the anticipated peak operating luminance is presented in the center of the display. The spot is measured by comparing its luminous profile with the graticule markings. As with the shrinking raster technique, determination of the line edge is subjective.

### TV Limiting Resolution Method

This technique involves the display of 2-dimensional high-contrast bar patterns or lines of various size, spacing, and angular orientation [5]. The observer subjectively determines the limiting resolution of the image by the smallest set of bars that can be resolved. Following are descriptions of potential errors with this technique:

- A phenomenon called *spurious resolution* can occur that leads the observer to overestimate the limiting resolution. Spurious resolution occurs beyond the actual resolution limits of the display. It appears as fine structures that can be perceived as line spacings closer than the spacing at which the bar pattern first completely blurs. This situation arises when the frequency-response characteristics fall to zero, then go negative, and perhaps oscillate as they die out. At the bottom of the negative trough, contrast is restored, but in reverse phase (white becomes black, and black becomes white).

- The use of test charts imaged with a video camera can lead to incorrect results because of the addition of camera resolution considerations to the measurement. Electronically generated test patterns are more reliable image sources.

The proper setting of brightness and contrast is required for this measurement. Brightness and contrast controls are adjusted for the desired peak luminance with an active raster background luminance (1 percent of peak luminance) using a stair-step video signal. Too much contrast will result in an inflated limiting resolution measurement; too little contrast will result in a degraded limiting resolution measurement.

Electronic resolution pattern generators typically provide a variety of resolution signals from 100 to 1000 *TV lines/picture height* (TVL/ph) or more in a given multiple (such as 100). Figure 23.3 illustrates an electronically generated resolution test pattern for high-definition video applications.

### Application Considerations

The subjective techniques discussed in this section, with the exception of TV limiting resolution, measure the resolution of the *display* [5]. The TV pattern test measures *image* resolution, which is quite different.

Consider as an example a video display in which the scan lines can just be perceived—about 480 scan lines per picture height. This indicates a *display* resolution of at least 960 TV lines, counting light *and* dark lines, per the convention. If a pattern from an electronic generator is displayed, observation will show the image beginning to deteriorate at about 340 TV lines. This characteristic is the result of beats between the image pattern and the raster, with the beat frequency decreasing as the pattern spatial frequency approaches the raster spatial frequency. This ratio of 340/480 = 0.7 (approximately) is known as the *Kell factor*. Although debated at length, the factor does not change appreciably in subjective observations.

## 23.2b Objective CRT Measurements

Four common types of objective measurements may be performed to assess the capabilities of a CRT display device:

- Half-power width
- Impulse Fourier transform
- Knife-edge Fourier transform
- Discrete frequency

Although more difficult to perform than the subjective measurements discussed so far, objective CRT measurement techniques offer greater accuracy and better repeatability. Some of the procedures require specialized hardware and/or software.

### Half-Power-Width Method

Under the half-power-width technique, a single horizontal line is activated with the brightness and contrast controls set to a typical operating level. The line luminance is equivalent to the highlight luminance (maximum signal level). The central portion of the line is imaged with a microscope in the plane of a variable-width slit. The open slit allows all the light from the line to pass

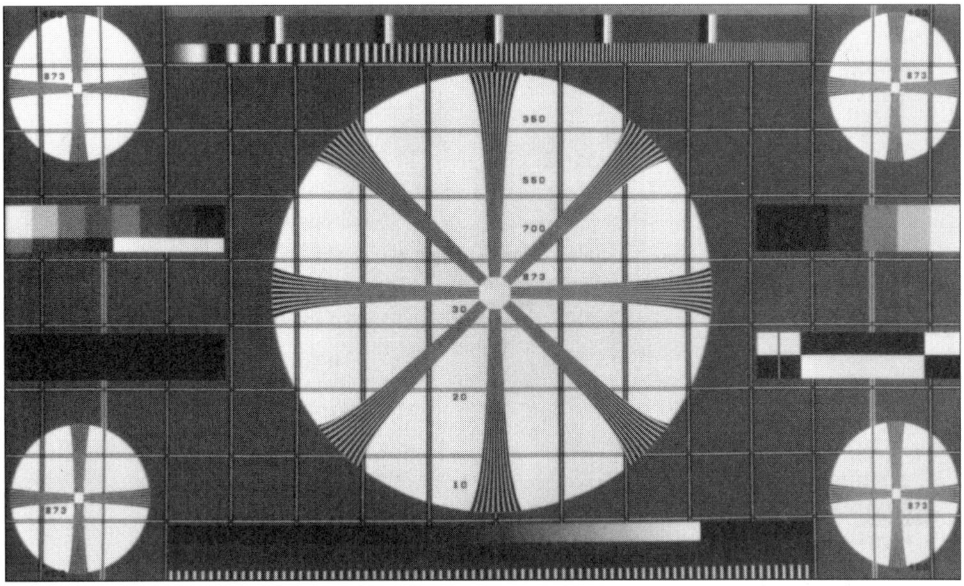


**Figure 23.3** Wide aspect ratio resolution test chart produced by an electronic signal generator. (*Courtesy of Tektronix.*)

through to a photodetector. The output of the photodetector is displayed on an oscilloscope. As the slit is gradually closed, the peak amplitude of the photodetector signal decreases. When the signal drops to 50 percent of its initial value, the slit width is recorded. The width measurement divided by the microscope magnification represents the *half-power width* of the horizontal scan line.

The half-power width is defined as the distance between symmetrical integration limits, centered about the maximum intensity point, which encompasses half of the total power under the intensity curve. The half-power width is not the same as the half-intensity width measured between the half-intensity points. The half-intensity width is theoretically 1.75 times greater than the half-power width for a Gaussian spot luminance distribution.

It should be noted that the half-power line-width technique relies on line width to predict the performance of the CRT. Many of the precautions outlined previously apply here also. The primary difference, however, is that line width is measured under this technique objectively, rather than subjectively.

## Fourier Transform Methods

The impulse Fourier transform technique involves measuring the luminance profile of the spot and then taking the Fourier transform of the distribution to obtain the MTF. The MTF, by definition, is the Fourier transform of the line spread function. Commercially available software may be used to perform these measurements using either an impulse or knife edge as the input waveform. Using the vertical spot profile as an approximation to the horizontal spot profile is not always appropriate, and the same reservations expressed in the previous section apply in this case as well.

The measurement is made by generating a single horizontal line on the display with the brightness and contrast set as discussed previously. A microphotometer with an effective slit-aperture width approximately one-tenth the width of the scan line is moved across the scan line (the long slit axis parallel to the scan line). The data taken is stored in an array, which represents the luminance profile of the CRT spot, distance vs. luminance. The microphotometer is calibrated for luminance measures and for distance measures in the object plane. Each micron step of the microphotometer represents a known increment in the object plane. The software then calculates the MTF of the display based on its line spread from the calibrated luminance and distance measurements. Finite slit-width corrections also may be made to the MTF curve by dividing it by a measurement-system MTF curve obtained from the luminance profile of an ideal knife-edge aperture or a standard source.

The knife-edge Fourier transform measurement may be conducted using a low-spatial-frequency vertical bar pattern (5 to 10 cycles) across the display with the brightness and contrast controls set as discussed previously. The frequency response of the square wave pattern generator and video pattern generator should be greater than the frequency response of the display system (100 MHz is typical). The microphotometer scans from the center of a bright bar to the center of a dark bar (left to right), measuring the width of the boundary and comparing it to a knife edge. The microphotometer slit is oriented vertically, with its long axis parallel to the bars. The scan usually is made from a light bar to a dark bar in the direction of spot movement. This procedure is preferred because waveforms from scans in the opposite direction may contain certain anomalies. When the beam is turned on in a square wave pattern, it tends to overshoot and oscillate. This behavior produces artifacts in the luminance profile of the bar edge as the beam moves from an off to an on state. In the on-to-off direction, however, the effects are minimal and the measured waveform does not exhibit the same anomalies that can corrupt the MTF calculations.

The bar-edge (knife-edge) measurement, unlike the other techniques discussed so far, uses the horizontal spot profile to predict display performance. All of the other techniques use the vertical profile as an approximation of the more critical horizontal spot profile. The bar-edge measurement will yield a more accurate assessment of display performance because the displayed image is being generated with a spot scanned in the horizontal direction.

## Discrete Frequency Method

The discrete sine wave frequency-response measurement technique provides the most accurate representation of display performance. With this approach, there are no assumptions implied about the shape of the spot, the electronics bandwidth, or low-frequency light scatter. Discrete spatial frequency sine wave patterns are used to obtain a discrete spatial frequency MTF curve that represents the signal-in to luminance-out performance of the display.

The measurement is begun by setting the brightness and contrast as discussed previously, with black-level luminance at 1 percent of the highlight luminance. A sine wave signal is produced by a function generator and fed to a pedestal generator where it is converted to an RS-170-A or RS-343 signal, then applied to the CRT. The modulation and resulting spatial frequency pattern are measured with a scanning microphotometer. The highlight and black-level measurements are used to calculate the modulation constant for each spatial frequency from 5 cycles/display width to the point that the modulation constant falls to less than 1 percent. The modulation constant values then are plotted as a function of spatial frequency, generating a discrete spatial frequency MTF curve.

## 23.3 Viewing Environment Considerations

The environment in which a video display device is viewed is an important criterion for critical viewing situations. Applications in which color purity and adherence to set standards are important require a standardized (or at least consistent) viewing environment. For example, textile colors viewed on a display with a white surround will appear different than the same colors viewed with a black surround. By the same token, different types of ambient lighting will make identical colors appear different on a given display.

The SMPTE has addressed this issue with RP 166-1995, which specifies the environmental and surround conditions that are required in television or video program review areas for the "consistent and critical evaluation" of conventional television signals [6]. Additionally, the practice is designed to provide for repeatable color grading or correction. A number of important parameters are specified in RP 166-1995, including the following:

- The distance of the observer from the monitor screen should be 4 to 6 picture heights for SDTV displays.
- The observer should view the monitor screen at a preferred angle in both the horizontal and vertical planes of 0° ±5° and, in any event, no greater than ±15° from the perpendicular to the midpoint of the screen.
- The viewing area decor should have a generally neutral matte impression, without dominant colors.
- Surface reflectances should be nonspecular and should not exceed 10 percent of the peak luminance value of the monitor white.

The Recommended Practice suggests placing the monitor in a freestanding environment 2.5 to 5 screen heights in front of the wall providing the visual surround. Another acceptable approach is to mount the monitor in a wall with its face approximately flush with the surface of the wall. It is further recommended that all light sources in use during picture assessment or adjustment have a color quality closely matching the monitor screen at reference white (i.e., D65).

It is often necessary to have black-and-white monitors surrounding one or more color monitors in a studio control room. According to RP 166-1995, the black-and-white monitors should be the same color temperature as the properly adjusted color monitor(s), 6500 K. Black-and-white monitors are normally equipped with P4 phosphors, at about 9300 K. This cooler color temperature prevents the background surrounding the color monitors from remaining neutral. Most black-and-white monitors can be ordered with 6500 K phosphors.

### 23.3a Picture Monitor Alignment

The proper adjustment and alignment of studio picture monitors is basic to video quality control. Uniform alignment throughout the production chain also ensures consistency in color adjustment, which facilitates the matching of different scenes within a program that may be processed at different times and in different facilities. The SMPTE has addressed this requirement for conventional video through RP 167-1995. The Recommended Practice offers a step-by-step process by which color monitors can be set. Key elements of RP 167-1995 include the following [7]:

- **Initial conditions.** Setup includes allowing the monitor to warm up and stabilize for 20 to 30 minutes. The room ambient lighting should be the same as it is when the monitor is in normal service, and several minutes must be allowed for visual adaptation to the operating environment.
- **Initial screen adjustments.** The monitor is switched to the setup position, in which the red, green, and blue screen controls are adjusted individually so that the signals are barely visible.
- **Purity.** Purity, the ability of the gun to excite only its designated phosphor, is checked by applying a low-level flat-field signal and activating only one of the three guns at a time. The display should have no noticeable discolorations across the face.
- **Scan size.** The color picture monitor application establishes whether the *overscan* or *underscan* presentation of the display will be selected. An underscanned display is one in which the active video (picture) area, including the corners of the raster, is visible within the screen mask. Normal scan brings the edges of the picture tangent to the mask position. Overscan should be no more than 5 percent.
- **Geometry and aspect ratio.** Display geometry and aspect ratio are adjusted with the crosshatch signal by scanning the display device with the green beam only. Correct geometry and linearity are obtained by adjusting the pincushion and scan-linearity controls so that the picture appears without evident distortions from the normal viewing distance.
- **Focus.** An ideal focus target is available from some test signal generators; if it is unavailable, multiburst, crosshatch, or white noise can be used as tools to optimize the focus of the displayed picture.
- **Convergence.** Convergence is adjusted with a crosshatch signal; it should be optimized for either normal scan or underscan, depending upon the application.
- **Aperture correction.** If aperture correction is used, the amount of correction can be estimated visually by ensuring that the 2T $\sin^2$ pulse has the same brightness as the luminance bar or the multiburst signal when the 3 and 4.2 MHz bursts have the same sharpness and contrast.
- **Chrominance amplitude and phase.** The chrominance amplitude and phase are adjusted using the SMPTE color bar test signal and viewing only the blue channel. Switching off the comb filter, if it is present, provides a clear blue channel display. Periodically, the red and green channels should be checked individually in a similar manner to verify that the decoders are working properly. A detailed description of this procedure is given in [7].
- **Brightness, color temperature, and gray scale tracking.** The 100-IRE window signal is used to supply the reference white. Because of typical luminance shading limitations, a centrally placed PLUGE [8] signal is recommended for setting the monitor brightness control. The black set signal provided in the SMPTE color bars also can be used for this purpose.
- **Monitor matching.** When color matching two or more color monitors, the same alignment steps should be performed on each monitor in turn. Remember, however, that monitors cannot be matched without the same phosphor sets, similar display uniformity characteristics, and

similar sharpness. The most noticeable deviations on color monitors are the lack of uniform color presentations and brightness shading. Color matching of monitors for these parameters can be most easily assessed by observing flat-field uniformity of the picture at low, medium, and high amplitudes.

For complete monitor-alignment procedures, see [7].

As more experience is gained with DTV-based systems, operating parameters such as those detailed in this section will no doubt be updated to take into consideration the unique attributes and requirements of HDTV.

## 23.4 References


1. MacAdam, D. L.: "Visual Sensitivities to Color Differences in Daylight," *J. Opt. Soc. Am.*, vol. 32, pp. 247–274, 1942.
2. Bender, Walter, and Alan Blount: "The Role of Colorimetry and Context in Color Displays," *Human Vision, Visual Processing, and Digital Display III*, Bernice E. Rogowitz (ed.), *Proc. SPIE* 1666, SPIE, Bellingham, Wash., pp. 343–348, 1992.
3. Tannas, Lawrence E., Jr.: *Flat Panel Displays and CRTs*, Van Nostrand Reinhold, New York, pg. 18, 1985.
4. Standards and definitions committee, Society for Information Display.
5. Verona, Robert: "Comparison of CRT Display Measurement Techniques," *Helmet-Mounted Displays III*, Thomas M. Lippert (ed.), Proc. SPIE 1695, SPIE, Bellingham, Wash., pp. 117–127, 1992.
6. "Critical Viewing Conditions for Evaluation of Color Television Pictures," SMPTE Recommended Practice RP 166-1995, SMPTE, White Plains, N.Y., 1995.
7. "Alignment of NTSC Color Picture Monitors," SMPTE Recommended Practice RP 167-1995, SMPTE, White Plains, N.Y., 1995.
8. Quinn, S. F., and C. A. Siocos: "PLUGE Method of Adjusting Picture Monitors in Television Studios—A Technical Note," *SMPTE Journal*, SMPTE, White Plains, N.Y., vol. 76, pg. 925, September 1967.

Chapter

# 24

# Application of the Zone Plate Signal

**Jerry C. Whitaker, Editor**

## 24.1 Introduction

The increased information content of advanced high-definition display systems requires sophisticated processing to make recording and transmission practical [1]. This processing uses various forms of bandwidth compression, scan-rate changes, motion-detection and motion-compensation algorithms, and other techniques. Zone plate patterns are well suited to exercising a complex video system in the three dimensions of its signal spectrum: horizontal, vertical, and temporal. Zone plate signals, unlike most conventional test signals, can be complex and dynamic. Because of this, they are capable of simulating much of the detail and movement of actual video, exercising the system under test with signals representative of the intended application. These digitally generated and controlled signals also have other important characteristics needed in test waveforms for video systems.

A signal intended for meaningful testing of a video system must be carefully controlled, so that any departure from a known parameter of the signal is attributable to a distortion or other change in the system under test. The test signal also must be predictable, so that it can be accurately reproduced at other times or places. These constraints usually have led to test signals that are electronically generated. In a few special cases, a standardized picture has been televised by a camera or monoscope—usually for a subjective, but more detailed, evaluation of overall performance of the video system.

A zone plate is a physical optical pattern, which was first used by televising it in this way. Now that electronic generators are capable of producing similar patterns, the label "zone plate" is applied to the wide variety of patterns created by video test instruments.

Conventional test signals, for the most part limited by the practical considerations of electronic generation, have represented relatively simple images. Each signal is capable of testing a narrow range of possible distortions; several test signals are needed for a more complete evaluation. Even with several signals, this method may not reveal all possible distortions or allow study of all pertinent characteristics. This is true especially in video systems employing new forms of sophisticated signal processing.

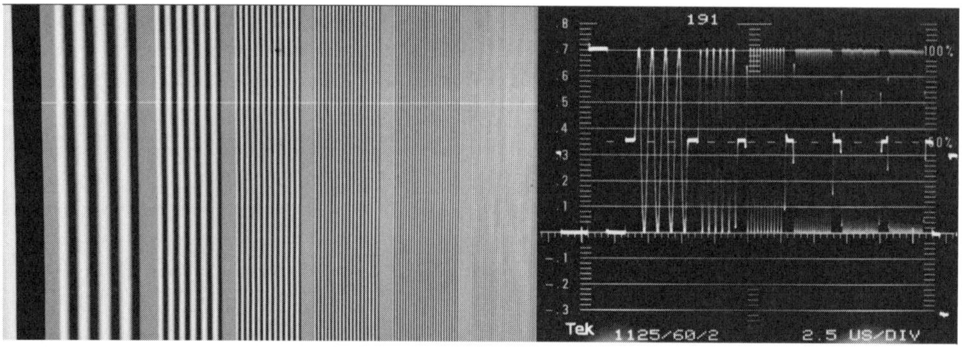


**Figure 24.1** Multiburst video test waveform: (*a, left*) picture display, (*b, right*) multiburst signal as viewed on a waveform monitor (1H). (*Courtesy of Tektronix.*)

## 24.1a Simple Zone Plate Patterns

The basic testing of a video communication channel historically has involved the application of several single frequencies—in effect, spot-checking the spectrum of interest [1]. A well-known and quite practical adaptation of this idea is the multiburst signal, shown in Figure 24.1[1]. This test waveform has been in use since the earliest days of video. The multiburst signal provides several discrete frequencies along a TV line.

The frequency-sweep signal is an improvement on multiburst. Although harder to implement in earlier generators, it was easier to use. The frequency-sweep signal, illustrated in Figure 24.2, varies the applied signal frequency continuously along the TV line. In some cases, the signal is swept as a function of the vertical position (field time). Even in these cases, the signal being swept is appropriate for testing the spectrum of the horizontal dimension of the picture.

Figure 24.3 shows the output of a zone plate generator configured to produce a horizontal single-frequency output. Figure 24.4 shows a zone plate generator configured to produce a frequency-sweep signal. Electronic test patterns, such as these, may be used to evaluate the following system characteristics:

- Channel frequency response
- Horizontal resolution
- Moiré effects in recorders and displays
- Other impairments

Traditionally, patterns that test vertical (field) response have been less frequently used. As new technologies implement conversion from interlaced to progressive scan, line-doubling display techniques, vertical antialiasing filters, scan conversion, motion detection, or other pro-

1. Figure 24.1(*a*) and other photographs in this section show the "beat" effects introduced by the screening process used for photographic printing. This is largely unavoidable. The screening process is quite similar to the scanning or sampling of a television image—the patterns are designed to identify this type of problem.

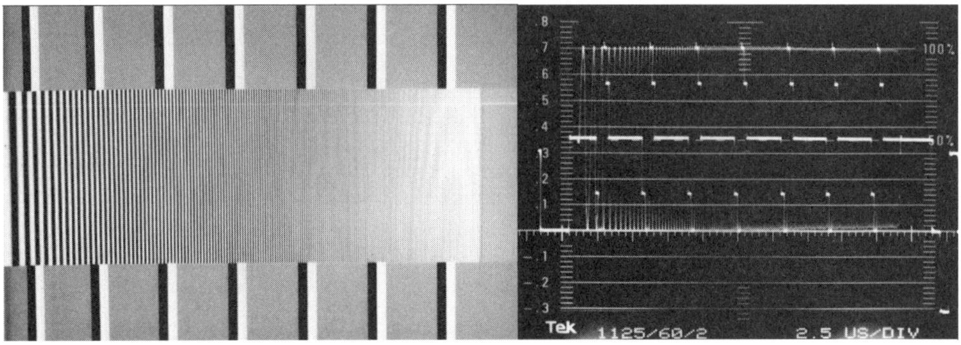


**Figure 24.2** Conventional sweep-frequency test waveform: (*a, left*) picture display, (*b, right*) waveform monitor display, with markers (1H). (*Courtesy of Tektronix.*)

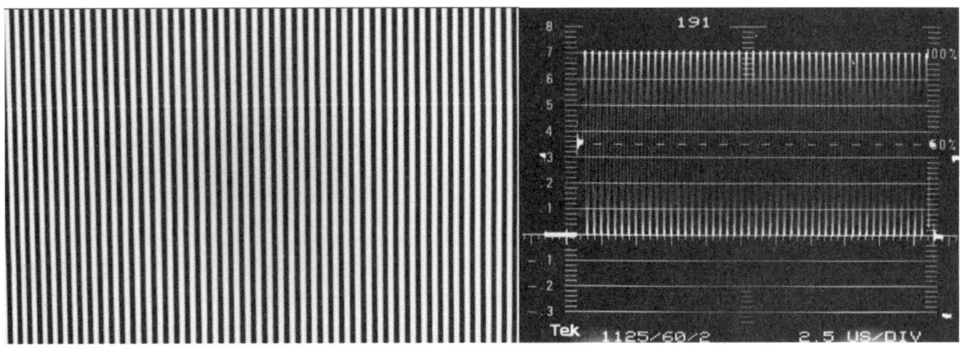


**Figure 24.3** Single horizontal frequency test signal from a zone plate generator: (*a, left*) picture display, (*b, right*) waveform monitor display (1H). (*Courtesy of Tektronix.*)

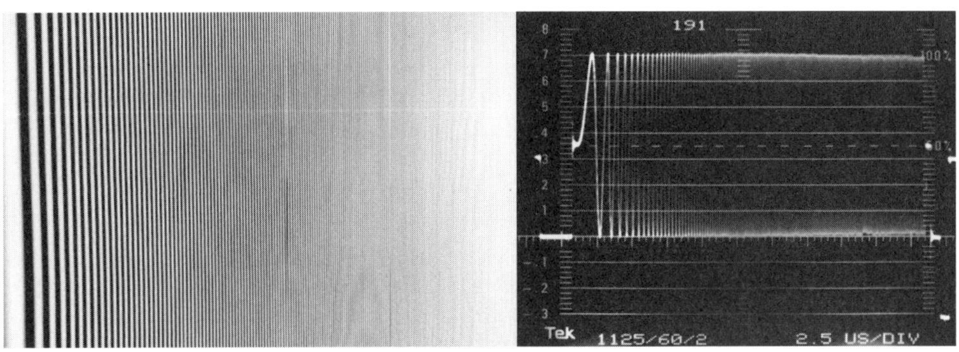


**Figure 24.4** Horizontal frequency-sweep test signal from a zone plate generator: (*a, left*) picture display, (*b, right*) waveform monitor display (1H). (*Courtesy of Tektronix.*)

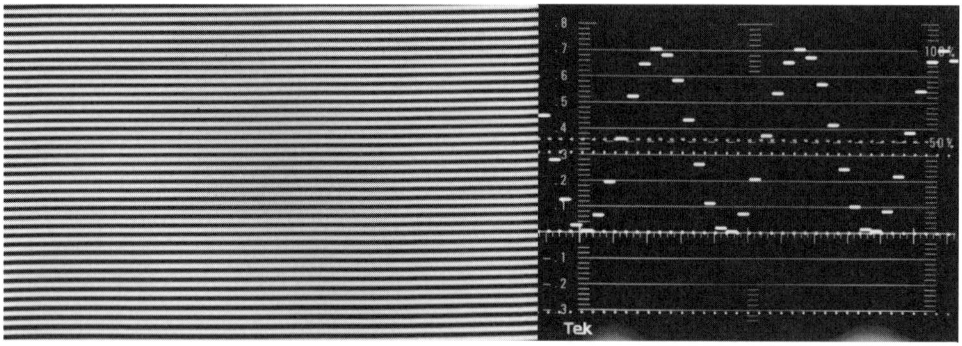


**Figure 24.5** Single vertical frequency test signal: (*a, left*) picture display, (*b, right*) magnified vertical-rate waveform, showing the effects of scan sampling. (*Courtesy of Tektronix.*)

**Figure 24.6** Vertical frequency-sweep picture display. (*Courtesy of Tektronix.*)

cesses that combine information from line to line, vertical testing patterns will be more in demand.

In the vertical dimension, as well as the horizontal, tests can be done at a single frequency or with a frequency-sweep signal. Figure 24.5 illustrates a magnified vertical-rate waveform display. Each "dash" in the photo represents one horizontal scan line. Sampling of vertical frequencies is inherent in the scanning process, and the photo shows the effects on the signal waveform. Note also that the signal voltage remains constant during each line, changing only from line to line in accord with the vertical-dimension sine function of the signal. Figure 24.6 shows a vertical frequency-sweep picture display.

The horizontal and vertical sine waves and sweeps are quite useful, but they do not use the full potential of a zone plate signal source.

## Producing the Zone Plate Signal

A zone plate signal is created in real time by a test signal generator [1]. The value of the signal at any instant is represented by a number in the digital hardware. This number is incremented as the

**Figure 24.7** Combined horizontal and vertical frequency-sweep picture display. (*Courtesy of Tektronix.*)

scan progresses through the three dimensions that define a point in the video image: horizontal position, vertical position, and time.

The exact method by which these dimensions alter the number is controlled by a set of coefficients. These coefficients determine the initial value of the number and control the size of the increments as the scan progresses along each horizontal line, from line to line vertically, and from field to field temporally. A set of coefficients uniquely determines a pattern, or a sequence of patterns, when the time dimension is active.

This process produces a sawtooth waveform; overflow in the accumulator holding the signal number effectively resets the value to zero at the end of each cycle of the waveform. Usually, it is desirable to minimize the harmonic energy content of the output signal; in this case, the actual output is a sine function of the number generated by the incrementing process.

## 24.1b Complex Patterns

A pattern of sine waves or sweeps in multiple dimensions may be produced, using the unique architecture of the zone plate generator [1]. The pattern shown in Figure 24.7, for example, is a single signal sweeping both horizontally and vertically. Figure 24.8 shows the waveform of a single selected line (line 263 in the 1125/60/2 HDTV system). Note that the horizontal waveform is identical to the one shown in Figure 24.4(*b*), even though the vertical-dimension sweep is now also active. Actually, different lines will give slightly different waveforms. The horizontal frequency and sweep characteristics will be identical, but the starting phase must be different from line to line to construct the vertical signal.

Figure 24.9 shows a 2-axis sweep pattern that is most often identified with zone plate generators, perhaps because it quite closely resembles the original optical pattern. In this circle pattern, both horizontal and vertical frequencies start high, sweep to zero (in the center of the screen), and sweep up again to the end of their respective scans. The concept of 2-axis sweeps actually is more powerful than it might first appear. In addition to purely horizontal or vertical effects, there are possible distortions or artifacts that are apparent only with simultaneous excitation in both axes. In other words, the response of a system to diagonal detail may not be predictable from information taken from the horizontal and vertical responses.

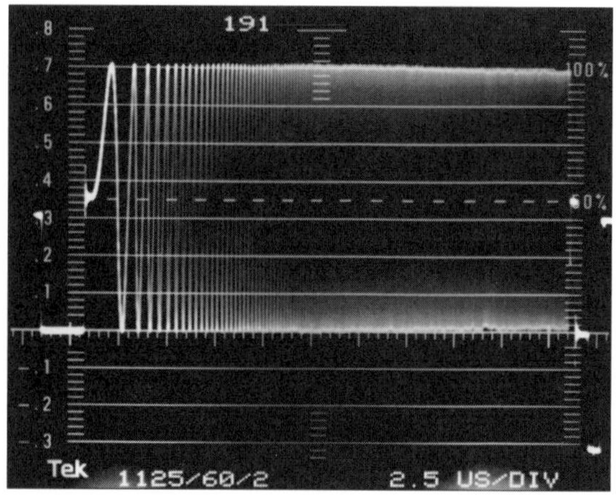


**Figure 24.8** Combined horizontal and vertical frequency sweeps, selected line waveform display (1H). This figure shows the maintenance of horizontal structure in the presence of vertical sweep. (*Courtesy of Tektronix.*)

**Figure 24.9** The best-known zone plate pattern, combined horizontal and vertical frequency sweeps with zero frequency in the center screen. (*Courtesy of Tektronix.*)

## 24.1c The Time (Motion) Dimension

Incrementing the number in the accumulator of the zone plate generator from frame to frame (or field to field in an interlaced system) creates a predictably different pattern for each vertical scan [1]. This, in turn, creates apparent motion and exercises the signal spectrum in the temporal dimension. Analogous to the single-frequency and frequency-sweep examples given previously, appropriate setting of the time-related coefficients will create constant motion or motion sweep (acceleration).

Specific motion-detection and interpolation algorithms in a system under test may be exercised by determining the coefficients of a critical sequence of patterns. These patterns then may be saved for subsequent testing during development or adjustment. In an operational environment, appropriate response to a critical sequence could ensure expected operation of the equipment or facilitate fault detection.

Although motion artifacts are difficult to portray in the still-image constraints of a printed book, the following example gives some idea of the potential of a versatile generator. In Figure 24.10, the vertical sweep maximum frequency has been increased to the point where it is zero-beating with the scan at the bottom of the screen. (Note that the cycles/ph of the pattern matches

**Figure 24.10** Vertical frequency-sweep picture display. (*Courtesy of Tektronix.*)

**Figure 24.11** The same vertical sweep as shown in Figure 12.32, except that appropriate pattern motion has been added to "freeze" the beat pattern in the center screen for photography or other analysis. (*Courtesy of Tektronix.*)

the lines/ph per field of the scan.) Actually, in direct viewing, there is another noticeable artifact in the vertical center of the screen: a harmonic beat related to the gamma of the display CRT. Because of interlace, this beat flickers at the field rate. The photograph integrates the interfield flicker, thereby hiding the artifact, which is readily apparent when viewed in real time.

Figure 24.11 is identical to the previous photo, except for one important difference—upward motion of ½-cycle per field has been added to the pattern. Now the sweep pattern itself is integrated out, as is the first-order beat at the bottom. The harmonic effects in center screen no longer flicker, because the change of scan vertical position from field to field is compensated by a change in position of the image. The resulting beat pattern does not flicker and is easily photographed or, perhaps, scanned to determine depth of modulation.

A change in coefficients produces hyperbolic, rather than circular 2-axis patterns, as shown in Figure 24.12. Another interesting pattern, which has been used for checking complex codecs, is shown in Figure 24.13. This is also a moving pattern, which was altered slightly to freeze some aspects of the movement for the purpose of taking the photograph.

**Figure 24.12** A hyperbolic variation of the 2-axis zone plate frequency sweep. (*Courtesy of Tektronix.*)

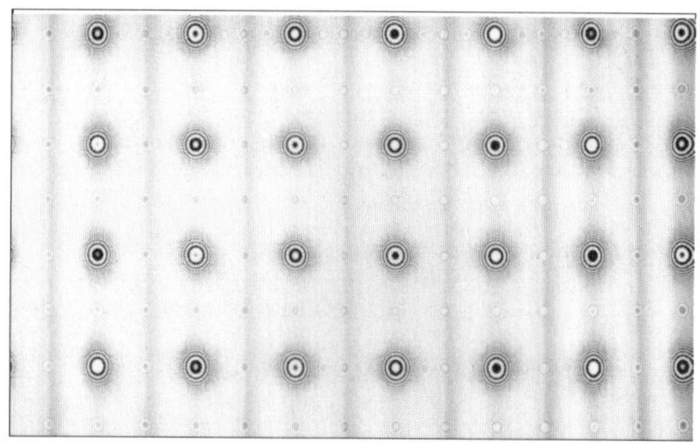

**Figure 24.13** A 2-axis frequency sweep in which the range of frequencies is swept several times in each axis. Complex patterns such as this may be created for specific test requirements. (*Courtesy of Tektronix.*)

## 24.2 References

1. "Broadening the Applications of Zone Plate Generators," Application Note 20W7056, Tektronix, Beaverton, Oreg., 1992.

Chapter

# 25

# Picture Quality Measurement

Jerry C. Whitaker, Editor

## 25.1 Introduction

Picture-quality measurement methods include subjective testing, which is always used—at least in an informal manner—and objective testing, which is most suitable for system performance specification and evaluation [1]. A number of types of objective measurement methods are possible for digital television pictures, but those using a human visual system model are the most powerful.

As illustrated in Figure 25.1, three key testing layers can be defined for the modern television system:

- **Video quality.** This consists of signal quality and picture quality (discussed in some detail previously in Section 18).
- **Protocol analysis.** Protocol testing is required because the data formatting can be quite complex and is relatively independent of the nature of the uncompressed signals or the eventual conversion to interfacility transmission formats. Protocol test equipment can be both a source of signals and an analyzer that locates errors with respect to a defined standard and determines the value of various operational parameters for the stream of data.
- **Transmission system analysis.** To send the video data to a remote location, one of many possible digital data transmission methods can be used, each of which imposes its own analysis issues.

Table 25.1 lists several dimensions of video-quality measurement methods. Key definitions include the following:

- *Subjective* measurements: the result of human observers providing their opinions of the video quality.
- *Objective* measurements: performed with the aid of instrumentation, manually with humans reading a calibrated scale or automatically using a mathematical algorithm.
- *Direct* measurements: performed on the material of interest, in this case, pictures (also known as *picture-quality* measurements).

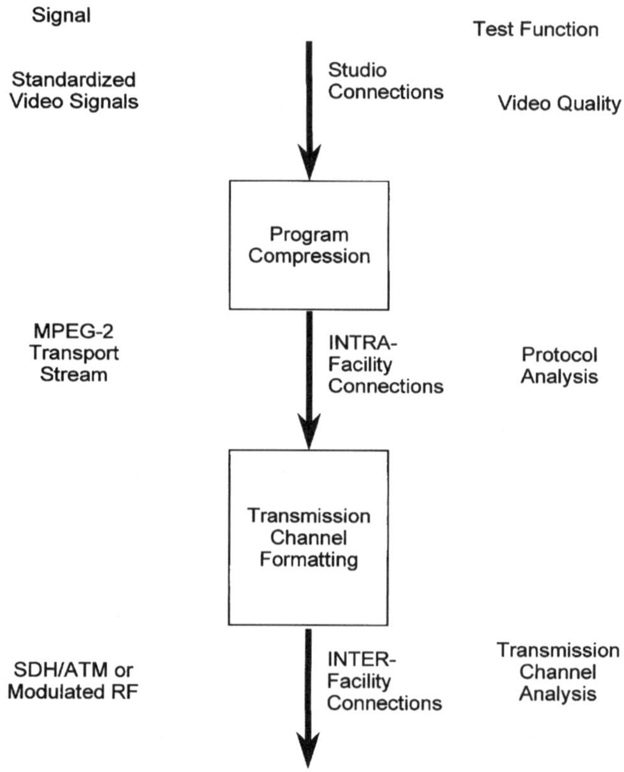


**Figure 25.1** Video testing layers for digital television. (*After* [1].)

**Table 25.1 Video Quality Definitions** (*After* [1].)

| Parameter | In-Service | Out-of-Service |
|---|---|---|
| **Indirect measurement** | | |
| Objective signal quality | Vertical interval test signals | Full-field test signals |
| **Direct measurement** | | |
| Subjective picture quality | Program material | Test scenes |
| Objective picture quality | Program material | Test scenes |

- *Indirect* measurements: made by processing specially designed test signals in the same manner as the pictures (also known as *signal-quality* measurements). Subjective measurements are performed only in a direct manner because the human opinion of test signal picture quality is not particularly meaningful.

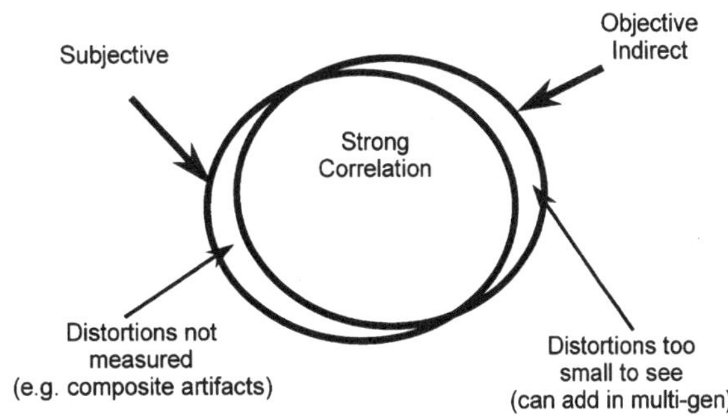


**Figure 25.2** Functional environment for traditional video measurements. (*After* [1].)

- *In-service* measurements: made while the program is being displayed, directly by evaluating the program material or indirectly by including test signals with the program material.
- *Out-of-service*: appropriate test scenes are used for direct measurements, and full-field test signals are used for indirect measurements.

In the mixed environment of compressed and uncompressed signals, video-quality measurements consist of two parts: signal quality and picture quality.

## 25.1a Signal/Picture Quality

Signal-quality measurements are made with a suite of test signals as short as one line in the vertical interval [1]. In a completely uncompressed system, such testing will give a good characterization of picture quality. This is not true, however, for a system with compression encoding/decoding because picture quality will change based on the data rate, complexity, and encoding algorithm. Picture-quality measurements, instead, require natural scenes (or some equivalent thereof) that are much more complex than traditional test signals. These complex scenes stress the capabilities of the encoder, resulting in nonlinear distortions that are a function of the picture content.

Out-of-service picture-quality measurements are similar to indirect signal-quality measurements in one aspect: the determination of the system response to a high-quality reference. However, they actually measure the degradation in reference picture quality rather than that of a synthetic test signal. In-service, such signals determine the response to program material and its degradation through the system. If the program material is "easy," the measurement may not have a great deal of practical value.

Objective indirect signal-quality measurements are a reasonably good way to determine the picture quality for uncompressed systems. That is, there is a good correlation between subjective measurements made on pictures from the system and objective measurements made on a suite of test signals using the same system. (See Figure 25.2.) The correlation is not perfect for all tests, however. There are distortions in composite systems, such as false color signals caused by poorly filtered high-frequency luminance information being detected as chroma. These distortions are not easily measured by objective means. Also, there are objective measurements that are so sensitive they do not directly relate to subjective results. However, such objective results often are

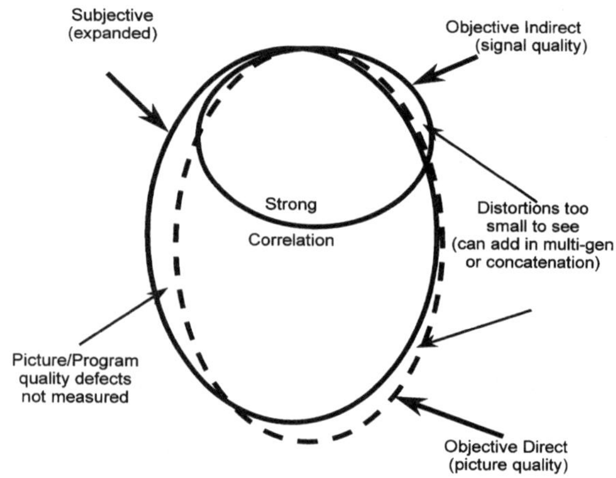


**Figure 25.3** Functional environment for digital video measurements. (*After* [1].)

useful because their effect will be seen by a human observer if the pictures are processed in the same way a number of times.

The use of digital compression has expanded the types of distortions that can occur in the modern television system. Because signal-quality measurements will not do the job, objective picture-quality measurements are needed, as illustrated in Figure 25.3. The total picture-quality measurement space has increased because of subjective measurements that now include multiminute test scenes with varying program material and variable picture quality. The new objective measurement methods must have strong correlation with subjective measurements and cover a broad range of applications.

Even with all the objective testing methods available for analog and full-bandwidth digital video, it is important to have human observation of the pictures. Some impairments are not easily measured, yet are obvious to a human observer. This situation certainly has not changed with the addition of modern digital compression. Therefore, casual or informal subjective testing by a reasonably expert viewer remains an important part of system evaluation and/or monitoring.

## 25.1b Automated Picture-Quality Measurement

Objective measurements can be made automatically with an instrument that determines picture degradation through the system [1]. Two somewhat mutually exclusive ways are available for classifying objective picture-quality measurement systems. Although there are several practical methods with a variety of algorithmic approaches, they may be generally divided into the following classes:

- *Feature extraction*, which does an essentially independent analysis of input and output pictures.
- *Picture processing*, where the complete input and output pictures are directly compared in some manner and must be available at the measurement instrument (also known as *picture differencing*).

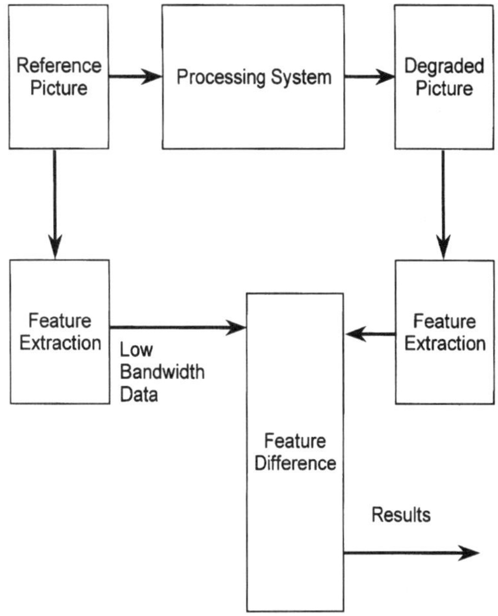


**Figure 25.4** Block diagram of the feature-extraction method of picture-quality analysis. (*After* [1].)

From a system standpoint, the *glass box* approach utilizes knowledge of the compression system to measure degradation. An example would be looking for blockiness in a DCT system. The *black box* approach makes no assumptions about operation of the system.

In feature extraction, analysis time is reduced by comparing only a limited set of picture characteristics. These characteristics are calculated, and the modest amount of data is embedded into the picture stream. Examples of compression impairments would be block distortion, blurring/ smearing, or "mosquito noise." At the receiver, the same features of the degraded picture are calculated, providing a measure of the differences for each feature (Figure 25.4). The major weakness of this approach is that it does not provide correlation between subjective and objective measurements across a wide variety of compression systems or source material.

For the picture-processing scheme, the reference and degraded pictures are filtered in an appropriate manner, resulting in a data set that may be as large as the original pictures. The difference between the two data sets is a measure of picture degradation, as illustrated in Figure 25.5.

The need for a standard for objective measurements has been addressed in ANSI T1.801.03 [2]. Several variations and improvements to the basic ANSI toolbox are listed in [1]. Because of the limitations of the ANSI method, work continues with the goal of refining objective measurement methods.

A number of approaches have been proposed by researchers using the human visual system (HVS) model as a basis. Such a model provides an image-quality metric that is independent of video material, specific types of impairments, and the compression system used. The study of the HVS has been going on for decades, investigating such properties as contrast sensitivity, spatio-temporal response, and color perception. Perhaps one of the best-known derivatives of this

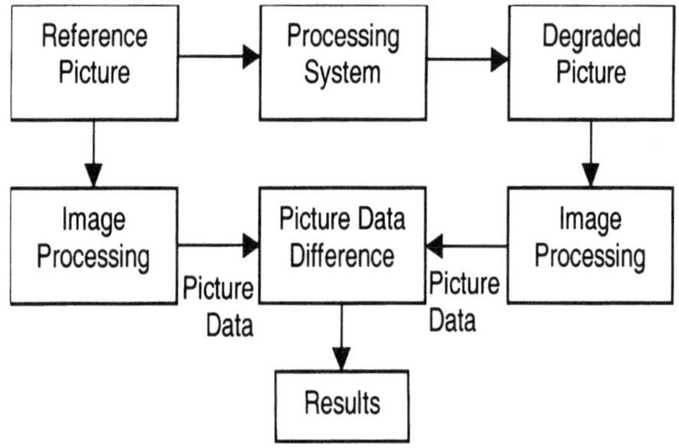


**Figure 25.5** Block diagram of the picture-differencing method of picture-quality analysis. (*After* [1].)

work is the *JNDmetrix* (Sarnoff/Tektronix), described in [3]. Other work in this area involves the *Three-Layered Bottom-Up Noise Weighting Model* described in [4].

## In-Service Measurements

In-service measurements are made simultaneously with regular program transmissions [5]. For traditional analog signal quality measurements, this is easily accomplished using vertical blanking interval test signals. In-service measurements also are important for systems incorporating bit-rate reduction and other forms of digital processing. Picture quality analysis must use the actual program material analyzed at the source and the processed output for the measurement.

Both picture differencing and feature extraction can be used for in-service measurements. The availability of the source sequence may not be a limitation for in-service monitoring in most applications, given that picture quality is limited primarily by the (transmission) compression process. Presuming this allows a defect monitoring solution to be designed to look for trace artifacts, rather than a loss of image fidelity.

Although this approach would seemingly relegate double-ended picture quality testing to the systems evaluations lab, it is conceivable that program producers may want to have absolute image quality confirmed, or even guaranteed.

Real-time measurements are desirable for in-service operation. In this context, *real-time* means that measurements are continuously available as the sequence passes through the system under test. There may be a nominal delay for processing, much as is done with the averaging function in sophisticated analog measurement systems. Generally speaking, every video field would be evaluated.

*Non-real-time* is the condition where the source and processed sequence information are captured, with results available after some nominal time for computation, perhaps a minute or two.

In-service picture quality measurement using the picture differencing method can be accomplished as shown in Figure 25.6. At the program source, a system-specific decoder is used to provide the processed sequence to a picture quality evaluation system. The results would be available in non-real-time, sampled real-time, or real-time, depending on the computational power of the measurement instrument. The system-specific decoder provides the same picture quality output as the decoder at the receive end of the transmission system.

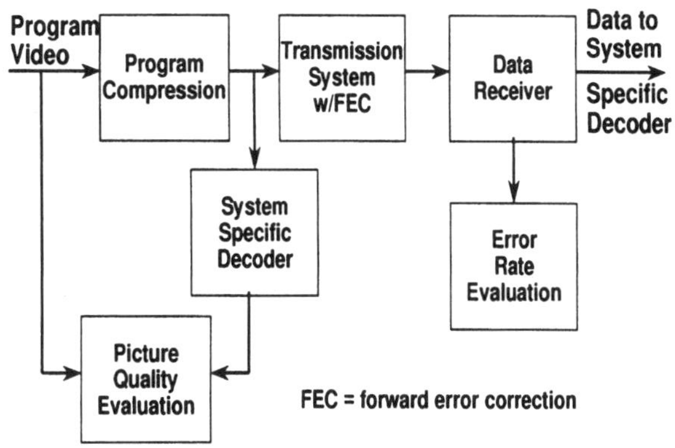


**Figure 25.6** In-service picture quality measurement system. (*After* [5].)

Practical transmission systems for broadcast applications operate either virtually error-free (resulting in no picture degradation) or with an unrecoverable number of errors (degradation so severe that quality measurements are not particularly useful). Therefore, bit-error-rate evaluation based on calculations in the receiver may provide sufficient monitoring of the transmission system.

The foregoing is based on the assumption that the transmission path is essentially fixed. Depending upon the application, this may not always be the case. Consider a terrestrial video link, which may be switched to a variety of paths before reaching its ultimate destination. The path also can change at random intervals depending on the network arrangement. The situation gives rise to compression concatenation issues for which there are no easy solutions.

As identified by Uchida [6], video quality deterioration accumulates with each additional coding device added to a cascade connection. This implies that, although picture deterioration caused by a single codec may be imperceptible, cascaded codecs, as a whole, can cause serious deterioration that cannot be overlooked. Detailed measurements of concatenation issues are documented in [6].

## 25.2 References


1. Fibush, David K.: "Picture Quality Measurements for Digital Television," *Proceedings of the Digital Television '97 Summit*, Intertec Publishing, Overland Park, Kan., December 1997.

2. ANSI Standard T1.801.03-1996, "Digital Transport of One-Way Video Signals: Parameters for Objective Performance Assessment," ANSI, Washington, D.C., 1996.

3. Fibush, David K.: "Practical Application of Objective Picture Quality Measurements," *Proceedings IBC '97*, IEE, pp. 123–135, Sept. 16, 1997.

4. Hamada, T., S. Miyaji, and S. Matsumoto: "Picture Quality Assessment System by Three-Layered Bottom-Up Noise Weighting Considering Human Visual Perception," *SMPTE Journal*, SMPTE, White Plains, N.Y., pp. 20–26, January 1999.

5. Bishop, Donald M.: "Practical Applications of Picture Quality Measurements," *Proceedings of Digital Television '98*, Intertec Publishing, Overland Park, Kan., 1998.

6. Uchida, Tadayuki, Yasuaki Nishida, and Yukihiro Nishida: "Picture Quality in Cascaded Video-Compression Systems for Digital Broadcasting," *SMPTE Journal*, SMPTE, White Plains, N.Y., pp. 27–38, January 1999.

# Listing of Figures and Tables

## Tables

# References Cited in this Book


A. DeVries et al, "Characteristics of Surface-Wave Integratable Filters (SWIFS)," *IEEE Trans.*, vol. BTR-17, no. 1, p. 16.

"Advanced Television Enhancement Forum Specification," Draft, Version 1.1r26 updated 2/2/99, ATVEF, Portland, Ore., 1999.

Aiken, J. A.: "A Thin Cathode Ray Tube," *Proc. IRE*, vol. 45, pg. 1599, 1957.

"Alignment of NTSC Color Picture Monitors," SMPTE Recommended Practice RP 167-1995, SMPTE, White Plains, N.Y., 1995.

Alkin, Oktay: "Digital Coding Schemes," *The Electronics Handbook*, Jerry C. Whitaker (ed.), CRC Press, Boca Raton, Fla., pp. 1252–1258, 1996.

Allison, J.: *Electronic Engineering Semiconductors and Devices*, 2nd ed., McGraw-Hill, London, pg. 308–309, 1990.

Altman, F. J., and W. Sichak: "A Simplified Diversity Communication System," *IRE Trans.*, vol. CS-4, March 1956.

Amm, D. T., and R. W. Corrigan: "Optical Performance of the Grating Light Valve Technology," Projection Displays V Symposium, *SPIE Proceedings*, SPIE, San Jose, Calif., vol. EI 3634-10, February 1999.

Amm, D. T., and R.W. Corrigan: "Grating Light Valve Technology: Update and Novel Applications," *SID Symposium—Anaheim*, SID, San Jose, Calif., May 1998.

ANSI Standard T1.801.03-1996, "Digital Transport of One-Way Video Signals: Parameters for Objective Performance Assessment," ANSI, Washington, D.C., 1996.

Applebaum. S. P.: "Adaptive Arrays," *IEEE Trans.*, vol. AP-24, pg. 585, September 1976.

Ashizaki, S., Y. Suzuki, K. Mitsuda, and H. Omae: "Direct-View and Projection CRTs for HDTV," *IEEE Transactions on Consumer Electronics*, vol. 34, no. 1, pp. 91–98, February 1988.

Ashizaki, S., Y. Suzuki, K. Mitsuda, and H. Omae: "Direct-View and Projection CRTs for HDTV," *IEEE Transactions on Consumer Electronics*, IEEE, New York, N.Y., 1988.

ATSC, "Guide to the Use of the ATSC Digital Television Standard," Advanced Television Systems Committee, Washington, D.C., Doc. A/54, Oct. 4, 1995.

ATSC: "Performance Assessment of the ATSC Transmission System, Equipment, and Future Directions," ATSC Task Force on RF System Performance, Advanced Television Systems Committee, Washington, DC, revision: 1.0, April 12, 2001.

Baldwin, T. F., and D. S. McVoy: *Cable Communications*, Prentice-Hall, Englewood Cliffs, N.J., 1983.

Barbin, R., and R. Hughes: "New Color Picture Tube System for Portable TV Receivers," *IEEE Trans. Broadcast TV Receivers*, vol. BTR-18, no. 3, pp. 193–200, August 1972.

Barkow, W. H., and J. Gross: "The RCA Large Screen 110× Precision In-Line System," ST-5015, RCA Entertainment, Lancaster, Pa.

Bates, W., P. Gelinas, and P. Recuay: "Light Valve Projection System for HDTV," *Proceedings of the ITS*, International Television Symposium, Montreux, Switzerland, 1991.

Bauman, E.: "The Fischer Large-Screen Projection System," *SMPTE Journal*, SMPTE, White Plains, N.Y., vol. 60, pg. 351, 1953.

Baur, G.: *The Physics and Chemistry of Liquid Crystal Devices*, G. J. Sprokel (ed.), Plenum, New York, N.Y., pg. 62, 1980.

Bean, B. R., and E. J. Dutton: "Radio Meteorology," National Bureau of Standards Monograph 92, March 1, 1966.

Bender, Walter, and Alan Blount: "The Role of Colorimetry and Context in Color Displays," *Human Vision, Visual Processing, and Digital Display III*, Bernice E. Rogowitz (ed.), Proc. SPIE 1666, SPIE, Bellingham, Wash., pp. 343–348, 1992.

Bendov, O.: "On the Validity of the Longley-Rice (50,90/10) Propagation Model for HDTV Coverage and Interference Analysis," *Proceedings of the Broadcast Engineering Conference*, National Association of Broadcasters, Washington, D.C., 1999.

Benson, K. B., and D. G. Fink: *HDTV: Advanced Television for the 1990s*, McGraw-Hill, New York, 1990.

Beverage, H. H., and H. O. Peterson: "Diversity Receiving System of RCA Communications, Inc., for Radiotelegraphy," *Proc. IRE*, vol. 19, pg. 531, April 1931.

Bishop, Donald M.: "Practical Applications of Picture Quality Measurements," *Proceedings of Digital Television '98*, Intertec Publishing, Overland Park, Kan., 1998.

Black, H. S.: U.S. Patent 2,102,671.

Blacker, A., et al.: "A New Form of Extended Field Lens for Use in Color Television Picture Tube Guns," *IEEE Trans. Consumer Electronics*, pp. 238–246, August 1966.

Blaha, R.: "Degaussing Circuits for Color TV Receivers," *IEEE Trans. Broadcast TV Receivers*, vol. BTR-18, no. 1, pp. 7–10, February 1972.

Blaha, Richard J.: "Large Screen Display Technology Assessment for Military Applications," *Large-Screen Projection Displays II*, William P. Bleha, Jr., (ed.), Proc. SPIE 1255, SPIE, Bellingham, Wash., pp. 80–92, 1990.

Bleha, W. P.: "Image Light Amplifier (ILA) Technology for Large-Screen Projection," *SMPTE Journal*, SMPTE, White Plains, N.Y., pp. 710–717, October 1997.

Bleha, William P., Jr., (ed.): *Large-Screen Projection Displays II*, Proc. SPIE 1255, SPIE, Bellingham, Wash., 1990.

Bonang, C., and C. Auvray-Kander: "Next Generation Broadband Networks for Interactive Services," in *Proceedings of IBC 2000*, International Broadcasting Convention, Amsterdam, 2000.

Brennan, D. G.: "Linear Diversity Combining Techniques," *Proc. IRE*, vol. 47, pg. 1075, June 1959.

"Broadening the Applications of Zone Plate Generators," Application Note 20W7056, Tektronix, Beaverton, Oreg., 1992.

Bullington, K.: "Radio Propagation at Frequencies above 30 Mc," *Proc. IRE*, pg. 1122, October 1947.

Bullington, K.: "Radio Propagation Variations at VHF and UHF," *Proc. IRE*, pg. 27, January 1950.

Burgmans, A., et. al.: *Information Display*, pg. 14, April/May 1998.

Burrows, C. R., and M. C. Gray: "The Effect of the Earth's Curvature on Groundwave Propagation," *Proc. IRE*, pg. 16, January 1941.

Buzak, Thomas S.: "Recent Advances in PALC Technology," *Proceedings of the 18th International Display Research Conference,* Society for Information Display, San Jose, Calif., Asia Display '98, pp. 273–276, 1998.

"Cable Television Information Bulletin," Federal Communications Commission, Washington, D.C., June 2000.

Carpenter, C.: et al., "An Analysis of Focusing and Deflection in the Post-Deflection Focus Color Kinescope," *IRE Trans. Electron Devices*, vol. 2, pp. 1–7, 1955.

Casteloano, Joseph A.: *Handbook of Display Technology*, Academic, New York, N.Y., 1992.

Chambers, J. A., S. Tantaratana, and B. W. Bomar: "Digital Filters," *The Electronics Handbook*, Jerry C. Whitaker (ed.), CRC Press, Boca Raton, Fla., pp. 749–772, 1996.

Chang, I.: "Recent Advances in Display Technologies," *Proc. SID*, Society for Information Display, San Jose, Calif., vol. 21, no. 2, pg. 45, 1980.

Chen, H., and R. Hughes: "A High Performance Color CRT Gun with an Asymmetrical Beam Forming Region," *IEEE Trans. Consumer Electronics*, vol. CE-26, pp. 459–465, August, 1980.

Chen, K. C., W. Y. Ho and C. H. Tseng: "Invar Mask for Color Cathode Ray Tubes," in *Display Technologies*, Shu-Hsia Chen and Shin-Tson Wu (eds.), Proc. SPIE 1815, SPIE, Bellingham, Wash., pp.42–48, 1992.

Cheng, J. B., and Q. H. Wang: "Studies on YAG Phosphor Screen for HDTV Projector," *Proc. SPIE 2892*, SPIE, Bellingham, Wash., pg. 36, 1996.

Cheng, Jia-Shyong, et. al.: "The Optimum Design of LCD Parameters in Projection and Direct View Applications," *Display Technologies*, Shu-Hsia Chen and Shin-Tson Wu (eds.), Proc. SPIE 1815, SPIE, Bellingham, Wash., pp. 69–80, 1992.

Ciciora, Walter S.: "Cable Television," in *NAB Engineering Handbook*, 9th ed., Jerry C. Whitaker (ed.), National Association of Broadcasters, Washington, D.C., pp. 1339–1363, 1999.

Ciciora, Walter, et. al.: "A Tutorial on Ghost Canceling in Television Systems," *IEEE Transactions on Consumer Electronics*, IEEE, New York, vol. CE-25, no. 1, pp. 9–44, February 1979.

Clapp, R., et al.: "A New Beam Indexing Color Television Display System," *Proc. IRE*, vol. 44, no. 9, pp. 1108–1114, September 1956.

Clement, Pierre, and Eric Gourmelen: "Internet and Television Convergence: IP and MPEG-2 Implementation Issues," *Proceedings of the 33rd SMPTE Advanced Motion Imaging Conference*, SMPTE, White Plains, N.Y., February 1999.

Cohen, C.: "Sony's Pocket TV Slims Down CRT Technology," *Electronics*, pg. 81, February 10, 1982.

Collocott, T. C., A. B. Dobson, and W. R. Chambers (eds.): *Dictionary of Science & Technology.*

Compton, R. T., Jr., R. J. Huff, W. G. Swarner, and A. A. Ksienski: "Adaptive Arrays for Communication Systems: An Overview of Research at the Ohio State University," *IEEE Trans.*, vol. AP-24, pg. 599, September 1976.

Cook, James H., Jr., Gary Springer, Jorge B. Vespoli: "Satellite Earth Stations," in *NAB Engineering Handbook*, Jerry C. Whitaker (ed.), National Association of Broadcasters, Washington, D.C., pp. 1285–1322, 1999.

Corrigan, R. W., B. R. Lang, D. A. LeHoty, and P. A. Alioshin: "An Alternative Architecture for High Performance Display," Silicon Light Machines, Sunnyvale, Calif., 1999. Presented at the 141st SMPTE Technical Conference (paper 141-25).

Credelle, T. L., et al.: "Cathodoluminescent Flat Panel TV Using Electron Beam Guides," *SID Int. Symp. Digest*, Society for Information Display, San Jose, Calif., pg. 26, 1980.

Credelle, T. L., et al.: *Japan Display '83*, pg. 26, 1983.

Credelle, T. L.: "Modular Flat Display Device with Beam Convergence," U.S. Patent 4,131,823.

Critical Viewing Conditions for Evaluation of Color Television Pictures," SMPTE Recommended Practice RP 166-1995, SMPTE, White Plains, N.Y., 1995.

CRT Control Grid Having Orthogonal Openings on Opposite Sides," U.S. Patent 4,242,613, Dec. 30, 1980.

"CRTs: Glossary of Terms and Definitions," Publication TEP92, Electronic Industries Association, Washington, 1975.

Davis, C., and D. Say: "High Performance Guns for Color TV—A Comparison of Recent Designs," *IEEE Trans. Consumer Electronics*, vol. CE-25, August 1979.

de Lisle, E. W.: "Computations of VHF and UHF Propagation for Radio Relay Applications," RCA, Report by International Division, New York, N.Y.

Di Toro, M. J.: "Communications in Time-Frequency Spread Media," *Proc. IEEE*, vol. 56, October 1968.

Dickson, F. H., J. J. Egli, J. W. Herbstreit, and G. S. Wickizer: "Large Reductions of VHF Transmission Loss and Fading by the Presence of a Mountain Obstacle in Beyond-Line-of-Sight Paths," *Proc. IRE*, vol. 41, no. 8, pg. 96, August 1953.

Digital HDTV Grand Alliance System: Record of Test Results, October 1995.

"Documents of the XVth Plenary Assembly," CCIR Report 238, vol. 5, Geneva, 1982.

"Documents of the XVth Plenary Assembly," CCIR Report 563, vol. 5, Geneva, 1982.

"Documents of the XVth Plenary Assembly," CCIR Report 881, vol. 5, Geneva, 1982.

Donofrio, R.: "Image Sharpness of a Color Picture Tube by MTF Techniques," *IEEE Trans. Broadcast TV Receivers*, vol. BTR-18, no. 1, pp. 1–6, February 1972.

Dougherty, H. T., and E. J. Dutton: "The Role of Elevated Ducting for Radio Service and Interference Fields," NTIA Report 81–69, March 1981.

Dressler, R.: "The PDF Chromatron—A Single or Multi-Gun CRT," *Proc. IRE*, vol. 41, no. 7, July 1953.

Dworsky, Lawrence N., and Babu R. Chalamala: *Field-Emission Displays*, Society for Information Display, San Jose, Calif., pp. F-1/3–F1/66, 1999.

Dye, D. W.: *Proc. Phys. Soc.*, Vol. 38, pp. 399–457, 1926.

Eccles, D. A., and Y. Zhang: "Digital-Television Signal Processing and Display Technology," *SID 99 Digest*, Society for Information Display, San Jose, Calif., pp. 108–111, 1999.

Eckersley, T. L.: "Ultra-Short-Wave Refraction and Diffraction," *J. Inst. Elec. Engrs.*, pg. 286, March 1937.

Einolf, Charles: "DTV Receiver Performance in the Real World," *2000 Broadcast Engineering Conference Proceedings*, National Association of Broadcasters, Washington, D.C., pp. 478–482, 2000.

"Electron Gun with Astigmatic Flare-Reducing Beam Forming Region," U.S. Patent 4,234,814, Nov. 18, 1980.

Elliot, R. S.: *Antenna Theory and Design*, Prentice-Hall, Englewood Cliffs, N.J., pg. 64, 1981.

Engelson, M., and J. Herbert: "Effective Characterization of CDMA Signals," *Microwave Journal*, pg. 90, January 1995.

Epstein, J., and D. Peterson: "An Experimental Study of Wave Propagation at 850 Mc," *Proc. IRE*, pg. 595, May 1953.

Erez, Beth: "Protecting content in the Digital Home," *Proceedings of the 33rd SMPTE Advanced Motion Imaging Conference*, SMPTE, White Plains, N.Y., pp. 231–238, February 1999.

FCC Regulations, 47 CFR, 15.65, Washington, D.C.

FCC/ACATS SSWP2-1306: Grand Alliance System Test Procedures, May 18, 1994.

Fewer, D. R.: "Design Principles for Junction Transistor Audio Power Amplifiers," *Trans. IRE PGA*, AU-3(6), November–December 1955.

Fibush, David K.: "Picture Quality Measurements for Digital Television," *Proceedings of the Digital Television '97 Summit*, Intertec Publishing, Overland Park, Kan., December 1997.

Fibush, David K.: "Practical Application of Objective Picture Quality Measurements," *Proceedings IBC '97*, IEE, pp. 123–135, Sept. 16, 1997.

Fink, D. G. (ed.): *Television Engineering Handbook*, McGraw-Hill, New York, N.Y., 1957.

Fink, D. G., and D. Christiansen (eds.): *Electronic Engineer's Handbook*, 2nd ed., McGraw-Hill, New York, 1982.

Fink, D. G., et al.: "The Future of High-Definition Television," *SMPTE Journal*, SMPTE, White Plains, N.Y., vol. 89, February/March 1980.

Fink, D. G.: *Color Television Standards*, McGraw-Hill, New York, N.Y., 1986.

Fiore, J., and S. Kaplin: "A Second Generation Color Tube Providing More Than Twice the Brightness and Improved Contrast," *IEEE Trans. Consumer Electronics*, vol. CE-28, no. 1, pp. 65–73, February 1982.

Flechsig, W.: "CRT for the Production of Multicolored Pictures on a Luminescent Screen," French Patent 866,065, 1939.

Florence, J., and L. Yoder: "Display System Architectures for Digital Micromirror Device (DMD) Based Projectors," *Proc. SPIE*, SPIE, Bellingham, Wash., vol. 2650, Projection Displays II, pp. 193–208, 1996.

Fockens, P., and C. G. Eilers: "Intercarrier Buzz Phenomena Analysis and Cures," *IEEE Trans. Consumer Electronics*, vol. CE-27, no. 3, pg. 381, August 1981.

Frerking, M. E.: *Crystal Oscillator Design and Temperature Compensation*, Van Nostrand Reinhold, New York, N. Y., 1978.

Fritz, Victor J.: "Full-Color Liquid Crystal Light Valve Projector for Shipboard Use," *Large Screen Projection Displays II*, William P. Bleha, Jr. (ed.), Proc. SPIE 1255, SPIE, Bellingham, Wash., pp. 59–68, 1990.

Fujio, T., J. Ishida, T. Komoto, and T. Nishizawa: "High-Definition Television Systems—Signal Standards and Transmission," *SMPTE Journal*, SMPTE, White Plains, N.Y., vol. 89, August 1980.

Garner, L. H.: "High-Power Solid State Amplifiers," *Trans. IRE PGA*, 15(4), December 1967.

Garrod, Susan A. R.: "D/A and A/D Converters," *The Electronics Handbook*, Jerry C. Whitaker (ed.), CRC Press, Boca Raton, Fla., pp. 723–730, 1996.

Garrod, Susan, and R. Borns: *Digital Logic: Analysis, Application, and Design*, Saunders College Publishing, Philadelphia, pg. 919, 1991.

Gerhard-Multhaupt, R.: "Light Valve Technologies for HDTV Projection Displays: A Summary," *Proceedings of the ITS*, International Television Symposium, Montreux, Switzerland, 1991.

Gibson, E. D.: "Automatic Equalization Using Time-Domain Equalizers," *Proc. IEEE*, vol. 53, pg. 1140, August 1965.

Glenn, W. E., C. E. Holton, G. J. Dixon, and P. J. Bos: "High-Efficiency Light Valve Projectors and High-Efficiency Laser Light Sources," *SMPTE Journal*, SMPTE, White Plains, N.Y., pp. 210–216, April 1997.

Glenn, William E.: "Large Screen Displays for Consumer and Theater Use," *Large Screen Projection Displays II*, William P. Bleha, Jr., (ed.), Proc. SPIE 1255, SPIE, Bellingham, Wash., pp. 36–43, 1990.

Glenn, William. E.: "Principles of Simultaneous Color Projection Using Fluid Deformation," *SMPTE Journal*, SMPTE, White Plains, N.Y., vol. 79, pg. 788, 1970.

Godfrey, R., et al.: "Development of the Permachrome Color Picture Tube," *IEEE Trans. Broadcast TV Receivers*, vol. BTR-14, no. 1, 1968.

Good, W.: "Projection Television," *IEEE Trans.*, vol. CE-21, no. 3, pp. 206–212, August 1975.

Good, W.: "Recent Advances in the Single-Gun Color Television Light-Valve Projector," *Soc. Photo-Optical Instrumentation Engrs.*, vol. 59, 1975.

Gove, R. J., V. Markandey, S. Marshall, D. Doherty, G. Sextro, and M. DuVal: "High-Definition Display System Based on Digital Micromirror Device," International Workshop on HDTV (HDTV '94), International Institute for Communications, Turin, Italy (October 1994).

Gow, J., and R. Door: "Compatible Color Picture Presentation with the Single-Gun Tri Color Chromatron," *Proc. IRE*, vol. 42, no. 1, pp. 308–314, January 1954.

Gretag AG: "What You May Want to Know about the Technique of Eidophor," Regensdorf, Switzerland.

Grinberg, J. et al.: "Photoactivated Birefringent Liquid-Crystal Light Valve for Color Symbology Display," *IEEE Trans. Electron Devices*, vol. ED-22, no. 9, pp. 775–783, September 1975.

Grossner, N.: *Transformers for Electronic Circuits*, 2nd ed., McGraw-Hill, New York, pp. 344–358, 1983.

Hamada, T., S. Miyaji, and S. Matsumoto: "Picture Quality Assessment System by Three-Layered Bottom-Up Noise Weighting Considering Human Visual Perception," *SMPTE Journal*, SMPTE, White Plains, N.Y., pp. 20–26, January 1999.

*Handbook of Physics*, McGraw-Hill, New York, N.Y., 1958.

Hardy, A. C., and F. H. Perrin: *The Principles of Optics*, McGraw-Hill, New York, N.Y., 1932.

Harper, C. A. (ed.): *Handbook of Components for Electronics*, McGraw-Hill, New York, N.Y., 1977.

Hasker, J.: "Astigmatic Electron Gun for the Beam Indexing Color TV Display," *IEEE Trans. Electron Devices*, vol. ED-18, no. 9, pg. 703, September 1971.

Hayashi, M., N. Yamada, and B. Sastra: "Development of a 42-in. High-Definition Plasma-Addressed LCD," in *SID International Symposium Digest of Technical Papers*, Society for Information Display, San Jose, Calif., pp.280–284, 1999.

Herold, E.: "A History of Color TV Displays," *Proc. IEEE*, vol. 64, no. 9, pp. 1331–1337, September 1976.

Hietala, Alexander W., and Duane C. Rabe: "Latched Accumulator Fractional-N Synthesis With Residual Error Reduction," United States Patent, Patent No. 5,093,632, March 3, 1992.

Hockenbrock, Richard: "New Technology Advances for Brighter Color CRT Displays," *Display System Optics II*, Harry M. Assenheim (ed.), Proc. SPIE 1117, SPIE, Bellingham, Wash., pp. 219-226, 1989.

Hoff, L. E., and A. R. King: "Skywave Communication Techniques," Tech. Rep. 709, Naval Ocean Systems Center, San Diego, Calif, March 30,1981.

Hoffman, Gary A.: "IEEE 1394: The A/V Digital Interface of Choice," 1394 Technology Association Technical Brief, 1394 Technology Association, Santa Clara, Calif., 1999.

Hornbeck, Larry J.: "Digital Light Processing for High-Brightness, High-Resolution Applications," *Projection Displays III*, Electronic Imaging '97 Conference, SPIE, Bellingham, Wash., February 1997.

Hoskoshi, K., et al.: "A New Approach to a High Performance Electron Gun Design for Color Picture Tubes," 1980 IEEE Chicago Spring Conf. Consumer Electronics.

Howald, R.: "Understand the Mathematics of Phase Noise," *Microwaves & RF*, pg. 97, December 1993.

Howe, R., and B. Welham: "Developments in Plastic Optics for Projection Television Systems," *IEEE Trans.*, vol. CE-26, no. 1, pp. 44–53, February 1980.

Hu, C., Y. Yu and K. Wang: "Antiglare/Antistatic Coatings for Cathode Ray Tube Based on Polymer System," in *Display Technologies*, Shu-Hsia Chen and Shin-Tson Wu (eds)., Proc. SPIE 1815, SPIE, Bellingham, Wash., pp.42–48, 1992.

Hubel, David H.: *Eye, Brain and Vision*, Scientific American Library, New York, N.Y., 1988.

Hufford, G. A: "A Characterization of the Multipath in the HDTV Channel," *IEEE Trans. on Broadcasting*, vol. 38, no. 4, December 1992.

Hulst, G. D.: "Inverse Ionosphere," *IRE Trans.*, vol. CS-8, pg. 3, March 1960.

Hutter, Rudolph G. E.: "The Deflection of Electron Beams," *Advances in Image Pickup and Display*, B. Kazen (ed.), vol. 1, Academic Press, New York, pp. 212-215, 1974.

IEEE Guide for Surge Withstand Capability, (SWC) Tests, ANSI C37.90a-1974/IEEE Std. 472-1974, IEEE, New York, 1974.

Ilcisen, K. J., et. al.: *Eurodisplay '96*, pg. 595, 1996.

Itah, N., et al.: "New Color Video Projection System with Glass Optics and Three Primary Color Tubes for Consumer Use," *IEEE Trans. Consumer Electronics*, vol. CE-25, no. 4, pp. 497–503, August 1979.

Jasik, H.: *Antenna Engineering Handbook*, McGraw-Hill, New York, Chapter 24, 1961.

Johnson, A.: "Color Tubes for Data Display—A System Study," Philips ECG, Electronic Tube Division.

Johnson, J. B:, "Thermal Agitation of Electricity in Conduction," *Phys. Rev.*, vol. 32, pg. 97, July 1928.

Judd, D. B., and G. Wyszecki: *Color in Business, Science and Industry*, 3rd ed., John Wiley and Sons, New York, N.Y.

Judd, D. B.: "The 1931 C.I.E. Standard Observer and Coordinate System for Colorimetry," *Journal of the Optical Society of America*, vol. 23, 1933.

Kahn, L. R.: "Ratio Squarer," *Proc. IRE*, vol. 42, pg. 1704, November 1954.

Kase, C. A., and W. L. Pritchard: "Getting Set for Direct-Broadcast Satellites," *IEEE Spectrum*, vol. 18, no. 8, pp. 22–28, 1981.

Kaufman, Ed: *IES Illumination Handbook*, Illumination Engineering Society.

Kikuchi, M., et al.: "A New Coolant-Sealed CRT for Projection Color TV," *IEEE Trans.*, vol. CE-27, IEEE, New York, no. 3, pp. 478-485, August 1981.

King, Nigel J. R.: "Phase Locked Loop Variable Frequency Generator," United States Patent, Patent No. 4,204,174, May 20, 1980.

Kingslake, Rudolf (ed.): *Applied Optics and Optical Engineering*, vol. 1, Chap. 6, Academic, New York, N.Y., 1965.

Kirchner, R. J.: "Properties of Junction Transistors," *Trans. IRE PGA*, AU-3(4), July-August 1955.

"Kodak Filters for Scientific and Technical Uses," Eastman Kodak Co., Rochester, N.Y.

Kraus, J. D.: *Antennas*, McGraw-Hill, New York, N. Y., Chapter 12, 1950.

Kurahashi, K., et al.: "An Outdoor Screen Color Display System," *SID Int. Symp. Digest 7*, Technical Papers, vol. XII, Society for Information Display, San Jose, Calif., pp. 132–133, April 1981.

Lakatos, A. I., and R. F. Bergen: "Projection Display Using an Amorphous-Se-Type Ruticon Light Valve," *IEEE Trans. Electron Devices*, vol. ED-24, no. 7, pp. 930–934, July 1977.

Lapedes, D. N. (ed.): *The McGraw-Hill Encyclopedia of Science & Technology*, 2nd ed., McGraw-Hill, New York, N.Y.

Law, H.: "A Three-Gun Shadowmask Color Kinescope," *Proc. IRE*, vol. 39, pp. 1186–1194, October 1951.

Ledoux, B, P. Bouchard, S. Laflèche, Y. Wu, and B. Caron: "Performance of 8-VSB Digital Television Receivers," *Proceedings of the International Broadcasting Convention*, IBC, Amsterdam, September 2000.

Lee, E. A., and D. G. Messerschmitt: *Digital Communications*, 2nd ed., Kluwer, Norell, Mass., 1994.

Lim, G. S., et. al.: "New Driving Method for Improvement of Picture Quality in 40-in. AC PDP," in *Asia Display '98—Proceedings of the 18th International Display Research Conference*, Society for Information Display, San Jose, Calif., pp. 591–594, 1998.

Lo, Y. T.: "TV Receiving Antennas," in *Antenna Engineering Handbook*, H. Jasik (ed.), McGraw-Hill, New York, N.Y., pp. 24–25, 1961.

Longley, A. G., and P. L. Rice: "Prediction of Tropospheric Radio Transmission over Irregular Terrain—A Computer Method," ESSA (Environmental Science Services Administration), U.S. Dept. of Commerce, Report ERL (Environment Research Laboratories) 79-ITS 67, July 1968.

Lucchesi, B., and M. Carpenter: "Pictures of Deflected Electron Spots from a Computer," *IEEE Trans. Consumer Electronics*, vol. CE-25, no. 4, pp. 468–474, 1979.

Lucky, R. W.: "Automatic Equalization for Digital Communication," *Bell Sys. Tech. J.*, vol. 44, pg. 547, April 1965.

Luxenberg, H., and R. Kuehn: *Display Systems Engineering*, McGraw-Hill, New York, 1968.

Lynn, D. K., C. S. Meyer, and D. C. Hamilton (eds.): *Analysis and Design of Integrated Circuits*, McGraw-Hill, New York, N.Y., 1967.

MacAdam, D. L.: "Visual Sensitivities to Color Differences in Daylight," *J. Opt. Soc. Am.*, vol. 32, pp. 247–274, 1942.

Mack, C. L.: "Diversity Reception in UHF Long-Range Communications," *Proc. IRE*, vol. 43, October 1955.

Maeda, M.: *Japan Display '83*, pg. 2, 1971.

Maseo, Imai, et. al.: "High-Brightness Liquid Crystal Light Valve Projector Using a New Polarization Converter," *Large Screen Projection Displays II*, William P. Bleha, Jr., (ed.), Proc. SPIE 1255, SPIE, Bellingham, Wash., pp. 52–58, 1990.

Masterson, W., and R. Barbin: "Designing Out the Problems of Wide-Angle Color TV Tube," *Electronics*, pp. 60–63, April 26, 1971.

Masuda, T., et. al: Conference Record—International Display Resolution Conference, pg. 357, 1994.

McPetrie, J. S., and L. H. Ford: "An Experimental Investigation on the Propagation of Radio Waves over Bare Ridges in the Wavelength Range 10 cm to 10 m," *J. Inst. Elec. Engrs.*, pt. 3, vol. 93, pg. 527, 1946.

Mears, N., "Method and Apparatus for Producing Perforated Metal Webs," U.S. Patent 2,762,149, 1956.

Megaw, E. C. S.: "Some Effects of Obstacles on the Propagation of Very Short Radio Waves," *J. Inst. Elec. Engrs.*, pt. 3, vol. 95, no. 34, pg. 97, March 1948.

Megawave Corp.: "Megawave/NAB Joint Technology Development Improved Antennas For NTSC Off-The-Air TV Reception," *NAB Engineering Conference Proceedings*, National Association of Broadcasters, Washington, D.C., 1996.

Mikoshiba, Shigeo: *Color Plasma Displays*, Society for Information Display, San Jose, Calif., pp M-4/3–M-4/68, 1999.

Mitsuhashi, Tetsuo: "HDTV and Large Screen Display," *Large-Screen Projection Displays II*, William P. Bleha, Jr. (ed.), Proc. SPIE 1255, SPIE, Bellingham, Wash., pp. 2-12, 1990.

Miyazawa, H.: "Evaluation and Measurement of Airplane Flutter Interference," *IEEE Trans. on Broadcasting*, vol. 35, no. 4, pp 362–367, December 1989.

Mokhoff, N.: "A Step Toward Perfect Resolution," *IEEE Spectrum*, IEEE, New York, N.Y., vol. 18, no. 7, pp. 56–58, July 1981.

Monsen, P.: "Fading Channel Communications," *IEEE Commun.*, vol. 18, pg. 16, January 1980.

Morizono, M.: "Technological Trends in High Resolution Displays Including HDTV," *SID International Symposium Digest*, paper 3.1, Society for Information Display, San Jose, Calif., May 1990.

Morrell, A., et al.: *Color Picture Tubes*, Academic Press, New York, pp. 91-98, 1974.

Morrell, A.: "Color Picture Tube Design Trends," *Proc. SID*, Society for Information Display, San Jose, Calif., vol. 22, no. 1, pp. 3–9, 1981.

Morris, James E.: "Liquid-Crystal Displays," in *The Electrical Engineering Handbook*, Richard C. Dorf (ed.), CRC Press, Boca Raton, Fla., 1993.

Moss, H.: *Narrow Angle Electron Guns and Cathode Ray Tubes*, Academic, New York, N.Y., 1968.

Na, Young-Sun, et. al.: "A New Data Driver Circuit for Field Emission Display," in *Asia Display '98—Proceedings of the 18th International Display Research Conference*, Society for Information Display, San Jose, Calif., pp. 137–140, 1998.

NAB TV TechCheck: "Canadians Perform Indoor DTV Reception Tests," National Association of Broadcasters, Washington, D.C., October 9, 2000.

NAB TV TechCheck: "CEA Establishes Definitions for Digital Television Products," National Association of Broadcasters, Washington, D.C., September 1, 2000.

NAB TV TechCheck: "Digital Cable–DTV Receiver Compatibility Standard Announced," National Association of Broadcasters, Washington, D.C., November 15, 1999.

NAB TV TechCheck: "FCC Adopts Rules for Labeling of DTV Receivers," National Association of Broadcasters, Washington, D.C., September 25, 2000.

NAB TV TechCheck: "New Digital Receiver Technology Announced," National Association of Broadcasters, Washington, D.C., August 30, 1999.

Nasibov, A., et al.: "Electron-Beam Tube with a Laser Screen," *Sov. J. Quant. Electron.*, vol. 4, no. 3, pp. 296–300, September 1974.

National Bureau of Standards Circular 462, "Ionospheric Radio Propagation," June 1948.

Neal, C. B., and S. Goyal, "Frequency and Amplitude Phase Effects in Television Broadcast Systems," *IEEE Trans.*, vol. CE-23, no. 3, pg. 241, August 1977.

Nishio, T., and K. Amemiya: "High-Luminance and High-Definition 50-in.-Diagonal Co-Planar Color PDPs with T-Shaped Electrodes," in *SID International Symposium Digest of Technical Papers*, Society for Information Display, San Jose, Calif., pp.268–272, 1999.

Norgard, John: "Electromagnetic Spectrum," *NAB Engineering Handbook*, 9th ed., Jerry C. Whitaker (ed.), National Association of Broadcasters, Washington, D.C., 1999.

Norgard, John: "Electromagnetic Spectrum," *The Electronics Handbook*, Jerry C. Whitaker (ed.), CRC Press, Boca Raton, Fla., 1996.

Norton, K. A.: "Ground Wave Intensity over a Finitely Conducting Spherical Earth," *Proc. IRE*, pg. 622, December 1941.

Norton, K. A.: "The Propagation of Radio Waves over a Finitely Conducting Spherical Earth," *Phil. Mag.*, June 1938.

Nyquist, H.: "Certain Factors Affecting Telegraph Speed," *Bell System Tech. J.*, vol. 3, pp. 324–346, March 1924.

Nyquist, H.: "Thermal Agitation of Electrical Charge in Conductors," *Phys. Rev.*, vol. 32, pg. 110, July 1928.

Oess, F.: "CRT Considerations for Raster Dot Alpha Numeric Presentations," *Proc. SID, Society for Information Display*, San Jose, Calif., vol. 20, no. 2, pp. 81–88, second quarter, 1979.

Ohkoshi, A., et al.: "A New 30V-in Beam Index Color Cathode Ray Tube," *IEEE Trans. Consumer Electronics*, vol. CE-27, p. 433, August 1981.

Palac, K.: Method for Manufacturing a Color CRT Using Mask and Screen Masters, U.S. Patent 3,989,524, 1976.

Parks, T. W., and J. H. McClellan: "A Program for the Design of Linear Phase Infinite Impulse Response Filters," *IEEE Trans. Audio Electroacoustics*, AU-20(3), pp. 195–199, 1972.

Pease, Richard W.: "An Overview of Technology for Large Wall Screen Projection Using Lasers as a Light Source," *Large Screen Projection Displays II*, William P. Bleha, Jr. (ed.), Proc. SPIE 1255, SPIE, Bellingham, Wash., pp. 93-103, 1990.

Pease, Richard W.: "An Overview of Technology for Large Wall Screen Projection Using Lasers as a Light Source," *Large Screen Projection Displays II*, William P. Bleha, Jr., (ed.), Proc. SPIE 1255, SPIE, Bellingham, Wash., pp. 93–103, 1990.

Petersen, A., and D. B. Sinclair: "A Singled-Ended Push-Pull Audio Amplifier," *Proc. IRE*, vol. 40, January 1952.

Peterson, O. H., H. H. Beverage, and J. B. Moore: "Diversity Telephone Receiving System of RCA Communications, Inc.," *Proc. IRE*, vol. 19, pg. 562, April 1931.

Peterson, R., R. Ziemer, and D. Borth: *Introduction to Spread Spectrum Communications*, Prentice-Hall, Englewood Cliffs, N. J., 1995.

Pfahnl, A.: "Aging of Electronic Phosphors in Cathode Ray Tubes," *Advances in Electron Tube Techniques*, Pergamon, New York, N.Y., pp. 204–208.

Phillips, Thomas E., et. al.: "1280 × 1024 Video Rate Laser-Addressed Liquid Crystal Light Valve Color Projection Display," *Optical Engineering*, Society of Photo-Optical Instrumentation Engineers, vol. 31, no. 11, pp. 2300–2312, November 1992.

Pitts, K., and N. Hurst: "How Much do People Prefer Widescreen (16 × 9) to Standard NTSC (4 × 3)?," *IEEE Transactions on Consumer Electronics*, vol. 35, no. 3, pp. 160–169, August 1989.

Pleasant, D.: "Practical Simulation of Bit Error Rates," *Applied Microwave and Wireless*, pg. 65, Spring 1995.

Plonka, Robert: "Can ATV Coverage be Improved with Circular, Elliptical or Vertical Polarized Antennas?," *NAB Engineering Conference Proceedings*, National Association of Broadcasters, Washington, D.C., 1996.

Pohlmann, Ken: *Principles of Digital Audio*, McGraw-Hill, New York, N.Y., 2000.

Pointer, R. M.: "The Gamut of Real Surface Colors," *Color Res. App.*, vol. 5, 1945.

Poorter, T., and F. W. deVrijer: "The Projection of Color Television Pictures," *SMPTE Journal*, SMPTE, White Plains, N.Y., vol. 68, pg. 141, 1959.

Price, R., and P. E. Green, Jr.: "A Communication Technique for Multipath Channels," *Proc. IRE*, vol. 46, ph. 555, March 1958.

Proakis, J. G.: "Advances in Equalization for Intersymbol Interference," in *Advances in Communication Systems: Theory & Application*, A. V. Balakrishnan and A. J. Viterbi (eds.), Academic Press, New York, N.Y., 1975.

Quinn, S. F., and C. A. Siocos: "PLUGE Method of Adjusting Picture Monitors in Television Studios—A Technical Note," *SMPTE Journal*, SMPTE, White Plains, N.Y., vol. 76, pg. 925, September 1967.

Qureshi, S.: "Adaptive Equalization," *IEEE Commun.*, vol. 20, pg. 9, March 1982.

Qureshi, Shahid U. H.: "Adaptive Equalization," *Proceedings of the IEEE*, IEEE, New York, vol. 73, no. 9, pp. 1349–1387, September 1985.

*Radio Amateur's Handbook*, American Radio Relay League, Newington, Conn., 1983.

"Radio Wave Propagation," Summary Technical Report of the Committee on Propagation of the National Defense Research Committee, Academic Press, New York, N.Y., 1949.

"Receiver Planning Factors Applicable to All ATV Systems," Final Report of PS/WP3, Advanced Television Systems Committee, Washington, D.C., Dec. 1, 1994.

"Recommended Practice for Measurement of X-Radiation from Direct View TV Picture Tubes," Publication TEP 164, Electronics Industries Association, Washington, D.C., 1981.

"Report of the Ad Hoc Committee, Federal Communications Commission," vol. 1, May 1949; vol. 2, July 1950.

Riegler, R. L., and R. T. Compton, Jr.: "An Adaptive Array for Interference Rejection," *Proc. IEEE*, vol. 61, June 1973.

Riley, Thomas A. D.: "Frequency Synthesizers Having Dividing Ratio Controlled Sigma-Delta Modulator," United States Patent, Patent No. 4,965,531, October 23, 1990.

Robbins, J., and D. Mackey: "Moire Pattern in Color TV," *IEEE Trans. Consumer Electronics*, vol. CE-28, no. 1, pp. 44–55, February 1982.

Robertson, A.: "Projection Television—1 Review of Practice," *Wireless World*, vol. 82, no. 1489, pp. 47–52, September 1976.

Robinder, R., D. Bates, P. Green: "A High Brightness Shadow Mask Color CRT for Cockpit Displays," *SID Digest*, Society for Information Display, vol. 14, pp. 72-73, 1983.

Rohde, Ulrich L. Rohde and David P. Newkirk: *RF/Microwave Circuit Design for Wireless Applications*, John Wiley & Sons, New York, N.Y., 2000.

Rohde, Ulrich L.: "Key Components of Modern Receiver Design—Part 1," *QST*, pg. 29, May 1994.

Rohde, Ulrich L.: *Digital PLL Frequency Synthesizers*, Prentice-Hall, Englewood Cliffs, N.J., 1983.

Rohde, Ulrich L.: *Microwave and Wireless Synthesizers: Theory and Design*, John Wiley & Sons, New York, N.Y., pg. 209, 1997.

Rossweiler, G. C., F. Wallace, and C. Ottenhoff: "Analog versus Digital Null-Steering Controllers," ICC '77 Conf. Rec., 1977.

Rublack, W.: "In-Line Plural Beam CRT with an Aspherical Mask," U.S. Patent 3,435,668, 1969.

Sakamoto, Y.: and E. Miyazaki, *Japan Display '83*, pg. 30, 1983.

Sarma, Kalluri R.: "Active-Matrix LCDs," in *Seminar Lecture Notes*, Society for Information Display, San Jose, Calif., vol. 1, pp. M3/3–M3/45, 1999.

Say, D.: "Picture Tube Spot Analysis Using Direct Photography," *IEEE Trans. Consumer Electronics*, vol. CE-23, pp. 32–37, February 1977.

Say, D.: "The High Voltage Bipotential Approach to Enhanced Color Tube Performance," *IEEE Trans. Consumer Electronics*, vol. CE-24, no. 1, pg. 75, February 1978.

Schiecke, K.: "Projection Television: Correcting Distortions," *IEEE Spectrum*, IEEE, New York, N.Y., vol. 18, no. 11, pp. 40–45, November 1981.

Schwartz, J.: "Electron Beam Cathodoluminescent Panel Display," U.S. Patent 4,137,486.

Sears, F. W.: *Principles of Physics*, III, Optics, Addison-Wesley, Cambridge, Mass., 1946.

Selvidge, H.: "Diffraction Measurements at Ultra High Frequencies," *Proc. IRE*, pg. 10, January 1941.

Sextro, G., I. Ballew, and J. Lwai: "High-Definition Projection System Using DMD Display Technology," *SID 95 Digest*, Society for Information Display, San Jose, Calif., pp. 70– 73, 1995.

Sgrignoli, Gary: "Preliminary DTV Field Test Results and Their Effects on VSB Receiver Design," ICEE '99.

Sherr, S.: *Electronic Displays*, Wiley, New York, N.Y., 1979.

Sherr, S.: *Fundamentals of Display System Design*, Wiley-Interscience, New York, N.Y., 1970.

Shockley, W: "A Unipolar Field-Effect Transistor," *Proc. IRE*, vol. 40, November 1952.

Shockley, W: "The Theory of P-N Junctions in Semiconductors and P-N Junction Transistors," *Proc. JRE*, vol. 41, June 1953.

Sinclair, C.: "Small Flat Cathode Ray Tube," *SID Digest*, Society for Information Display, San Jose, Calif., pg. 138, 1981.

Sklar, B.: *Digital Communications: Fundamentals and Applications*, Prentice-Hall, Englewood Cliffs, N. J., 1988.

Smith, E. E., and E. W. Davis: "Wind-induced Ions Thwart TV Reception," *IEEE Spectrum*, pp. 52—55, February 1981.

Standards and definitions committee, Society for Information Display.

"Standards Testing: Bit Error Rate," application note 3SW-8136-2, Tektronix, Beaverton, OR, July 1993.

Stanley, T.: "Flat Cathode Ray Tube," U.S. Patent 4,031,427.

Stemson, A: *Photometry and Radiometry for Engineers*, John Wiley and Sons, New York, N.Y.

Swartz, J.: "Beam Index Tube Technology," *SID Proceedings*, Society for Information Display, San Jose, Calif., vol. 20, no. 2, p. 45, 1979.

Takeuchi, Kazuhiko, et. al.: "A 750-TV-Line Resolution Projector Using 1.5 Megapixel a-Si TFT LC Modules," *SID 91 Digest*, Society for Information Display, pp. 415–418, 1991.

Takeuchi, Kazuhiko, et. al.: "A 750-TV-Line Resolution Projector using 1.5 Megapixel a-Si TFT LC Modules," *SID 91 Digest*, Society for Information Display, San Jose, Calif., pp. 415–418, 1991.

Taneda, T., et al.: "A 1125-Scanning Line Laser Color TV Display," *SID 1973 Symp. Digest Technical Papers*, Society for Information Display, San Jose, Calif., vol. IV, pp. 86–87, May 1973.

Tannas, Lawrence E., Jr.: *Flat Panel Displays and CRTs*, Van Nostrand Reinhold, New York, pg. 18, 1985.

Tate, Jeffrey P., and Patricia F. Mead: "Crystal Oscillators," in *The Electronics Handbook*, Jerry C. Whitaker (ed.), CRC Press, Boca Raton, Fla., pp. 185–199, 1996.

"Television Receivers and Video Products," UL 1410, Sec. 71, Underwriters Laboratories, Inc., New York, 1981.

*The Cambridge Encyclopedia*, Cambridge University Press, 1990.

*The Columbia Encyclopedia*, Columbia University Press, 1993.

"The Propagation of Radio Waves through the Standard Atmosphere," Summary Technical Report of the Committee on Propagation, vol. 3, National Defense Research Council, Washington, D.C., 1946, published by Academic Press, New York, N.Y.

Thomas, Gomer: "ATSC Datacasting—Opportunities and Challenges," *Proceedings of the 33rd SMPTE Advanced Motion Imaging Conference*, SMPTE, White Plains, N.Y., pp. 307–314, February 1999.

TMS320C55x DSP Functional Overview, Texas Instruments, Dallas, TX, literature No. SRPU312, June 2000.

Tomioka, M., and Y. Hayshi: "Liquid Crystal Projection Display for HDTV," *Proceedings of the International Television Symposium*, ITS, Montreux, Switzerland, 1991.

Tong, Hua-Sou: "HDTV Display—A CRT Approach," *Display Technologies*, Shu-Hsia Chen and Shin-Tson Wu (eds.), Proc. SPIE 1815, SPIE, Bellingham, Wash., pp. 2–8, 1992.

Trent, R. L.: "Design Principles for Transistor Audio Amplifiers," *Trans. IRE PGA*, AU-3(5), September–October 1955.

Tsuruta, Masahiko, and Neil Neubert: "An Advanced High Resolution, High Brightness LCD Color Video Projector," *SMPTE Journal*, SMPTE, White Plains, N.Y., pp. 399–403, June 1992.

Uba, T., K. Omae, R. Ashiya, and K. Saita: "16:9 Aspect Ratio 38V-High Resolution Trinitron for HDTV," *IEEE Transactions on Consumer Electronics*, vol. 34, no. 1., pp. 85–89, February 1988.

Uchida, Tadayuki, Yasuaki Nishida, and Yukihiro Nishida: "Picture Quality in Cascaded Video-Compression Systems for Digital Broadcasting," *SMPTE Journal*, SMPTE, White Plains, N.Y., pp. 27–38, January 1999.

Ungerboeck, G.: "Trellis-Coded Modulation with Redundant Signal Sets," parts I and II, *IEEE Comm. Mag.*, vol. 25 (Feb.), pp. 5-11 and 12-21, 1987.

Ungerboeck, Gottfried: "Fractional Tap-Spacing Equalizer and Consequences for Clock Recovery in Data Modems," *IEEE Transactions on Communications*, IEEE, New York, vol. COM-24, no. 8, pp. 856–864, August 1976.

"Using Vector Modulation Analysis in the Integration, Troubleshooting and Design of Digital RF Communications Systems," Product Note HP89400-8, Hewlett-Packard, Palo Alto, Calif., 1994.

van der Pol, Balth, and H. Bremmer: "The Diffraction of Electromagnetic Waves from an Electrical Point Source Round a Finitely Conducting Sphere, with Applications to Radiotelegraphy and to Theory of the Rainbow," pt. 1, *Phil. Mag.*, July, 1937; pt. 2, *Phil. Mag.*, November 1937.

Van, M. W., and J. Von Esdonk: "A High Luminance High-Resolution Cathode-Ray Tube for Special Purposes," *IEEE Trans Electron Dev.*, IEEE, New York, N.Y., ED-30, pg. 193, 1983.

Verona, Robert: "Comparison of CRT Display Measurement Techniques," *Helmet-Mounted Displays III*, Thomas M. Lippert (ed.), Proc. SPIE 1695, SPIE, Bellingham, Wash., pp. 117–127, 1992.

Walker, P. J.: "A Current Dumping Audio Power Amplifier," *Wireless World*, December 1975.

Wang, Q. H., J. B. Cheng, and Z. L. Lin: "A New YAG Phosphor Screen for Projection CRT," *Electron Lett.*, vol. 34, no. 14, pg. 1420, 1998.

Wang, Qionghua, Jianbo Cheng, Zulun Lin, and Gang Yang: "A High-Luminance and High-Resolution CRT for Projection HDTV Display," *Journal of the SID*, Society for Information Display, San Jose, Calif., vol. 7, no. 3, pg 183–186, 1999.

Wang, S., et al.: "Spectral and Spatial Distribution of X-Rays from Color Television Receivers," *Proc. Conf. Detection and Measurement of X-radiation from Color Television Receivers*, Washington, D.C., pp. 53–72, March 28–29, 1968.

Watson, R.: "Receiver Dynamic Range; Pt. 1, Guidelines for Receiver Analysis," *Microwaves & RF*, vol. 25, pg. 113, December 1986.

"Waveform Analysis: Noise and Jitter," application note 3SW8142-2, Tektronix, Beaverton, OR, March 1993.

Weber, Larry F.: "Plasma Displays," in *The Electrical Engineering Handbook*, Richard C. Dorf (ed.), CRC Press, Boca Raton, Fla., 1786–1798, 1993.

*Webster's New World Encyclopedia*, Prentice Hall, 1992.

Wedding, Donald K., Sr.: "Large Area Full Color ac Plasma Display Monitor," *Large Screen Projection Displays II*, William P. Bleha, Jr., (ed.), Proc. SPIE 1255, SPIE, Bellingham, Wash., pp. 29–35, 1990.

Weinberg, L.: *Network Analysis and Synthesis*, McGraw-Hill, New York, N.Y., 1967.

Wells, John Norman: "Frequency Synthesizers," European Patent, Patent No. 0125790B2, July 5, 1995.

Widlar, R. J.: "A Unique Current Design for a High Performance Operational Amplifier Especially Suited to Monolithic Construction," *Proc. NEC*, 1965.

Widrow, B., and J. M. McCool: "A Comparison of Adaptive Algorithms Based on the Methods of Steepest Descent and Random Search," *IEEE Trans.*, vol. AP-24, pg. 615, September 1976.

Widrow, B., P. E. Mantey, L. J. Griffiths, and B. B. Goode: "Adaptive Antenna Systems," *Proc. IEEE*, vol. 55, pg. 2143, December 1967.

Williams, Charles S., and Becklund, Orville A.: Optics: *A Short Course for Engineers and Scientists*, Wiley Interscience, New York, N.Y., 1972.

Wilson, E.: "Evaluate the Distortion of Modular Cascades," *Microwaves*, vol. 20, March 1981.

Wilson, J., and J. F. B. Hawkes: *Optoelectronics: An Introduction*, Prentice-Hall, London, pg. 145, 1989.

Woodhead, A., et al.: *1982 SID Digest*, Society for Information Display, San Jose, Calif., pg. 206, 1982.

Wugofski, T. W.: "A Presentation Engine for Broadcast Digital Television," *International Broadcasting Convention Proceedings*, IBC, London, England, pp. 451–456, 1999.

Wyszecki, G., and W. S. Stiles: *Color Science, Concepts and Methods, Quantitative Data and Formulae*, 2nd ed., John Wiley and Sons, New York, N.Y.

X-Radiation Measurement Procedures for Projection Tubes," TEPAC Publication 102, Electronic Industries Association, Washington, D. C.

Yamada and Uematsu, "New Color TV with Composite SAW IF Filter Separating Sound and Picture Signals," *IEEE Trans.*, vol. CE-28, no. 3, p. 193. Copyright 1982 IEEE.

Yamamoto, Y., Y. Nagaoka, Y. Nakajima, and T. Murao: "Super-compact Projection Lenses for Projection Television," *IEEE Transactions on Consumer Electronics*, IEEE, New York, N.Y., August 1986.

Yoshida, S., et al.: "25-V Inch 114-Degree Trinitron Color Picture Tube and Associated New Development," *Trans. BTR*, pp. 193-200, August 1974.

Yoshida, S., et al.: "A Wide Deflection Angle (114°) Trinitron Color Picture Tube," *IEEE Trans. Electron Devices*, vol. 19, no. 4, pp. 231–238, 1973.

Yoshida. S.: et al., "The Trinitron—A New Color Tube," *IEEE Trans. Consumer Electronics*, vol. CE-28, no. 1, pp. 56–64, February 1982.

Yoshizawa, T., S. Hatakeyama, A. Ueno, M. Tsukahara, K. Matsumi, K. Hirota: "A 61-in High-Definition Projection TV for the ATSC Standard," *SID 99 Digest*, Society for Information Display, San Jose, Calif., pp. 112–115, 1999.

Younse, J. M.: "Projection Display Systems Based on the Digital Micromirror Device (DMD)," *SPIE Conference on Microelectronic Structures and Microelectromechanical Devices for Optical Processing and Multimedia Applications*, Austin, Tex., SPIE Proceedings, SPIE, Bellingham, Wash., vol. 2641, pp. 64–75, Oct. 24, 1995.

Zborowski, R. W.: “Application of On-Channel Boosters to Fill Gaps in DTV Broadcast Coverage,” *NAB Engineering Conference Proceedings*, National Association of Broadcasters, Washington, D.C., 2000.

Ziemer, R., and W. Tranter: *Principles of Communications: Systems, Modulation, and Noise*, 4th ed., Wiley, New York, 1995.

Ziemer, Rodger E.: “Digital Modulation,” *The Electronics Handbook*, Jerry C. Whitaker (ed.), CRC Press, Boca Raton, Fla., pp. 1213–1236, 1996.

# Subject Index

## E



## F

## G



## H



## I

### J



### K



### L



### M

## S

# Appendix: Television/Video Standards

## 1) Advanced Television Systems Committee

The Advanced Television Systems Committee (ATSC) is an international, non-profit organization developing voluntary standards for digital television. The ATSC has over 200 member organizations representing the broadcast, broadcast equipment, motion picture, consumer electronics, computer, cable, satellite, and semiconductor industries. The following is a partial list of ATSC Standards and technical activities.

### ATSC Digital Television Standard, *Document A/53*

The Digital Television Standard describes the system characteristics of the advanced television (ATV) system. The document and its normative annexes provide detailed specification of the parameters of the system including the video encoder input scanning formats and the pre-processing and compression parameters of the video encoder, the audio encoder input signal format and the pre-processing and compression parameters of the audio encoder, the service multiplex and transport layer characteristics and normative specifications, and the VSB RF/transmission subsystem. The system is modular in concept and the specifications for each of the modules are provided in the appropriate annex. This document includes Amendment No. 1 to Doc. A/53.

### Guide to the use of the ATSC Digital Television Standard, *Document A/54*

This guide provides an overview and tutorial of the system characteristics of the advanced television (ATV) system defined by ATSC Standard A/53, *ATSC Digital Television.*

### Digital Audio Compression (AC-3*), Document A/52*

This document specifies coded representation of audio information and the decoding process, as well as information on the encoding process. The coded representation specified is suitable for use in digital audio transmission and storage applications, and may convey from 1 to 5 full bandwidth audio channels, along with a low frequency enhancement channel. A wide range of encoded bit-rates is supported by this specification. Typical applications of digital audio compression are in satellite or terrestrial audio broadcasting, delivery of audio over metallic or optical cables, or storage of audio on magnetic, optical, semiconductor, or other storage media.

### Standard for Coding 25/50 Hz Video, *Document A/63*

This document describes the characteristics for the video subsystem of a digital television system operating at 25 Hz and 50 Hz frame rates.

### Transmission Measurement and Compliance Standard for DTV, *Document A/64 Rev. A*

This document describes methods for testing, monitoring, and measurement of the transmission subsystem intended for use in the digital television (DTV) system, including specifications for maximum out-of-band emissions, parameters affecting the quality of the inband signal, symbol error tolerance, phase noise and jitter, power, power measure, frequency offset, and stability. In addition, it describes the condition of the RF symbol stream upon loss of MPEG packets. (The ATSC approved a revision to this document on May 30, 2000, that includes the revised FCC DTV emission mask.

### PSIP for Terrestrial Broadcast and Cable, *Document A/65 Rev A with Amendment No. 1*

The Program and System Information Protocol Standard provides a methodology for transporting digital television system information and electronic program guide data. The standard includes an amendment that provides new functionality known as *Directed Channel Change* (DCC). This new feature will allow broadcasters to tailor programming or advertising based upon parameters defined by the viewer such as: postal, zip or location code, program identifier, demographic category, and content subject category. Potential applications include customized programming services, commercials based upon demographics, and localized weather and traffic reports.

### Conditional Access For Terrestrial Broadcast, *Document A/70*

This document defines a standard for the Conditional Access system for ATSC terrestrial broadcasting to enable broadcasters to fully utilize the capabilities of digital broadcasting. This standard is based, whenever possible, on existing open standards and defines the building blocks necessary to ensure interoperability. The ATSC CA module is replaceable; to ensure that ATSC hosts are protected against obsolescence as security is upgraded. This standard applies to all CA vendors that supply CA service on behalf of an ATSC service provider. An overview of the CA standard is given in Annex C. (This document includes an Amendment that the ATSC approved on May 30, 2000.)

### Modulation and Coding Requirements for DTV Applications Over Satellite, *Document A/80*

This document defines a standard for modulation and coding of data delivered over satellite for digital television contribution and distribution applications. The data can be a collection of program material including video, audio, data, multimedia, or other material. It includes the ability to handle multiplexed bit streams in accordance with the MPEG-2 systems layer, but it is not limited to this format and makes provision for arbitrary types of data as well. QPSK, 8PSK and 16 QAM modulation modes are included, as well as a range of forward error correction techniques.

### Data Broadcast, *Document A/90*

The ATSC Data Broadcast Standard defines protocols for data transmission compatible with digital multiplex bit streams constructed in accordance with ISO/IEC 13818-1 (MPEG-2 systems). The standard supports data services that are both TV program related and non-program related. Applications may include enhanced television, webcasting, and streaming video services. Data broadcasting receivers may include PCs, televisions, set-top boxes, or other devices. The standard provides mechanisms for download of data, delivery of datagrams, and streaming data.

### DTV Application Software Environment (DASE)

The DASE standard will define a software layer (middleware) that allows programming content and applications to run on a common receiver platform. Interactive and enhanced applications need access to common receiver features in a platform-independent manner. The standard will provide enhanced and interactive content creators the specifications necessary to ensure that their applications and data will run uniformly on all brands and models of receivers. Manufacturers will be able to choose hardware platforms and operating systems for receivers, but provide the commonality necessary to support applications made by many content creators.

### Interactive Services

This standard defines session level protocols carried over interaction channels associated with interactive services. The interaction channel may be one or two way and connects a user (operating through a DTV receiver) with some service provider. The ATSC Interactive Services Protocols are intended to operate on a variety of physical networks, by focusing on higher layer protocols but not addressing specific applications

### VSB Enhancements

Responding to the evolving needs of broadcasters, ATSC has initiated a standards activity aimed at enhancing the VSB modulation specifications that are a part of the ATSC Digital Television Standard (A/53). The new standards proceeding is intended to give broadcasters additional flexibility, including the ability to transmit programming and data to portable and mobile receivers. Just getting underway as this book went to press, it will move forward in parallel with the DTV implementation process already well underway.

### Implementation Subcommittee

The ATSC Implementation Subcommittee (IS) evaluates technical requirements, operational impacts, preferred operating methods, time frames, and cost impacts of various DTV implementation issues. Based upon this analysis, the Subcommittee identifies potential requirements for standards, recommended practices, or guidelines.

For additional information on the Advanced Television Systems Committee, visit the web site www.atsc.org.

## 2) CEA Standards

The following documents relating to television broadcasting have been adopted by the Consumer Electronics Association. For additional information contact the CEA at www.ce.org.

## a) Antennas

### CEB6-C

"TV Receiving Antenna Manufacturers Guide to Categorizing Antennas for Use With the CEA TV Antenna Sector Map Program, Antenna Types and Characteristics, Minimum Performance Requirements, Packaging, and Marking Specifications"

This bulletin was prepared to provide manufacturers of television receiver antennas with appropriate guidance on determining antenna categories and minimum performance requirements to comply with the CEA TV Antenna Selector Map program. Essential elements of this program include color coding of various television reception environments in a market, with corresponding color-coded marking of antennas to provide reception. See EIA-774 for test and measurement procedures. The CEA copyrighted logo for marking antennas and logo use guidelines are available from CEA.

### CEB7

"TV Receiving Antenna Manufacturers Guide to Indoor Antennas for Use with the CEA Indoor TV Antenna Certification Program, Indoor Antenna Characteristics, Packaging and Marking Specifications, and Minimum Performance Requirements"

This bulletin was prepared by the CEA R-5 Antennas Committee to provide manufacturers of indoor television receiver antennas with appropriate guidance on determining antenna characteristics and minimum performance requirements to comply with the CEA Indoor TV Antenna Certification program. Essential elements of this program include indoor TV antenna technical specifications and the use of a CEA copyrighted certification logo. Test and measurement procedures are contained in the EIA-774 standard, and the CEA copyrighted logo for marking antennas and logo use guidelines are available from CEA.

### EIA-774

"TV Receiving Antenna Performance Presentation and Measurement"

This standard is intended to provide television receive antenna manufacturers with appropriate test and measurement procedures to examine antenna performance parameters necessary to comply with elements of the CEMA TV Antenna Selector Map program, specifically EIA CEB-6. Essential elements include procedures to determine antenna gain, front-to-back ratio, directivity, and distortion performance of active antennas with integrated amplifiers.

## b) Television

### CEB1-A

"Recommended Practice for Content Advisories"

This bulletin provided guidance to receiver manufacturers and designers concerning expected receiver response to content advisory information transported in the EIA-744-A U.S. or Canadian content advisory (program rating) packet. EIA/CEA-608-B incorporates this information.

## CEB2

"Recommended Practice for Expansion of Extended Data Service (XDS) to Include Cable Channel Mapping System Information"

This bulletin provided guidance to receiver designers and manufacturers concerning expected receiver response to channel allocation data packets. EIA/CEA-608-B incorporates this information. EIA-CEB-2 is withdrawn. See EIA/CEA-608-B.

## CEB4

"Recommended Practice for VCR Specifications"

The purpose of this recommended practice is to: 1) include essential information for the VHS VCR user, and 2) standardize the format for the presentation of the information.

## CEB5

"Recommended Practice for DTV Receiver Monitor Mode Capability"

CEB5 is intended to provide recommendations to digital television (DTV) designers/manufacturers concerning a "monitor" mode capability. See EIA-762 for minimum specifications for a DTV remodulator.

## CPEB1

"Standard Method of Measurement of Ionizing Radiation from Television Receivers for Factory Quality Assurance"

Establishes a standard method of obtaining ionizing radiation characteristics from production television receivers.

## CPEB2

"Definition of Normal Operating Conditions for Television Receivers"

Specifies normal operating conditions of broadcast television receivers for the purpose of X-radiation measurements.

## CPEB3

"Measurement Instrumentation for X-Radiation from Television Receivers"

Specifies instrumentation for the measurement of X-radiation emission from television receivers.

## EIA/CEA-608-B

"Line 21 Data Services"

Serves as a technical guide for those providing encoding equipment and/or decoding equipment to produce material with encoded data embedded in Line 21 of the vertical blanking interval of the NTSC video signal. It is also a usage guide for those who will produce material using such equipment. Revision incorporates content advisory.

## EIA/CEA-775.1

"Web Enhanced DTV 1394 Interface Specification"

This standard includes mechanisms to allow a source of MPEG service to utilize the MPEG decoding and display capabilities in a DTV.

## EIA/CEA-775.2

"Service Selection Information for Digital Storage Media Interoperability"

A digital storage device such as a D-VHS or hard disk digital recorder may be used by the DTV or by another source device such as a cable set-top box to record or time-shift digital television signals. This standard supports the use of such storage devices by defining Service Selection Information (SSI), methods for managing discontinuities that occur during recording and playback, and rules for management of partial transport streams.

## EIA/CEA-805

"Data Services on the Component Video Interfaces"

This standard specifies how data services are carried on component video interfaces (CVI), as described in EIA-770.1-A (for 2H 480p signals only), EIA-770.2-A (for 2H 480p signals only) and EIA-770.3-A. This standard applies to all CE devices carrying data services on the CVI vertical blanking interval (VBI). This standard does not apply to signals which originate in 1H 480I, as defined in EIA-770.1-A and EIA-770.2-A. The first data service defined is Copy Generation Management System (CGMS) information, including signal format and data structure when carried by the VBI of standard definition progressive and high definition $YP_bP_r$ type component video signals. It is also intended to be usable when the $YP_bP_r$ signal is converted into other component video interfaces including RGB and VGA.

## EIA/CEA-818-B

"Cable Compatibility Requirements"

This standard defines the minimum requirements that shall be met by digital cable TV systems and digital TV receivers such that the receivers may be connected directly to the RF output of the cable system to provide selected baseline services.

## EIA/CEA-819

"Cable Compatibility Requirements for Two-Way Digital Cable TV Systems"

This standard defines the minimum requirements that shall be met by two-way digital cable TV systems and two-way digital TV receivers such that the receivers may be connected directly to the RF output and input of the two-way cable system to provide specified services.

## EIA/CEA-849

"Application Profiles for EIA-775A Compliant DTVs"

This standard specifies profiles for various applications of the EIA-775A standard. The application areas covered here include digital streams compliant with ATSC terrestrial broadcast, direct-broadcast satellite (DBS), OpenCable™, and standard definition Digital Video (DV) camcorders.

## EIA/CEA-863

"Connection Color Codes for Home Theater Systems"

This standard defines the colors for marking connections commonly used for electronic devices in a home theater system. This standard adds continuity to installation information and assures consistency of information to installers.

## EIA/CEA-CEB-8

"Consideration of EIA-608-B Data Within the DTV Closed Captioning (EIA-708-B) Construct"

EIA/CEA CEB-8 provides guidance on the use and processing of the EIA/CEA-608-B data stream embedded within the ATSC MPEG-2 video elementary transport stream. EIA/CEA CEB-8 augments EIA-708-B (addressing DTV Closed Captioning) Sections 4.3 and 9.23.

## EIA/IS-105.1

"Decoder Interface Standard"

Specifies an interconnection method for attaching a cable decoder to a piece of consumer electronics equipment such as a TV or a VCR. Two ports—an audio/video/control port and an IF port—are covered under this interim standard. This specifies the physical characteristics of the interface between the decoder and the receiver.

## EIA/IS-16-A

"Immunity of Television Receivers and Video Cassette Recorders (VCRs) to Direct Radiation from Radio Transmissions, 0.5 to 30 MHz"

This Interim Standard establishes as a performance guideline for the immunity of TV receivers and VCR's to direct radiation from radio transmissions below 30 MHz. It provides the recommended procedure for measuring the immunity of TV receivers and VCRs over the frequency range of 0.5 to 30 MHz.

## EIA/IS-31

"Recommended Design Guideline, Rejection of Educational Interference to Ch. 6 Television Reception"

Establishes the design guideline for a color television receiver or VCR in the play-through mode to provide rejection of educational (and non-commercial) FM interference to Ch. 6 television reception equal to or exceeding that of the FM median receiver. It defines the signal and operation of the TV product, and the evaluation criteria to be used in implementing the interference immunity guideline. It also establishes guidelines for signal levels at which TV products should coexist with other RF sources, without material performance degradation.

## EIA/IS-702

"Copy Generation Management System (Analog)"

This standard included packet description data relating to the Copy Generation Management System (Analog) (CGMS-A). EIA/CEA-608-B incorporates this information. EIA-702 is withdrawn. See EIA/CEA-608-B.

## EIA/TIA-250-C

"Electrical Performance for Television Transmission Systems (ANSI/EIA/TIA-250-C-89)"

This Standard specifies the minimal transmission performance characteristics, consistent with good engineering practice, of television transmission of 525-line NTSC color or monochrome video and associated audio signals suitable for television broadcasting or for similar application. These limits are used for the acceptance of new systems or restoration of existing systems after maintenance. It should be noted that transmission systems utilize analog, digital, or a mixture of analog and digital techniques. Definitions, standards, and methods of measurement are given for both the video and related audio signals being carried from a few hundred feet to thousands of miles, including satellite transmission.

## EIA-105.2

"Decoder Interface Control Standard"

The EIA/IS-105 specification is designed to allow consumer electronic devices such as television, VCRs, and other receiving devices to be connected to various external devices via the Decoder Interface Multi-pin connector.

## EIA-170

"Electrical Performance Standards Monochrome Television Studio Facilities"

Establishes definitions, minimum standards, and methods of measurement for the electrical performance for monochrome television studio facilities. It is intended to apply only to locally generated signals; that is, signals generated in the studio itself or at a nearby point where control can be exercised over picture quality.

## EIA-189-A

"Encoded Color Bar Signal"

The EIA Standard Color Bar Signal is intended for use as a test signal for the following principal reasons: a) adjustment of color monitors, b) adjustment of color encoders, and c) rapid checks of color television transmission systems.

## EIA-23

"RF Interface Specification for Television Receiving and Cable Television Systems"

This specification is intended to apply to all cable systems and to all receiving devices which may be directly connected to a cable system residential outlet, including, but not limited to, television sets, video cassette recorders, and converters (whether furnished by cable operators or independently acquired by subscribers).

## EIA-378

"Measurement of Spurious Radiation from FM and TV Broadcast Receivers in the Frequency Range of 100 to 1000 MHz, Using the EIA Laurel Broadband Antenna"

Describes potential sources of spurious radiation from frequency modulation and television broadcast receivers and establishes measurement methods whereby the strength of some radiations may be determined.

## EIA-462

"Electrical Performance Standards for Television Broadcast Demodulators"

This Standard includes aural and visual performance standards for various functions and operating modes which may or may not be offered by a particular demodulator manufacturer, or may be offered by a manufacturer not in one instrument, but in two or more separate instruments.

## EIA-544

"Immunity of TV and VCR Tuners to Internally Generated Harmonic Interference from Signals in the Band 535 kHz to 30 MHz (ANSI/EIA-544-88)"

This Standard establishes performance guidelines for rejection of interference by television receivers, video cassette recorders, and tuners. It details a measurement procedure which determines the level of interfering signal which will generate harmonics in the tuner, causing interference 40 dB below a desired signal at the intermediate frequency (IF) output of a tuner under test. This standard covers interference immunity to CB, amateur radio, and other transmissions.

## EIA-679-B

"National Renewable Security Standard (NRSS)"

NRSS provides two physical designs. Part A defines a removable and renewable security element form factor that is an extension of the ISO-7816 standard. Part B defines a removable and renewable security element based on the PCMCIA ("C Card") form factor. The common attributes allow either an NRSS-A or NRSS-B device to provide security for applications involving pay and subscription cable or satellite television services, telephone, and all forms of electronic commerce.

## EIA-693

"Audio/Video Bus (AVBus) Physical Layer and Media Specification (ANSI/EIA-693-97)"

This standard contains the performance specifications necessary to implement an Audio/Video (AV) Bus to carry baseband audio, video and control signals for limited distances between consumer audio and video equipment in the home. The purpose of this standard is to present all the information necessary for the development of an AV physical network and devices to communicate and share information over the network.

## EIA-708-B

"Digital Television (DTV) Closed Captioning"

This document is intended as a definition of DTV Closed Captioning (DTVCC) and provides specifications and/or guidelines for caption service providers, DTVCC decoder and encoder manufacturers, DTV receiver manufacturers, and DTV signal processing equipment manufacturers. This specification includes: a) a description of the transport method of DTVCC data in the DTV signal, b) a description of DTVCC specific data packets and structures, c) a specification of how DTVCC information is to be processed, d) a list of minimum implementation recommendations for DTVCC receiver manufacturers, and e) a set of recommended practices for DTV encoder and decoder manufacturers.

### EIA-744-A

"Transport of Content Advisory Information Using Extended Data Service (XDS)"

The XDS Content Advisory (program rating) packet transports essential content advisory information viewers need to block selected programming. EIA/CEA-608-B incorporates this information. EIA-744-A is withdrawn. See EIA/CEA-608-B.

### EIA-745

"Transport of Cable Channel Mapping System Information Using Extended Data Service (XDS)"

This standard defined three new Miscellaneous Class data packets within XDS that transport information necessary to permit viewers to select a desired channel based on its broadcast channel assignment. EIA/CEA-608-B incorporates this information. EIA-745 is withdrawn. See EIA/CEA-608-B.

### EIA-746-A

"Transport of Internet Uniform Resource Locator (URL) Information Using Text-2 (T-2) Service"

This document was a proposed amendment to EIA-608 to insert Internet Uniform Resource Locators (URLs) within the line-21 data system using the Text-2 (T-2) service. EIA/CEA-608-B incorporates this information. EIA-746-A is withdrawn. See EIA/CEA-608-B.

### EIA-752

"Transport of Transmission Signal Identifier (TSID) Using Extended Data Service (XDS)"

This document was a proposed amendment to EIA-608 to include a unique 16-bit Transmission Signal Identifier in a new Extended Data Service (XDS) packet. EIA/CEA-608-B incorporates this information. EIA-752 is withdrawn. See EIA/CEA-608-B.

### EIA-761-A

"DTV Remodulator Specification with Enhanced OSD Capability"

This standard defines minimum specifications for a one-way data path utilizing an 8 VSB trellis or a 16 VSB remodulator in compliance with ATSC Standard A/53, Annex D. This standard also defines on-screen display (OSD) capabilities. This standard applies to any type of device used to connect to an ATSC compliant digital television receiver (DTV) receiver. Devices meeting this standard should interoperate with any ATSC compliant receiver that also supports "monitor mode." This standard addresses required RF output specifications, on-screen display (OSD) capabilities, and capability profiles for a DTV remodulator and recommendations concerning input to the remodulator. This standard does not address 8 VSB without OSD. For information concerning 8 VSB without OSD, see also EIA-762 and EIA-799.

### EIA-762

"DTV Remodulator Specification"

This standard defines a minimum specification for a one-way data path utilizing an 8-VSB trellis remodulator in compliance with ATSC A/53, Annex D. This standard applies to any type of device used to connect to an ATSC compliant digital television receiver (DTV) receiver. Devices meeting this standard should interoperate with any ATSC compliant receiver that also supports "monitor mode" (see EIA CEB-5.) This standard addresses both required RF output specifications for a DTV remodulator and recommendations concerning input to the remodulator.

## EIA-766-A

"U.S. and Canadian Region Rating Tables (RRT) and Content Advisory Descriptors for Transport of Content Advisory Information using ATSC A/65-A Program and System Information Protocol (PSIP)"

This standard augments ATSC Standard A/65-A and SCTE DVS-097 Rev. 7, both titled Program and System Information Protocol for Terrestrial Broadcast and Cable (PSIP). Along with the above two standards, this standard designates the RRT which provides the receiver with the definition of the rating system and the Content Advisory Descriptor which provides the receiver with the specific program rating for each program. Specifically, this standard specifies the exact syntax to be used to define the U.S. and Canadian Rating Region Tables (RRT) in accordance with A/65-A Section 6.4 as well as the exact syntax to be used in the Content Advisory Descriptors that convey the rating information for each program in accordance with A/65-A Section 6.7.4. Thus DTV receivers may block unwanted programs as determined by the user.

## EIA-770.1-A

"Analog 525 Line Component Video Interface—Three Channels"

This standard defines the physical characteristics of an interface and the parameters of the signals carried across the interface, using three parallel channels for the interconnection of equipment operating with analog component video signals. This standard includes specifications for two scanning structures: 1H - having 525 lines, 59.94 fields/second, 2:1 interlaced, and a horizontal scanning rate of 15.734 kHz; and 2H—for doubled scanned interfaces having 525 lines, 59.94 frames/second, progressively scanned, and having a horizontal scanning rate of 31.47 kHz. Both interfaces shall be capable of either 4:3 or 16:9 aspect ratios.

## EIA-770.2-A

"Standard Definition TV Analog Component Video Interface"

This standard defines the physical characteristics of an interface and the parameters of the signals carried across that interface, using three parallel channels for the interconnection of equipment operating with analog component video signals. The standard includes specifications for: 1) 480i video format defined by 480 active lines, 525 total lines, 2:1 interlaced at 59.94 or 60 fields/second; and 2) 480p video format defined by 480 active lines, 525 total lines, progressively scanned at 59.94 or 60 frames/second. Both video formats shall be capable of either 4:3 or 16:9 aspect ratios.

## EIA-770.3-A

"High Definition TV Analog Component Video Interface"

This standard defines two raster-scanning systems for the representation of stationary or moving two-dimensional images sampled temporally at a constant frame rate. The first image format specified is 1920 × 1080 samples (pixels) inside a total raster of 1125 lines. The second image format specified is 1280 × 720 samples (pixels) inside a total raster of 750 lines. Both image formats shall have an aspect ratio of 16:9.

## EIA-775-A

"DTV 1394 Interface Specification"

This standard defines a specification for a baseband digital interface to a DTV using the IEEE-1394 bus and provides a level of functionality that is similar to the analog system. It is designed to enable interoperability between a DTV compliant with this standard and various types of consumer digital audio/video sources including digital set-top boxes (STBs) and analog/digital hard disk or videocassette recorders (VCRs).

## EIA-796

"NRSS Copy Protection Systems"

The copy protection systems that have been included in EIA-796 are itemized for the purpose of identification. The systems outlined in EIA-796 all support the copy protection frameworks described in EIA-679-B, Parts A and B.

## EIA-799

"On-Screen Display Specification"

This standard specifies syntax semantics for bitmapped graphics data typically used for on-screen display (OSD). The standard is applicable whenever it is necessary to specify a standard method for delivery of bitmapped graphics data. The pixel formats include optional alpha-blend and transparency attributes to support composition of graphics over analog or digitally decoded video within the display.

## EIA-806

"Carriage of DTV PSIP Information Using the XDS Method"

This standard addresses transmission of DTV PSIP data using the XDS data carriage method. EIA/CEA-608-B incorporates this information. EIA-806 is withdrawn. See EIA/CEA-608-B.

## IETNTS1

"Industrial Electronic Tentative Standard No. 1 (IETNTS1), Color Television Studio Picture Line Amplifier Output Drawing"

Specifies timing parameters within the horizontal blanking interval of a color television signal. This document is based on EIA-170.

## TVSB5

"Multichannel TV Sound System BTSC System Recommended Practices"

Specifies the transmission of multichannel television sound (MTS) in accordance with the BTSC system and the FCC rules governing its use. This document is intended for both manufacturers and broadcasters.

## c) Television Receivers

### EIA-336

"Color Coding for Chassis Wiring (ANSI/EIA-336-68) (R73) (R79)"

The colors described in this Standard are as defined in EIA-359, "EIA Standard Colors for Color Identification and Coding."

## d) Home Networks

### ANSI/EIA-600.10

"Introduction to the CEBus Standard"

The EIA-600 Specification covers the overall topology of the EIA-600 network and the detailed topology for each individual medium used; the electrical and physical specifications for the media usable by EIA-600; the physical interface from a device to the medium and the signaling method specifications to be used on the medium; the protocol to be used for network access and the description of the control message format; and a command language that allows all devices to communicate a common set of functions to be performed. Aspects of the overall EIA-600 network that are not addressed in this specification are operation and maintenance of the network. This standard establishes a minimal set of rules for compliance. It does not rule out extended services that may be provided, as long as the rules of this standard are adhered to within the system. It is, in fact, the intention of the standards to permit extended services (defined by users) to exist.

### ANSI/EIA-600.31

"Power Line Physical Layer and Medium Specification (ANSI/EIA-600.31-97)"

This document is the preliminary specification for the CEBus Power Line (PL) Physical Layer and Media portion of the Physical Layer and Media Specifications of EIA-600. Its purpose is to present the information necessary for the development of a PL physical network and devices to communicate and share information over the network. This is one of a series of documents covering the various media that comprise the CEBus standard.

### ANSI/EIA-600.32

"Twisted Pair Physical Layer & Medium Specification (ANSI/EIA-600.32-97)"

This document is the specification for the CEBus Twisted Pair (TP) Physical Layer and Medium. Its purpose is to present all the information necessary for the development of a TP physical network and devices to communicate and share information over that network in an orderly manner. This is one of a series of documents covering the various media that comprise the CEBus standard.

### ANSI/EIA-600.33

"Coax Cable Physical Layer & Medium Specification (ANSI/EIA-600.33-97)"

This document is the preliminary specification for the CEBus Coax (CX) Physical Layer and Medium. Its purpose is to present all the information necessary for the development of a CX physical network and devices to communicate and share information over that network in an orderly manner. This is one of a series of documents covering the various media that comprise the CEBus standard.

### ANSI/EIA-600.34

"IR Physical Layer & Medium Specification (ANSI/EIA-600.34-97)"

This document is a preliminary specification for the CEBus Infrared (IR) Physical Layer and Medium portion of the Physical Layer and Medium specifications of EIA-600. Its purpose is to present all the information necessary for the development of a IR physical network and devices to communicate and share information over that network to and from IR and other CEBus media in an orderly manner. This is one of a series of documents covering the various media that comprise the CEBus standard.

### ANSI/EIA-600.35

"RF Physical Layer & Medium Specification (ANSI/EIA-600.35-97)"

This document is the preliminary specification for the CEBus Radio Frequency (RF) Physical Layer and Medium portion of the Physical Layer and Medium specifications of EIA-600. Its purpose is to present all of the information necessary for the development of a RF physical layer for the CEBus device. This is one of a series of documents covering various media that comprise the CEBus standard.

### ANSI/EIA-600.37

"Symbol-Encoding Sublayer"

This document describes the portion of the Node Physical Layer that interfaces to the Medium Access Control (MAC) Sublayer and to Layer System Management (LSM). This sublayer is called the Symbol Encoding (SE) Sublayer.

### ANSI/EIA-600.38

"Power Line/Radio Frequency Symbol Encoding Sublayer"

This document describes the portion of the Power Line or RF Physical Layer that interfaces to the Medium Access Control (MAC) Sublayer and to Layer System Management (LSM). This sublayer is called the Power Line/RF Symbol Encoding (PL/RF SE) Sublayer.

### ANSI/EIA-600.41

"Description of the Data Link Layer (ANSI/EIA-600.41-97)"

This document provides a prose description of the Data Link Layer Design for the CEBus Network. The intent of this document is to be descriptive, rather than provide a formal specification, and contains a discussion of the Data Link Layer interfaces to the Network Layer and Physical Layer, as well as a functional description of the Data Link Layer.

### ANSI/EIA-600.42

"Node Medium Access Control Sublayer (ANSI/EIA-600.42-97)"

This part of the CEBus standard is a technical specification of the services and protocol for the Node Medium Access Control Sublayer.

### ANSI/EIA-600.43

"Node Logical Link Control Sublayer (ANSI/EIA-600.43-97)"

This part of the CEBus standard is a technical specification of the services and protocol for the Node Logical Link Control Sublayer.

### ANSI/EIA-600.45

"Node Network Layer Specification"

This document is the CEBus Node Network Layer part of EIA-600.

### ANSI/EIA-600.46

"Node Application Layer Specification"

This document is the CEBus Node Application Layer part of EIA-600.

### ANSI/EIA-600.81

"Common Application Language (CAL) Specification (ANSI/EIA-600.81-97)"

This document describes the basic framework of CAL. It is intended as an introduction to CAL operation and syntax that stresses the object-oriented aspects of CAL. It is believed that the object-oriented methodology offers the best means of understanding the complex interaction between devices, controls, and controllers present in the CEBus environment.

### ANSI/EIA-600.82

"CAL Context Description (ANSI/EIA-600.82-97)"

This document describes the contexts, or main subsystems within a device, supported by the Common Application Language (CAL).

### ANSI/EIA-633.10

"Introduction to EIA-600 Conformance Specification"

This standard is concerned with conformance of an implementation to the associated protocol specifications contained in EIA-600. A dual audience is expected for this standard. The first is laboratories that may be interested in acting as conformance testing agencies. Such agencies are tasked with converting these requirements into a hardware/software test system. The second audience is the set of designers of EIA-600 compatible products. Any designer of compatible products should understand the importance of the tests described in this standard, as they relate to implementing EIA-600. The second group should read this standard while considering whether its hardware and/or software implementation is likely to pass the stated tests.

### ANSI/EIA-633.31

"Power Line Physical Layer Conformance Specification"

This portion of the conformance standard specifies tests to determine conformance of a Node's Power Line (PL) PL Physical Layer to IS-60. Part one of this standard provides an overview of the conformance philosophy. The reader is urged to review that material before attempting to use the details provided in this part.

### ANSI/EIA-633.32

"Twisted Pair Physical Layer Conformance (ANSI/EIA-633.32-97)"

This standard specifies tests to determine conformance of a device's Twisted Pair Physical Layer to EIA-600.

### ANSI/EIA-633.34

"Infrared Physical Layer Conformance (ANSI/EIA-633.34-97)"

This standard specifies tests to determine conformance of a Node's IR Physical Layer to EIA-600.

### ANSI/EIA-633.37

"Symbol Encoding Sublayer Physical Layer Conformance (ANSI/EIA-633.37-97)"

This standard specifies tests to determine conformance of a Node's Symbol Encoding Sublayer to EIA-600.

### ANSI/EIA-633.38

"PL and RF Symbol Encoding Physical Layer Conformance (ANSI/EIA-633.38-97)"

This standard specifies tests to determine conformance of a Node's Power Line or RF Symbol Encoding Sublayer to EIA-600.

### ANSI/EIA-633.46

"Node Application Layer Conformance Specification"

This portion of the conformance standard specifies tests to determine conformance of a Node's Application Layer to EIA-600.

### ANSI/EIA-633.81

"CAL Conformance Specification"

This portion of the conformance standard specifies tests to determine conformance of a Node's CAL to EIA-600.81. Part one of this standard provides an overview of the conformance philosophy. The reader is urged to review that material before attempting to use the details provided in this part.

### ANSI/EIA-709.1-A

"Control Network Protocol Specification (ANSI/EIA-709.1-A-99)"

This specification applies to a communication protocol for networked control systems. The protocol provides peer-to-peer communication for networked control and is suitable for implementing both peer-to-peer and master-slave control strategies. This specification describes services in layers 2-7. In the layer 2 specification, it also describes the data link layer and the MAC sub-layer interface to the physical layer. The physical layer provides a choice of transmission media. The interface described in this specification supports multiple transmission media at the physical layer.

## ANSI/EIA-709.2-A

"Control Network Power Line (PL) Channel Specification (ANSI/EIA-709.2-A-99)"

This document specifies the Control Network Power Line (PL) Channel and serves as a companion document to the EIA-709.1 Control Network Protocol Specification. Its purpose is to present the information necessary for the development of a PL physical network and nodes to communicate the share information over the network. This is one of a series of documents covering the various media that comprise the EIA-709 Standard.

## ANSI/EIA-709.3

"Free-Topology Twisted-Pair Channel Specification"

This document specifies the EIA-709.3 free-topology twisted-pair channel and serves as a companion document to the EIA-709.1 Control Network Protocol Specification. The channel supports communication at 78.125 kbps between multiple nodes, each of which consists of a transceiver, a protocol processor, and application processor, a power supply, and application electronics.

## ANSI/EIA-709.4

"Fiber-Optic Channel Specification"

In conjunction with ANSI/EIA-709.1-A Control Network Protocol Specification, EIA-709.4 defines a complete 7-layer protocol stack for communications on an EIA-709.4 single-fiber (half-duplex) fiber-optic channel. EIA-709.4 specifies the physical layer (OSI Layer 1) requirements for the EIA -709.4 fiber-optic channel which encompasses the interface to the Media Access Control (MAC) layer and the interface to the medium. The single-fiber channel implemented as specified in EIA-709.4 allows two nodes to communicate bidirectionally across a single piece of fiber cable.

## ANSI/EIA-721.1

"Generic Common Application Language (Generic CAL) Specification"

This document describes the basic framework of Generic CAL. It is intended as an introduction to Generic CAL operation and syntax that stresses the object-oriented aspects of Generic CAL. It is believed that the object-oriented methodology offers the best means of understanding the complex interaction between devices, controls, and controllers present in a Generic Network environment.

## ANSI/EIA-721.2

"Generic CAL Context Description"

This document describes the contexts, or main subsystems within a device, supported by the Generic Common Application Language (Generic CAL).

## ANSI/EIA-721.3

"Node Application Layer Specification"

This Application Layer consists of four main elements. The application Process is the interface to the Application Layer. Services are provided by the Generic Common Application Language (Generic CAL) Element to the User Element of the Application Process. Generic CAL is the language framework through which Resource Allocation and Control functions are executed. Services are provided by the Message Transfer Element to the Generic CAL Element. The Message Transfer Element interfaces to the lower layers of the Generic Network either directly or through the Association Control Element. The lower layers are representative of some home automation networks. Additional OSI layers may be included. An adaptation layer may be required between the Generic CAL Application Layer and the Generic Network lower layers.

## ANSI/EIA-721.4

"Generic Common Application Language Quality of Service"

This specification for Generic CAL consists of an Application Layer containing a command language and a Message Transfer Service Element. The specifications of the lower OSI layers are not within the scope of this standard. However, the services provided by the lower layers affect the performance and composition of messages issued from the Application Layer. These lower layer service options are collectively called the Quality-of-Service (QOS) available from the communications protocol. This portion of EIA-721 standard describes the lower layer QOS options that may impact the Application Layer. Recommended capabilities are specified. Also, a mechanism to convey these options to the Generic CAL Application Layer using Layer System Management functions is presented.

## ANSI/EIA-776.1

"CEBus-EIB Router Communications Protocol-Description of the CEBus-EIB Router"

This document describes the operation of a CEBus-EIB Router. This document is not intended to define how a router should operate, but to provide an overview of the operation and the coordination of various router elements.

## ANSI/EIA-776.2

"CEBus-EIB Router Communications Protocol—CEBus-EIB Router Medium Access Control Sublayer"

The CEBus-EIB Router Medium Control (MAC) Sublayer is almost identical to the CEBus or EIB Node MAC Sublayer corresponding to the "CEBus Side" or the Router. The differences are in the way the Router does address matching on a received packet and on the information exchanged in some of the service primitives. Rather than copy the Node MAC specification here and make minor changes, the Router MAC is specified by exception to the Node MAC for both the CEBus and EIB Specifications.

## ANSI/EIA-776.3

"CEBus-EIB Router Communications Protocol—CEBus-EIB Router Logical Link Control Sublayer"

This section specifies the CEBus-EIB Router Logical Link Control Sublayer interfaces to the Router Network Layer and to the Layer System Management. The interfaces are described in terms of service primitives which are abstract interfaces across a layer boundary. A service primitive represents an exchange of information into or out of a layer. Although service primitives are defined using a format similar to that of programming language procedure calls, no implementation technique is implied.

### ANSI/EIA-776.4

"CEBus-EIB Router Communications Protocol - CEBus-EIB Router Network Layer"

The CEBus-EIB Router Network Layer is conceptually divided into several elements, each performing distinct well-defined services. Each element may be thought of as an independent process that communicates with the other elements and protocol layers through specified interfaces.

### ANSI/EIA-776.5

"CEBus-EIB Router Communications Protocol - The EIB Communications Protocol"

EIB is a control system for related applications in homes and buildings. The EIB system offers standardized basic and system components, e.g., Bus Coupling Units (BCU), Power Supply Units (PSU), Bus Interface Modules (BIM), Routers, and RS-232 data interfaces. EIB offers the capability of constructing devices in a modular form using system devices like BCU or BIM that support communications-specific functions. A standardized interface called Physical External Interface (PEI) reduces the expense of developing EIB devices and allows them to be exchanged.

### EIA/CEA-844

"XML Encoding of Generic Common Application Language"

This standard specifies the encoding of Generic Common Application Language (CAL) into XML. It is based on ANSI/EIA-721 and EIA-851.

### EIA/CEA-851

"VHN Home Network"

The R-7.4 VHN Home Network Standard defines a flexible and open network architecture and communications protocol specification for digital devices in the home.

### EIA-600 CEBus SET

"EIA Home Automation System (CEBus)"

Provides the necessary specifications for the Consumer Electronic Bus (CEBus), a local communications and control network designed specifically for the home. The CEBus network provides a standardized communication facility for exchange of control information as data among devices and services in the home. The major motivations for its development were: a) to develop a universal low cost method for devices in the home to communicate regardless of manufacturer; b) to allow the introduction of new products and services to the home at minimal cost and confusion to consumers; c) to meet the majority of anticipated home control requirements with a single multi-media network standard; and d) to minimize the redundancy of control and operation methods among devices and equipment in the home. CEBus is intended to handle existing and anticipated control communication requirements at minimum practical costs consistent with a broad spectrum of residential applications. It is intended for such functions as remote control, status indication, remote instrumentation, energy management, security systems, and entertainment device coordination. These situations require economical connection to a shared local communication network carrying relatively short digital messages. A major objective of this specification is compatibility. It is intended that every implementation of CEBus be able to co-exist with every other implementation; that every device that meets this specification can communicate with all other CEBus devices; and the language used for control functions will be understood by all devices. This version includes portions of IS-60 that have not been revised as well as new ANSI approved updates.

## 3) SCTE Standards

The following documents relating to digital television have been adopted or are under development by the Society of Cable Telecommunications Engineers. For additional information contact the SCTE at www.scte.org.

### a) Digital Video

#### DVS 011

"Cable and Satellite Extensions to ATSC System Information Standards"

In conjunction with the Advanced Television Systems Committee document, ATSC A/56, Systems Information for Digital Television, this document defines a proposed Standard for System Information (SI) compatible with Moving Picture Experts Group (MPEG-2) compliant digital multiplex bitstreams constructed in accordance with ISO/IEC 13818-1 (MPEG-2) and transmitted on cable and satellite, and cable or satellite. The SI specifications here provides specific definitions and syntax for non-broadcast applications, extending the SI protocol by defining additional tables and descriptors applicable to cable and satellite, and cable or satellite transmission.

#### DVS 018

"ATSC Digital Television Standard"

This standard was prepared by the Advanced Television Systems Committee (ATSC) Technology Group on Distribution (T3). The Digital Television Standard describes the system characteristics of the U.S. advanced television (ATV) system. The document and its normative annex as provide detailed specifications and the parameters of the system including the video encoder, the audio encoder input signal format and the pre-processing and compression parameters of the audio encoders, the service multiplex and transport layer characteristics and normative specifications, and the VSB RF/Transmission subsystem.

#### DVS 019

"Digital Audio Compression (AC-3) Standard"

The normative portions of this standard specify a coded representation of audio information and specify the decoding process. Informative information on the encoding process is included. The coded representation specified herein is suitable for use in the digital audio transmission and storage applications. The coded representation may convey from 1 to 5 full bandwidth audio channels, along with a low frequency enhancement channel.

### DVS 020

"Guide to the Use of the ATSC Digital Television Standard"

This document provides a guide to understanding the system characteristics of the U.S. advanced television system as documented in the ATSC Digital Television Standard (DVS 018).

### DVS 022

"System Information for Digital Information"

This document defines a Standard for System Information (SI) compatible with digital multiplex bit streams constructed in accordance with ISO/IEC 13818 (MPEG-2). It defines the standard protocol that carries relevant System Information tables contained within packets carried in the transport multiplex. This document describes tables that are applicable to terrestrial (over-the-air), cable, SMATV, MMDS, and satellite broadcast signals. Only certain messages are applicable in each of the domains. All tables and messages defined in this document are carried in the Network Packet Identifier (PID).

### DVS 026

"Subtitling Method for Broadcast Cable"

This document defines a standard for a transmission protocol supporting multilingual subtitling services to augment video and audio within MPEG-2 multiplexes. This document is revised from a General Instrument standard entitled "Subtitles and Emergency Messages." Document Number STD-095-002, Revision 1.3, dated July 6, 1995.

### DVS 031

"Digital Video Transmission Standard for Cable Television"

This standard describes the framing structure, channel coding, and channel modulation for a digital multi-service television distribution system that is specific to a cable channel. The system can be used transparently with the distribution from a satellite channel, as many cable systems are fed directly from the satellite links. The cable channel, including optical fiber, is primarily regarded as a bandwidth-limited linear channel, with a balanced combination of white noise, interference, and multi-path distortion. The quadrature amplitude modulation (QAM) technique used, together with adaptive equalization and concatenated coding, is well-suited to this application and channel. The specification covers both the 64 and 256 QAM. Most features of both modulation schemes are the same. Where there are differences, the specific details for each modulation scheme are covered in the standard. The design of the modulation, interleaving and coding is based upon testing and characterization of cable systems in North America. The modulation is QAM with a 64-point signal constellation (64 QAM) and with a 256-point signal constellation (256 QAM), transfer selectable. The forward error correction (FEC) is based on a concatenated coding approach that produces high-coding gain at moderate complexity and overhead. Concatenated coding offers improved performance over a block code, at a similar overall complexity. The system FEC is optimized for quasi-error free operation at a threshold output error event rate of one error per 15 minutes. The data format input to the modulation and coding is assumed to be a Moving Picture Experts Group (MPEG-2) transport. However, the method used for MPEG synchronization is de-coupled from FEC synchronization. For example, this enables the system to carry Asynchronous Transfer Mode (ATM) packets easily without interfering with ATM synchronization. There are two modes supported: Mode 1 has a symbol rate of 5.057 Msps and Mode 2 has a symbol rate of 5.361 Msps. Typically, Mode 1 will be used for 64 QAM and the Mode 2 will be used for 256 QAM. The system will be compatible with future implementations of higher data rate schemes employing high-order QAM extensions. ITU-T/SCTE.

## DVS 033

"SCTE Video Compression Formats"

The standard format consists of four tables: Table 1, Standardized Video Input Formats; Table 2, Compression Format constraints—Tier 1; Table 3, Compression Format Constraints—Tier 2 (Additional Modes); and Table 4, Legend for MPEG-2 codes values in Table 3.

## DVS 043

"QPSK Tools for Forward and Reverse Data Paths"

The SCTE Digital Access Network can support forward and reverse communication paths. The forward path supports digital transmission channels (DTCs), and forward data channels (FDCs).

## DVS 046

"Specifications for Digital Transmission Technologies"

It is the goal of this document to promote the widespread availability of cost effective products and services, and the growth of digital video services without limiting technological change or innovation. This document provides the specification of digital video and related audio technologies in use or expected to be in use by the Cable TV industry in the U.S. In summary, it provides specifications to Channelization, Return Channel, Modulation, Forward Error Correction, Network, Transport, Conditional Access Interface, Video, Audio, Interactivity, and Cluster Interface.

## DVS 047

"National Renewable Security Standard (NRSS)"

This standard describes the physical and electrical parameters of the interface for Conditional Access (CA) devices for Consumer Electronics (CE). The purpose of this document is to specify the interface to a CA device that performs the signal access and, optionally, de-scrambling and entitlement keeping and key generation. Several CA devices could be interfaced to the same CE device simultaneously. This interface definition allows the entire CA system to be removable and replaceable. Very few constrains are placed on the conditional access system architecture.

## DVS 053

"VBI Extension for ATSC Digital Television Standards"

This document defines a standard for the carriage of Vertical Blanking Interval (VBI) services in MPEG-2 compliant bitstream constructed in accordance with ISO/IEC 13818-2. The approach builds upon a data structure defined in the ATSC A/53 Digital Television Standards, and is designed to be backward-compatible with the method.

## DVS 055

"EIA Interim Standard IS-679 B of National Renewable Security Standard (NRSS)"

A partial logical architecture has been assumed for a host in order to define the place in the host where the common interface can logically occur. The impact upon the freedom of choice for host designers in other respects has been minimized. The common interface consists of two components, the Transport Stream Interface and the Command Interface. Both are layered to make the overall interface design and implementation easier. The upper layers are common to all implementations but alternative lower-layer implementations are possible. This specification includes one based upon the PC Card standard but others may be included in future versions.

## DVS 064

"National Renewable Security Standards (NRSS) Part A and Part B"

NRSS provides a means for security for digital television receivers and digital VCR receivers. It has two physical designs, Part A and Part B. Both are removable and renewable security elements. Part A was from the ISO-7816 standard. It has 8 electrical contacts using a limited number of PC cards for security and in serial communication. This device costs less and can be smaller. Part B is more of interest to the cable television industry. It has 68 electrical contacts using parallel communications (one host and as many modules as one wants). It is more extensible and robust.

## DVS 068

"ITU-T Recommendation J.83"

The recommendation "Digital multi-programme systems for television, sound and data services for cable distribution" covers the definition of the frame structure channel coding and modulation for digital multi-program signals for television, sound and data services distributed by cable networks.

## DVS 071

"Digital Multi Programming Distribution by Satellite"

With the aim to promote the convergence on a world-wide standard for satellite digital multi-programme reception systems for television, sound and date services, the Common Functions Requirements for the reception of Digital Multi-programme Television emissions by satellite are described. These Common Functional Requirements configures the universal elements of the satellite Integrated Receiver Decoder (IRD). Although the present Annex addresses satellite operating in the 11/12 GHz frequency ranges, other ranges are not excluded.

### DVS 075

"Cue Command in Digital Systems"

The recommended practices was established by the Cable Television Advertising Bureau Technical Committee and the National Cable Television Association (NCTA) Cueing and Signaling Subcommittee in 1984, functions include Entry and Exit tones for analog systems. The digital versions of these indications are referred to as the Cue Command. This contribution provides a brief description of the various Cue functions, and cue commands that allow similar functionality in a digital system.

### DVS 076

"Digital Cable Ready Receivers: Practical Considerations"

This contribution has been prepared in ensuring that any ongoing efforts to define the Digital-Cable Ready Receiver (DCRR) takes into account digital cable system requirements. The DCRR has to meet: 1) backwards compatible with analog systems, and 2) compatible with digital cable systems.

### DVS 080

"Digital Broadband Delivery Phase 1.0 Functions"

This document provides an overview of the Digital Broadband Delivery System (DBDS). Included is a description of the system architecture, components and subsystems, and interfaces.

### DVS 082

"Broadband File System Product Description Release 1.2s"

The Broadband File System (BFS) provides a mechanism for delivering various types of data from a group of servers to a client which is attached to a network. This data may contain any type of information. Server and client can view the data as a file system similar to disk operating systems. BSF operates in one way or two way mode.

### DVS 084

"Common Interface for DVB Decoder Interface"

The first purpose of this document is to explain why the Common Interface Specification is designed the way it is. Second purpose is to guide how to implement and use the Common Interface. This includes recommendations for various design options where specific limits were not in the specifications. PCMCIA was chosen with suitable rationale for its Physical Layer.

### DVS 097

"Program and System Information Protocol for Terrestrial Broadcast and Cable"

This document defines a Standard for System Information (SI) and Program Guide (PG) data compatible with digital multiplex bit streams constructed in accordance with ISO/IEC 13818-1 (MPEG-2 Systems). The document defines the standard protocol for transmission of the relevant data tables contained within packets carried in the Transport Stream multiplex. The protocol defined herein will be referred to as Program and System Information Protocol (PSIP). Tables described in this document shall be applicable to terrestrial (over-the-air) and cable signals. Adopted.

### DVS 098

"IPSI Protocol for Terrestrial Broadcast with Examples"

The Integrated Program and System Information (IPSI) protocol is a small collection of tables designed to operate within every transport stream for terrestrial broadcast of digital television. Its purpose is to describe the information at the system and programs for all digital channels carried in a particular transport. Information for analog and digital channels from different transports can be incorporated. The relational hierarchy for the component tables is explained through typical applications.

### DVS 114

"SMPTE Splice point for MPEG-2 Transport"

This standard defines constraints on encoding of and syntax for MPEG-2 Transport Streams such that they may be spliced. The constraints here are applied individually to programs within Transport Streams. The standard enables splicing of programs within a multiprogram transport stream either simultaneously or independently. The standard may also be used with single program transport streams. The document specifies constraints for both seamless and non-seamless splice points. This standard specifies the technique for carrying cue information as a transport stream.

### DVS 147

"Revision to DVS 022"

This document defines a Standard System Information (SI) compatible with digital multiplex bit streams constructed in accordance with ISO/IEC 13818-1 (MPEG-2). The document defines the standard protocol that carries relevant System Information tables contained within packets carried in the transport multiplex.

### DVS 154

"Digital Program Insertion Control API"

This API describes the communication involved in inserting digital ads into a digital multiplex.

### DVS 159

"Optional Extensions for Carriage of NTSC VBI Data in Cable Digital Transport Streams"

This document defines a standard for the carriage of Vertical Blanking Interval (VBI) services in MPEG-2 compliant bit streams constructed in accordance with ISO/IEC 13818-2. The approach builds upon a data structure defined in the ATSC A/53 Digital Television Standard, and is designed to be backwards-compatible with this method.

### DVS 161

"ATSC Data Broadcast Specification"

This document defines a standard for data transmission compatible with digital multiplex bit streams constructed in accordance with ISO/IEC 13818-1 (MPEG-2 Systems).

### DVS 165

"DTV Interface Specification"

This standard defines a minimum specification for an analog and baseband digital interface to a digital TV receiver. This specification is designed to enable interoperability between a DTV compliant and various types of consumer audio/video devices including digital set-top boxes (STBs) and analog/digital hard disk or video cassette recorders (VCRs).

### DVS 168

"Emergency Alert System Interface to Digital Cable Network"

The purpose of this document is to describe the interface between an Emergency Alert Receiver Server (EARS) and an Emergency Alert Controller (EAC).

### DVS-181

"Service Protocol"

This document defines a standard wire line protocol that gives a cable operator a method for advertising services that do no conform to traditional broadcast content oriented services. These services may include e-mail, web browsing, e-commerce, and games. SP supports the notion of a "service" that is defined as an application and parameter data that is passed to the application upon activation. This expands on the existing content-centric definition of "service."

### DVS-192

"Splicer Application Programmer's Interface Definition Overview"

This document specifies extensions to MPEG-2 to allow the transmission of data intended to be transcoded into the VBI of MPEG-2 decoded video. This document is not intended to contradict or replace the ATSC A/53 and EIA 708 standards for digital television broadcast.

### DVS-211

"Service Information for Digital Television"

This document defines a Standard for Service Information (SI) compatible with digital multiplex bit streams constructed in accordance with ISO/IEC 13818-1 (MPEG-2). The previous standard protocol (System Information), as described in ATSC A/56, also in SCTE DVS/147, carries relevant System Information tables contained within packets in the transport multiplex for many installed digital cable systems. System Information represents system-wide information in the Network Packet Identifier (PID) described in SCTE DVS/147.

### DVS-213

"Copy Protection for POD Module Interface"

The content protection mechanism presented in this document is intended to protect the interface between the Point of Deployment (POD) Removable Security Module and the host device. National control centers and/or cable system head-ends have the ability to communicate with individual addressable POD Modules for authorizing and de-authorizing of digital services on a channel by channel basis. Secured video services are encrypted throughout the entire delivery chain from originator to subscriber and copy protection shall be used on the outputs of the host device as more fully defined below.

## b) Data Standards

### DSS 97-02

"Data-Over -Cable Radio Frequency (RF) Interface Specification"

This document defines the radio-frequency interface specifications for high speed data-over-cable systems that are developed by the Multimedia Cable Network System (MCNS). The intended service will allow transparent bi-directional transfer of Internet Protocol (IP) traffic, between the cable system headend and customer locations, over a hybrid fiber/coax (HFC) cable television system. The transmission path over the cable system is realized at he headend by a Cable Modem Termination System (CMTS), and at each customer location by a cable modem (CM). At he headend or hub, the interface to the data-over-cable system is the Cable Modem Termination System-Network-Side Interface (CMTS-NSI). At customer locations, it is the cable-modem-to-customer-premise-equipment interface (CMCI). The intent is for the MCNS operators to transparently transfer IP traffic between these interfaces, including but not limited to datagrams, Dynamic Host Configuration Protocol (DHCP), Internet Group Messaging Protocol (IGMP), and IP Group addressing (broadcast and multicast). ITU-T/SCTE.

# About the Editor

**Jerry C. Whitaker** is Technical Director of the Advanced Television Systems Committee (ATSC), Washington, D.C. He was previously President of Technical Press, a consulting company based in the San Jose (CA) area. Mr. Whitaker has been involved in various aspects of the electronics industry for over 25 years, with specialization in communications. Current book titles include the following:

- Editor-in-chief, *Standard Handbook of Video and Television Engineering*, 3rd ed., McGraw-Hill, 2000
- *DTV Handbook*, 3rd ed., McGraw-Hill, 2001
- *Video Displays*, McGraw-Hill, 2000
- Editor, *Interactive Television Demystified*, McGraw-Hill, 2001
- Editor, *Video and Television Engineers' Field Manual*, McGraw-Hill, 2000
- *Radio Frequency Transmission Systems: Design and Operation*, McGraw-Hill, 1990
- *Maintaining Electronic Systems*, CRC Press, 1991
- Editor-in-chief, *The Electronics Handbook*, CRC Press, 1996
- *AC Power Systems Handbook*, 2nd ed., CRC Press, 1999
- *Power Vacuum Tubes Handbook*, 2nd ed., CRC Press, 1999
- *The Communications Facility Design Handbook*, CRC Press, 2000
- *The Resource Handbook of Electronics*, CRC Press, 2000
- Co-author, *Television and Audio Handbook for Engineers and Technicians*, McGraw-Hill, 1989
- Co-author, *Communications Receivers*, 3rd ed., McGraw-Hill, 2000
- Co-editor, *Information Age Dictionary*, Intertec Publishing/Bellcore, 1992

Mr. Whitaker has lectured extensively on the topic of electronic systems design, installation, and maintenance. He is the former editorial director and associate publisher of *Broadcast Engineering* and *Video Systems* magazines, and a former radio station chief engineer and television news producer.

Mr. Whitaker is a Fellow of the Society of Broadcast Engineers and an SBE-certified professional broadcast engineer. He is also a fellow of the Society of Motion Picture and Television Engineers, and a member of the Institute of Electrical and Electronics Engineers.

Mr. Whitaker has twice received a Jesse H. Neal Award *Certificate of Merit* from the Association of Business Publishers for editorial excellence. He has also been recognized as *Educator of the Year* by the Society of Broadcast Engineers.

Mr. Whitaker resides in Morgan Hill, California.

## On-Line Updates

Additional updates relating to DTV in general, and this book in particular, can be found at the *Standard Handbook of Video and Television Engineering* web site:

**www.tvhandbook.com**

The tvhandbook.com web site supports the professional video community with news, updates, and product information relating to the broadcast, post production, and business/industrial applications of digital video.

Check the site regularly for news, updated chapters, and special events related to video engineering. The technologies encompassed by *Television Receivers* are changing rapidly, with new developments each month. Changing market conditions and regulatory issues are adding to the rapid flow of news and information in this area.

Specific services found at **www.tvhandbook.com** include:

- **Video Technology News**. News reports and technical articles on the latest developments in digital television, both in the U.S. and around the world. Check in at least once a month to see what's happening in the fast-moving area of digital television.
- **Television Handbook Resource Center**. Check for the latest information on professional and broadcast video systems. The Resource Center provides updates on implementation and standardization efforts, plus links to related web sites.
- **tvhandbook.com Update Port**. Updated material for *Television Receivers* is posted on the site regularly. Material available includes updated sections and chapters in areas of rapidly advancing technologies.
- **tvhandbook.com Book Store**. Check to find the latest books on digital video and audio technologies. Direct links to authors and publishers are provided. You can also place secure orders from our on-line bookstore.

In addition to the resources outlined above, detailed information is available on other books in the McGraw-Hill Video/Audio Series.